IV International Conference on Laser and Plasma Research and Technologies (LaPlas 2018)

Journal of Physics: Conference Series
Volume 1238

Moscow, Russia
30 January - 1 February 2018

ISBN: 978-1-5108-9174-6
ISSN: 1742-6588

Printed from e-media with permission by:

Curran Associates, Inc.
57 Morehouse Lane
Red Hook, NY 12571

Some format issues inherent in the e-media version may also appear in this print version.

Institute of Physics
Dirac House, Temple Back
Bristol BS1 6BE UK

Phone: 44 1 17 929 7481
Fax: 44 1 17 920 0979

techtracking@iop.org

Additional copies of this publication are available from:

Curran Associates, Inc.
57 Morehouse Lane
Red Hook, NY 12571 USA
Phone: 845-758-0400
Fax: 845-758-2633
Email: curran@proceedings.com
Web: www.proceedings.com

TABLE OF CONTENTS

IOP Conf. Series: Journal of Physics: Conf. Series **1238** (2019) 011001 doi:10.1088/1742-6596/1238/1/011001

Abstract

A.P. Kuznetsov and S.V. Genisaretskaya

IV Conference on Plasma and Laser Research and Technologies took place on January 30 until February 1, 2018 at the National Research Nuclear University "MEPhI". The Conference was organized by the Institute for Laser and Plasma Technologies and was supported by the Competitiveness Program of the National Research Nuclear University "MEPhI".

The conference program consisted of seven sections:

- Laser Physics and Laser Technologies
- Plasma Physics and Plasma Technologies
- Controlled Thermonuclear Fusion
- Modern Problems of Theoretical Physics and Superstrong Laser Fields
- Challenges in Physics of Solid State, Functional Materials and Nanosystems
- Modern Trends of Quantum Metrology
- Particle Accelerators and Radiation Technologies.

The conference is based on scientific fields as follows:

- Laser, plasma and radiation technologies in industry, energetic, medicine;
- High-temperature plasma physics, environmentally-friendly energetic based on controlled thermonuclear fusion;
- New metamaterials, "smart" alloys and quantum systems;
- Photonics, quantum metrology, optical information processing;
- Ultrahigh optical fields, high-power lasers, Mega Science facilities;
- Spectroscopic synchrotron, neutron, laser research methods, quantum mechanical calculation and computer modelling of condensed media and nanostructures;
- High energy density laser physics.

More than 300 specialists took part in the Conference. They represented leading Russian scientific research centers and universities (National Research Centre "Kurchatov Institute", P.N. Lebedev Physical Institute, A.M. Prokhorov General Physics Institute, Joint Institute for Nuclear Research, Moscow Institute of Physics and Tecnology and others) and leading scientific centers and universities from Germany, Italy, France, Canada, Japan.

We would like to thank heartily all of the speakers, participants, organizing and program committee members for their contribution to the conference.

IOP Conf. Series: Journal of Physics: Conf. Series **1238** (2019) 011002 doi:10.1088/1742-6596/1238/1/011002

Peer review statement

All papers published in this volume of *Journal of Physics: Conference Series* have been peer reviewed through processes administered by the proceedings Editors. Reviews were conducted by expert referees to the professional and scientific standards expected of a proceedings journal published by IOP Publishing.

IOP Conf. Series: Journal of Physics: Conf. Series **1238** (2019) 012001 doi:10.1088/1742-6596/1238/1/012001

Behaviour of Redeposited Tungsten Layers with Varying Impurity Content during Thermal and Radiation Loads

L.B. Begrambekov, O.A. Bidlevich, A.V. Grunin, N.A. Puntakov, A.N. Voityuk.

National Research Nuclear University MEPhI (Moscow Engineering Physics Institute), Kashirskoe shosse 31, 115409 Moscow, Russia.

lbb@plasma.mephi.ru

Abstract. Tungsten is going to be implemented as a material for plasma-facing divertor tiles in an ITER tokamak. Accumulation of dust particles due to exfoliation of redeposited tungsten layers can severely affect the operation of a fusion devices. Behaviour of tungsten films deposited on tungsten, including layers with added impurities of iron (0,2 at.% and 2 at.% concentration), as well as dust particles with Al impurity, has been studied. It is shown that a film with 2 at.% impurity of iron exhibits exfoliation and blistering when exposed to thermal and radiation loads, while tungsten dust particles with aluminium impurity undergo various structural changes depending on their Al content when heated.

1. Introduction

Tungsten will be used as a plasma-facing material for ITER divertor tiles. Behavior of tungsten during thermal cycling and plasma irradiation in different modes was investigated in a number of works [1-3]. Tungsten atoms sputtered by plasma irradiation would form redeposited layers on divertor which tiles are irradiated with lesser intensity [4,5]. Formation of dust particles might be the result of the formation of films with poor adhesion. Dust accumulation can severely affect the discharge parameters in future thermonuclear facilities and lead to the accumulation of dangerously high amounts of tritium. This work is dedicated to the investigation of processes in deposited layers during thermal cycling and irradiation by hydrogen ion flux of high power density. The experiments on generating tungsten and tungsten-aluminium dust composed of flake particles formed by destruction of layers formed on the walls of a plasma chamber by irradiating tungsten and aluminium targets with plasma ions are also presented.

2. Experimental Setup

Deposition of layers and irradiation by hydrogen ion flux was conducted in a CODMATT (Coating Deposition and MATerial Testing) facility [6]. Tungsten films were formed on polished tungsten plates 15×7.5×1 mm^3 in size by atoms sputtered from the surface of a tungsten(99.9 at%) target by argon plasma ions. Deposition rate was 2 μm/hr, layer thickness was 2 μm.

During the investigation of the behavior of redeposited tungsten films in ITER-like conditions, it is important to consider the possibility of impurities added to tungsten during redeposition, such as beryllium and, possibly, stainless steel sputtered from different parts of the facility. Due to safety reasons our laboratory is not permitted to work with beryllium, thus aluminium is used as a beryllium proxy and an impurity in tungsten films, as well as stainless steel atoms (Fe, Cr, Ni) deposited in a different experiment.

Thermal cycling of samples was conducted in a Multifunctional Research Mass-Spectrometry Analysis Complex (MIKMA) [7]. The samples were heated by radiation from a flat tungsten filament mounted on an opposite side of a sample under investigation and heated by a pulse current in a vacuum of no less than 10^{-6} Torr. The sample was heated to the temperatures of up to 1200 °C. Temperature gradient on a sample was no more than ±10°C at the sample size. The samples were heated at a rate of 5°C/s and cooled for 5 min to the temperature of 300°C. A number of samples were prepared for testing, with a small (0,2 at.%) and large (2 at.%) iron impurity content in the deposited film.

Amorphous tungsten dust was accumulated in a "DEKOR" plasma coating deposition facility [8] During the discharge, deposited films formed on stainless steel sheets that partially exfoliated as dust particles. Approx. 0,5 g of dust was collected after each discharge. The particles had a form of flakes ranging in size from a few microns up to 500 μm (Fig. 1a) and being 2-6 μm thick. As such, the size of flakes was comparable to those obtained in modern thermonuclear facilities [9]. Particle content measured using Energy-dispersive X-ray spectroscopy (EDS), included W (up to 90 at.%), Al (up to 60%) and O (50 at.%). An addition of a small (<10 at.%) amount of stainless steel impurity (Fe, Cr, Ni) leads to the consequence that oxide layers formed on the plasma chamber wall exfoliated along the tungsten flakes. The flakes were then separated by size using a 35×35 μm^2 net.

a b

Fig. 1. Flake particles: a) overview, b) thickness of a particle.

Analysis of the deposited layers' structure and profile before and after testing was conducted using a Tescan Vega 3 Scanning Electron Microscope (SEM), with a cross-section done using a FIE Quanta 200 3D system (scanning electron microscope with gallium ion beam).

3. Experiment Results

3.1. Thermal cycling of deposited coating

After 260 cycles, traces of restructurisation were observed on the surface of deposited coating with a small impurity content – formation of separate globules of 200-500 nm in size. (Fig. 1) No exfoliation or cracking of the coating was observed.

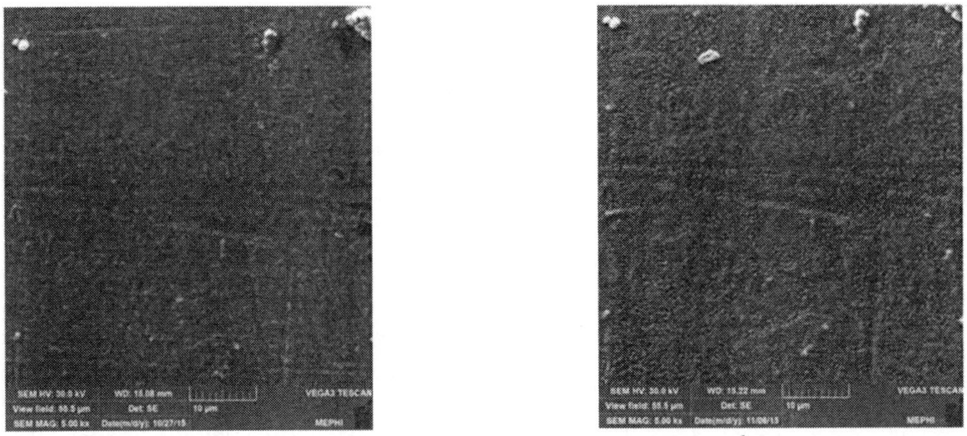

a b

Fig. 2. Deposited tungsten layer with a small impurity content: a) directly after deposition; b) after 260 heating cycles.

Similar testing was conducted on a tungsten sample with a large impurity content (2 at.%). Thermal cycling led to the formation of larger globules 1-3 μm in size (Fig. 3), which is supposedly connected to the recrystallisation processes on the deposited layer. After 250 heating cycles, formation of pores can be observed on the surface. After 360 cycles, pore formation continued. Additionally, exfoliation occurred on some parts of the sample, which may indicate the formation of a large number of pores in the bulk of a deposited layer.

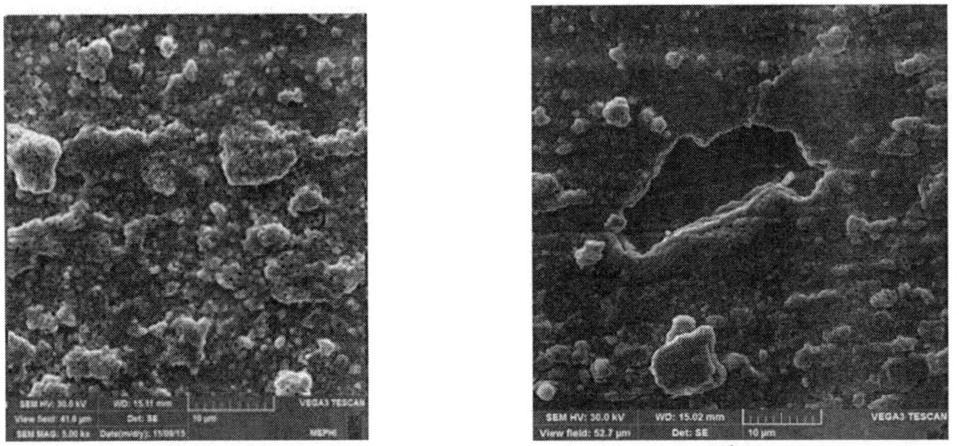

a b

Fig. 3. Depositel layer with a 2 at.% iron impurity content: a) pores on a surface after 260 heating cycles; b) exfoliation after 350 heating cycles

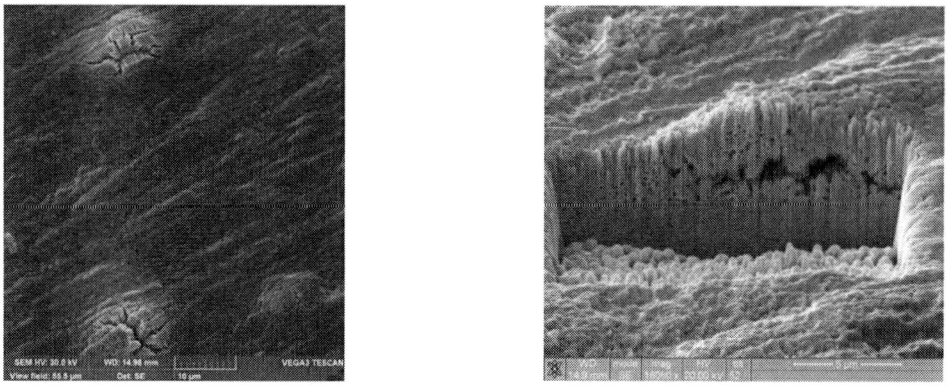

a b

Fig. 4. Deposited tungsten coating after irradiation by hydrogen plasma ions: a) separate blisters; b) blister cross-section.

3.2. Irradiation of deposited films by ions

During the irradiation of a sample with a 2 at.% impurity content by H_2^+ ions with a power density of 5 MW/m^2 (ion energy 5 keV/at., cyclic irradiation mode, 20 pulses with the duration of 5 seconds each), blisters formed on the surface of the sample. The temperature of the sample during the experiment did not exceed 1200 °C (the sample cooled between the pulses to the temperature of 600 °C). Fig. 4a shows the image of the film after irradiation. No blisters formed on the sample with a low impurity content. Fig. 4b shows the formation of a large number of microscopic gas bubbles near the interface between the layer and the substrate. Accumulation of hydrogen in these bubbles could lead to the formation of strains parallel to the surface in the deposited films and form blisters. It is interesting to note that pores formed in the samples heated without ion irradiation. That can be explained by pores formation leading to the gas trapped during deposition escaping the coating, with strains and blisters not forming as a result.

3.3. Investigation of structural changes in dust particles during heating

During the heating experiments, dust particles were places on a box-shaped heating element made of thin-sheet tungsten mounted between the electrodes in a VUP-2K station plasma chamber. The temperature was controlled using a tungsten-rhenium thermocouple attached to a heater. Dust particles were heated to temperatures of up to 2000 K with an increment of 100 K each time, changes in structure were investigated using SEM and EDS after each heating.

No changes in structure of amorphous particles were observed in a temperature range of up to 1073 K. Starting from 1473 K, which corresponds to the minimal temperature at which the recrystallization process for tungsten occurs, the formation of metal grains and pores at the grain boundaries occurred (Fig. 5a). With further temperature increase, the size of grains increased, and the oxygen content decreased due to the decomposition of oxides in the particles. If a particle included aluminium, then, when the content was high (> 55 at.%), molten aluminium droplets formed and spread on the surface of the heater at 1473 K (Fig. 5b). When the particle consisted of two layers, with differing contents for each layer (85 at.% W and 60 at.% Al, respectively), conical structures appeared when the particle was heated to temperatures of up to 1273 K (Fig. 5c), which can be indicative of the development of internal stress in the flake. For small aluminium contents (<25 at.%), no such effects were observed after heating to above temperatures. It is speculated that aluminium evaporates before forming droplets.

In case of crystal dust, large particles were baked along the edges at temparatures above 1173 K, and small particles forming conglomerates baked together (Fig. 5d, 5e).

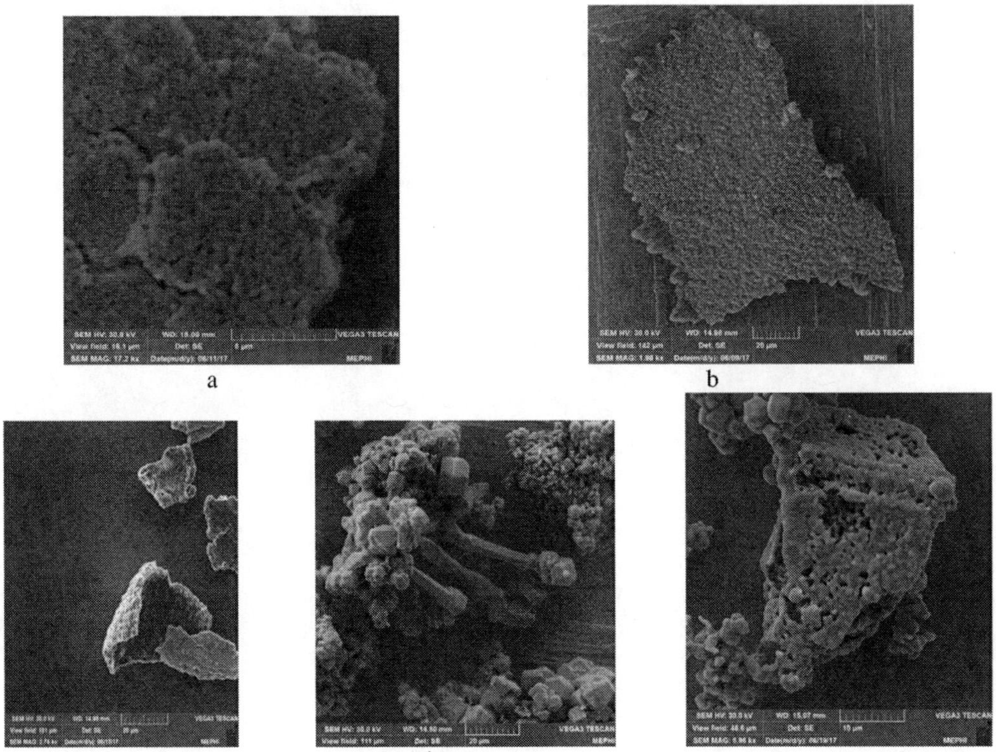

a b c d e

Fig. 5. Different structural changes in dust particles during heating: a) recrystallisation, b) cones, c) molten Al droplets, d) baking of large crystal particles along the edges, e) baking of small crystal particle conglomerates.

4. Conclusion

Thermal cycling and irradiation of tungsten samples with deposited tungsten films with different iron impurity concentration (0,2 at.% and 2 at.%) by hydrogen was conducted . During thermal cycling of tungsten films with impurities, globules formed on the surface of the film, with larger globules (1-3 μm) forming on a layer with a higher impurity concentration (2 at.%). Pores also formed on a layer with higher impurity content, which led to exfoliation of deposited film, which wasn't observed on a lower impurity (0,2 at.%) sample. During hydrogen ion irradiation, pore formation in coatings with a high impurity content resulted in exfoliation and blister formation in deposited films.

It is shown that, during heating of dust particles, a structural change occurs in them, such as: recrystallization and crystal grain growth, formation of cone-like structures due to internal strains, melting and evaporation of components with a low melting point (Al) in case of amorphous dust particles.

References

[1] Krieger K., et al., J. Nucl. Mater., 313—316(2003)327—332.
[2] Buzi L., et al., J. Nucl. Mater., 455(2014)316—319.
[3] Krasheninnikov S.I., et al., Plasma Phys. Control. Fusion, 2011, vol. 53, № 8, p. 083001.
[4] M. Balden, et al., J. Nucl. Mater., 438(2013)S220-S223.
[5] S. Krat, et al., J. Nucl. Mater., 438(2013)742-S745.
[6] E. Azizov, et al., J. Nucl. Mater., 463(2015)792-795
[7] A. Airapetov, et al., J. Nucl. Mater., 415, Issue 1, Supplement, 1(2011) S1042-S1045.
[8] Begrambekov L.B., Gordeev A. A., Grunin A.V., Evsin A.E., et al. Materials of VII

International School-Conference of young researchers and specialists IHISM'11 (24-28 October, 2011, Zvenigorod), p. 323-340
[9] S.I. Krasheninnikov et al., Plasma Phys. Controlled Fusion 53(2011) 083001.

Acknowledgments

This work was conducted with financial support from the Russian Science Foundation under the grant № 17-12-01575.

Electronic, local atomic structure of lutetium tetraphenylporfyrin: XPS and XAFS spectroscopy stadies

R N Mozhchil[1], A M Ionov[1], S I Bozhko[1], V S Bozhko[1], V D Rumyantseva[2], A L Trigub[3], A P Menushenkov[4]

[1] Institute of Solid State Physics RAS, Academician Ossipyan str. 2, 142432 Chernogolovka, Russia
[2] Moscow Technological University (MIREA), Vernadsky Avenue 78, 119571 Moscow, Russia
[3] "Kurchatov Institute", pl. Akademika Kurchatova 1, 123182 Moscow, Russia
[4] National Research Nuclear University MEPhI (Moscow Engineering Physics Institute), Kashirskoe sh. 31, 115409 Moscow, Russia

E-mail: mr_mozhzcil@mail.ru

Abstract. The electronic and local atomic structure of lutetium metalloporphyrins Lu(acac)TPP and precursor tetraphenylporphyrin (TPP) have been studied by X-ray photoelectron spectroscopy (XPS) and absorption spectroscopy (XAFS). The XPS spectroscopy data show changes in the electronic structure of tetraphenylporphyrins under incorporation of the lutetium atom: a redistribution of the electron density between the nitrogen atoms of the pyrrole- and aza- group and appearance of one broadened peak of the N1s state. The integer trivalent state of the rare-earth metal (Lu^{3+}) in metalloporphyrins has been confirmed by both XPS and XAFS methods.

1. Introduction

Because of nonlinear optical properties the metalloporphyrins are very promising compounds as an active substance for optical communications, optical and electrical signals processing, information storage [1]. Porphyrins are of interest as low-dimensional conductors [2], promising for nanotechnology and the creation of elements and devices based on sensitivity of individual macromolecules to electrical, magnetic and electromagnetic fields [1].

Molecules of porphyrins form stable complexes with rare-earth (RE) metals, which have intense absorption in the near-IR region of the spectrum [3]. Substation of various metals into the meso- and/or β-positions of the macrocycle makes it possible to change the physico-chemical properties of metalloporphyrins. This plays an important role in their use in medicine and photochemistry [4].

This paper is devoted to the study of the electron and local atomic structures of lutetium tetraphenylporphyrin.

2. Methods

The porphyrin complexes were synthesized in the laboratory of the Moscow Technological University (MIREA), the synthesis was described in details in [5].

The electronic structure and the chemical bonding in porphyrins were studied experimentally by X-ray photoemission spectroscopy (XPS). High resolution Lu4d, Br3d, N1s, C1s, O1s and valence band (VB) spectra were measured at Kratos AXIS Ultra DLD spectrometer with the photon energy 1486.69 eV, AlK_α mono, total resolution $Ag3d_{5/2}$ was about 0.48 eV. Spectra were calibrated using $Ag3d_{5/2}$ and $In3d_{5/2}$ lines. Samples for XPS spectroscopy were pressed into the indium substrate *ex-situ*. The surfaces of prepared samples were additionally cleaned by resistive heating up to ~ 400 K in ultra high vacuum (UHV) *in-situ*. The base pressure during measurements was in the range $5{\cdot}10^{-10}$-$2{\cdot}10^{-9}$ torr.

L_3-Lu -edge X-ray spectra were measured on the "Structural Materials" beamline of Kurchatov synchrotron radiation source [6]. The energy of electrons in the storage ring was 2.5 GeV, the typical current was 80-100 mA. The measurements were carried out at room temperature in transmission geometry (channel-cut Si(111) monochromator), the intensity of X-ray beam was measured before and after the sample by ionization chambers filled with N_2/Ar mixtures for optimal detection efficiency. The monochromator was calibrated according with maximum jump derivative absorption spectra of *K*-Cu absorption spectra (E_0= 8979 eV) of the metal foil

3. Results and discussion

3.1. The XPS study.

The analysis of wide XPS spectra of the investigated porphyrins demonstrated the presence of all the elements making up the compounds (Lu4d, N1s, C1s, O1s). It is worth to note that the relative concentrations of elements, according to XPS, in the first approximation, correspond to the chemical composition.

A characteristic two-peak structure with binding energies of 399.4 and 397.2 eV was observed, in the XPS spectra of the core levels of N1s tetraphenylporphyrin (Fig.1a). This structure, corresponding to nitrogen in the pyrro-group (sp^3 bond configuration) and aza-group (sp^2), respectively, agrees well with the results for thiolporphyrins and related phthalocyanine compounds.

There is a change in the state of N1s spectra of metalloporphyrins (Fig. 1b), indicating redistribution of the electron density between pyrro- and aza-N in porphyrins after addition of lutetium (deprotonation). Such a change is in a good agreement with the data obtained for RE and 3d metalloporphyrins (Me=RE, Fe, Co, Ni) and copper phthalocyanines [7,8,9]. The so-called "metallization" (introducing of a metal into the center of the TPP molecule) leads to the replacement of two hydrogen ions by a metal ion, which in turn equally interacts with all the nitrogen ions in the coordination cavity, thereby causing more uniform redistribution of the electron density between nitrogen and appearance of the single broadened peak N1s. It should be noted that splitting of the nitrogen line observed in the spectrum (Fig. 1b) can be explained by presence of the porphyrin, caused by the peculiarity of the synthesis.

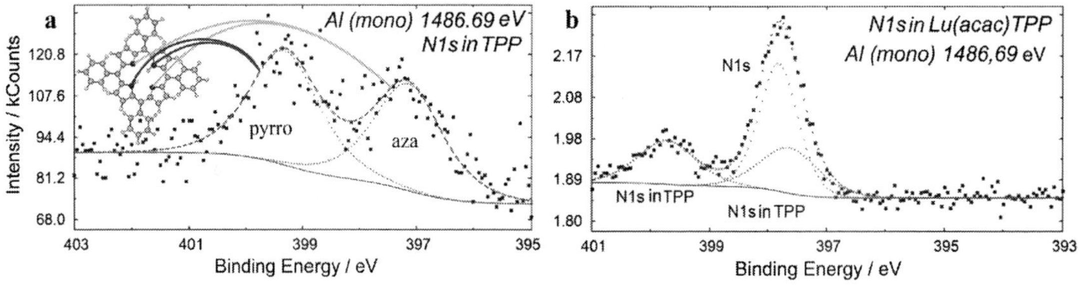

Figure 1. Decomposition of XPS spectrum of the N1s level for TPP (a) and Lu(acac)TPP (b).

Lutetium is a rare-earth element with a fully filled *f*-electron shell and the Lu4d spectrum is represented by a doublet splitting, with an area ratio of 3:2. A doublet of chlorine Cl2p is superimposed on the Lu4d line (Fig. 2), which is also related to the synthesis features. Parameters of the Lu4d line in oxide and metal were used in the decomposition of the lutetium spectrum in the porphyrin complex.

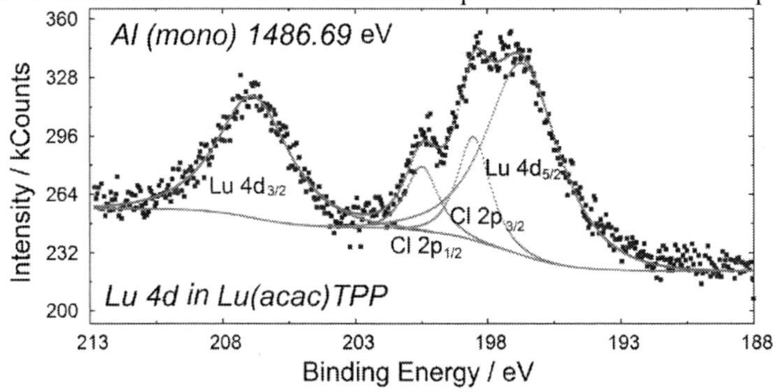

Figure 2. Decomposition of level Lu4d for Lu(acac)TPP

For all investigated lutetium complexes a wide peak in the XPS spectra of the core level C1s is observed. It corresponds to different carbon states in non-equivalent positions in the molecule whose peaks exhibit small chemical shifts (Fig. 3). Decomposition of O1s level in Lu(acac)TPP gives narrower lines responsible for oxygen in both the compound itself and in small amount of impurities. Decomposition of O1s and C1s level in Lu(acac)TPP is similar to ytterbium complex [8].

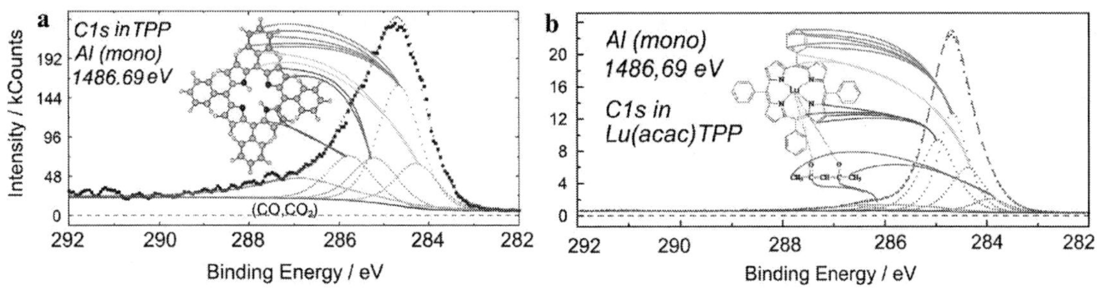

Figure 3. Decomposition of XPS spectrum of the C1s level for TPP (a) and Lu(acac)TPP (b).

The spectra of the VB for studied porphyrins (hv = 1486,69 eV) are compared in Figure 4 with that for metal-free porphyrins.

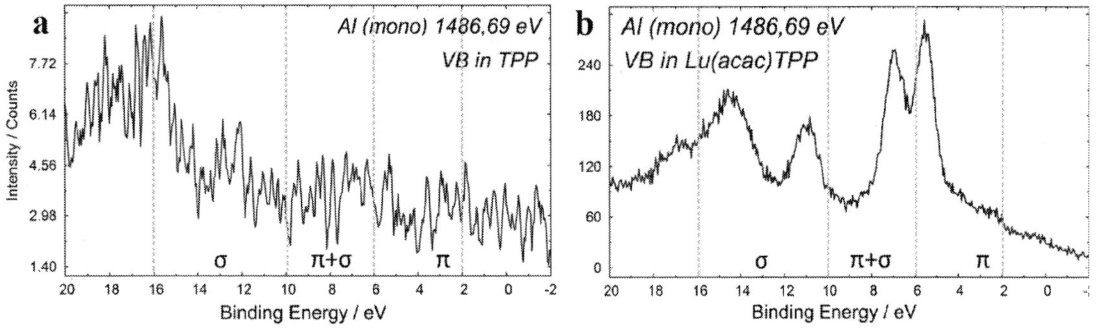

Figure 4. The spectra of the VB of porphyrins: TPP (a) and Lu(acac)TPP (b).

The analysis of the experimental and published data [4,10] allows us to assume that the valence bands of porphyrins metalloporphyrins of RE, 3d metals and tetraphenylporphyrins are similar. They are formed by π- (in a range 1-6 eV), π+σ- (5-10 eV), and σ-states (6-16 eV) of porphyrin macrocycles and semi-core levels of rare-earth elements (4f, 5p) and according to the published data, in the first approximation, can be described by a superposition of the spectra of the structural components, benzene and pyrrole.

3.2. XAFS research

The L_3-Er XANES (X-ray absorption near edge structure) spectrum in the macrocycle is shown in Fig.5a. The "white line" is observed as a single maximum at an energy of 9245 eV without any splitting for Lu(acac)TPP indicating an integer trivalent state of Lu in both compounds.

Fit of the calculated EXAFS spectrum to the experimental ones was carried out in the application of geometric optimization data (Fig.5b), by using the Artemis program [11].

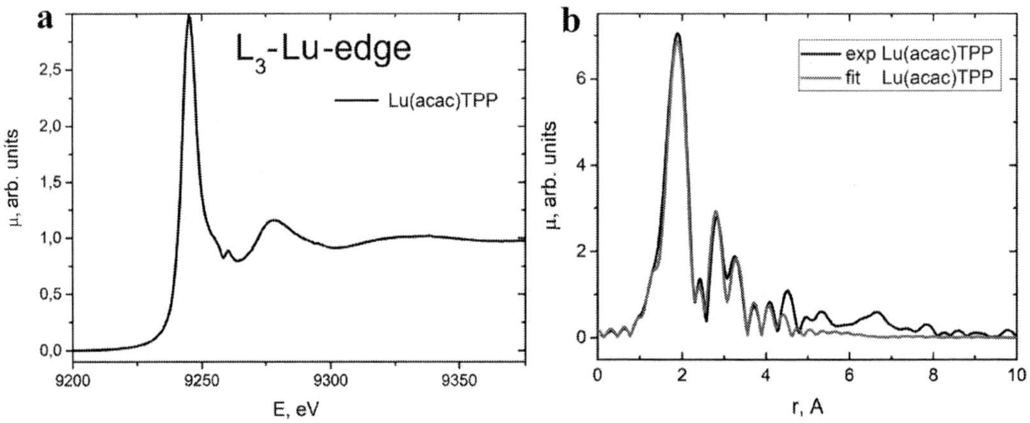

Figure 5. (a) XANES spectra of L_3-Lu-edge for Lu(acac)TPP. (b) Fourier transform modulus of the experimental (black) and model (red) EXAFS functions.

The results of processing EXAFS-spectra for Lu(acac)TPP (Table 1) and a systematic comparison of the results for complexes Yb(acac)TPP and Er(acac)TPP [12] indicate a small difference in the structure of the porphyrin core for various the rare earth element.

Table 1. Local structure parameters (bond length, coordination numbers and the Debye-Waller factors) for Lu(acac)TPP.

atomic bonds	R, N, σ^2
M-N	2.38, 4, 0.003
M-O	2.26, 4, 0.005
M-C_a	3.27, 4, 0.008
M-C_m	3.63, 4, 0.003
M-C_b	4.53, 8, 0.004

A further investigation of metalloporphyrins and their derivatives should give a better understanding of formation and distortion processes of the porphyrin macrocycle.

4. Conclusions

The change in the electronic state in porphyrins after introduction of rare earth into the central cavity of the molecule was demonstrated. In the XPS spectra of the investigated metalloporphyrins, single broadened peak of N1s states was observed after metallation. It indicates a small difference in the binding energy of pyrrole- and aza-N, while in the N1s spectrum of the tetraphenylporphyrin state a distinctive two-peak structure corresponding to nitrogen in the pyrrole (sp_3) and aza-group (sp_2), respectively, is observed. Thus, introduction of the RE atom into the central porphyrin area leads to a more uniform redistribution of the electron density between the pyrrole and aza-group nitrogen and the appearance of broadened N1s peak with a binding energy in a range 398.1-398.3 eV in the XPS spectra.

The integer trivalent state of lutetium in metalloporphyrins was established, which is indicated in Lu(acac)TPP by both the doublet of the 4d lutetium line in XPS spectra and the unsplitted “white line” in L_3-Lu XANES spectra.

The analysis of the valence band of porphyrins formed by π- and σ-states of porphyrin macrocycles and semi-core levels of rare-earth elements (4f, 5p) revealed its similarity with RE metalloporphyrins, which agrees with quantum mechanical calculations.

References

[1] Rosa A, Ricciardi G, Baerends E J, et. al. 2005 *Inorg. Chem.* **44(19)** 6609–22
[2] Ray P C, Leszczynski J 2006 *Chem. Phys. Lett.* **419** 4 578–83
[3] Li C, Ly J, Lei B, Fan W et. al. 2004 *J. Phys. Chem. B* **108** 28 9646–9
[4] Ghosh A, Moulder J, Bröring M, Voge E 2001 *Angew. Chem. Int. Ed.* **40** 2 431-4
[5] Gorshkova A S, Rumyantseva V D, et. al. 2014 *Apple Academic Press* **10** 375
[6] Chernyshov A A, Veligzhanin A A, et. al. 2009 *Methods Phys. Res. A.* **603** 95-8
[7] Park K T et al. 2003 *Surface Science* **529** 285-92
[8] Mozhchil R N, Ionov A M et al. 2017 *J. Surface investigation X-ray Synchrotron and Neutron,* **11** 3 517-22
[9] Diller K, Papageorgiou A C, Klappenberger F, Allegretti F, Bartha J V and Auwarter W 2016 *Chem. Soc. Rev.* **45** 1629-56
[10] Blyth R I R, Thompson J et. al. 2004 *Synthetic. Metals* **142** 293–8
[11] Ravel B, Newville M 2005 *J. Synchrotron Rad.* **12** 4 537–41
[12] Mozhchil R N, Menushenkov A P, Ionov A M , Bozhko S I, Rumyantseva V D, Trigub A L 2015 *Physics Procedia* **71** 318-22

IOP Conf. Series: Journal of Physics: Conf. Series **1238** (2019) 012003 doi:10.1088/1742-6596/1238/1/012003

Impact wear-resistance of Ti-Cr-B-N coatings produced by pulsed CAE of ceramic target

Ph V Kiryukhantsev-Korneev, K A Kuptsov

National University of Science and Technology 'MISIS', 4 Leninsky Prospekt, Moscow 119049, Russia

E-mail: kiruhancev-korneev@yandex.ru

Abstract. Ti–Cr–B–N coatings were successfully deposited using reactive pulsed cathodic arc evaporation (CAE) of ceramic TiCrB target. The influence of the substrate negative bias voltage on the coating impact wear resistance and mechanical properties has been studied. Structural investigations have been carried out using X-ray diffraction analysis, transmission and scanning electron microscopy, glow-discharge optical-emission spectroscopy, and optical profilometry. The coating properties have been estimated by impact tests, scratch-test, and nanoindentation. The obtained results show that the coatings consist of Ti(Cr)N fcc phase with crystallites 1–2 nm in size and amorphous BNx tissure. The coatings obtained at U_{bias}=-250 V demonstrated highest hardness of 24 GPa, whereas samples deposited at U_{bias}=-500V exhibited minimal wear rate (worn area of 12 μm^2) in impact conditions. Coatings obtained by pulsed CAE with intensive ion-bombardment possess high adhesion strength higher than 120 N.

1. Introduction

Hard wear resistant coatings for various applications can successfully be deposited by magnetron sputtering of composite targets fabricated using self-propagating high-temperature synthesis (SHS) [1]. Previously, magnetron sputtering have been used for sputtering of the SHS targets in the DC mode [2–5], in the pulsed DC [6-8], RF [9], and HIPIMS [10] modes. There are limited data [11] on preparing coatings by cathodic arc evaporation of SHS targets, although this method has several undoubted advantages. The main advantage of vacuum-arc evaporation, in comparison with magnetron sputtering, is the high energies of particles; dominance of ions in the flow transferred to the substrate; and, accordingly, better coating adhesion [12]. An undesirable feature of arc evaporation is the formation of the drop phase, which degrades coating properties. To exclude drop-phase formation, various techniques can be applied: mechanical separation, electromagnetic filtration, cathode rotation, and pulsed mode of CAE [13–15]. The pulse mode provides more arc spots with a higher motion velocity [15]. The increase in the arc-spot motion velocity reduces the probability of drop-phase formation and leads to a more uniform distribution of the heat flow over the target surface and volume, thus reducing the arising thermal stresses. The latter circumstance is very important for SHS ceramic cathode materials that are sensitive to thermal impacts. The aim of present work was to study coatings obtained by pulsed CAE of SHS targets and analyze the influence of the substrate negative bias voltage on the coating properties including adhesion, and impact wear resistance. The objects of study were Ti–Cr–B–N coatings showed excellent combination of mechanical, tribological properties, thermal stability, wear- and corrosion resistance [16].

2. Experimental part

The TiB–38%Cr_4Ti_9B–4%Cr_2Ti cathode 18 mm in diameter and 7 mm thick for CAE was fabricated from Ti, Cr, and B powders using SHS force pressing technology. The coatings were deposited using a UVN-2M pumping system and a Sokol 50/20 pulsed-arc evaporator. A scheme and photograph of the evaporator are shown in Fig. 1. The parameters of CAE were fixed on level: ignition frequency 20 Hz, ignition voltage 7 kV, discharge voltage and current 200 V and 1.2 A, respectively, total pressure 0.07 Pa, atmosphere 100% N_2, substrate-cathode/anode distance (*Ls-k/a*) of 20/10 mm. Negative bias voltage (*U*bias) was varied between 0 and -500V (Table 1). The typical pulse configuration is shown on Fig. 1.

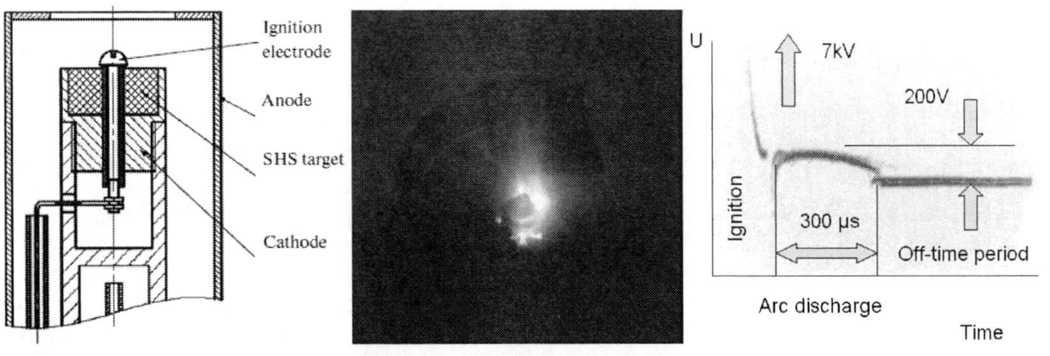

a b c

Figure 1 - Pulsed CAE of TiCrB target: scheme of evaporator (a), photo of process (b), and pulse configuration (c)

Table 1. Structure and properties of Ti-Cr-B-N coatings.

№	U_{bias} (V)	Concentrations (at.%)				D_{111} (nm)	H (Gpa)	E (Gpa)	W (%)	Ra (nm)	S (μm^2)
		Ti	Cr	B	N						
1	0	39	13	27	21	1.8	22	213	52	24	43
2	-250	-	-	-	-	1.6	24	226	53	21	70
3	-500	38	13	37	12	1.1	21	268	50	19	12

The substrate materials were Si plates (deposition time of 20 min), Ni foils (20 min), and WC-6%Co discs (1h). Before depositing coatings, the substrates were subjected to ultrasonic treatment in isopropyl alcohol. The element depth profiles were obtained by glow discharge optical emission spectroscopy (GDOES) using a Profiler 2 instrument (Horiba Jobin Yvon, France) [17]. Using these profiles, we calculated depth-averaged elemental concentrations. The coating structures were analyzed by Raman spectroscopy (LabRam Horiba JY spectrometer), transmission and scanning electron microscopy (TEM and SEM) with JEM-200CX Jeol and S-4800 Hitachi microscopes. X-ray diffraction (XRD) analysis was carried out with a Geigerflex Rigaku diffractometer using Co*Kα* radiation. The hardness, modulus of elasticity, and elastic recovery of the coatings were determined at load of 8 mN using a Nano Hardness Tester nanoindenter (CSM Instruments) equipped with a Berkovich indenter. The indentation curves were calculated by the Oliver-Pharr method [18]. Adhesion strength was estimated by means of a scratch tester (CSM Instruments, Switzerland). Impact wear were studied using "ball-on-plate" impact tester (Cemecon, Germany). The coatings deposited onto cemented carbide substrates were subjected to a range of impacts at a constant impact frequency of 50 Hz using a cemented carbide ball with 5-mm diameter. The deformation zone of a sample was cooled by compressed air. Each sample was subjected to tests at 250 N for 10^5 cycles. The coating roughness and impact cavities were studied using a Wyko 1100NT optical profiler.

3. Results and Discussion

The GDOES analysis showed that all elements were distributed uniformly through the coating thickness. Average concentrations are listed in the Table 1. Results of SEM showed that the coatings possess dense homogeneous structure (figure 2).

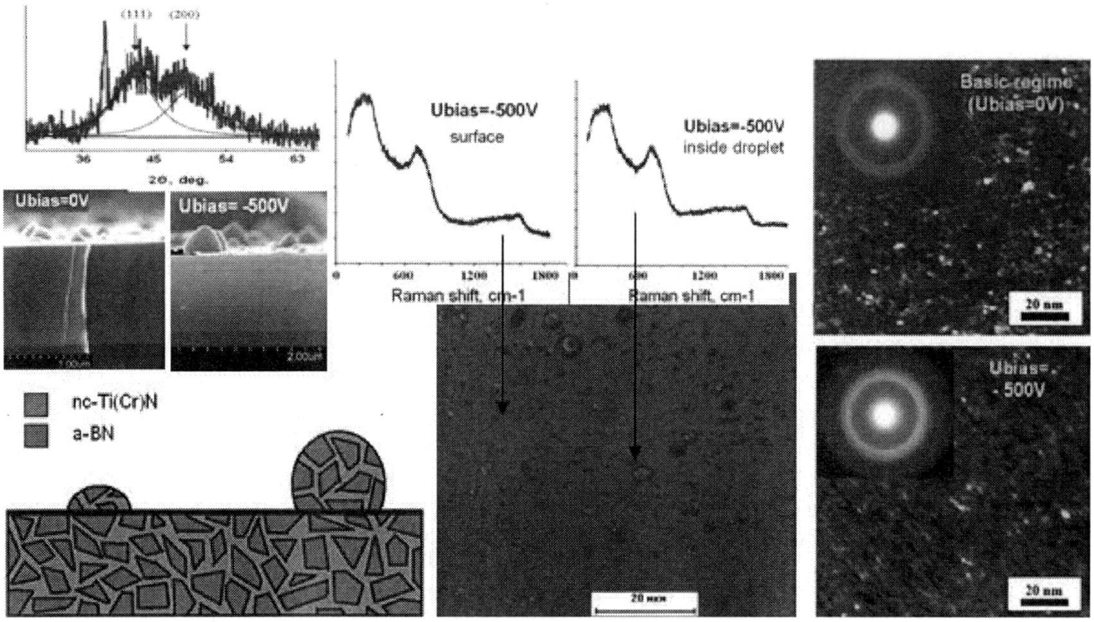

Figure 2 – Structure of Ti-Cr-B-N coatings deposited by Pulsed CAE

It can be seen that there are no columnar-structure elements that are typical for CAE deposited reference TiN coatings. Particles of crystallized drop phase, which determines the coating roughness, can be seen on the surface. The Raman data revealed, along with the TiN-based phase, boron–nitrogen bonds (peaks in the spectra at ~1380 and ~1580 cm^{-1}), which can be explained by the presence of amorphous phase and/or a small amount of hexagonal phase (with a composition similar to BN) in the coatings [19-20]. Note that according to the Raman study the flat regions and droplets have the same structure (Fig. 2). The roughness measurements for Ti–Cr–B–N coatings showed that the *Ra* parameter decreases from 24 to 19 nm with an increase in the bias from 0 to –500 V. Note that obtained results are close or slightly better compare to the results for Ti–B–N CAE coatings obtained before [19] and possess *Ra*=22–56 nm. Thus, Ti–Cr–B–N coatings have a satisfactory roughness. Decreasing of Ra parameter with U_{bias} can be explained by intensification of surface ion bombardment, sputtering/evaporation of deposited droplets, and decreasing of the droplet concentration.

Selected area diffraction patterns (Fig. 2) showed that fcc phase based on titanium nitride is main component of coatings 1-3. Comparison of dark-field images obtained by TEM showed that the crystallite size decreased from 1.8 to 1.1 nm when U_{bias} was raised from 0 up to -500V. According to the XRD results there are strongly broadened low-intensity peaks in the coating-diffraction patterns between the positions of the (111) and (200) lines of titanium nitride, as well as narrow high-intensity lines from the silicon substrate at ~38°. The mean crystallite size (D_{111}) estimated by Scherrer formula decreases from 1.8 to 1.1 nm with an increase in the bias magnitude from 0 to –500 V, which can be explained by the more intense ion bombardment of the growing coating, which leads to a higher surface energy, stronger crystallite imperfection, and larger number of crystallization centers.

Scratch-test of the Ti-Cr-B-N coatings deposited by pulsed CAE method with intensive preliminary ion etching showed that samples has excellent adhesion to the cemented carbide substrate (fig. 3).

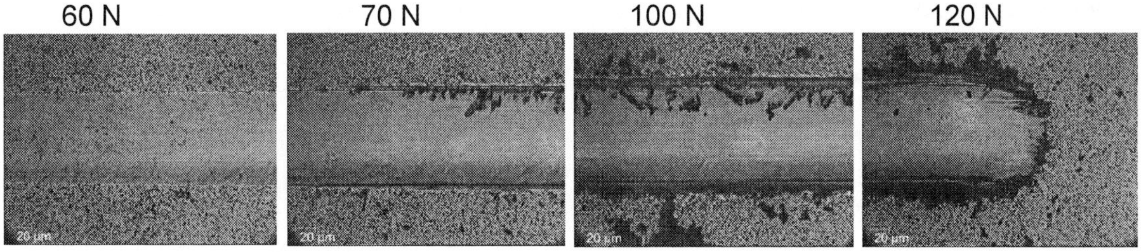

Figure 3 – Adhesion strength of Ti-Cr-B-N coatings deposited by Pulsed CAE

Microscopic examination of scar regions has not revealed the formation of any cracks and serious delaminations of the coatings up to maximal load of 120 N. Some fragments of materials around deformation zone can be attributed to the wear of the coating or substrate during test. Note that Ti-Cr-B-N coatings deposited by magnetron sputtering demonstrated critical load of failure in range 30-40 N [21]. Nanoindentation experiments showed that coating 2 has the highest hardness, H=24 GPa and elastic recovery, W=53%, whereas coating 3 exhibited maximal elastic modulus, E=268 GPa (Table 1). The internal stresses, crystallite size and the volume fraction of amorphous phase are known to strongly affect the mechanical properties of nanocomposite coatings. Previously, the hardness values for magnetron-sputtered Ti–Cr–B–N coatings obtained using a similar SHS target were found to be ~17 GPa for coatings with a crystallite size of 1–2 nm and high amorphous-phase content [21]. Optical 2D and 3D images of impact cavities obtained after the impact-tests at 250N for 10^5 cycles are shown in Fig. 4.

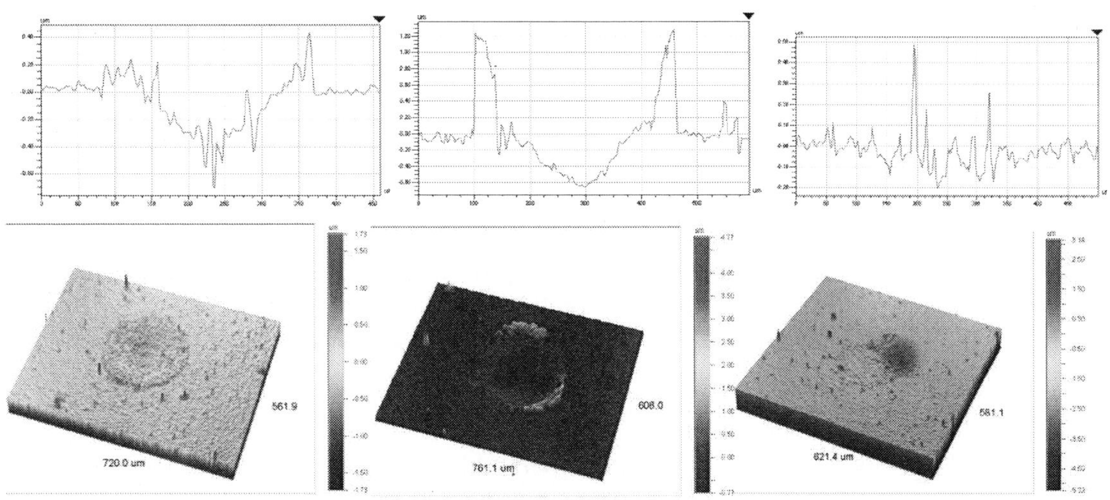

Figure 4 – 2D and 3D images of impact cavities on the surface of Ti-Cr-B-N coatings deposited by Pulsed CAE at Ubias of 0 V (a), -250 V (b), and -500 V (c)

The Ti-Cr-B-N coatings obtained at Ubias=-500V exhibited the smallest deformation zone size and impact wear area (S) (Table 1) compared to the coatings deposited with smaller bias. The high wear rate of coating 2 can be due to its high hardness. In the process of testing the microparticles break away from the coating and play role of abrasive. Similar data were obtained in the work dedicated to the Cr-B-N coatings [22]. It should also be mentioned that although the plasticity index, H/E and resistance of a coating to plastic deformation H^3/E^2, were proposed to be a reliable indicator of wear resistance in a coating [23], it does not explain the difference in behavior of the coatings during impact testing. The 3 coatings exhibited minimal H/E=0.08 and also minimal H^3/E^2 = 0.13 GPa. Perhaps internal stresses, crack resistance, crystallite size, and volume of amorphous phase are more significant parameters for this type of test.

4. Conclusion

The nanocomposite nc-TiCrN/a-BN coatings were obtained by pulsed CAE technology at substrate bias voltages of 0, -250, and -500V. The coatings demonstrated dense, non-porous structure, relative low concentration of the condensed droplet phase and low roughness (Ra=19-24 nm), hardness between 21 and 24 GPa, high critical load of adhesive failure (>120 N), good impact wear resistance (worn areas between 12 and 70 μm^2). The combination of high hardness, adhesion strength, and wear resistance makes Ti–Cr–B–N CAE deposited coatings promising candidates for protective coatings to be used in various tribological applications

Acknowledgments

We thank M.Ya. Bychkova for scratch-test. This work was carried out with financial support from the Ministry of Education and Science of the Russian Federation in the framework of state assignment No.11.7172.2017/8.9.

References

[1] Levashov E A, Mukasyan A S, Rogachev A S and Shtansky D V 2017 *International Materials Reviews* **62** 203
[2] Shtansky D V, Lyasotsky I V, D'yakonova N B, Kiryukhantsev-Korneev F V, Kulinich S A, Levashov E A and Moore J J 2004 *Surface and Coatings Technology* **182** 204
[3] Kiryukhantsev-Korneev Ph V , Iatsyuk I V, Shvindina N V, Levashov E A, Shtansky D V 2017 *Corrosion Science* **123** 319
[4] Paternoster C, Fabrizi A, Cecchini R, Spigarelli S, Kiryukhantsev-Korneev Ph V, Sheveyko A 2008 *Surface and Coatings Technology* **203** 736
[5] Shtansky D V, Sheveyko A N, Sorokin D I, Lev L C, Mavrin B N, Kiryukhantsev-Korneev Ph V 2008 *Surface and Coatings Technology* **202** 5953
[6] Audronis M, Leyland A, Matthews A, Kiryukhantsev-Korneev F V, Shtansky D V, Levashov E A 2007 *Plasma Processes and Polymers* **4** 687
[7] Lin J, Mishra B, Moore J J, Pinkas M and Sproul W D 2008 *Surface and Coatings Technology* **203** 588
[8] F. V. Kiryukhantsev-Korneev, K. A. Kuptsov, A. N. Sheveiko, E. A. Levashov, D. V. Shtansky 2013 *Russian Journal of Non-Ferrous Metals* **54** 330
[9] Zhong D, Moore J J, Mishra B M, Ohno T, Levashov E A and Disam J 2003 *Surface and Coatings Technology* **163-164** 50
[10] Kiryukhantsev-Korneev Ph V , Horwat D, Pierson J F and Levashov E A 2014 *Technical Physics Letters* **40** 614
[11] Werner Z , Stanisławski J, Piekoszewski J, Levashov E A and Szymczyk W 2003 *Vacuum* **70** 263
[12] Kiryukhantsev-Korneev F V, Shirmanov N A, Sheveiko A N, Levashov E A , Petrzhik M I, Shtanskii D V 2010 *Russian Engineering Research* **30** 909
[13] Martin P J and Bendavid A 2001 Thin Solid Films **394** 1
[14] Leng Y X Chen J Y Wang J Wan G J Sun H Yang P Huang N 2006 *Surface and Coatings Technology* **201** 157.
[15] Witke T, Schuelke T, Schultrich B, Siemroth P, Vetter J 2000 *Surface and Coatings Technology* **126** 81
[16] Kiryukhantsev-Korneev F V, Sheveiko A N, Komarov V A, Blanter M S, Skryleva E A, Shirmanov N A, Levashov E A and Shtansky D V 2011 *Russian Journal of Non-Ferrous Metals* **52** 311

[17] Kiryukhantsev-Korneev Ph V 2012 *Protection of Metals & Physical Chemistry of Surfaces* **48** 585
[18] Pharr G M, Oliver W C, Brotzen F R 1992 *Journal of materials research* **7** 613
[19] Ben el Mekki M, Djouadi M A, Guiot E, Mortet V, Pascallon J, Stambouli V, Bouchier D, Mestres N and Nouet G 1999 *Surface and Coatings Technology* **116–119** 93.
[20] Ben el Mekki M, Mestres N and Pascual J 2006 *Surface and Coatings Technology* **201** 2553
[21] Kiryukhantsev-Korneev F V, Petrzhik M I, Sheveiko A N, Levashov E A and Shtanskii D V 2007 *The Physics of Metals and Metallography* **104** 167
[22] Kiryukhantsev-Korneev Ph V, Pierson J. F Bychkova M. Y. . Manakova O. S, Levashov E. A. Shtansky D. V. 2016 *Tribology Letters* 63(3):44
[23] Leyland A and Matthews A 2000 *Wear* **246** 1

Modelling of multicomponent polydisperse systems using small-angle scattering data

P V Konarev[1,2], A E Kryukova[1,2], V V Volkov[1] and D I Svergun[3]
[1]Laboratory of Reflectometry and Small-angle Scattering, A.V. Shubnikov Institute of Crystallography of FRC "Crystallography and photonics", Russian Academy of Sciences, Leninsky pr. 59, 119333, Moscow, Russia
[2]NRC "Kurchatov Institute", pl. Kurchatova 1, 123098 Moscow, Russia
[3]European Molecular Biology Laboratory (EMBL), Hamburg Outstation, Notkestrasse 85, 22607, Hamburg, Germany

E-mail: konarev@ns.crys.ras.ru

Abstract. Algorithms for modelling of multicomponent polydisperse systems using small-angle scattering data have been developed. They cover a wide range of particle form-factors and take into account interparticle interactions. The algorithms are implemented in computer programs MIXTURE and POLYSAS, and their performance is tested on a number of simulated and experimental examples. The stability of the solution search is explored for the case of a two-component polydisperse system of spherical particles.

1. Introduction

Structural characterization of nanoscale objects is important for a better understanding of various physical, chemical and biological systems. The methods like X-ray crystallography, nuclear magnetic resonance and electron microscopy provide valuable information about particle shape with high resolution up to a few Å. However, these methods need special sample preparations, which are not always easy, and upon crystallization, freezing or isotope labelling artefacts or differences compared to the native state of the object can be observed. In contrast, small-angle scattering (SAS) does not require special treatment of the sample and permits to study materials in their native conditions yielding particle envelopes at low resolution of 1-2 nm [1]. The use of SAS in combination with other structural methods yields complementary information and improves the reliability of the structural results.

Depending on the level of structural organization one can divide the nanostructured systems into the following categories: monodisperse systems consisting of identical non-interacting particles (e.g. diluted solutions of macromolecules, metal nanoparticles, self-assembled polymer clusters etc.), polydisperse multicomponent systems containing interacting particles of different sizes and shapes (mixtures of proteins, nanoparticles and microemulsions) and partially oriented systems (liquid crystals, crystallized polymers etc.). For monodisperse objects, SAS allows one to determine the geometrical characteristics (radius of gyration, maximum size, excluded volume of the particle) and particle shape. For multicomponent systems the main structural task of SAS analysis is to determine the number and the type of components, and estimate their volume fractions, average sizes and polydispersity. For the partially oriented systems, the scattering patterns contain diffraction peaks, and their positions and widths determine the lattice plain distances, the sizes of crystallites and the degree of disorder.

During last decades effective algorithms on low resolution ab initio shape determination for monodisperse solutions have been developed and are widely used [2]. Unfortunately, they cannot be directly applied for polydisperse multicomponent systems since the simultaneous restoration of the particle shape and the size distributions is a highly ambiguous task. To overcome this complexity the shape of the particle is often approximated by simple geometrical body (sphere, cylinder, ellipsoid etc.) and the size distribution functions of the particles can be restored in this case.

In the present paper we describe the functionality and potential applications of the program MIXTURE [3] and its interactive graphical version implemented in POLYSAS [4] designed to analyze multicomponent systems containing various types of polydisperse particles and take into account the interparticle interactions. These programs belongs to the package ATSAS [2] freely available to the academic users. The performance of the programs will be illustrated by several practical examples. The tests of the stability of the restored solutions for two-component systems will be presented and the improvements in the search algorithms will be discussed.

2. Modelling of polydisperse mixtures using small-angle scattering data

The scattering intensity from a system containing distinct components is a sum of the corresponding intensities which are weighted by their volume fractions:

$$I(s) = \sum_{k=1}^{N} v_k I_k(s), \tag{1}$$

where the sum goes over different N components, v_k and $I_k(s)$ are the volume fraction and the scattering intensity from the k–th component, respectively, s is the momentum transfer $s = (4\pi/\lambda)\sin\theta$, where 2θ is the scattering angle, λ is the X-ray or neutron wavelength.

Each component (e.g. particles of certain shape) in a dilute mixture can have its own polydispersity in particle size and certain type of interparticle interactions, and its intensity in the general case can be calculated according to the following expression:

$$I_k(s) = S_k(s) \int D_k(R) V_k(R) [\Delta\rho_k(R)]^2 i_{0k}(s,R) dR, \tag{2}$$

where $D_k(R)$, $V_k(R)$ and $i_{0k}(s,R)$ denote the volume size distribution of the particles, the effective volume of the particle with size R and the normalized form-factor for the k-th component, respectively. Here, $\Delta\rho_k(R)$ is the scattering contrast of k-th component and $S_k(s)$ its structural factor, the latter responsible for the interparticle interactions (for dilute systems $S_k(s)$ is simply equal to 1).

The use of equations (1) and (2) permits one to analyze complicated cases of polydisperse systems using SAS data. A program MIXTURE [3] was developed to characterize mixtures of particles with different shapes (using geometrical bodies or high resolution models of macromolecules) containing up to ten components (types of particles).

Presently, MIXTURE contains a wide range of form-factors for the components, different versions of the program including the following types of particles: solid, hollow and core-shell spheres, solid, hollow and core-shell cylinders, solid, hollow and core-shell dumb-bells and dimers of dumb-bells, two-axes ellipsoids, high resolution models of macromolecules supplied by the user, bilayer lipid shells, mixtures of multilayer spherical lipid vesicles. Several components of the same type are allowed, e.g. one can design a system containing small and large spherical particles, long and short cylinders etc. Each component is characterized by its relevant structural parameters (e.g. the average size R, the polydispersity degree dR from Gauss or Schulz type of distributions, the volume fraction etc.) and the non-linear minimization procedures (e.g. quasi Newton method due to Broyden-Fletcher-Goldfarb-Shanno [5]) with simple bounds on the variables are employed to find the mixture with the best overall fit to the experimental data.

MIXTURE can be run automatically using a command file with model specifications and initial values of the parameters as well as their upper and lower limits. Alternatively, POLYSAS provides a

IOP Conf. Series: Journal of Physics: Conf. Series **1238** (2019) 012004 doi:10.1088/1742-6596/1238/1/012004

convenient interface where MIXTURE can be launched from the pop-up menus and each parameter of the model can be changed interactively using sliders or spin buttons. After the minimization, the best fit to the experimental data and the partial intensities from each components are stored as separate files, and the optimized parameter values are written in a log file after each run to build history of running MIXTURE under different conditions.

3. Practical applications to proteins and nanoparticles

After validation using simulated data, MIXTURE was applied to a number of experimental X-ray data sets. Some examples are presented below to illustrate the capacity of the method to quantitatively describe the structural composition of multicomponent systems. The X-ray synchrotron scattering data were recorded in collaborative user projects on the X33 beamline of the EMBL [6] on the storage ring DORIS-III (DESY, Hamburg). Figure 1 presents the X-ray scattering data from native and mutant forms of lumazine synthase from *Aquifex aeolicus* (LASQ) and *Bacillus subtilis* (LSBS) [7]. The latter enzyme catalyzes the reaction of riboflavin biosynthesis (vitamin B_2), an essential element for living organisms. In solution it forms small and large icosahedral capsids with a diameter of 16 and 30-32 nm, respectively. The relative abundance of small and large capsids strongly depends on buffer composition and on pH. Besides, incomplete and deformed capsids can also be present in solution. MIXTURE permitted to estimate volume fractions and polydispersity degree of above components at different conditions, the obtained results being in agreement with cryo-EM measurements [7] and indicate that the ability of lumazine synthase to form multiple assembly states has an important biological and physiological meaning in catalytic processes of riboflavin biosynthesis.

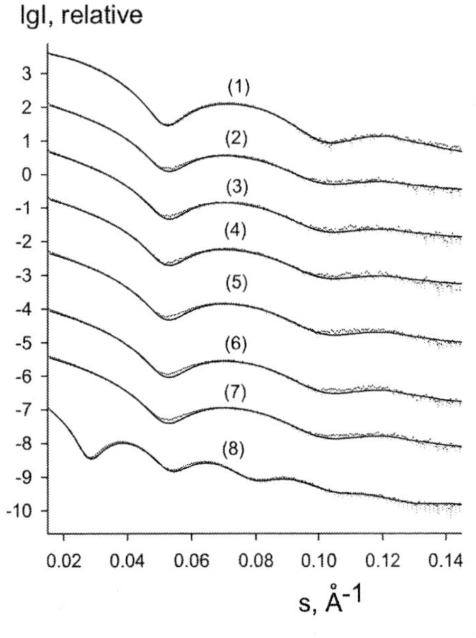

Figure 1 Experimental SAXS data (dots) and MIXTURE fits (solid lines) of lumazine synthase. (1) LSBS in borate buffer (pH 7); (2-4) LSAQ in phosphate buffer (pH 6,7,8); (5-7) LSAQ in Tris buffer (pH 7,8,9); (8) LSAQ mutant in Tris buffer (pH 8).

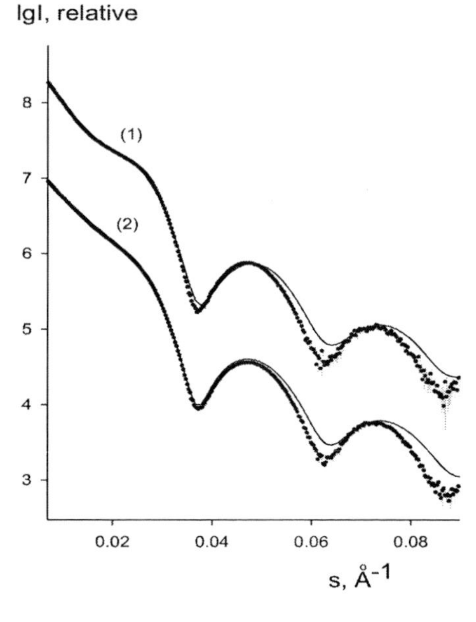

Figure 2 SAXS scattering patterns (dots with experimental errors bars) and fits from MIXTURE (solid lines) for iron oxide nanoparticles (NP) (curve 1) and NPs embedded into viral capsid of brome mosaic virus (BMV) (curve 2).

Figure 2 displays the X-ray scattering data from iron oxide nanoparticles (NP) embedded into the viral capsid of brome mosaic virus (BMV) [8]. The FeO/Fe_2O_3 nanoparticles prepared of iron oleate by thermal decomposition were stabilized by the coating with poly(maleic anhydride-alt-1-octadecene) (PMAOD) modified with poly (ethylene glycol) (PEG) tails. The encapsulation of NPs into the virus shells BMV was reached by mixing and subsequent dialysis of the solution. The experimental data were fitted by MIXTURE with a model containing mixtures monomeric virus-like nanoparticles (VNP), dimeric and tetrahedron-like "associates" of VNPs that can be seen from EM experiments [8]. It was found that the functionalized NPs tend to form significant fraction (around 50%) of these associates, while the BMV VNPs stay largely (80% volume fraction) monomeric. This result can be explained either by the screening of attractive magnetic NP interactions with the BMV shells or by the limited PEG tails penetration into the shells. The average distance between the individual NPs inside the clusters (around 28 nm) exceeds the NP sizes (25 nm) due to the PEG tails. The formation of mainly single particles for BMV VNPs is important for development of bioprobes with predicted and well-defined properties.

Both these examples demonstrate the capability of the program MIXTURE in elucidating the structural organization of multicomponent polydisperse systems.

4. Stability boundaries for the parameters of the restored solutions for two component systems of spherical particles

For complicated multi-component systems, ambiguous results may be obtained largely depending on the initial values of the parameters for the minimization. The stability of restored solutions using MIXTURE was assessed for a two-component systems of spherical particles [9, 10] with two well-separated size distributions at different noise levels. It was shown that the presence of moderate noise of Poisson type (up to 25%) does not significantly worsen the ranges of solution stability for the model parameters. At the same time the method may be highly sensitive to each individual pair of model parameters, and the contour plots showing the areas of finding the correct solution have complicated fractal-like boundaries with separated "islands" of instabilities. The latter ones can be explained both by the multimodal behaviour of the minimized functional and by the internal ambiguity of the given structural model. Similar trends were observed for the cases with partially overlapped size distributions. Three model systems were selected, which contain two types of spherical nanoparticles characterized by their radii R_1 and R_2, polydispersities dR_1 and dR_2 and volume fractions v_1 and v_2 (see Table 1).

Table 1. The "true" parameters of three theoretical model systems.

№, model number	v_1, relative	v_2, relative	R_1, nm	R_2, nm	dR_1, nm	dR_2, nm
1	0.75	0.25	6.0	8.0	0.1	1.0
2	0.75	0.25	7.0	8.0	0.3	0.3
3	0.50	0.50	6.0	8.0	0.7	0.7

The SAS curves for these systems without noise and with added Gaussian and Poisson noise (at a relative level of 25%) are displayed in Figure 3 (left panel). The corresponding volume distributions ($D_V(R)$ functions) are shown in Figure 3 (right panel). The contour plots of solution stability ranges *versus* the pair of parameters (R_1, R_2) presented in Figure 4 reveal that a wider range for the initial guess of the parameters can be chosen in the case of partially overlapped size distributions compared to the systems with two well-separated size distributions.

An improved search strategy may thus be suggested based on the above results. In future, such analysis will be expanded to real experimental data to help defining distinct model cases with higher probabilities for finding the 'true' solution for the system.

IOP Conf. Series: Journal of Physics: Conf. Series **1238** (2019) 012004 doi:10.1088/1742-6596/1238/1/012004

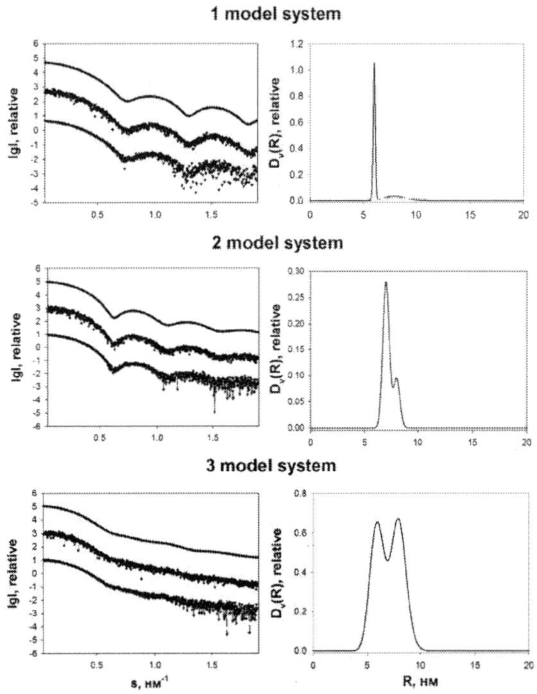

Figure 3. The left panel contains the simulated theoretical SAS curves with the parameters from Table 1 (curve 1 - with relative noise level 0%, curve 2 - Gaussian noise 25%, curve 3 - Poisson noise 25%). The curves are vertically displaced for better visualization. The right panel displays the volume size distribution functions $D_V(R)$.

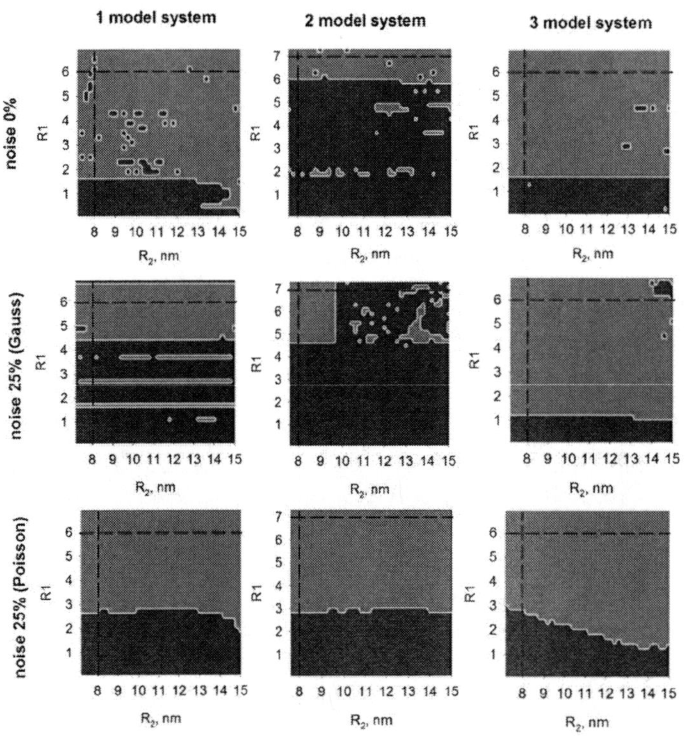

Figure 4. The contour plots of solution stability ranges for three model systems depending on the pair of parameters (R_1, R_2). The left column corresponds to the model 1, the central column – the model 2, the right one – the model 3. The plots for the curves without noise (relative noise level 0%) are located in the first row, with Gaussian noise (relative noise level 25%) - in the second row, with Poisson noise (relative noise level 25%) - in the third row. The interception of the dotted lines corresponds to the theoretical ('true') solution. If the restored parameters are located within 5% in the vicinity of 'true' solution, then the solution is treated as successful (red areas), otherwise the solution is treated as false (blue areas).

5. Conclusions

A program MIXTURE designed for the structural characterization of multicomponent polydisperse systems using SAS data was developed and successfully applied to a wide range of systems, including microemulsions [11], biological macromolecules [7] and nanoparticles [8]. The analysis of solution stability using the program on two-component systems [9, 10] showed a strong dependence of finding the 'true' solution on the type of the model parameters and on the experimental noise level and provided hints for the improved search strategy.

Acknowledgements

The authors acknowledge the support from the Federal Agency of Scientific Organizations (Agreement No 007-ГЗ/Ч3363/26).

References

[1] Feigin L A and Svergun D I *Structure Analysis by Small-Angle X-Ray and Neutron Scattering* 1987 (New York: Plenum Press)
[2] Franke D, Petoukhov M V, Konarev P V, Pankovich A, Tuukkanen A, Mertens H D T, Kikhney A G, Hajizadeh N R, Franklin J M, Jeffries C M and Svergun D I 2017 *J. Appl. Cryst.* **50** 1212-25
[3] Konarev P V, Volkov V V, Sokolova A V, Koch M H J and Svergun D I 2003 *J. Appl. Cryst.* **36** 1277-82
[4] Konarev P V, Volkov V V and Svergun D I 2016 *J. Phys. Conf. Ser.* **747** 012036
[5] Gill P.E., Murray W., Wright M.H. *Practical Optimization* 1981 (London: Academic Press)
[6] Blanchet C E, Zozulya A V, Kikhney A G, Franke D, Konarev P V, Shang W, Klaering R, Robrahn B, Hermes C, Cipriani F, Svergun D I and Roessle M 2012 *J. Appl. Cryst.* **45** 489-95
[7] Zhang X, Konarev P V, Petoukhov M V, Svergun D I, Xing L, Cheng R H, Haase I, Fischer M, Bacher A, Ladenstein R and Meining W 2006 *J. Mol. Biol.* **362** 753-70
[8] Malyutin A G, Cheng H, Sanchez-Felix O R, Carlson K R, Stein B D, Konarev P V, Svergun D I, Dragnea B and Bronstein L M 2015 *ACS Appl. Mater. Interfaces* **7** 12089-98
[9] Kryukova A E, Konarev P V and Volkov V V 2017 *J. Phys. Conf. Ser.* **941** 012069
[10] Kryukova A E, Konarev P V and Volkov V V 2018 *Cryst. Reports* **63** 26-31
[11] Svergun D I, Konarev P V, Volkov V V, Koch M H J, Sager W F C, Smeets J and Blokhuis E M 2000 *J. Chem. Phys.* **113** 1651-65

IOP Conf. Series: Journal of Physics: Conf. Series **1238** (2019) 012005 doi:10.1088/1742-6596/1238/1/012005

Comparison of levitation forces of the cc-tapes stacks from different manufacturers

M A Osipov, A S Starikovskii, I A Rudnev

National Research Nuclear University MEPhI (Moscow Engineering Physics Institute), Kashirskoe sh. 31, 115409 Moscow, Russia

E-mail: MAOsipov1@mephi.ru

Abstract. High-temperature superconductors (HTSC) have a significant potential for application in systems based on magnetic levitation. The wide spread of second-generation HTSC tapes (CC-tape from coated conductors) allows the use of stacks of CC-tapes as an alternative to bulk superconductors, particularly in levitation systems. The paper presents the results of research of CC-tapes of various manufacturers - SuNAM, THEVA and SuperOx. The tapes were cut into square 12 mm × 12 mm fragments and stacked in stacks, the number of tapes varying from 10 to 100. The measured dependences of the levitation force on the distance between the magnet and the stack have similar patterns for tapes of all manufacturers - the force of interaction increases with increasing thickness of stacks. However, the influence of the thickness of the stack on the levitation force for different CC-tapes is of a different nature.

1. Introduction

High-temperature superconductors (HTS) have the significant potential for magnetic-force-based applications. Currently, the widely used material in this application is bulk REBa2Cu3O7-x (REBCO, wherein RE corresponds to the rare earth elements). This material is capable of trapping magnetic fields and sustaining stable levitation a above a permanent magnet without active control. All of this opens up a wide range of applications: levitation bearings, maglev (magnetic levitation) trains, flywheel energy storage, motors and generators [1, 2]. However, wide spread of CC-tapes enables the use of HTS stacks as alternative to bulk superconductors. Using stacks of CC-tapes makes it possible to overcome some of the difficulties associated with the bulk drawbacks such as brittleness, low thermal conductivity.

Nowadays, researchers actively investigate the possibility of using HTS tapes in levitation applications. The stack, formed from pieces of industrial tape, is superior to the bulk superconductor on the tensile strength and thermal stability at low temperatures [3]. A detailed comparison of the advantages and disadvantages of stacks of tapes and bulk HTS for levitation applications is given in our previous work [4]. In our previous papers we studied the possibility of capturing the magnetic field by stacks of tapes [5, 6] and the levitation force of the stack of HTS tapes above the permanent magnet with respect to the lateral displacement, vibrations and crossed magnetic fields [4, 7, 8]. Also, we have studied the levitation force and trapped flux for various magnetizations of the stack [9].

For the development of an HTS bearing and transportation systems, it is important to predict the levitation performance of a stack of CC-tapes based on tape characteristics, such as critical current and engineering critical current density.

In our work we present new results on the vertical levitation force of HTS stacks of various thickness, from 10 to 100 tapes provided by different manufacturers: Theva, Sunam, SuperOx.

2. Experimental Techniques

2.1. Samples

Samples of different manufacturers were tested: Theva, Sunam, SuperOx. In the experiments on measuring the levitation force we used 12 mm wide tapes, coated with silver, without copper coating. The main characteristics of the tapes are presented in Table 1.

The Theva manufacturer provided two tapes with serial numbers 17617 and 17618, which are samples of the commercially produced TPL1100 tape. The minimum critical current of these tapes, measured by the manufacturer by use of TapeStar at 77K, self-field, are 360 and 518 A, the average critical current is 429 and 670. The substrate of the tapes consists of Hastelloy c276, 90-100 μm thick. The tape produced by Sunam has a substrate of similar thickness of about 100 microns, the substrate material is non-magnetic steel. The average critical current of the tape measured by the four-contact method is about 850 A. Tapes manufactured by SuperOx differ from all the above samples with a thinner substrate, and as a result, with thinner tapes and higher engineering current density. The average critical current of the tape is about 550 A according to the manufacturer specification.

Table 1. Comparative characteristics of CC-tapes of different manufacturers.

Tape	Average critical current	Thickness of stack of 100 tapes, mm	Substrate
Theva 617	429	11	Hastelloy C276 90 – 100 μm
Theva 618	670	11	Hastelloy C276 90 – 100 μm
Sunam	850	11,5	Non-magnetic SUS : thickness >80 μm
SuperOx	550	5,5	Hastelloy C276 40 – 50 μm

2.2. Experimental setup

Measurements of levitation force were conducted by use of home-made experimental setup. A basic experiment details can be found in [9]. A stack of CC-tapes was placed in a brass holder, which, in turn, is placed in a cryostat with liquid nitrogen (Fig.1a). The cryostat with the sample holder (4 on Fig.1b) was mounted on a motion system (1 on Fig.1) capable of vertical and lateral movements with a minimal step of 2 μm

In our experiment we used a stack of four permanent magnets Nd-Fe-B (5 in Fig.1a) for supplying an applied magnetic field. Each magnet was 30 mm in diameter, 10 mm high, made of N45 grade material providing a surface remnant flux density of 0.30–0.33 T. The whole stack of magnets had a surface flux density of 0.58 T. The stacked magnets were placed on top of the load cell (6 in Fig. 1a) to measure the vertical levitation force.

All measurements were carried out in the zero field cooling mode. The stack was placed at a vertical distance of 80 mm away from the permanent magnet, where external magnetic field could be neglected. After that, the cryostat was filled with liquid nitrogen to cool the sample. Further, the sample was lowered down with a step of 0.5 mm with repulsive force measurements on each step. After that, the sample was lowered down in 0.5 mm steps with measurements of the repulsive force at each step. During measuring procedure, the sample displacement was stopped for 500 milliseconds on each movement step. The distance between the stack and the permanent magnets in the final position was 5.5 mm. The stray magnetic field provided by the stack of magnets at this distance was 0.35 T.

IOP Conf. Series: Journal of Physics: Conf. Series **1238** (2019) 012005 doi:10.1088/1742-6596/1238/1/012005

The sample was descended from an initial position to a test final position. The distance between the stack and the permanent magnets in the final position was 7 mm. At the final distance the sample was moved laterally in order to evaluate the effect of the lateral displacement on the repulsive or lateral forces. In field cooling (FC) mode, the sample was first descended to its final position, and then it was cooled by liquid nitrogen. Examples of descending curve measurements for stacks of 100 tapes are shown in Fig. 1b.

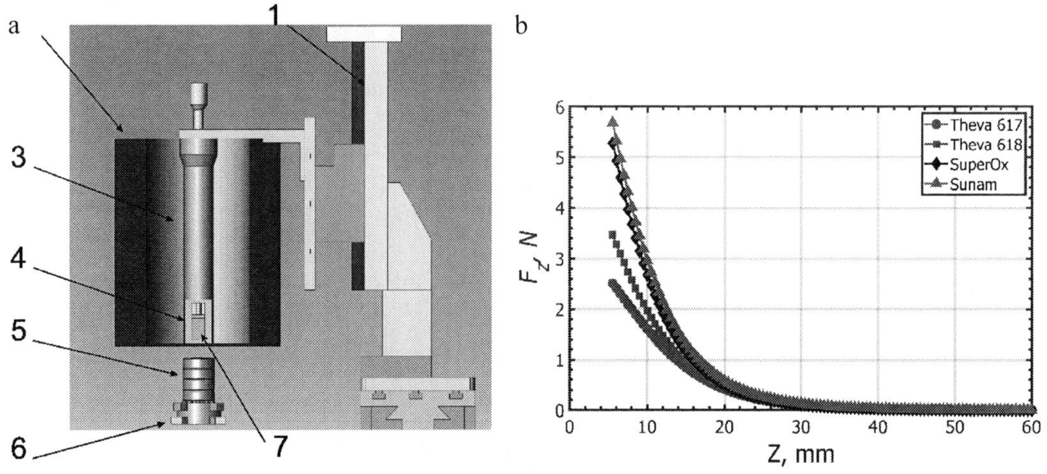

Figure 1. a) Experimental setup for levitation force measurements. 1 – motion system, 2 – cryostat, 3 – liquid nitrogen, 4 – brass holder, 5 – stack of Nd–Fe–B permanent magnets, 6 – load cell, 7 – stack of tapes.

b) The dependence of repulsive force FZ between the stack and the permanent magnet on the vertical coordinate Z during descending of the stack. Results for stacks of 100 layers are shown. Legend displays the manufacturer of the tape. The temperature was 77 K.

3. Results and Discussion

From the descending curves in Fig. 1b, it is seen that samples of tapes with a higher critical current have a higher levitation force. A sample of SuperOx is standing out. Despite the lower critical current of the tapes compared to Sunam, SuperOx shows a comparable to Sunam levitation force. This can be explained by the smaller thickness of SuperOx tapes, which leads to higher values of the engineering current density. It also stands out that with the decrease in distance, and hence with the growth of the field, the differences between the tapes levitation force only increase. This may indicate that in higher fields it becomes more relevant to use CC-tapes with a higher critical current.

Let us consider in more detail the maximum repulsive force - the repulsive force measured at the minimum distance to the magnet. The dependence of this force on the number of tapes in the stack for different manufacturers is presented in Figure 2a. It can be seen from the figure that for the tapes of Sunam and Theva companies stacks of tapes with a higher critical current demonstrate higher levitation force. Although the critical current of the SuperOx tape is comparable to the Theva 618's critical current, the levitation force in this case is higher and comparable in magnitude to Sunam. This can be explained, as mentioned above, by the smaller thickness of SuperOx tapes, which leads to higher values of the engineering current density. It can be seen from the graphs that curves with a higher value of levitation force saturate faster. The difference in the levitation force at first increases with the number of tapes. However, for a higher number of tapes in the stack, the advantage of tapes with a higher critical current is reduced. Under the experimental conditions, it is most effective to use tapes with a higher critical current value for stacks of about 50-60 tapes.

Fig. 2b shows the dependence of the maximum repulsive force on the number of tapes in the stack normalized to the engineering density of the critical current in one tape (77K, self-field). It can be seen from the graphs that the curves for the samples of tapes produced by SuperOx and Theva coincide well with each other. However, Sunam's tapes differ greatly in their levitation characteristics from other tapes. This suggests that the engineering current density of tapes can not be a universal factor that allows predicting the levitation properties of a stack of tapes. The difference in the behavior of stacks may be due to the difference in the type of pinning and is explained by the different dependancies of critical current on external magnetic field and on the angle between field and tape plane.

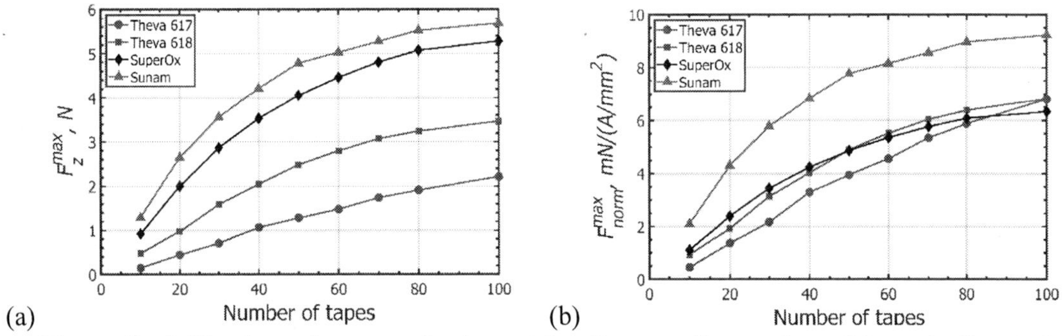

Figure 2. a) The dependence maximal repulsive force on the number of tapes in the stack for different manufacturers. The measurement temperature was 77 K.

b) The dependence maximal repulsive force on the number of tapes in the stack for different manufacturers, normalized to the engineering density of the critical current in one tape (77K, self-field). The measurement temperature was 77 K.

4. Conclusions

We have carried out an investigation of the influence of the number of the elements in the stack on levitation force for stacks of 10–100 GdBCO tapes provided by different manufacturers. All measurements were performed in ZFC mode in liquid nitrogen. Based on the obtained results, we can conclude that:

1) As a rule, with increasing engineering current density in a single tape, the levitation force of a stack of CC-tapes increases.

2) The engineering current density of CC-tapes cannot be a universal factor that allows to accurately predict the levitation properties of a stack of tapes.

3) For large values of the engineering density of the critical current, the levitation force goes to saturation faster.

Acknowledgments

The study was carried out at the expense of a grant from the Russian Science Foundation (Project No. 17-19-01527).

References

[1] Sotelo G G, et al. 2014 *J. Phys. Conf. Ser.* **507(3)** 032017
[2] Werfel F N, et al. 2012 *Supercond. Sci. Technol.* **25(1)** 014007
[3] Patel A, et al. 2013 *Appl. Phys. Lett.* **102** 102601
[4] Osipov M, Abin D, Pokrovskii S, Rudnev I 2016 *IEEE Trans. Appl. Supercond.* **26(4)** 3601704
[5] Osipov M, et al. I 2015 *Prog. Supercond. Cryog.* **17(1)** 21–4
[6] Rudnev I, et al. 2015 *Phys. Procedia* **65** 141–4
[7] Abin D, Osipov M, Pokrovskii S, Rudnev I 2016 *IEEE Trans. Appl. Supercond.* **26(3)** 8800504
[8] Pokrovskii S, Abin D, Osipov M, Rudnev I 2016 *IEEE Trans. Appl. Supercond.* **26(3)** 8201304.
[9] Osipov M, Abin D, Pokrovskii S, Rudnev I 2017 *IEEE Trans. Appl. Supercond.* **27(4)** 7851052

IOP Conf. Series: Journal of Physics: Conf. Series **1238** (2019) 012006 doi:10.1088/1742-6596/1238/1/012006

Structure and magnetic hysteresis properties of $NdFe_{11}Ti$ based alloys and their nitrides after different methods of obtaining

M V Zheleznyi, I V Shchetinin, M V Gorshenkov, A I Bazlov, A V Kamynin, P Aggrey, A G Savchenko
National University of Science and Technology «MISiS» (NUST «MISiS»), Leninskyi pr. 4, 119049 Moscow, Russia

E-mail: markiron@mail.ru

Abstract. We have studied the effect of synthesis methods on the structural state and magnetic hysteresis properties of the $NdFe_{11}Ti$ compound with $ThMn_{12}$ structure obtained by arc-melting, annealing, melt spinning and nitriding. We found prolonged high-temperature annealing and melt spinning form a single-phase state. However, the grain size of the alloy after melt spinning is 1000 times as less which leads to high magnetic properties as compared to annealing. Nitriding increases the magnetic hysteretic properties after both annealing and melt spinning.

1. Introduction

Compounds based on Fe and rare-earth elements with the $ThMn_{12}$ structure have been known for more than 30 years. RFe_{12} (R: rare-earth element) compounds have fundamental magnetic properties comparable to the compound $Nd_2Fe_{14}B$. $NdFe_{12}$ after nitriding has higher magnetic properties: saturation magnetization 1.66 T, Curie temperature 550 ° C and anisotropy field 6.4 MA / m [1]. However, binary compounds RFe_{12} are stable only in the thin films forms. The stabilizing elements replacing Fe in the compounds $RFe_{12-x}M_x$ (M = Al, Cr, V, Ti, Mo, W, Si or Nb) are used to obtain a phase with the $ThMn_{12}$ [2-7]. Nitriding of the compounds $RFe_{12-x}M_x$ increases saturation magnetization, the Curie temperature and anisotropy field [8]. However, now these alloys are impractical for application because of small magnetic hysteretic properties compared to the Nd-Fe-B system. Therefore, studying of structure's formation and magnetic properties of $NdFe_{11}Ti$ alloys after melt spinning and nitriging is an actual task.

2. Experimental

Alloys of the composition $NdFe_{11}Ti$ with a mass 150 g were obtained by arc melting in an argon atmosphere. The ingot was divided into two parts. One part was annealed in an argon atmosphere at a temperature of 1100 ° C for 168 hours. Another part of the ingot was quenched from the liquid state in an argon atmosphere on a rotating copper wheel of a DVX-II equipment. The linear rotation speed of the copper wheel was 30, 35 and 47 m/s. After both annealing and melt spinning the samples were exposed to nitriding at pressure of 2 atm and temperature of 500 °C for 4 hours. Their chemical composition was defined by a X-ray fluorescence wave-dispersion spectrometer Rigaku ZSX Primus II. X-ray diffraction studies were made using a Rigaku Ultima IV diffractometer (CoK_α radiation, λ = 1.79021 Å). Spectra were processed by the Rietveld method via the Rigaku PDXL 2 software. Quantity of amorphous phase was defined by the method by Savchenko et al. [9]. The microstructure of the cast alloy was studied by a Tescan Vega 3SB SEM with a Oxford Instruments EDX detector for elemental microanalysis The microstructure of the quenched samples was studied by a TEM JEOL JEM 2100 with EDX detector for elemental microanalysis. The magnetic hysteresis properties of samples were measured by a Quantum Design PPMS system in a magnetic field up to 11.2 MA/m at room temperature.

3. Results and discussion

Table 1 shows chemical composition of the as-cast sample which is very close to nominal compound $NdFe_{11}Ti$. Fig. 1 shows the XRD results for samples after arc melting and annealing. The results of phase analysis are shown in Table 2. Sample after arc-melting was characterized by multi-phase state: $NdFe_{11}Ti$ (70%), α-Fe (20%), Fe_2Ti (10 %).

Table 1. Chemical composition of $NdFe_{11}Ti$ alloy after arc-melting

Sample	Chemical composition (mass. %)			Total content of impurity elements (mass. %)
	Nd	Fe	Ti	Mo, Cu, Ni, Cr, Si, Al
$NdFe_{11}Ti$	17.89	76.17	5.94	-
Experiment	17.86 ± 0.05	75.33 ±0.03	6.14 ±0.02	≤ 0.53

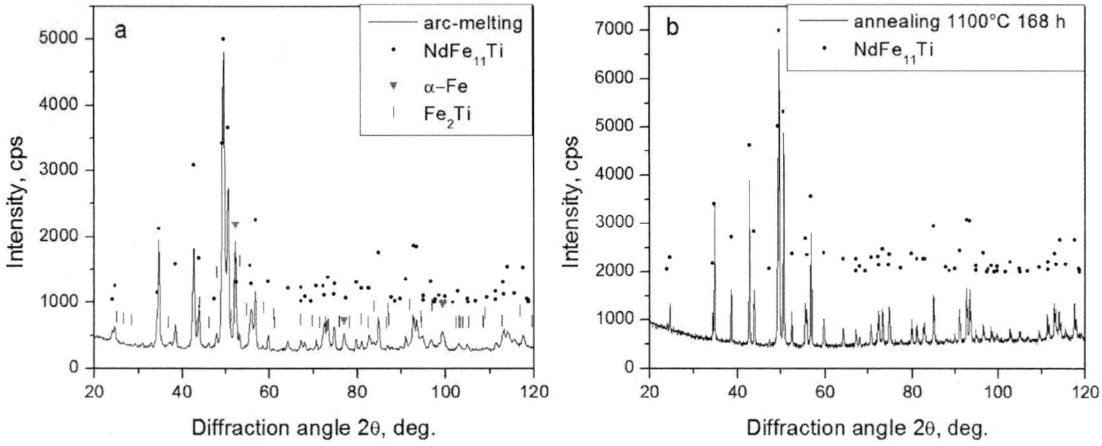

Figure 1. X-ray diffraction patterns of $NdFe_{11}Ti$ alloy after arc-melting (a) and annealing (b)

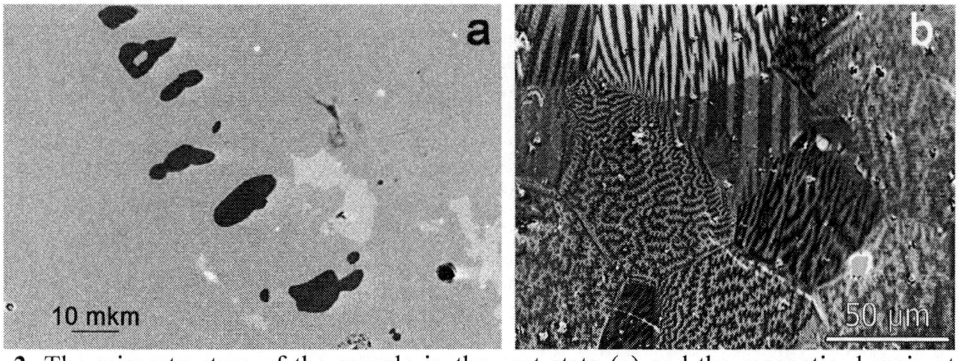

Figure 2. The microstructure of the sample in the cast state (a) and the magnetic domain structure after annealing (b)

The results obtained are confirmed by scanning electron microscopy in contrast of the back-scattered electrons. Three phases are found on the image of the microstructure of the alloy (Fig. 2a).

The cast alloy after annealing was characterized by a single-phase state of the intermetallic compound $NdFe_{11}Ti$. Sample after annealing has grains with an average size of about 150 μm that are shown in image obtained by the Kerr method (Fig. 2b). The magnetic domain structure corresponds to a ferro-magnetic material with the uniaxial type of magnetocrystalline anisotropy [10]. The magnetic domains at room temperature were oriented in each grains along the axis of easy magnetization (*c*-axis of the crystal lattice of the $ThMn_{12}$ structure).

Table 2. The XRD results of the $NdFe_{11}Ti$ alloy in the cast state and after annealing

Phase composition (Space group)	Content (vol. %)	Lattice parameters			The volume of a unit cell (V, Å^3)
		a, Å	*c*, Å	*c/a*	
arc-melting state					
$NdFe_{11}Ti$ (*I4/mmm*)	70.0(3)	8.589(2)	4.796(9)	0.556	353.87(17)
α-Fe (*Im-3m*)	20.0(8)	2.879(9)	2.879(9)	1	23.86(15)
Fe_2Ti (*P6₃/mmc*)	10.0(7)	4.787(5)	7.815(9)	1.633	155.09(16)
annealing state					
$NdFe_{11}Ti$ (*I4/mmm*)	≈ 100.0(5)	8.587(4)	4.793(2)	0.558	353.41(14)

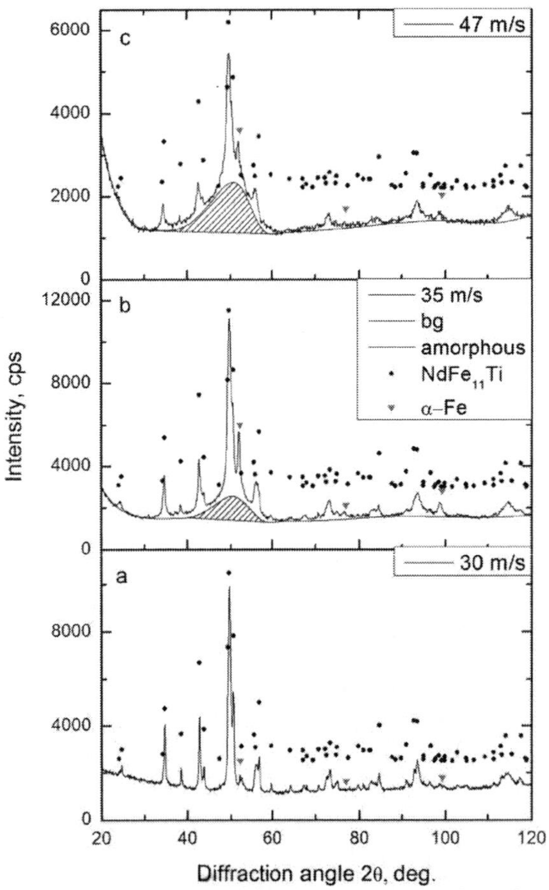

Figure 3. X-ray diffraction patterns for the $NdFe_{11}Ti$ alloy after melt spinning with v_s = 30 m/s (a), 35 m/s (b) and 47 m/s (c)

Figure 4. The results of transmission electron microscopy of samples after melt spinning with v_s = 30 m/s

Fig. 3 shows diffraction patterns of samples after melt spinning with different wheel speeds v_s = 30, 35 and 47 m/s, respectively. Results of the XRD analysis are shown in Table 3. The results of X-ray diffraction analysis show that during melt spinning a formation of an amorphous phase is possible. The volume fraction of the amorphous phase increases from 0 % to 30 % with increasing the wheel speed from 30 to 47 m/s, respectively. Melt spinning with wheel speed v_s = 30 m/s leads to the formation of $NdFe_{11}Ti$ (97 %) compound and a few amount of α-Fe (3%). Fig. 4 shows results of transmission electron microscopy study of the $NdFe_{11}Ti$ alloy after melt spinning with a wheel speed v_s = 30 m/s. The microstructure of the alloy is the $NdFe_{11}Ti$ phase in a equiaxed polycrystalline state with an average grain size of 120-150 nm. An amorphous phase along the grain boundaries of the main phase was not found.

Thus, using annealing at a temperature of 1100 °C for 168 hours and melt spinning with a wheel speed v_s = 30 m/s the single-phase state of $NdFe_{11}Ti$ compound were obtained. The average grain sizez of the alloy microstructure after melt spinning and after annealing differed by 1000 times (120-150 nm and 150 μm, respectively).

Table 3. The XRD results of the $NdFe_{11}Ti$ alloy after melt spinning with various wheel speed

Phase (Space group)	Content (vol. %)	Lattice parameters			The volume of a unit cell (V, Å^3)
		a, Å	*c*, Å	*c/a*	
v_s = 47 m / s					
$NdFe_{11}Ti$ (*I4/mmm*)	61.0(9)	8.542(3)	4.878(4)	0.571	355.90(14)
α-Fe (*Im-3m*)	6.0(4)	2.890(4)	2.890(4)	1	24.13(15)
Amorphous	33(4)	-	-	-	-
v_s = 35 m / s					
$NdFe_{11}Ti$ (*I4/mmm*)	75.0(9)	8.528(5)	4.809(6)	0.564	349.78(18)
α-Fe (*Im-3m*)	5.0(4)	2.884(9)	2.884(9)	1	24.00(14)
Amorphous	20(3)	-	-	-	-
v_s = 30 m / s					
$NdFe_{11}Ti$ (*I4/mmm*)	97.0(9)	8.555(7)	4.806(5)	0.562	351.72(15)
α-Fe (*Im-3m*)	3.0(8)	2.879(3)	2.879(3)	1	23.87(14)

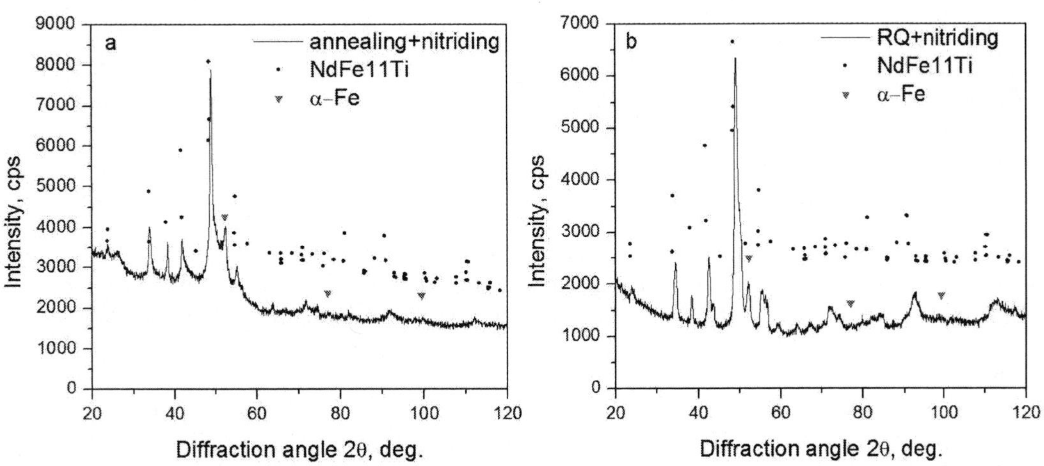

Figure 5. X-ray diffraction patterns of the $NdFe_{11}Ti$ alloy in the heat-treatment state after nitriding (a); in the melt-spun state at v_s = 30 m/s after nitriding (b)

The XRD results for the different samples in the heat-treated and as-spun (a wheel speed 30 m/s) states after nitriding are shown in Fig. 5 and Table 4. After nitriding the volume of the unit cell of the $NdFe_{11}Ti$ phase in the heat-treated state increased by 6%. Nitriding also leads to release of a small amount of α-Fe (5%) from the matrix phase. These results are inconsistent with previously published literature data. The relative increase in the unit cell volume (the volume effect ΔV/V) for the $NdFe_{11}Ti$ → $NdFe_{11}TiN$ transition was 3-4%. The volume effect of the $NdFe_{11}Ti$ phase for the alloy in the melt-spun state at v_s = 30 m / s after nitriding was 3%. Also the fraction of α-Fe was increased from 3 to 10%.

Table 4. The XRD results for $NdFe_{11}Ti$ alloy after nitriding

Treatment	Phase composition (Space group)	Content (vol. %)	Lattice parameters			The volume of a unit cell (V, Å^3)
			a, Å	*c*, Å	*c/a*	
Annealing + N_2	$NdFe_{11}TiN$ (*I4/mmm*)	95.0(3)	8.734(9)	5.022(9)	0.575	383.06(14)
	α-Fe (*Im-3m*)	5.0(8)	2.876(4)	2.876(4)	1	23.79(20)
Melt spinning + N_2	$NdFe_{11}TiN$ (*I4/mmm*)	90.0(3)	8.645(4)	4.838(3)	0.560	361.60(14)
	α-Fe (*Im-3m*)	10.0(8)	2.870(3)	2.870(3)	1	23.64(15)

Magnetic hysteresis properties of samples in the heat-treatment and as-spun at v_s = 30 m/s states after nitriding are shown in Table 5. The alloys after annealing and nitriding were in a large-grain state and had low coercive force $_iH_c$ (5.2 and 29.6 kA/m) and remanence magnetization σ_r (3 and 20 A·m^2/kg). Saturation magnetization σ_s had increased from 138 to 143 A·m^2/kg which corresponds to the literature data. Coercive force for sample in the as-spun state after nitriding had increased from 23.2 to 84.2 kA/m. This fact is linked with increase of magnetocrystalline anisotropy field after nitriding [8]. Similarly, remanent magnetization and saturation magnetization after nitriding had respectively increased from 43 to 46 A·m^2/kg and from 122 to 139 A·m^2/kg which is linked with volume effect. Saturation magnetization of the sample after melt spinning because of defect crystalline structure was σ_s = 122 A·m^2/kg.

Table 5. Magnetic hysteresis properties of the alloy after various treatments at room temperature

Treatment	Coercive force ($_iH_c$, Oe (kA/m))	Remanence magnetization (σ_r, A·m^2/kg)	Saturation magnetization (σ_s, A·m^2/kg)
Annealing	65 (5.2)	3	138
Annealing + N_2	370 (29.6)	20	143
Quenching from liquid	290 (23.2)	43	122
Quenching from liquid + N_2	1053 (84.2)	46	139

The observed differences in the magnetic characteristics are associated with the features of the microstructure and substructure (dispersity and microdeformations), as well as with different the volume effect for the samples in the heat-treatment and as-spun states both before and after nitriding.

4. Conclusion

We have successfully made magnetic material from $NdFe_{11}Ti$-$NdFe_{11}TiN$ alloys by arc melting, homogenizing annealing, melt spinning and nitriding. The sample after homogenizing annealing at a temperature of 1100 °C for 168 hours contains the hard magnetic phase with $ThMn_{12}$ structure as majority phase. Sample in an almost single-phase state (97% of 1:12 phase) was produced by melt spinning with the average grain size about 150 nm. Nitriding of the alloy leads to an increase in main magnetic hysteresis properties. The maximum magnetic hysteresis properties were obtained by a combination of melt spinning and nitriding: $_iH_c$ = 84.2 kA/m, σ_r = 46 A·m^2/kg, σ_s = 139 A·m^2/kg.

Acknowledgements

The work was financially supported by the Ministry of Education and Science of the Russian Federation in the framework of the grant of the President of the Russian Federation to support young Russian scientists (Grant No. MK-3616.2017.2).

References

[1] Hirayama Y, et al. 2015 *Scr. Mater.* **95** 70-2 DOI: doi.org/10.1016/j.scriptamat.2014.10.016
[2] Ken Ohashi, et. al. 1988 *J. of the Less Com. Met.* **139(2)** L1-L5 DOI: doi.org/10.1016/0022-5088(88)90020-3
[3] De Mooij D B, Buschow K H J 1988 *J. of the Less Com. Met.* **136** 207-15 DOI: 10.1016/0022-5088(88)90424-9.
[4] Fuquana B, et al. 2005 *J. Magn. Magn. Mater.* **290–291** 1192
[5] Suzuki S, et. al., 2016 *J. of Mag. and Magn. Mat.* **401** 259 -68
[6] Suzuki S, et al. 2014 *AIP Adv* **117** 131
[7] Hirayama Y, et al. 2015 *Scr. Mater* **95** 70 - 2
[8] Akayama A, Fujii H, Yamamoto K, Tatami R 1994 *J. of Magn. and Magn. Mat.* **130(1–3)** 99-107. DOI: 10.1016/0304-8853(94)90662-9
[9] Savchenko A G , Medvedeva T M, Shchetinin I V, et.al. 2017 *J. of All. and Comp.* **707** 205-9. DOI: 10.1016/j.jallcom.2017.01.002.
[10] Bo-Ping Hu, Hong-Shuo Li, Gavigan J P et. al. 1989 *J. Phys.: Condens. Matter* **1** 755-70

Formation of ultrathin MoS_2 films using laser-based methods

V Yu Fominski, R I Romanov, V N Nevolin, D V Fominski, O V Komleva, V V Popov

National Research Nuclear University MEPhI (Moscow Engineering Physics Institute), Kashirskoe sh. 31, 115409 Moscow, Russia

E-mail: vyfominskij@mephi.ru

Abstract. A comparative analysis of the abilities of several novel methods to produce ultrathin molybdenum disulphide (MoS_2) films containing from 1 to 10 molecular layers was carried out. To deposit MoS_x films and MoO_x precursor films, the atomic flux was formed by laser ablation of Mo, MoS_2, and MoO_3 targets. Saturation with sulphur of the deposited layers was performed using a reactive gas (hydrogen sulphide) or by thermally activated treatment of thin-film precursors in a sulphur vapor. It has been established that the use of hydrogen sulphide makes it possible to obtain ultrathin MoS_2 films at relatively low temperatures ~ 350 °C. However, these films contained local defects which were absent in the films prepared by the treatment of thin film MoO_x precursors in sulphur vapours at higher temperatures (≥ 650°C).

1. Introduction

Transition metal dichalcogenides (TMDs), such as MoS2, are of particular interest to researchers, since the specific structure the TMDs allowed to form ultrathin films (2D and quasi-2D) containing the required number of mutually oriented molecular planes [1,2]. Studies of 2D-TMD materials continue to develop rapidly that is due to the discovery of new interesting electronic and optical properties. In many cases, the electronic, photonic, and mechanical properties of the 2D materials are fundamentally different from those of bulk TMD materials. Therefore, these 2D materials have great prospects for a number of new applications, including flexible electronics, spintronics, optoelectronics, catalysis etc. [3,4].

The success of the application of these materials largely depends on the success of the development of technology for obtaining them on sufficiently large areas of substrate. It is necessary to have technologies for reproducible deposition of various 2D materials with specified properties. For the preparation of heterostructures based on these thin-film materials, it is important to reconcile the temperature regimes of formation of an individual layer. At present, the most widely used method is chemical vapor deposition (CVD) which requires sufficiently high temperatures (~ 700°C) to produce MoS_2 films [5]. The steam flow of reagents is formed by heating solid precursors of sulfur and molybdenum oxide and transferring their vapors by an argon stream. Chemical interaction of vapor flux with the surface of the substrate held at appropriate temperature causes the deposition of MoS_2 film.

In the case of pulsed laser deposition (PLD) of MoS_2 films from the MoS_2/S target, a physical deposition of the flux of activated Mo and S atoms is realized [6,7]. The temperature regime is determined by the conditions for obtaining high-quality ultrathin films. The aim of the work was to compare the MoS_2 films, to produce which several variations of the PLD were used. At the same time, physical PLD deposition of MoS_x and MoO_x thin film precursors was combined with temperature activated chemical synthesis of MoS_2 using the reactive H_2S gas or vapor flux which was formed by laser ablation of S target.

2. Experimental methods

To produce ultrathin MoS_2 films, the PLD methods using nanosecond laser pulses were developed. Pulsed laser deposition in the traditional configuration (PLD/MoS_2) was applied. Laser ablation of the target MoS_2 was carried out in vacuum or hydrogen gas at an energy density of laser irradiation near the plasma formation threshold (1.5 J/cm^2). In this case, a congruent transfer of the target composition to the film was possible [8]. The substrate temperature was 650°C. In the second case, the PLD of molybdenum oxide thin films was carried out. For this purpose, the targets of molybdenum and molybdenum oxide were used. Ablation of the metal target was carried out in a residual gas (reactive PLD). The metal oxide films were subjected to sulfurization at 650°C in sulfur vapors arising from the laser ablation of the solid target S in hydrogen. Also, the thin film of molybdenum oxide was reduced in hydrogen sulfide at a pressure of 10 Pa at 350°C. For these methods, the following notation were accepted: PLD/ MoO_x/S and PLD/MoO_x/H_2S.

The prepared films were studied by scanning electron microscope (SEM), X-ray photoelectron spectroscopy (XPS), micro-Raman spectroscopy (MRS).

3. Results and discussion

Figure 1 shows the characteristic result of the investigation of an ultrathin MoS_2 film deposited on a sapphire substrate by the PLD/MoO_x/S method. The prepared film uniformly covered the surface of the substrate. On the surface, the film defects characteristic of the PLD method are visible. These are submicron particles of Mo formed during laser ablation of the Mo target in O_2-containing residual gas.

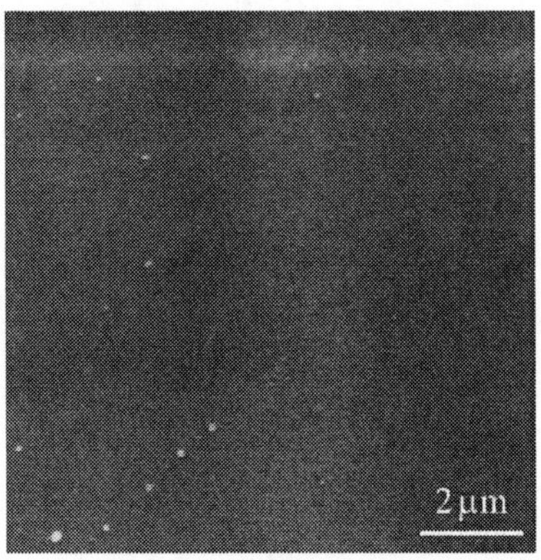

Figure 1. SEM image of an ultrathin MoS_2 film which was obtained on a sapphire substrate by sulfurization of molybdenum oxide film previously deposited by the reactive PLD method.

For the films prepared by the PLD/MoS_2 method, XPS studies showed that a characteristic effect of the formation of the sub-stoichiometric composition of MoS_x films (x <2) was observed. The lack of sulfur was compensated by the introduction of oxygen and the formation of Mo-O bonds (figure 2 (a)) together with the Mo-S bonds. It should be noted that the MRS studies of these films showed that the oxygen atoms did not have a significant effect on a local packing of Mo and S atoms in the MoS_x films. Oxygen could be introduced already in air into the layered structure of a film containing many defects (sulfur vacancies). No characteristic peaks corresponding to Mo-O bonds were detected on the MRS spectra (results not presented).

Because of using PLD/MoS_2 method, the MoS_x films with a layered structure were obtained, and the number of molecular layers depended on the number of laser pulses. When 100 pulses were used for target ablation, the films possessing 2 layers were obtained (figure 3(a)). This was indicated by the results of measuring the distance between the main peaks in the MRS spectra for MoS_2. The number of layers n was determined from the empirical dependence Δf=26.45 – 15.42/(1+1.44$n^{0,9}$) (cm^{-1}), where Δf is the distance between the line centers corresponding to E_{2g}^{1} and A_{1g} oscillations [7]. The positions of these lines in the MRS spectrum for bulk MoS_2 are shown in figure 3 (a) by dashed lines. In the same figure, squares indicate lines corresponding to the sapphire substrate.

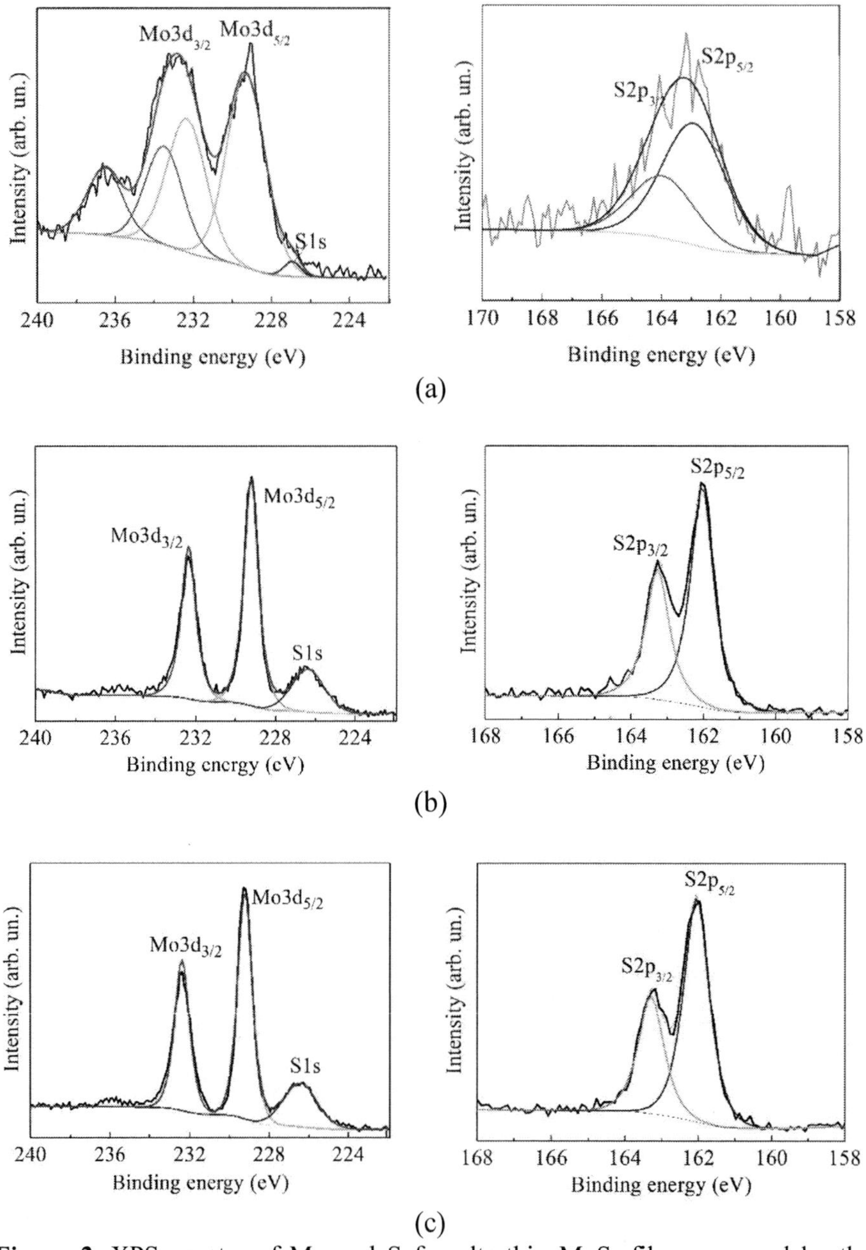

Figure 2. XPS spectra of Mo and S for ultrathin MoS_x films prepared by the different laser-based methods: PLD/MoS_2 (a), PLD/MoO_x/S (b), and PLD/MoO_x/H_2S (c).

According to the XPS studies the sulfurization of the MoO_x film at higher temperature using sulphur vapor and hydrogen sulphide gas proceeded quite efficiently (figure 2 (b,c)). The XPS spectra contained only lines corresponding to the chemical bonds of MoS_2. The use of H_2S could cause the formation of an increased concentration of molybdenum vacancies, since for these films the S/Mo ratio is 2.27. In the case of using sulphur vapor, S/Mo was ~ 2.1. The number of molecular layers formed in the MoS_2 films depended on the thickness of the precursor MoO_x. Figure 3 (a) shows MRS spectra for films possessing 5–10 layers.

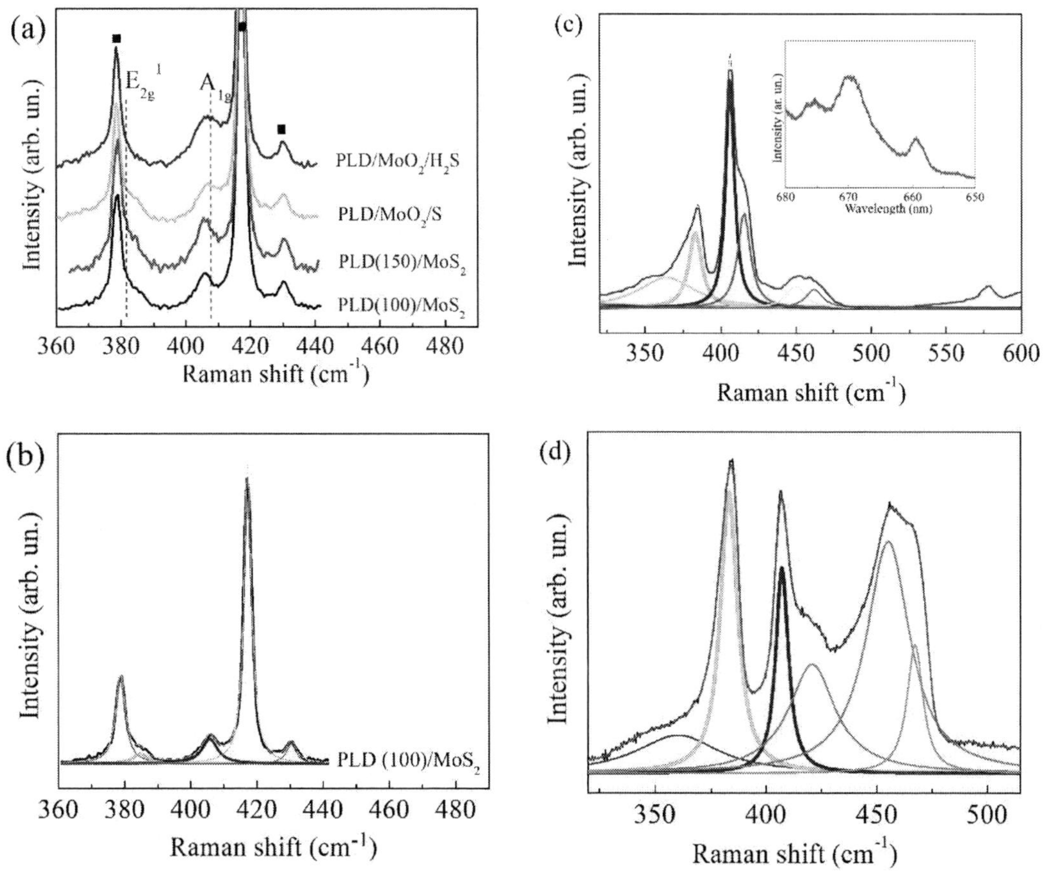

Figure 3. (a,b) MRS spectra for ultrathin MoS_x films prepared by the different laser-based methods at elevated temperatures; (c,d) MRS spectra of 2D MoS_2 films obtained by PLD/MoO_x/S on sapphire and PLD/MoO_x/H_2S on NaCl substrate at moderate temperature, respectively. The inset to (c) shows the luminescence spectrum of the film. The laser wavelength for MRS studies was 632.8 nm (a,b,d) and 532 nm (c).

Precision selection of the thickness of the metal-oxide film precursor made it possible to obtain MoS_2 films containing 1–2 molecular layers. Figure 3 (c) shows the MRS spectrum for such a film obtained by PLD/MoO_x/S on the sapphire at a sulfurization temperature of 650° C. In addition to the lines from the sapphire substrate, lines at 385.3 and 405.5 cm^{-1} was observed. The distance between these lines corresponds to a monolayer film. The inset to figure 3 (c) shows the luminescence spectrum which indicates the possibility of direct-gap transitions in a quite perfect lattice of prepared monolayer MoS_2 film.

Attempts to reduce the temperature of sulfurization of the metal oxide film using H_2S gas, which is known as strong reducing agent, have shown that this process may be realized at ~ 350°C. Figure 3 (d) shows the MRS spectrum for a two-layer MoS_2 film obtained on a NaCl substrate by the PLD/MoO_x/H_2S method at lower temperature. In addition to the typical for MoS_2 film E_{2g}^1 and A_{1g} lines, a peak at about 460 cm^{-1} appeared. One of the possible causes of this peak was a violation of the perfection of the local structure of MoS_2. A line with such a shift is observed in clustered MoS_3 films, and it corresponds to vibrations of apical sulfur atoms bound to three Mo atoms [9]. Also, the intensity of this peak is usually large in a quite perfect MoS_2 when using excitation radiation in the wavelength range 600–700 nm.

4. Conclusions.

The use of pulsed laser radiation of nanosecond duration makes it possible to realize several modes of formation of ultrathin MoS_2 films. Deposition of the laser-initiated flux of Mo and S atoms during ablation of the MoS_2 target makes it possible to obtain films with a controlled number of molecular layers, but they contain a high concentration of S vacancies. The result of sulfurization of thin film MoO_x precursor prepared by PLD method depended on the temperature of the process and the nature of the reducing agent. At sufficiently high temperatures (≥650°C), qualitative monolayer films can be formed using sulphur vapor and H_2S gas at low pressure. The use of H_2S gas allowed to lower the sulfurization temperature, however this caused the formation of local disturbances in the structure of MoS_2 films. It should be noted that the regulation of the type and concentration of defects in ultrathin TMD films is of great importance in several practical applications, in particular, in the use of these films for the catalytic activation of hydrogen evolution reaction.

Acknowledgments

The reported study was funded by RFBR according to the research project № 18-02-00191.

References

[1] LeRoy B J and Robinson J A 2018 *APL Mater.* **6** 026301.
[2] Choi W, Choudhary N, Han G H, Park J, Akinwande D and Lee Y H 2017 *Materials Today* **20** 116–30
[3] Choi K, Lee Y T and Im S 2016 *Nano Today* **11** 626–43.
[4] Das T and Ahn J-H 2017 *Flat Chem* **3** 43–63.
[5] Brent J R, Savjani N and O'Brien P 2017 *Prog. Mater. Sci.* **89** 411–78.
[6] Yang Z and Hao J 2016 *J. Mater. Chem.* C **4** 8859.
[7] Siegel G, Subbaiah Y P V, Prestgard M C and Tiwari A 2015 *APL Mater.* **3** 056103.
[8] Fominski V Yu, Nevolin V N, Romanov R I and Smurov I 2001 *J. Appl. Phys.* **89** 1449–57.
[9] Fominski V Yu, Romanov R I, Fominski D V and Shelyakov A V 2017 *Thin Solid Films* **642** 58–68.

IOP Conf. Series: Journal of Physics: Conf. Series **1238** (2019) 012008 doi:10.1088/1742-6596/1238/1/012008

Influence of nanosecond laser irradiation on the structure and conductivity of BC_x films

P V Zinin[1], R I Romanov[2], V Yu Fominski[2], V P Filonenko[3], I A Troyan[4], K M Bulatov[1], I S Vasil'evskii[2], D A Safonov[2], D V Fominski[2], A A Soloviev[2]

[1] Scientific and Technological Center of Unique Instrumentation, Russian Academy of Sciences, Butlerova str. 15, 117342, Moscow, Russia
[2] National Research Nuclear University MEPhI (Moscow Engineering Physics Institute), Kashirskoe sh. 31, 115409 Moscow, Russia
[3] Institute for High Pressure Physics, Russian Academy of Sciences, 142190, Troitsk, Moscow, Russia
[4] Federal Scientific Research Center "Crystallography and Photonics" of the Russian Academy of Sciences, Leninsky pr. 59, 119333, Moscow, Russia

E-mail: vyfominskij@mephi.ru

Abstract. Thin-film precursors of BC_x were formed by pulsed laser codeposition of boron and carbon. Targets made of pressed boron and carbon powders with an equal element content (B/C = 1/1) and an increased carbon content (B/C = 1/3) were used. The films were deposited on sapphire substrates at elevated temperature (700°C) which determined the initial properties of the precursor BC_x films. Irradiation of the films was carried out by laser pulses of nanosecond duration with varying intensity. The films obtained by laser annealing of BC_x (Q-BC_x) were studied by scanning electron microscopy and micro-Raman spectroscopy. Irradiation under optimal conditions made allowed to realize pulsed melting of the films and partial preservation of their continuity on the substrate. The local structure of Q-BC_x films and the nature of the changes in their electrophysical properties depended on the composition of the precursor films and the laser irradiation regimes.

1. Introduction

Thin films of BC_x attract the interest of researchers from various fields of science and technology, since they can have a whole complex of interesting and useful properties [1,2]. Particularly, the interest in this thin-film material increased after the discovery of superconductivity which is established for the metastable modification of Q-BC_x [3,4]. This modification is formed by pulsed laser melting of the films that initially consisted of alternating layers of boron and diamond-like carbon. The mixing of the film components proceeded in highly non-equilibrium conditions of the superheated liquid phase, and the superconducting phase was formed as a result of quenching of a highly supercooled liquid phase/alloy.

Due to the appearance of new important information on the superconducting properties of Q-BC_x films, there is an increasing interest to find more controlled and reproducible methods for obtaining continuous and smooth films of BC_x with superconducting properties. The aim of the work was to study the influence of the nature of the thin-film precursor BC_x on the character of structural and

IOP Conf. Series: Journal of Physics: Conf. Series **1238** (2019) 012008 doi:10.1088/1742-6596/1238/1/012008

electrophysical changes under the action of laser pulses of nanosecond duration. For this research, BC_x films of various structure and composition were created by pulsed laser deposition (PLD) at elevated temperature.

2. Experimental methods

Pulsed laser deposition of BC_x films was carried out from targets containing microparticles of diamond and boron, up to 5 μm in size. The atomic ratio of the elements in one B-C target was B/C = 1/1 (the target with a high concentration of B) and another B-C target was enriched with carbon (B/C=1/3). Targets were ablated by laser pulses with a wavelength of 266 nm in a vacuum chamber at a pressure of ~ 10^{-3} Pa. The radiation energy of laser pulses of nanosecond duration (~ 10 ns) was 40 mJ. After focusing the radiation on the targets, the fluence reached 7 J/cm^2. The films were deposited on sapphire substrates, the temperature of which was ~700°C. The thickness of the prepared films was approximately 100 nm.

The obtained BC_x films were irradiated by the pulses of the same laser in air at room temperature of the substrate. The intensity of irradiation was reduced, and it was varied in the range 0.1–1 J/cm^2. The structure and composition of the films before and after irradiation was studied by scanning electron microscopy (SEM), energy dispersive X-ray spectroscopy (EDXS), and micro-Raman spectroscopy (MRS). The wavelength of the analyzing laser radiation was 532 nm. The registration time varied from 2 to 10 min. To obtain additional information on the structure of the precursor BC_x films obtained by the PLD, very thin films were studied by transmission electron microscopy (TEM) and electron energy loss spectroscopy (EELS). Very thin films were deposited on NaCl crystals and transferred to metal grids for TEM analysis.

The electrical properties of the films were measured by a four-contact method in the van der Pauw geometry. Contacts to the sample with a circular mesa were formed from the InSn alloy, the linearity of the volt-ampere characteristics of all contacts was monitored. The assembled samples were placed on the insert in the cryostat, the resistance of the sample was measured by successive heating in the temperature range 4–300 K. The temperature was measured by a calibrated Lakeshore sensor. During the measurements, the direction of the current was switched to eliminate the effects of thermoelectric power. The resistivity was calculated by averaging the values from all pairs of contacts.

3. Results and discussion

Figure 1 (a) shows results of the SEM study of the morphology of BC_x film obtained from the target B/C=1/3. For this target, relatively smooth films with a small number of particles of submicron sizes were formed. Laser irradiation of the deposited films caused the surface to melt (figure 1 (b)) that manifested itself in a change in the topography of the sample surface. For an area of reduced fluence

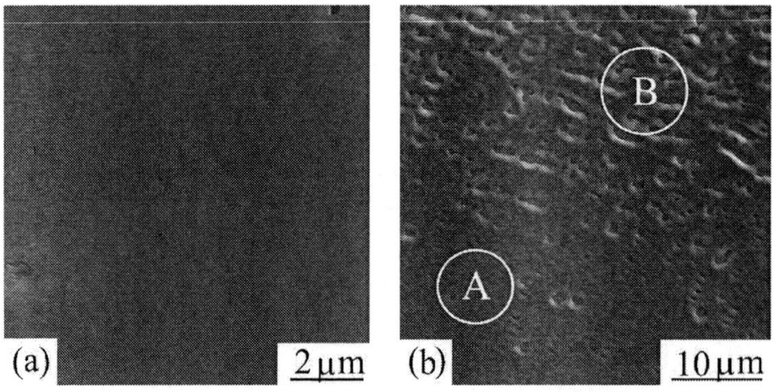

Figure 1. SEM images of BC_x film prepared from the target B/C=1/3: (a) – after preparation by PLD; (b) – after pulsed laser irradiation. Zones A and B are discussed in the text.

irradiation, the film remained continuous (Zone A), and for an area of increased fluence of irradiation, film breaks (Zone B) were detected. EDXS studies showed no change in the B/C ratio after laser irradiation.Figure 2 (a) shows the results of the SEM studies for the film obtained by ablation of the target B/C = 1/1. When using a target with a higher concentration of B, the deposited films contained many particles of submicron and micron sizes. Laser irradiation caused melting and mixing of submicron-sized particles with the main material of the film (figure 2 (b)). However, micron-sized particles retained a rounded shape after laser irradiation and were partially incorporated into the film matrix. According to the EDXS studies (scanning was carried out over a sufficiently large area), the chemical composition of the films was preserved after laser irradiation.

TEM studies of a very thin BC_x films showed (results not presented) that they had a homogeneous amorphous structure. According to EELS analysis, the concentration of diamond-like sp^3-bonds reached 80%. The boron atoms were embedded in an amorphous matrix of the film, forming a solid solution of B-C.

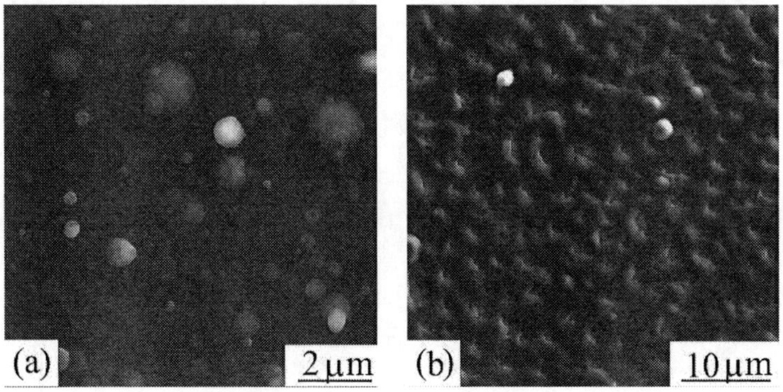

Figure 2. SEM images of the BC_x films prepared from the target B/C=1/1: (a) – after deposition by PLD; (b) – after pulsed laser irradiation.

Figure 3 shows the MRS spectra measured before and after laser irradiation of the BC_x films obtained from the two B-C targets. Spectra, which corresponded to areas irradiated with low (LF) and high (HF) fluence, differ only a little. The spectra of the as-deposited and laser irradiated films strongly depended on their composition. The appearance of relatively narrow lines on the spectrum of the film enriched in carbon (figure 3 (a)) indicated that after irradiation the structural state of the film became more equilibrium than immediately after the PLD. For the BC_x film possessing approximate equal concentrations of B and C, laser irradiation did not cause strong changes in the MRS spectra (figure 3 (b)).

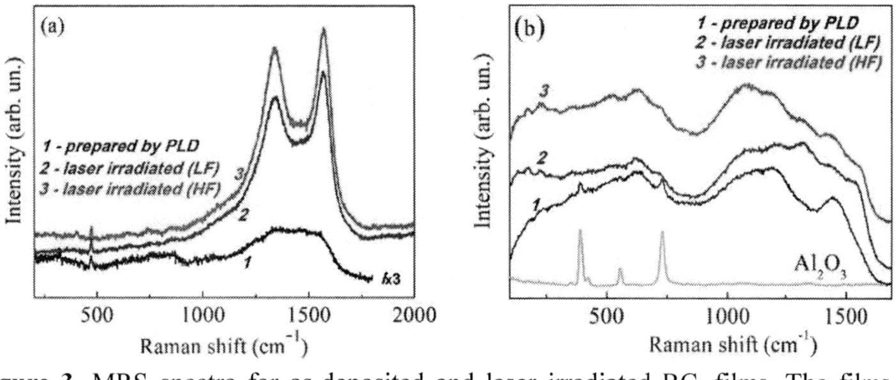

Figure 3. MRS spectra for as-deposited and laser irradiated BC_x films. The films were prepared by PLD from B/C=1/3 (a) and B/C=1/1 (b) targets.

IOP Conf. Series: Journal of Physics: Conf. Series **1238** (2019) 012008 doi:10.1088/1742-6596/1238/1/012008

Figure 4 shows the results of a rigorous mathematical processing of some of the MRS spectra, based on published data on the analysis of B-C materials by the MRS method (for example, [5-7]). The model spectrum of as-deposited film enriched in carbon (the target B/C=1/3) contained several lines, which could be due to the following factors (figure 4 (a)). The lines at 270, 645, 810, and 1040 cm^{-1} were determined by singularities in the phonon density of states. The lines at 1325 and 1560 cm^{-1} corresponded to the D and G states and were shifted from the characteristic positions due to disordering and introduction of boron. The line at 1470 cm^{-1} was due to hydrocarbons. After irradiation with high fluence, the spectrum contained two lines at 1337 and 1570 cm^{-1}, which corresponded to D and G lines displaced due to the structural features of the film. The lines at 1470 and 1135 cm^{-1} were caused by hydrocarbons.

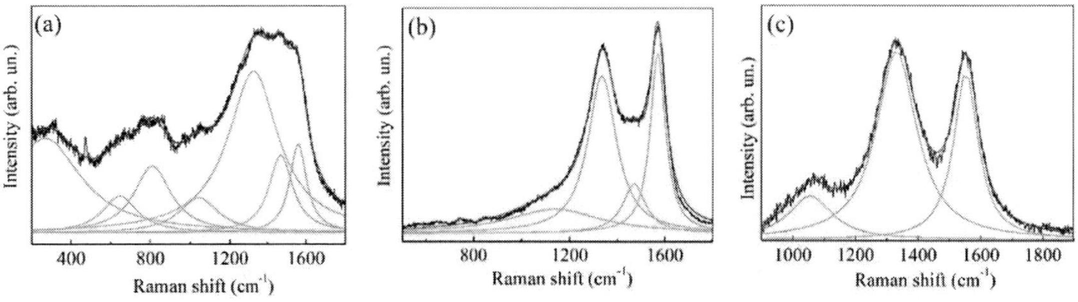

Figure 4. (a, b) – deconvolution of the MRS spectra for the BC_x film enriched with carbon before (a) and after (b) pulsed laser irradiation; (c) – deconvolution of MRS spectrum for a boron enriched BC_x film, illustrating the states which arise after the pulsed laser irradiation of the film.

Figure 4 (c) shows the result of processing the MRS spectra for an enriched in boron BC_x film (the target B/C=1/1). For the difference spectrum obtained by subtracting the spectrum of the as-deposited film from the spectrum of the irradiated (HF) film, three lines were revealed. There were G (1552 cm^{-1}) and D lines (1331 cm^{-1}), as well as a line at 1054 cm^{-1} which characterized the changes in the phonon density of states.

The characteristic results of measuring the resistance of BC_x films before and after laser irradiation are shown in figure 5. As-deposited BC_x films possessed semiconductor–type conductivity. The resistance of the films increased with decreasing temperature. After laser irradiation of a carbon-enriched film, the temperature dependence of the resistance was generally preserved, however, at very low temperatures, the effect of decreasing resistance was observed. After laser irradiation of the film enriched with boron, a metallic type of temperature-resistance dependence appeared in the range of 50–300 K. At temperatures below 50 K, a sharp increase in the resistance of the film was observed.

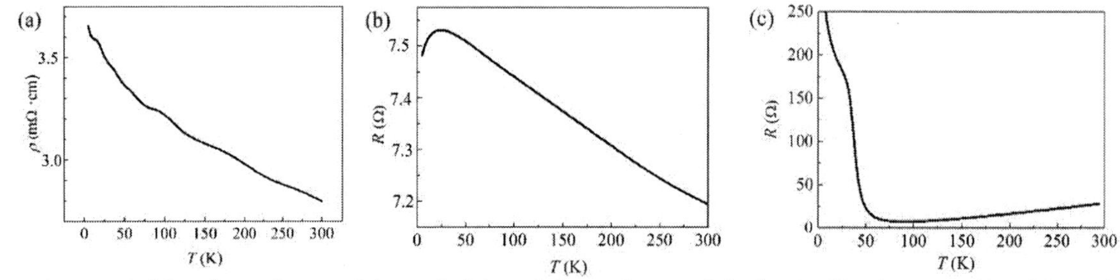

Figure 5. (a) – dependence of the resistivity of the carbon enriched BC_x film (target B/C=1/3) on the temperature; (b, c) – the temperature dependence of the surface resistance after laser irradiation of the films BC_x obtained from the targets B/C=1/3 and B/C =1/1, respectively.

4. Conclusions

Pulsed laser irradiation of amorphous BC_x films caused a decrease in the degree of nonequilibrium of the state which was formed by pulsed laser deposition on a sapphire substrate heated to 700°C. This is manifested in an increase in the contribution of graphite-like carbon bonds in both the enriched in carbon BC_x and containing high concentration of boron BC_x. For the BC_x films enriched in carbon, the semiconductor type of conductivity is retained after pulsed laser irradiation. However, for them, a drop of resistance is possible at very low temperatures. The BC_x films enriched with boron showed a metallic type of conductivity after pulsed laser melting. Such behaviour was broken at 50 K and manifested itself in a sharp increase in film resistance.

Acknowledgments

The reported study was funded by RSF according to the research project № 17-12-01535. Fominski V Yu thanks the Ministry of Science and Education of the Russian Federation for its support in the frame of the work "Organization of Scientific Research" (No. 3.4639.2017/6.7). Samples have been partially investigated in Shared Research Center “Heterostructure microwave electronics and Physics of wideband semiconductors”.

References

[1] Song W, Kim Y, Jung D S, Lee S, Jung W, Kwon O, Kim H K, Kim M S, An K S and Park C Y 2013 *Appl. Surf. Sci.* **284** 53–8.
[2] Zinin P V, Nozhkina A V, Romanov R I, Filonenko V P, Titov S A, Trojan I, Fominski V Y, Bulatov K M, Bykov A A, Kutuza I B, Anokhin A and Lomonosov A M 2018 *MRS Advances* **3** 45–52.
[3] Bhaumik A, Sachan R, Gupta S and Narayan J 2017 *ACS Nano* 2017 **11** 11915–22.
[4] Gupta S, Bhaumik A, Sachan R and Narayan J 2018 *JOM* **70** 450–5.
[5] Polyakov S N, Denisov V N, Mavrin V N, Kirichenko A N, Kuznetsov M S, Martyushov S Yu, Terentiev S A and Blank V D 2016 *Nanoscale Research Lett.* **11** 11.
[6] Zinin P V, Ming L C, Ishii H A, Jia R, Acosta T and Hellebrand E 2012 *J. Appl. Phys.* **111**. 114905.
[7] Zhang S, He Z, Ji X, Lu W, Wang C, Shen Q and Zhang L 2014 *J. Appl. Phys.* **115** 154906.

IOP Conf. Series: Journal of Physics: Conf. Series **1238** (2019) 012009 doi:10.1088/1742-6596/1238/1/012009

Influence of synthesis conditions on the crystal and local structures of WO_3 powders.

V V Popov[1,2], A P Menushenkov[1], A A Pisarev[1], M M Berdnikova[1], Ya V Zubavichus[2], R D Svetogorov[2], I V Shchetinin[3], M V Zheleznyi[3], N A Tsarenko[4], L A Arzhatkina[4], A A Yastrebtsev[1], B R Gaynanov[1], K V Ponkratov[5]

[1] National Research Nuclear University MEPhI (Moscow Engineering Physics Institute), Kashirskoe sh. 31, 115409 Moscow, Russia
[2] NRC "Kurchatov Institute", pl. Akademika Kurchatova 1, 123182 Moscow, Russia
[3] National University of Science and Technology MISiS (Moscow Institute of Steel and Alloys), Leninskiy pr. 4, 119049 Moscow, Russia
[4] JSC "Scientific Research Institute of Chemical Technology", Kashirskoe sh. 33, 115409 Moscow, Russia
[5] Renishaw plc, Kantemirovskaya st. 58, 115477 Moscow, Russia

E-mail: victorvpopov@mail.ru

Abstract. Influence of synthesis conditions on the crystal and local structures of WO_3 powders prepared by thermal decomposition of ammonium paratungstate and precipitation of tungstates aqueous solutions in strong acid conditions has been investigated. Combination of X-ray powder diffraction, X-ray absorption fine structure spectroscopy, IR- and Raman-spectroscopy, and scanning electron microscopy was used. The calcination of all initial compounds at temperatures $\geq 500^{\circ}C$ led to formation of the monoclinic γ-WO_3 single phase. It was concluded that the neutral octahedral complex $[W{=}O(OH)_4(H_2O)]$ can be a structural unit of the precursors prepared in acidic suspensions. The local structure of synthesized tungsten oxides consists of edge-shared and corner-shared distorted octahedral WO_6 species linked together.

1. Introduction

Transition element oxides are of great interest for both fundamental science and technological applications due to their multifunctional properties. Tungsten oxides WO_x , where $2 \leq x \leq 3$, are among the most important substances from this large group. Nowadays they are commonly used as photocatalysts, gas sensors, and electrochromic materials. There are also some new applications of WO_x for dye-sensitized solar cells, optical data storage and field-emission displays. Besides, cation-doped WO_x is considered as high-T_c superconductor [1].

Oxides WO_x and their hydrates can be synthesized using various chemical and physical methods, such as wet chemical precipitation, sol-gel hydrolysis of tungsten alkoxide, ion exchange and thermal decomposition of tungsten salts. Films of WO_x can be deposited on various substrates using various vacuum techniques [1-4]. Different crystalline structures as well morphologies occur depending on the processing conditions. The WO_x phase transitions are rather complex.

The aim of the present work is to study the influence of synthesis conditions on the crystal and local structures of WO_3 powders prepared by wet chemical precipitation and thermal decomposition of tungsten salts by means of X-ray diffraction, X-ray absorption fine structure spectroscopy, IR- and Raman-spectroscopy, and scanning electron microscopy.

2. Experimental

The starting materials were sodium tungstate (ST) $Na_2WO_4·2H_2O$ (98.5%), ammonium paratungstate (APT) $(NH_4)_{10}W_{12}O_{41}·5H_2O$ (99.0%). Precursors (tungsten trioxide hydrates or tungstic acids) were prepared by acidification of tungstates salts solutions with hydrochloric HCl (analytical grade) or nitric HNO_3 (analytical grade) acids until the final pH ~ 0 [2]. Tungsten trioxide WO_3 powders were prepared by calcination of synthesized precursors or APT at 300 - 1000°C for 2 - 6 h in air.

Crystal structures of synthesized powders were studied by X-ray powder diffraction at "Center for X-ray structural research and diagnostics of materials" MISiS using a Rigaku MiniFlex 600 diffractometer with Cu K_α radiation. The Rietveld full-profile analysis of X-ray diffraction patterns was performed with the Jana2006 software [5].

The Raman spectra were collected on a inVia Reflex confocal Raman microscope (Renishaw) (laser λ_1=532 nm and λ_2=785 nm) and a Nicolet iS50 FT-IR spectrometer (Thermo Fisher Scientific Inc.) with Thermo Scientific iS50 Raman module (λ=1064 nm) at room temperature. Infrared (IR) spectra of the synthesized powders were recorded on Nicolet iS50 FT-IR spectrometer with built-in all reactive, mid- and far-IR diamond iS50 ATR module in the range of 400-4000 cm^{-1}.

X-ray absorption fine structure (EXAFS) spectra were measured above W L_1- and L_3- edges at "Structural Materials Science" beamline of the Kurchatov synchrotron radiation source in transmission mode at room temperature. The processing and simulation of the EXAFS spectra were performed using IFEFFIT [6] program packages. The back scattering amplitudes and phases were calculated with FEFF-8.20 [7] based on the known crystal structure parameters and the diffraction data. The XANDA program [8] was used to fit the near-edge region of X-ray absorption spectra.

The particle morphology and elemental composition of synthesized powders were studied by Scanning Electron Microscopy coupled with Energy Dispersive X-ray Spectroscopy (SEM/EDS) using an X-Act energy dispersive detector Oxford Instruments with a spectral resolution of 125 eV mounted on a Vega 3 Tescan scanning electron microscope. The characteristic X-ray radiation was automatically processed with the AZtec program.

3. Results and discussion

It was found that acidic precipitation (pH 0; 80°C) of both ST and APT leads to formation of bright lemon yellow powders of tungsten trioxide monohydrates (tungstic acid) $WO_3·H_2O$ with orthorhombic structure (JCPDS № 43-0679). Particles of all synthesized precursors (both ST and APT) had size up to several tens of microns and were aggregates of smaller particles (Figure 1).

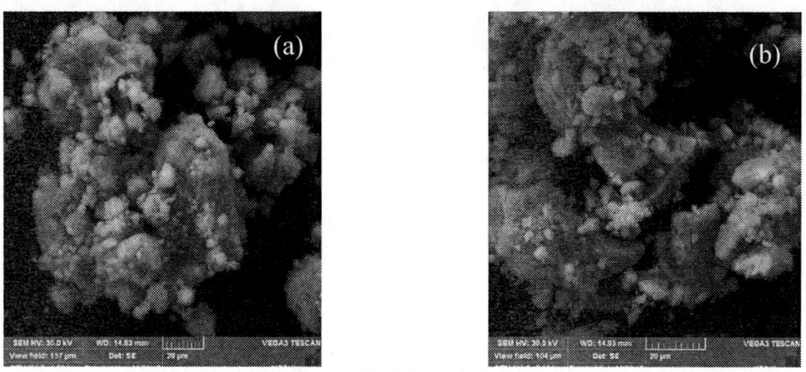

Figure 1 SEM images of (ST+HCl) (a) and (APT+HNO_3) (b) precursors

In the precursor samples annealed at 300°C we observed the appearance of the monoclinic phase I γ-WO_3 (JCPDS № 43-1035) and probably some amount of residual precursor (Figure 2a, 2b). In the sample prepared by APT at 300°C, we can see the presence of some amorphous phase and hexagonal ammonium tungsten oxide bronze $(NH_4)_{0.25}WO_3$ (JCPDS № 73-1084) (Figure 2c). Further increase of calcination temperature above 500°C led to formation of the of monoclinic I γ-WO_3 single phase (Figure 2). This is in a good agreement with data reported in [2, 4]. The XRD study showed that increase of calcination temperature in the range of 500-1000°C leads to increase of crystallite sizes (from tens nm at 300°C to a several hundreds nm at 700°C) and decrease of microstrains in the WO_3 powders.

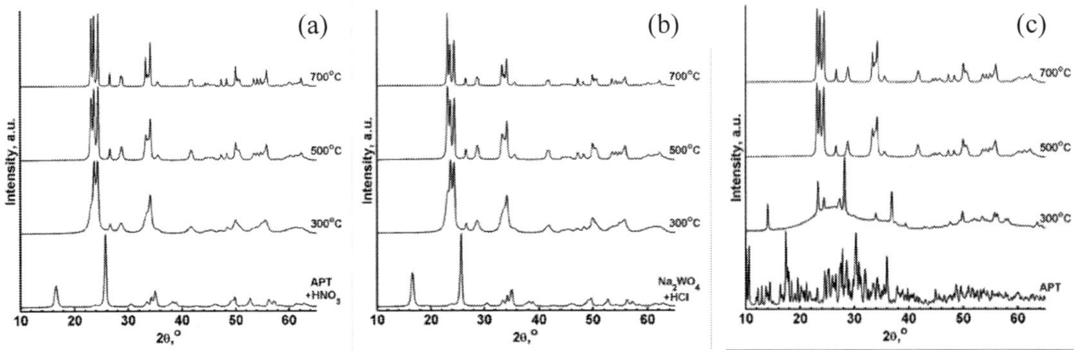

Figure 2 XRD patterns of powders prepared by calcination of ($APT+HNO_3$) (a), (ST+HCl) (b) precursors and APT (c)

Figure 3 shows SEM images of WO_3 particles prepared by calcitation of APT at 500°C for 2 h in air. Agglomerates with sizes in the range from a few to tens microns were observed. These agglomerates consist of primary particles with dimensions of a few hundreds nm which is close to the crystallite sizes determined from XRD data.

Figure 3 SEM images (at different magnification) of the WO_3 particles prepared by calcitation of APT at 500°C for 2 h in air

It was found that the IR spectra of $WO_3 \cdot H_2O$ precursor powders, synthesized from both ST and APT, containe a strong band corresponding to the stretching vibrations of OH groups (3406 cm^{-1}) and structural water molecules (1620 cm^{-1}), an IR band at 947 cm^{-1} corresponding to stretching vibrations of ν(W=O) and very strong and broad band at 667 cm^{-1} indicating the presence of stretching vibrations of ν(O-W-O) [3] (Figures 4a and 4b). The FT-IR spectrum of APT (Figure 4c) indicates N-H bending and stretching vibrations from NH_4^+ at wavenumbers of 1429 and 3163 cm^{-1}, respectively; O-H bond vibrations at 1640 and 3434 cm^{-1}, stretching vibrations of ν(W=O) at 952 cm^{-1} and W-O bond vibrations at 878, 719 and 522 cm^{-1} [4]. It should be noted that calcination of all initial compounds led to a significant decrease the intensity of the IR bands of both OH and N-H groups.

However their presence in the IR-spectra at 300°C indicated that both $WO_3 \cdot H_2O \rightarrow WO_3$ and APT → WO_3 phase transitions are incomplete. It should be noted that calcination of precursors (prepared from solutions) above 300°C caused the appearance of additional IR bands at 811 cm^{-1} and 730 cm^{-1} corresponding to the stretching vibrations of the W=O and O-W-O bonds, respectively, and indicating the formation of γ-WO_3 phase [3] (Figures 4a and 4b). A further increase of the temperature to 500°C led to complete disappearance of IR bands of both OH and N-H groups (Figure 4). IR spectra containing only W-O bond vibrations indicate completion of WO_3 formation. This is in a good agreement with the XRD data.

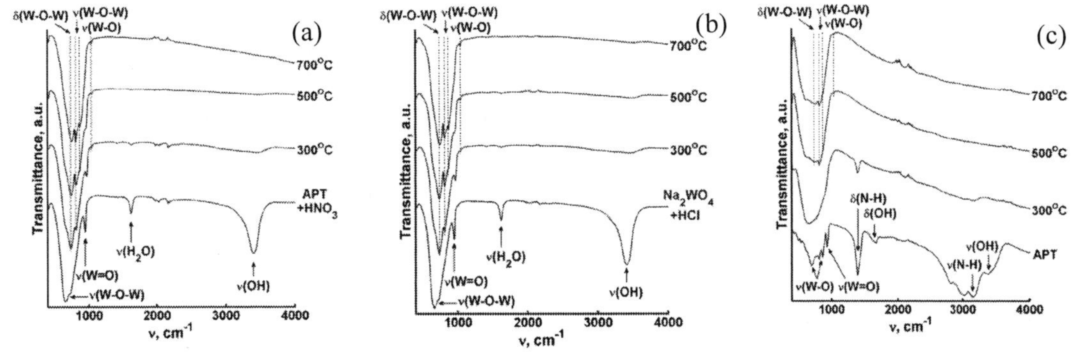

Figure 4 IR spectra of powders prepared by calcinations of ($APT+HNO_3$) (a), (ST+HCl) (b) precursors and APT (c)

The structural information was confirmed by Raman spectroscopy. As seen in Figure 5a, the precursor (ATP + HNO_3) contained a sharp mode at wavenumber of 942 cm^{-1} (corresponding to the stretching vibrations of W=O) and strong broad band at 630 cm^{-1} (corresponding to the stretching vibrations of O-W-O). The calcination of the precursor at 300°C led to the appearance of additional intense modes at 718 and 807 cm^{-1}, which indicates the presence of stretching vibrations ν(O-W-O), as well as mode at 274 cm^{-1} mode indicating bending vibrations δ(O-W-O) in γ-WO_3 [3]. It should be noted that the mode at 630 cm^{-1} was still visible in the sample prepared at 300°C for 2 h. This indicates the presence of the residual $WO_3 \cdot H_2O$ phase, which practically disappeared after annealing at the temperature of 700°C (Figure 5a). The Raman spectra of powders prepared by calcinations of (ST+HCl) precursor were similar (Figure 5b). The presence of W=O and O-W-O bonds as well as the formation of an intermediate substance at 300°C and γ-WO_3 at ≥ 500°C was found upon calcination of APT (Figure 5c).

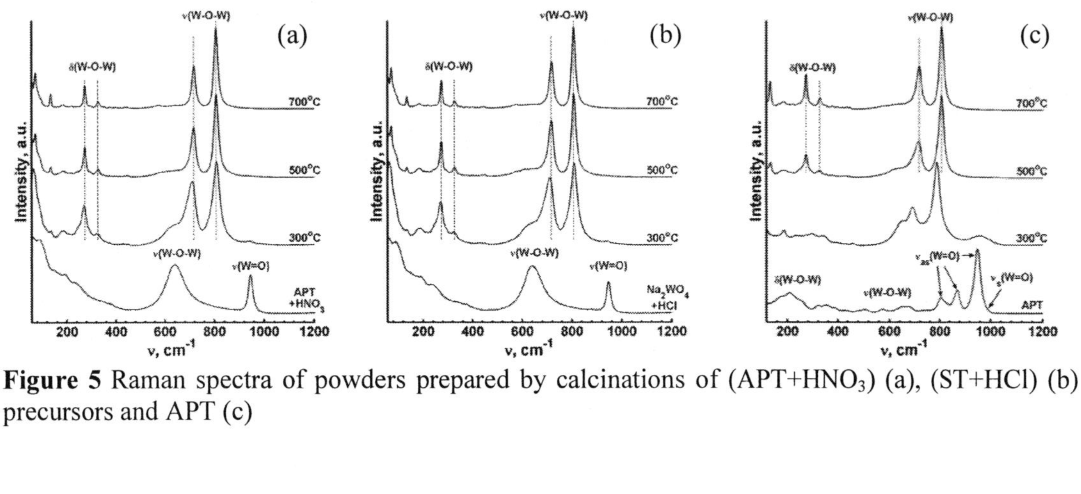

Figure 5 Raman spectra of powders prepared by calcinations of ($APT+HNO_3$) (a), (ST+HCl) (b) precursors and APT (c)

IOP Conf. Series: Journal of Physics: Conf. Series **1238** (2019) 012009 doi:10.1088/1742-6596/1238/1/012009

An analysis of IR- and Raman-data made it possible to conclude that the neutral octahedral complex [W=O(OH)$_4$(H$_2$O)] may be a structural unit of the precursors (both APT+HNO$_3$ and ST+HCl) prepared in acidic suspensions.

We performed XAFS-spectra measurements to understand the local atomic structure realignment during the heat treatment of precursors and prepared WO_3 powders. The W L$_3$-edge white line mostly is derived from electron transitions from the 2p$^{3/2}$ state to a vacant 5d state [9]. The W L$_3$-edge XANES spectra of the WO_3 powders synthesized from (APT+HNO$_3$) precursor are shown in Figure 6a. There is some splitting of W L$_3$-edge white line (Figure 6b), which indicates the presence of distorted octahedral structure units (WO_6) in investigated samples.

The W L$_1$-edge XANES spectrum provides information on the electronic state and the geometry of the tungsten species. The pre-edge peak is mainly attributed to forbidden electron transitions from 2s orbitals to 5d orbitals [9]. Figure 6c shows W L$_1$-edge XANES spectra of (APT+HNO$_3$) precursor and the synthesized WO_3 powders. The observed moderate pre-edge peaks confirmes the presence of distorted octahedral WO_6 species.

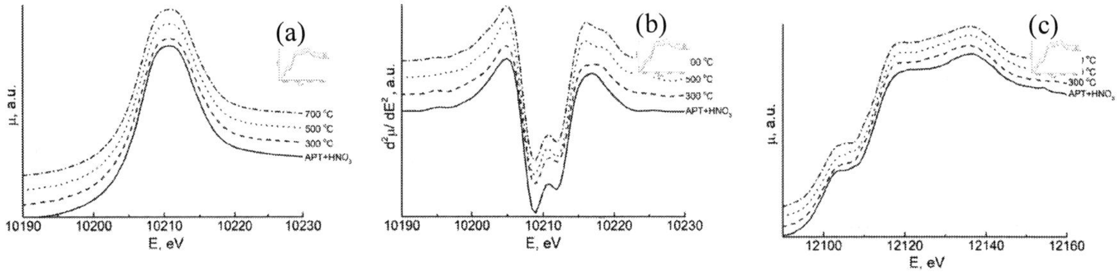

Figure 6 XANES of powders prepared by calcinations of (APT+HNO$_3$) precursor W L$_3$-edge (a), second derivation (b) and W L$_1$-edge (c)

The FT-EXAFS spectra of W L$_3$-edge of (APT+HNO$_3$) precursor, initial APT and the synthesized WO_3 powders are shown in Figure 7. The peak in the range of 1-2 Å, which is due to the W-O shell [9, 10], appeared in all FT spectra. In Figure 7 one can see that the peak of W-O shell for the precursor consists of a few components, which may be attributed to W=O, W-O and W···OH$_2$. Conrary, for WO_3 samples, the W-O shell had only single peak corresponding to W-O distance in WO_6 unit. These results are in a good agreement with IR (see Figure 4) and Raman spectroscopy data (see Figure 5).

The group of peaks in the range of 3-4 Å is due to W-W shell [9, 10]. One can see in Figure 7 that the W-W shell consists of a few components, which can be attributed to the edge-shared octahedral (shorter distance ~ 3.2 Å) and to the corner-shared octahedra WO_6 (longer distance ~ 3.7 Å). This is in a good agreement with published data [10].

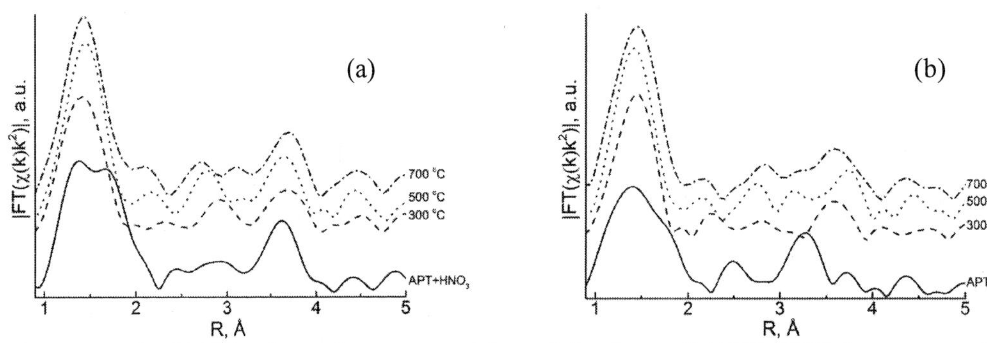

Figure 7 FT-EXAFS of powders prepared by calcinations of (APT+HNO$_3$) precursor (a) and APT (b)

4. Conclusion

A combination of X-ray powder diffraction, XAFS spectroscopy, Raman spectroscopy, IR-spectroscopy, and scanning electron microscopy was used for investigation of the influence of synthesis conditions on crystal and local structures of WO_3 powders prepared by thermal decomposition of ammonium paratungstate and precipitation of tungstates aqueous solutions in strong acid conditions. The calcination of all initial compounds at temperatures above 500°C led to formation of single phase of monoclinic I γ-WO_3. It was concluded that the neutral octahedral complex $[W{=}O(OH)_4(H_2O)]$ can be a structural unit of the precursors prepared in acidic suspensions. The local structure of synthesized tungsten oxides consist of edge-shared and corner-shared distorted octahedral WO_6 species linked together.

Acknowledgements

The work was supported by the Ministry of Education and Science of the Russian Federation (Grant № 3.9750.2017/8.9). The scanning electron microscopy was possible due to Grant № 14.Y26.31.0008 from the Ministry of Education and Science of the Russian Federation.

References

[1] Zheng H, Ou J Z, Strano M S, Kaner R B, Mitchell A and Kalantar-zadeh K 2011 *Adv. Funct. Mater.* **21** 2175
[2] Supothina S, Seeharaj P, Yoriya S and Sriyudthsak M 2007 *Ceram. Inter.* **33** 931
[3] Gotic M, Ivanda M, Popovic S and Music S 2000 *Mater. Sci. Eng. B* **77** 193
[4] Kalpakli A O, Arabaci A, Kahruman C, and Yusufoglu I 2013 *Int. J. Refract. Met. Hard Mater.* **37** 106
[5] Petricek V, Dusek M and Palatinus L 2014 *Z. Kristallogr.* **229** 345
[6] Newville M 2001 *J. Synchrotron Rad.* **8** 322
[7] Ankudinov A, Bouldin C, Rehr J and Conradson S 1998 *Phys. Rev. B* **58** 7565
[8] Klementiev K V 2001 *J. Phys. D: Appl. Phys.* **34** 209
[9] Yamazoe S, Hitomi Y, Shishido T and Tanaka T 2008 *J. Phys. Chem. C* **112** 6869
[10] Carrier X, Marceau E, Carabineiro H, Rodríguez-González V and Che M 2009 *Phys. Chem. Chem. Phys* **11** 7527

IOP Conf. Series: Journal of Physics: Conf. Series **1238** (2019) 012010 doi:10.1088/1742-6596/1238/1/012010

Self-consistent and Maxwell approximations to describe the excess conductivity anisotropy in FeSe above superconducting transition temperature

K K Kesharpu[1 †], P D Grigoriev[1 2 3], D I Lazeva[1] and T I Mogilyuk[4]

[1]National University of Science and Technology "MISiS", 119049 Moscow, Russia
[2]L. D. Landau Institute for Theoretical Physics, 142432 Chernogolovka, Russia
[3]P. N. Lebedev Physical Institute, RAS, 119991 Moscow, Russia
[4]National Research Center "Kurchatov Institute", 123098 Moscow, Russia

E-mail: †m1708377@edu.misis.ru

Abstract. Using the self-consistent approximation we calculate conductivity in an anisotropic heterogeneous media with superconducting inclusions and compare the results with those obtained previously using the Maxwell approximation and with available experimental data on excess conductivity above $T_c = 8$K in FeSe. Advantages and drawbacks of these two approximations are discussed. The obtained analytical formulas can be applied to various other anisotropic heterogeneous superconductors, including high-T_c layered superconductors.

1. Introduction

Increasing the transition temperature (T_c) in high-T_c superconducting materials is one of the major challenges in condensed matter physics. Most of high-T_c superconductors are usually metal alloys or complex oxide ceramics. The copper oxides [1] and iron based superconductors [2] are of special interests because T_c in these materials is highest at ambient pressure and can be controlled in some range by chemical composition and doping level. During the doping process, the dopant does not deposit uniformly, which transforms the initial spatially homogeneous compound to a heterogeneous material. Because of this inhomogeneity, superconducting (SC) properties are often observed first in spatially disconnected parts of the materials at $T > T_c$ [3]. As the temperature T approaches T_c, small superconducting areas become larger and their phases become coherent. In the end, at $T = T_c$, the whole volume becomes superconducting. This type of inhomogeneous development of superconducting properties above T_c was called *Gossamer Superconductivity* [4]. It may also be named *Heterogeneous Superconductivity*. Existence of such heterogeneous phase has been corroborated by magnetic response experiments using *scanning SQUID microscopy* method [5]. Spatial inhomogeneity in superconductor may increase T_c, as shown theoretically [3, 6]. It is also responsible for superconducting effects above T_c. The understanding of relation between spatial inhomogeneity and high-T_c superconductivity is very important for synthesizing new materials with higher T_c.

Recently it was proposed that the excess conductivity of a layered anisotropic heterogeneous compound due to isolated superconducting islands is also very anisotropic with the maximal effect along the least conducting direction [4, 7]. The quantitative description of this effect using the *Maxwell approximation (MA)* was proposed and used to analyze experimental data on conductivity and diamagnetic response in FeSe [4, 7]. In the present work we use *self-consistent approximation (SCA)* (See Refs. [8, 9] for short review) to derive analytical expression for conductivity and its anisotropy in such a heterogeneous material with two different phases. Then we compare the results obtained using SCA and Maxwell approximations with the experimental data in bulk FeSe. Finally, we discuss the advantages and drawbacks of these two approximations.

FeSe has the simplest chemical composition among the large family of iron-based high-T_c superconductors, but its electronic structure and properties are very interesting. Under pressure its transition temperature raises to 40 K [10]. Very promising is the observation of superconductivity with $T_c > 100$ K in FeSe monolayer on $SrTiO_3$ substrate [11]. The bulk FeSe has inhomogeneous microscopic structure and T_c [12]. This makes FeSe very convenient for the comparison with our model of anisotropic heterogeneous conductor.

2. Theory

Conductivity of inhomogeneous media under SCA is found by making 3 primary assumptions: (i) volume fractions of different phases are distributed randomly, (ii) spatial average of the effective field is equal to the applied field E, (iii) all the included phases are symmetrically treated. For simplicity we will derive the relation for a sample containing only two phases. Let us imagine a sample of heterogeneous material containing two regions (region-I and region-II as shown in Fig. 1). Let the conductivity of region-I, which is completely homogeneous, be σ_e. Contrary to region-I, region-II is inhomogeneous and contains two phases, i.e. phase-I and phase-II, with conductivity σ_1 and σ_2 respectively. Since both phases in region-II are treated symmetrically, the conductivity outside the boundary of a phase does not depend on which phase inclusion is considered. For example, conductivity outside the shaded part in region-II should be the same (*i.e., effective conductivity* (σ_e)), whether the shaded part belongs to phase-I or phase-II. This idea is incorporated in the mathematical derivation of σ_e by thinking that islands of conductivity σ_1 or σ_2 are embedded inside a large homogeneous material of conductivity σ_e.

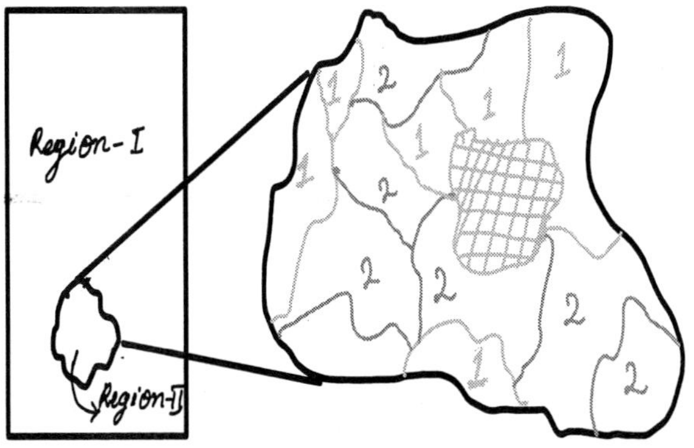

Figure 1. Square bounded region is the full sample containing region-I (*Conductivity* → σ_e; *Electric field* → E_I) and inside it, the embedded small region-II. The blow out of region-II contains phase-1 (*Conductivity* → σ_1; *Electric field* → E_{II}; *Color* → *Yellow*; *Notation* → 1) denoted as notation *1* and phase-2 (*Conductivity* → σ_2; *Electric field* → E_{II}; *Color* → *Red*; *Notation* → 2). The shaded region is of phase-1 surrounded by arbitrary phase.

Let E_I and E_{II} be the electric field inside regions I and II respectively on Fig. 1. These two electric fields should be equal, $E_I = E_{II}$, because SCA assumes that the effective conductivity value across the material on average is the same, and that the current flowing in both regions I and II is also the same, $j_I = j_{II}$. These two assumptions, combined with the relation $j = \sigma E$, imply that $E_I = E_{II}$. First we take spherical islands of different phases. Then the dipole moment of n_1 spherical islands of radius a_1 and of conductivity σ_1 embedded inside a material of conductivity σ_e can be written as

$$P_1 = \frac{n_1 E_{\rm II} 4\pi a_1^3(\sigma_1 - \sigma_e)}{3(\sigma_1 + 2\sigma_e)} = \frac{\phi_1 E_{\rm II}(\sigma_1 - \sigma_e)}{(\sigma_1 + 2\sigma_e)} . \quad (1)$$

Similarly the dipole moment of n_2 spherical islands of radius a_2 and of conductivity σ_2 embedded inside a material of conductivity σ_e is given by

$$P_2 = \frac{n_2 E_{\rm II} 4\pi a_2^3(\sigma_2 - \sigma_e)}{3(\sigma_2 + 2\sigma_e)} = \frac{\phi_2 E_{\rm II}(\sigma_2 - \sigma_e)}{(\sigma_2 + 2\sigma_e)} . \quad (2)$$

The presence of islands of both phases inside region-II creates a polarization which disturbs the electric field locally. But on average their contributions have to cancel each other to satisfy the condition $E_{\rm I} = E_{\rm II}$, and the total polarization of region-II is zero. Mathematically it can be written as $P_1 + P_2 = 0$, which with the help of Eqs. (1) and (2) gives the relation for σ_e:

$$\frac{\phi_1(\sigma_1 - \sigma_e)}{(\sigma_1 + 2\sigma_e)} + \frac{\phi_2(\sigma_2 - \sigma_e)}{(\sigma_2 + 2\sigma_e)} = 0 . \quad (3)$$

For superconducting inclusion inside a metallic phase we need to substitute $\sigma_1 \to \sigma_s$; $\sigma_2 \to \sigma_m$; $\phi_1 \to \phi_s$; $\phi_2 \to \phi_m$ in Eq. (3).

For the islands of ellipsoidal shape instead of single depolarization value, as in spherical case, we have 3 depolarization values for 3 main axes of the ellipsoid. We consider prolate spheroidal inclusions with $a_x = a_y$, $a_z^2/a_x^2 \equiv \gamma > 1$ and the eccentricity $\chi \equiv \sqrt{1 - 1/\gamma}$. After algebraic operation with depolarization matrix (See Eq. (18.19) of Ref. [9] for the whole procedure) the equation for effective conductivity σ_e can be written as

$$\phi_s(\boldsymbol{\sigma_s} - \boldsymbol{\sigma_e})R^{se} + \phi_m(\boldsymbol{\sigma_m} - \boldsymbol{\sigma_e})R^{me} = 0 , \quad (4)$$

where $R^{je} = \left[I + A^*\boldsymbol{\sigma_e}^{-1}(\boldsymbol{\sigma_j} - \boldsymbol{\sigma_e})\right]^{-1}$, I is the identity matrix, σ_j is the conductivity in phase j, $\boldsymbol{\sigma_j} = I\sigma_j$, and the effective conductivity tensor $\boldsymbol{\sigma_e}$ is diagonal:

$$\boldsymbol{\sigma_e} = \begin{bmatrix} \sigma_{xx} & 0 & 0 \\ 0 & \sigma_{yy} & 0 \\ 0 & 0 & \sigma_{zz} \end{bmatrix} . \quad (5)$$

The depolarization tensor A^* for a prolate ellipsoid is given by (see section 17.1.2 of Ref. [9])

$$A^* = \begin{bmatrix} Q & 0 & 0 \\ 0 & Q & 0 \\ 0 & 0 & 1-2Q \end{bmatrix} , \quad (6)$$

where

$$2Q = 1 + \frac{1}{\gamma - 1}\left[1 - \frac{1}{2\chi}\ln\left(\frac{1+\chi}{1-\chi}\right)\right] . \quad (7)$$

We substitute the values of σ_m, A^*, σ_e and $\phi_m = (1 - \phi_s)$ in Eq. (4), and simplify using that in the superconducting phase $\sigma_s \to \infty$. Then solving for σ_e we get

$$\frac{\sigma_{xx}}{\sigma_m^x} = \frac{\sigma_{yy}}{\sigma_m^y} = \frac{Q}{Q - \phi_s} , \quad \frac{\sigma_{zz}}{\sigma_m^z} = \frac{1 - 2Q}{1 - 2Q - \phi_s} , \quad (8)$$

which is valid only for the isotropic metallic conductivity $\sigma_m^x = \sigma_m^y = \sigma_m^z$.

The anisotropic case, when $\sigma_m^x \neq \sigma_m^y \neq \sigma_m^z$, can be transformed to isotropic by dilation with different coefficient along different axes. For layered compounds with in-plane isotropy $\sigma_m^x = \sigma_m^y > \sigma_m^z$, similar the procedure in Refs. [4, 7], we dilate only along the z-axis as $z^* = z/\sqrt{\eta}$, where $\eta = \sigma_m^z/\sigma_m^x$, so that after dilatation $\sigma_m^x = \sigma_m^y = \sigma_m^z$. The effective conductivity for this case can be obtained from Eq. (8) by making the substitution $\gamma \to \gamma^* \equiv \gamma/\eta$. For the highly anisotropic case, i.e when $\sigma_{zz} \ll \sigma_{xx}$ and $\gamma^* \gg 1$, Eq. (7) simplifies to

$$Q = \frac{1}{2} + \frac{2 - \ln(4\gamma^*)}{4\gamma^*}. \tag{9}$$

Then for prolate spheroid with ($a_z \gg a_x = a_y$) Eq. (8) gives

$$\begin{aligned} \frac{\sigma_{xx}}{\sigma_m^x} &= \frac{2\gamma^* + 2 - \ln(4\gamma^*)}{2\gamma^* + 2 - \ln(4\gamma^*) - 4\gamma^*\phi_s}, \\ \frac{\sigma_{zz}}{\sigma_m^z} &= \frac{\ln(4\gamma^*) - 2}{\ln(4\gamma^*) - 2 - 2\gamma^*\phi_s}, \end{aligned} \tag{10}$$

which remains valid also after the inverse dilatation to the original geometry. The value of $\gamma = a_z^2/a_x^2$ is determined by the shape of spheroids and can be taken arbitrarily, provided $\gamma^* = (a_z^2\sigma_{xx})/(a_x^2\sigma_{zz}) > 1$ (for $\gamma^* < 1$ one should use Eq. (17.31) of Ref. [9] instead of Eq. (7)). Eq. (10) differs strongly from the expression for conductivity in such a heterogeneous system obtained previously using the Maxwell approximation[4, 7]:

$$\frac{\sigma_{xx}}{\sigma_m^x} \approx \frac{1}{1-\phi_s} + \phi_s, \quad \frac{\sigma_{zz}}{\sigma_m^z} \approx \frac{1}{1-\phi_s} + \frac{2\gamma^*\phi_s}{\ln(4\gamma^*) - 2}. \tag{11}$$

In particular, conductivity in Eqs. (10) and (11) diverges at completely different ϕ_s values.

3. Comparison with experiment and with Maxwell approximation

In this section we compare the above theoretical calculation with the experimental data from Refs. [4] and [7] on resistivity in FeSe along the conducting layers *a-b* (ρ_{ab}) and perpendicular to them (ρ_c). There are several reasons for choosing FeSe. First, FeSe is a heterogeneous[12] quasi-2D superconductor whith large conductivity anisotropy: its resistivity along the z axis is about 400 times greater than along the conducting *x-y* planes, but the in-plane conductivity σ_{ab} is isotropic. Hence, FeSe is a perfect system to apply the prolate spheroid case as described in the previous section. Second, the excess conductivity above T_c due to superconducting inclusions in the Maxwell approximations has already been calculated, and the corresponding volume ratio (ϕ_s) of superconducting phase for the same system has already been extracted, which allows us to compare both the results. The ϕ_s values from Eq. (10) can be found as

$$\begin{aligned} \phi_s^x &= \frac{(\sigma_{xx} - \sigma_m^z)Q}{\sigma_{xx}} \approx \frac{(\sigma_{xx} - \sigma_m^x)\left[2\gamma^* + 2 - \ln(4\gamma^*)\right]}{4\gamma^*\sigma_{xx}}, \\ \phi_s^z &= \frac{(\sigma_{zz} - \sigma_m^z)(1 - 2Q)}{\sigma_{zz}} \approx \frac{(\sigma_{zz} - \sigma_m^z)\left[\ln(4\gamma^*) - 2\right]}{2\gamma^*\sigma_{zz}}, \end{aligned} \tag{12}$$

where ϕ_s^x and ϕ_s^z are the superconducting volume ratios calculated from experimental conductivity along the x and z axes respectively. In Ref. [7] it was shown experimentally that conductivity below $T = 40 - 50$ K has starts to deviate from the linear temperature dependence, hence we suspect that the SC islands starts to appear around $T = 40$ K. We find the metallic conductivities (σ_m^x, σ_m^z) by taking the inverse of linear extrapolated resistivity from high temperatures. In FeSe the conductivity along z-axis is around 400 less than along *x-y* plane for temperature below 50K. Hence, we use the value $\eta \approx 1/400$ for FeSe to calculate $\gamma^* \equiv \gamma/\eta$.

We use the conductivity along z-axis to find ϕ_s^z and then use ϕ_s^z to calculate conductivity along the xy plane. We choose z-axis to find the volume ratio ϕ because the effect of SC islands is most significant

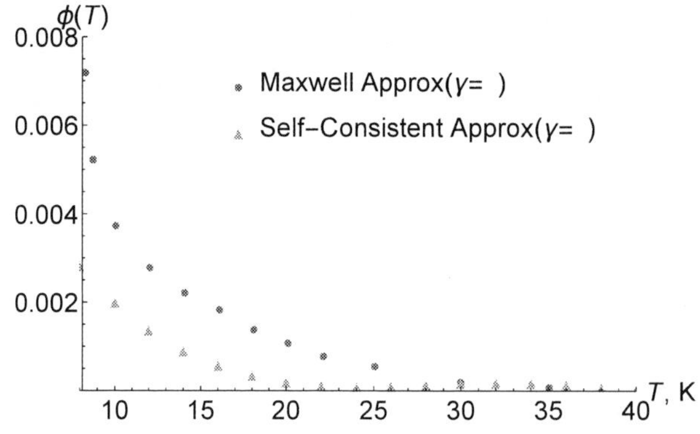

Figure 2. The volume ratio (ϕ_s^z) extracted from experimental data on resistivity from Ref. [4] using Eq. (12) under self-consistent approximation (orange triangles) and extracted from Fig. 4 of Ref. [7] under Maxwell approximation (blue circles). We used the value $\gamma = 1$ (spherical islands) and anisotropy ratio $\eta = \sigma_m^z/\sigma_m^x = 1/400$.

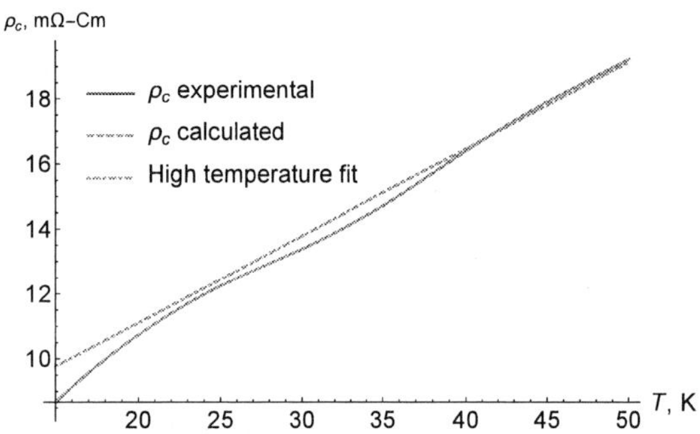

Figure 3. Comparison of experimental interlayer resistivity from Fig. 2c of Ref. [4] and theoretical result in Eq. (10). High temperature fit $\rho_c = 6.5 + 0.3T(\mathrm{m}\Omega\cdot\mathrm{cm})$ is found by extrapolating ρ_c above 50 K.

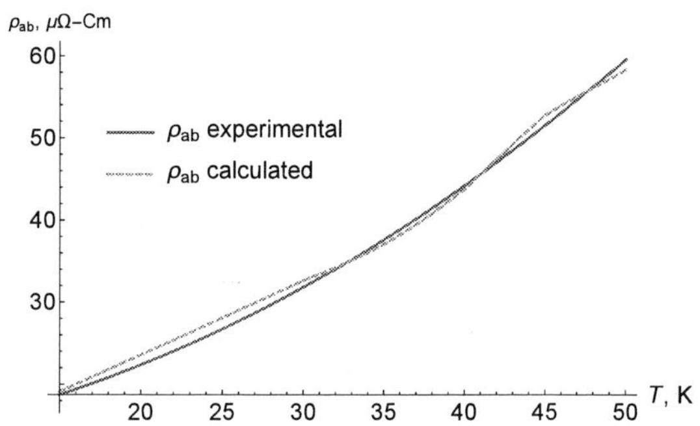

Figure 4. Comparison of experimental in-plane resistivity from Fig. 2c of Ref. [4] with theoretical result in Eq. (10). High temperature fit $\rho_{ab} = 12.3 + 0.21T(\mu\Omega\cdot\mathrm{cm})$ is found by extrapolating above 50 K.

in this direction, as it is the least conducting axis. In Fig. 2 we plot the superconducting volume ratio calculated using Eq. (12). We extracted resistivity data from Fig. 2c of Ref. [4]. Metallic conductivity $\sigma_m^z = 1/[0.0065 + 0.0003T](\mathrm{m}\Omega\cdot\mathrm{cm})^{-1}$ is found by extrapolating the ρ_c from 50 K upwards. The hump in the volume ratio is the due to decrease in the slope of ρ_c around temperature of $30 < T < 35$. For comparison, the SC volume fraction $\phi(T)$ extracted from Fig. 4 of Ref. [7] using Maxwell approximation for $\gamma = 1$ is also plotted in Fig. 2. Here we should mention that as we don't have exact metallic resistivity, we have made the linear approximation for ρ_c which might include some error of small percentage. In Figs. 4 and 3 the comparison of experimental and theoretical resistivity is shown. The complete agreement of prediction and experiment in Fig. 3 is because we calculated the ϕ_s using the same resistivity along the z axis.

4. Discussion

Above we derived the analytical expressions for conductivity in the heterogeneous superconductor using the self-consistent approximation (SCA) and compare the results with those[4, 7] obtained using Maxwell approximation and with experimental data in FeSe. The Maxwell and self-consistent approximations are the simplest to describe conductivity in heterogeneous materials [9]. They allow obtaining analytical results convenient for physical predictions and for comparison with experiment. The more accurate methods do not give simple analytical expressions and require the knowledge of spatial and size distribution of the second-phase inclusions [9]. The limitations and drawbacks of the SCA and Maxwell approximations differ, therefore their comparison is helpful for a reliable description of real materials.

The Maxwell approximation is valid only at small volume fraction $\phi \ll 1$ of the second phase, i.e. of superconducting inclusions in our case, and it gives incorrect percolation threshold $\phi = 1$. However, it is strictly substantiated in the limit $\phi \ll 1$ and coincides with the lower bound for the effective conductivity of media with superconducting inclusions at arbitrary ϕ. The SCA sometimes works well even at $\phi \sim 1$ and gives almost correct percolation thresholds in isotropic 2D and 3D cases. However, in anisotropic case these percolation thresholds differ for different directions, which is incorrect for most superconductors. Thus, in Eqs. (8) and (10) the conductivity along z and x axes diverges at different values of ϕ_s. The stronger is anisotropy of metallic phase, the larger is the difference between percolation thresholds. This is generally incorrect, but it may somewhat reflect physical situation if the length of SC islands or clusters is comparable to the size of the whole sample. In real superconductors, taking into account the Josephson coupling and proximity effect, the effects of anisotropic percolation thresholds may also appear, but in a different way. Thus, it is not a priori clear whether the SCA or Maxwell approximation is better for strongly anisotropic heterogeneous conductors. However, one may say that while the Maxwell approximation gives the lower bound, the SCA gives the upper bound for the effective conductivity along the least conducting axis. It is reflected in Fig. 2, where the same excess conductivity in the SCA is obtained for much smaller volume fraction of superconducting inclusions than in the Maxwell approximation. Hence, both these approximations are helpful in analyzing experimental data.

Acknowledgments

The work was carried out with financial support from the Ministry of Education and Science of the Russian Federation in the framework of increase Competitiveness Program of NUST ''MISIS'', implemented by a governmental decree dated 16th of March 2013, No 211. Sec. III was supported by the RSF grant # 16-42-01100. P.G. thanks RFBR grant #18-02-00280. T.M. thanks RFBR grant # 18-02-01022.

References

[1] Keimer B, Kivelson S A, Norman M R, Uchida R, and Zaanen J 2015 *Nature.* **518** 965
[2] Si Q, Yu R, and Abrahams E 2016 *Nat. Rev. Mater.* **1** 16017
[3] Kresin V Z, Ovchinnikov Y N, and Wolf S A 2006 *Phys. Rep.* **431** 231
[4] Sinchenko A A, Grigoriev P D, Orlov A P, et al. 2017 *Phys. Rev. B* **95** 165120
[5] Iguchi I, Yamaguchi T, and Sugimoto A 2001 *Nature* **412** 420
[6] Martin I, Podolsky D, and Kivelson S A 2005 *Phys. Rev. B* **72** 060502
[7] Grigoriev P D, Sinchenko A A, Kesharpu K K et al. 2017 *JETP Lett.* **105** 786
[8] Landauer R 1978 *AIP Conf. Proc.* **40** 2
[9] Torquato S 2002 *Random heterogeneous materials : microstructure and macroscopic properties* (New York: Springer-Verlag)
[10] Medvedev S, McQueen T M, Troyan I A et al. 2009 *Nat. Mater.* **8** 630
[11] Ge J F et al. 2015 *Nat. Mater.* **14** 285
[12] Naidyuk Yu G, Fuchs G, Chareev D A, and Vasiliev A N 2016 *Phys. Rev. B* **93** 144515

IOP Conf. Series: Journal of Physics: Conf. Series **1238** (2019) 012011 doi:10.1088/1742-6596/1238/1/012011

Selection of pulsed laser deposition conditions for preparation of perfect thin-film MoS_x hydrogen evolution catalysts

V N Nevolin[1], D V Fominski[2], R I Romanov[2], M I Esin[2], V Y Fominski[2], P F Kartsev[2]

[1]P N Lebedev Physical Institute of the Russian Academy of Science, Leninskii pr. 53, 119991 Moscow, Russia
[2]National Research Nuclear University MEPhI (Moscow Engineering Physics Institute), Kashirskoe sh. 31, 115409 Moscow, Russia

E-mail: vyfominskij@mephi.ru

Abstract. The influence of pulsed laser deposition conditions in the geometry "off-axis" on the catalytic properties of MoS_x films in the hydrogen evolution reaction is investigated. For the deposition of MoS_x films from MoS_2 target, pulsed laser radiation from the IR and UV wavelength ranges was used. The angle of incidence of the laser-induced plume on the surface of the substrate in a buffer gas was varied to check the influence of large in-size Mo-enriched particles. The efficiency of the catalyst was estimated from the results of the turnover frequency (TOF) measurement, which made it possible to minimize the influence of the "loading" of the catalyst on its characteristics. The effects of the chemical composition, local structure, and properties of the catalyst–substrate interface on the efficiency of the hydrogen evolution reaction are analysed. The regime of pulsed laser deposition of more effective thin-film MoS_x catalysts is determined.

1. Introduction

Recent studies have shown that thin-film forms of transition metal dichalcogenides (TMD), in particular molybdenum sulfide (MoS_x), can exhibit high electrocatalytic activity in the hydrogen evolution reaction (HER) [1,2]. Catalytically active sites were found on the edges of ultrathin layered MoS_2 structures, on the defect areas of the basal planes of such films, as well as on certain atoms of amorphous modifications of MoS_x [3,4]. Amorphous films of molybdenum sulfide attract interest because it is not necessary to solve the problem of forming clusters with a certain orientation to the substrate surface to obtain high density of catalytically active sites on the catalysts surface [5].

Pulsed laser deposition (PLD) allows to obtain effective TMD containing catalysts with different chemical and structural states, in which both amorphous regions and nanocrystalline inclusions can be present [6,7]. The realized structural state (local atomic packages) largely depends on the conditions for obtaining and depositing a laser plume. In the case of laser ablation of the target MoS_2 by laser pulses with a wavelength from the IR range (1.06 μm), relatively large particles enriched in molybdenum are inevitably formed [8]. In the case of the deposition of the plume along the normal to the surface of the substrate, these particles have an important influence on the morphology of the catalyst, since they increase the real area of its surface. In the case of the application of the deposition geometry by a sliding plume, large particles practically did not settle on the substrate [9].

The aim of the study was to investigate the influence of the conditions of laser plume generation and deposition on the catalytic activity of MoS_x films, which were created under the geometry of the off-axis PLD onto a substrate of glassy carbon (GC). In the case of deposition of relatively large particles enriched in Mo, the chemical nature of the catalyst support can be changed, and two types of MoS_x/Mo and MoS_x/GC interfaces may be realized. It is important to separate the effects in the catalytic hydrogen evolution associated with 1) an increase in the surface area; 2) the formation of catalytically active sites of various nature, including the appearance of pure Mo on the surface and 3) the appearance of interfaces of MoS_x/Mo and MoS_x/GC.

2. Experimental methods

For ablation of the target MoS_2, nanosecond pulses with a wavelength of 1064 and 266 nm were used. For both wavelengths, the laser fluence of the target irradiation was chosen from the condition for the formation of a plasma plume with a low concentration of sparks which could be caused by the formation of large particles. The resulting laser plume propagated along the direction of the normal to the surface of the target in argon at a pressure of 5 Pa. GC substrates were located at some distance from the axis of the laser plume expansion (figure 1). Thus, the direction of movement of the plume fraction was chosen, in which the sulfur concentration could be quite high [10]. In this case, the substrates were installed at different angles (θ) to the direction of propagation of the laser plume. This angle was varied in the interval 0°–90°. Film deposition time at any conditions was 8 min. The PLD was made at room temperature of GC substrate.

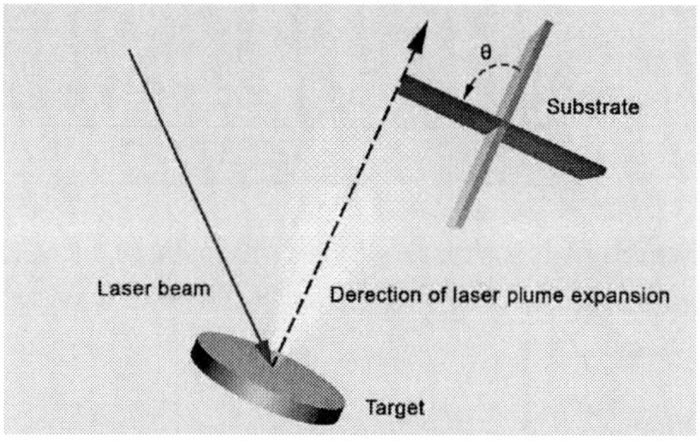

Figure 1. PLD of MoS_x films with various configuration of the off-axis deposition from MoS_2 target.

Structural and morphological characteristics of MoS_x films were studied by scanning electron (SEM) and transmission electron microscopy and microdiffraction (TEM and MD) methods, supplemented with energy-dispersive X-ray spectroscopy (EDXS). To analyze the catalytic activity of films, a traditional set of methods for electrochemical measurements in 0.25 M H_2SO_4 was used. The turnover frequency (TOF) was chosen as the main criterion of comparison, which made it possible to reduce the dependence of the catalytic characteristics of the prepared MoS_x films on the loading (thickness) of the catalytic material.

For estimation of TOF, a technique was used in which a cyclic voltammogram was measured in the range from -50 to +1150 mV after applying the cathodic polarization [11]. Integration of current peaks resulted from the oxidation of active sites allows estimating the number of active sites on the catalyst surface. The normalization of the current density, which is measured by linear voltammograms, by the number of catalytically active sites allows us to calculate the TOF of the catalyst.

3. Results and discussions

Figure 2 shows the characteristic morphology of MoS_x films, which were obtained by the off-axis PLD method with a wavelength of 1064 and 266 nm on the GC substrates. The substrates were placed at different angles to the axis of expansion of the laser plume. It can be seen that the wavelength did not have a significant effect on the morphology of the film, which for θ=90 ° contained a high surface

IOP Conf. Series: Journal of Physics: Conf. Series **1238** (2019) 012011 doi:10.1088/1742-6596/1238/1/012011

concentration of rounded particles of micron and submicron sizes. It was previously established that these particles are enriched in molybdenum [9].

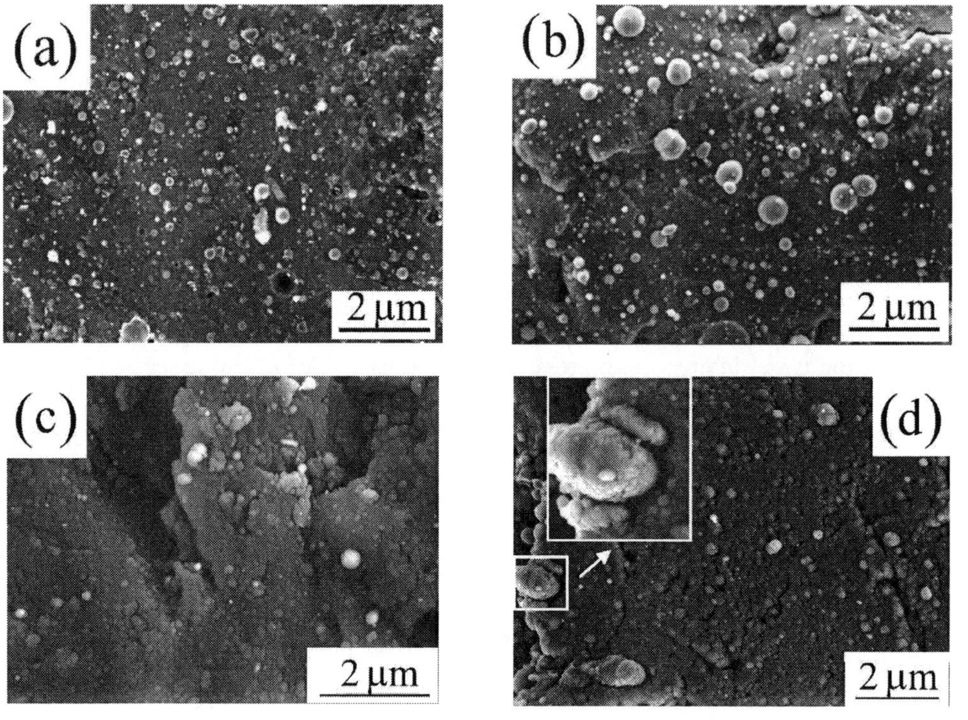

Figure 2. SEM images of MoS_x films, which was prepared by off-axis PLD at various θ: (a,b) θ=90° and wavelength 1064 and 266 nm, respectively; (c) θ=0° and wavelength 1064 nm; (d) θ=30° and wavelength 1064 nm.

When the substrate was turned parallel to the axis of expansion of the plume (θ=0°), the Mo particles were deposited in a much smaller amount, and the morphology of the film consisted of densely packed nanoparticles up to 50 nm in size. In the intermediate position (angle) of the substrate, the submicron-sized Mo particles are deposited onto the substrate, and the larger particles are deformed due to spreading over the GC surface. As a result, the height of such particles above the surface decreased noticeably (figure 2).

Measurements of the ratio x=S/Mo by the EDXS method have shown that at normal incidence of the plume x was ≤2. A decrease of an angle of the substrate tilt caused a noticeable increase in the sulfur concentration ($x \geq 3$).

The results of TEM studies of prepared films are shown in figure 3. For MoS_x films obtained at 90°, regions with relatively high-quality laminar packing of Mo and S atoms in a hexagonal 2H-MoS_2 lattice were observed (figure 3 (a)). The crystals sizes were no more than 15 nm, and the distances between the basal planes were ~ 0.6 nm. In addition to such local areas with an ordered structure, regions with amorphous atomic packing, which corresponded to diffuse reflections in the microdiffraction pattern, were also found for this film (results not presented). Films obtained at other angles of deposition of the laser plume had a more disordered structure [10]. However, high resolution TEM studies revealed the presence of nanoclusters with laminar packing of molecular planes in these films. The distance between the planes was about 0.6 nm (figure 3(b)).

IOP Conf. Series: Journal of Physics: Conf. Series **1238** (2019) 012011 doi:10.1088/1742-6596/1238/1/012011

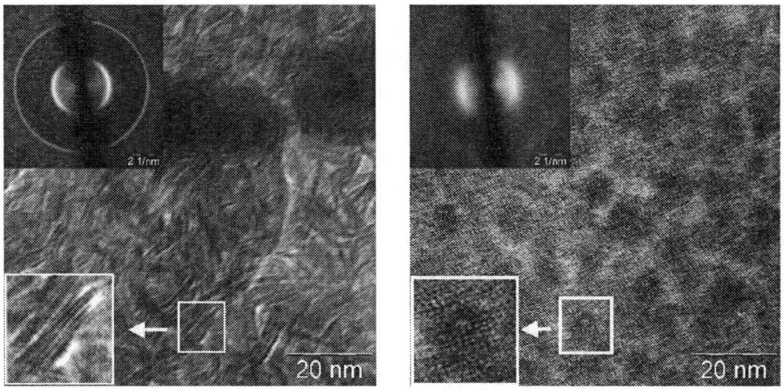

(a) (b)

Figure 3. High resolution TEM and MD patterns for the MoS_x, films prepared by off-axis PLD on the substrates at θ=90° (a) и θ=0° (b)

Figure 4 shows the results of electrochemical studies of the prepared catalytic MoS_x films. The deposition conditions significantly influenced the results of all the test types carried out. For the sample located at 90°, the most intense peaks of oxidation of the active sites were observed. The peaks could correspond to Mo (630/678 mV) and molybdenum sulfide (700 mV). Other samples had peaks of 600 and 850 mV, which could correspond to oxidation of amorphous Mo-3S clusters. The results of the TOF calculations for the prepared MoS_x films are given in table 1. According to calculations, the largest values of TOF were found in samples unfolded at an angle of 30° to the direction of expansion of the laser plume. At the same time, this sample demonstrated larger value of resistance to current transport during catalysis.

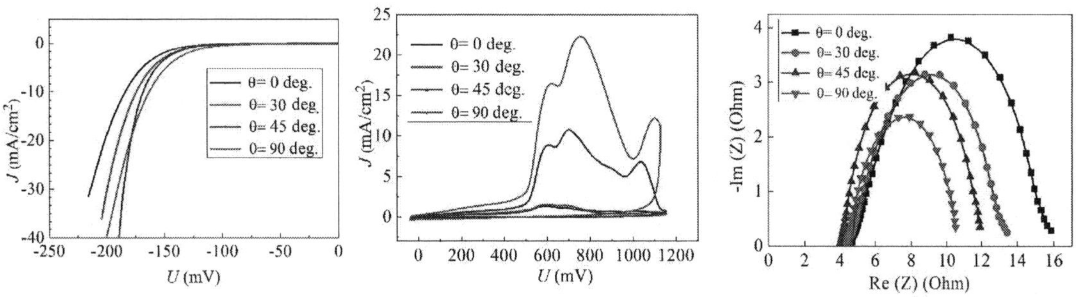

(a) (b) (c)

Figure 4. Linear (a) and cyclic (b) voltammograms, as well as Nyquist plots (d) for MoS_x films, which were deposited at different angles between the direction of the laser plume expansion and the substrate surface.

The deposition of a large number of molybdenum particles actually could change the chemical nature of the thin film MoS_x catalyst support. The influence of the nature of the interface of MoS_x film with such supports as glassy carbon and molybdenum on the catalytic activity of MoS_x was studied by calculating the change in the Gibbs binding energy (ΔG) during the formation of a hydrogen molecule. The calculations were performed within the framework of the density functional theory (DFT) using the software package for quantum-mechanical calculations Quantum ESPRESSO in the basis of plane waves. Figure 5 shows two models that have been analyzed. In one model, the MoS_3 chain of atoms was located above a single graphite plane that mimicked a local region of glassy carbon, for which a graphite-like short-range order of atomic packing is characteristic. In another model, the MoS_3 chain was located above the atomic plane of Mo. The reliability of the chain model of the MoS_x amorphous phase was considered in detail by Hibble and Wood [12]. Calculations have shown that absolute values of ΔG for the selected models differ only a little and the real values are 0.29 eV and -0.31 eV for the MoS_3/Mo(110) and MoS_3/GC contact regions, respectively

Table 1. Important catalytic characteristics of the obtained MoS_x films.

Deposition angle	Overvoltage, mV (J=10 mA/cm^2)	Tafel slope (mV)	TOF (s^{-1})
θ=0°	184	42.3	0.090
θ=30°	175	43.9	0.130
θ=45°	168	39.0	0.043
θ=90°	162	46.6	0.015

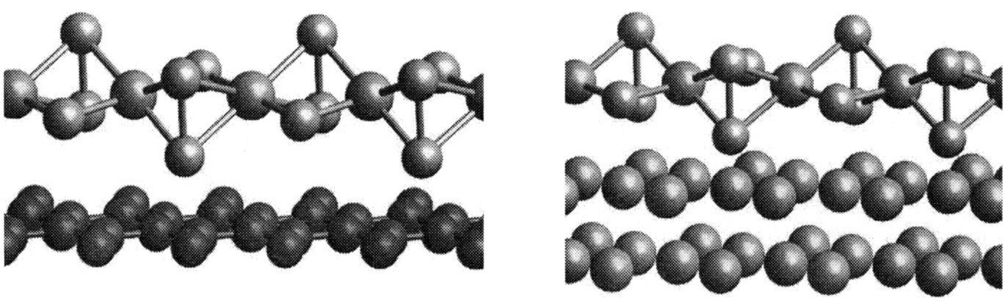

Figure 5. Atomic configurations, which were used for analysis of HER on the MoS_3 catalysis film under the influence of MoS_3/GC (a) and MoS_3/Mo (b) interfaces (Mo – blue, C – black, S – yellow).

Thus, the modification of the chemical nature of the support (Mo instead of GC) could not have a significant effect on the catalytic properties of MoS_x films. It should be noted that DFT calculations were carried out for the chain modification of molybdenum sulfide. In the real case, the MoS_x film contained amorphous and nanocrystalline phases.

4. Conclusions

In the case of production of thin-film MoS*x* catalysts by the PLD method from the MoS_2 target, the wavelength of the laser radiation had no appreciable effect on the morphology of the film. When the laser plume was deposited along the normal to the substrate surface, the prepared films consisted of micro- and submicro-meter Mo particles and a nanostructured matrix. When the substrate was oriented along the direction of plume motion, the contribution of the Mo particles to the film morphology was changed, since the Mo particles spread over the surface (θ=30°) or do not adhere to the surface at all (θ=0°). Deposition of Mo particles could increase the catalytic activity of MoS_x films by increasing the surface area and improving the current transport. However, a more important factor is the local packing in the MoS_x film, which largely depended on the angle of incidence of the laser plume. The most perfect catalytic MoS_x films were obtained when the substrate was slightly tilted (θ=30°) to the direction of plume expansion and a combination of propagated and scattered atomic flows was realized for film growth.

Acknowledgments
The reported study was funded by RFBR according to the research project № 18-02-00191.

References

[1] Seo B and Joo S H 2017 *Nano Convergence* **4**:19.
[2] He Z and Que W 2016 *Appl. Mater. Today* **3** 23–56.
[3] Voiry D, Salehi M, Silva R, Fujita T, Chen M, Asefa T, Shenoy V B, Eda G and Chhowalla M 2013 *Nano Lett. 13* 6222–27.
[4] Lin Z, Carvalho B R, Kahn E, Lv R, Rao R, Terrones H, Pimenta H A and Terrones M 2016 *2D Mater.* **3** 022002.
[5] Morales-Guio G G and Hu X 2014 *Acc. Chem. Res*. **47** 2671–81.
[6] Fominski V Yu, Romanov R I, Fominski D V and Shelyakov A V 2018 *Nucl. Instr. Meth. Phys. Res***.** B **416** 30-40.
[7] Grigoriev S N, Fominski V Y, Romanov R I, Volosova M A and Fominski D V *2018 J. Phys. Conf. Ser.* **941(1)** 012065
[8] Walck S D, Zabinski J S, Donley M S and Bultman J E 1993 *Surf. Coat. Technol.* **62** 412–16.
[9] Fominski V Yu, Romanov R I, Fominski D V, Dzhumaev P S and Troyan I A 2018 *Opt. Laser Technol***.** **102** 74-84.
[10] Fominski V Yu, Nevolin V N, Romanov R I and Smurov I 2001 *J. Appl. Phys*. **89** 1449–57.
[11] Ting L R L, Deng Y, Ma L, Zhang Y-J, Peterson A A and Yeo B S 2016 *ACS Catal.* **6** 861–67.
[12] Hibble S J and Wood G B 2004 *J. Am. Chem. Soc.* **126** 959–65.

IOP Conf. Series: Journal of Physics: Conf. Series **1238** (2019) 012012 doi:10.1088/1742-6596/1238/1/012012

The detection of nitro compounds vapor using sensors based on MIS-structures manufactured by laser deposition

M O Etrekova[1,2], A V Litvinov[1], A A Mikhailov[2]

[1] National Research Nuclear University MEPhI (MEPhI), Kashirskoe shosse 31, 115409, Moscow, Russia
[2] Scientific production company "INKRAM", Lyublinskaya Street 151, office 222, 109341, Moscow, Russia

E-mail: AVLitvinov@mephi.ru

Abstract. The conditions of the detection of nitro compounds vapors was investigated using the trinitrotoluene as an example. The detection method consists of pyrolysis and later registration by the MIS-sensor of gaseous products of thermal decomposition . It was shown that the offered method allows to achieve the detection limit of $1 \cdot 10^{-12}$ g/cm^3 (particles – 1 ng) with respect to the vapors of trinitrotoluene.

1. Introduction
This article is devoted to a new gas-analytical sensory method of detection of trace amounts of nitro compounds, allowing to create a device with a limit of detection of trinitrotoluene vapors not worse than $1 \cdot 10^{-12}$ g/cm^3 (particles – 1 ng).

The sensor is based on the metal-insulator-semiconductor (MIS) structure, developed in the National Research Nuclear University MEPhI [1]. The design and principle of operation of the sensor are described in details in [2].

2. Previous experience and statement of the problem
For the first time, the gas-analytical method of vapor detection of nitro-containing substances using a semiconductor sensor based on MIS-structure was proposed in 2006 by a group of researchers from Moscow Engineering Physics Institute [3].

The essence of the method was as follows. A gas sample with nitro compounds (trinitrotoluene, dinitrotoluene, nitroglycerin, hexogen, etc.) was heated to a temperature of 400 ° C and above. Then, the gaseous products of thermal decomposition (pyrolysis) were fed to the MIS-sensor and a useful signal was recorded. The nitrogen dioxide was used as a reference product of pyrolysis in this case.

It has been established [2] that MIS-sensors based on Pd-SiO_2-Si-Pd type structures obtained by laser deposition (Pd electrode thickness 30 nm, diameter 3 mm) have demonstrated the best sensitivity to NO_2 with a detection threshold of 1 ppb.

According to [4], at room temperature, the vapor pressure of TNT does not exceed 5...7 ppb or 60 pg per 1 cm^3. Thus, the sensitivity of the MIS-sensor to NO_2 is enough for its use in a vapor detector of nitro compounds.

This article presents the results of the research on the following tasks:

- studying the theoretical basis of the thermal decomposition reaction of nitro compounds,

- development of a procedure for preparing samples of nitro compounds of the required mass,
- development and creation of an experimental testing workbench,
- determination of optimal operating parameters of the experimental testing workbench,
- calibration of the experimental stand by trace amounts of nitro compounds.

3. The theoretical basis of the thermal decomposition reaction of nitro compounds

An experimental study of the regularities of the process of thermal destruction of molecules of nitro compounds has been conducted for more than half a century. However, the accumulated data are very contradictory. This can be explained by the competition of various channels of the primary act of the reaction of gas-phase monomolecular decomposition.

To describe the process of thermal decomposition of nitro compounds, it is customary to use the Arrhenius hypothesis, according to which a reaction involving molecules occurs only if they have some excess energy [5]. According to the kinetic theory of gases, the number of active molecules per unit volume is equal to their total number multiplied by the factor $e^{-\frac{E_a}{R \cdot T}}$, where E_a is the excess energy of the active molecule, J/mol (or activation energy); T is the temperature, K; R is the universal gas constant, $\frac{J}{mol \cdot K}$. Thus, if the reaction proceeds, the equation holds:

$$k = B \cdot e^{-\frac{E_a}{RT}}, \quad \text{or} \quad \ln k = -\frac{E_a}{R} \cdot \frac{1}{T} + \ln B, \tag{1}$$

where k is the rate constant of thermal decomposition, c^{-1}; B is the pre-exponential factor characterizing the probability of a chemical reaction.

According to the modern theory of monomolecular decomposition mechanisms of 2,4,6-trinitrotoluene [6], the most likely scenarios of the primary reaction act are the following:
a) breaking the C-N bond with the abruption of the NO_2 group (E_a = 258-280 kJ/mol);
b) nitro-nitrite rearrangement (E_a = 190 kJ/mol);
c) aciform formation (E_a = 105-133 kJ/mol).
As seen, the process of intramolecular hydrogen transfer from the CH_3 methyl group to the NO_2 group with the formation of aciforms is energetically less expensive.

However, due to the larger pre-exponential factor B at high temperatures, the rate of the TNT thermal decomposition reaction by the radical mechanism of the C-NO_2 bond breaking increases more rapidly with respect to alternative reactions and begins to predominate.

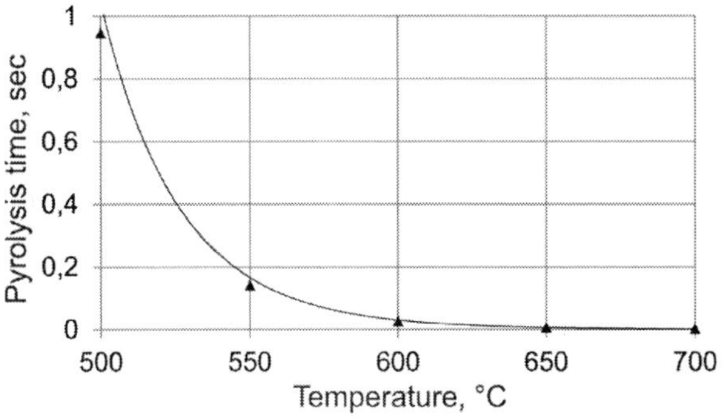

Figure 1. The dependence of the duration of thermal decomposition of nitrotoluenes from temperature

Using equation (1) and experimental data on the activation parameters of the thermal decomposition of nitrotoluenes [6] at temperatures of 230 °C and higher, in Figure 1 we show the dependency duration of the pyrolysis reaction of TNT (sec) from the temperature T, (C).

Thus, it is possible to estimate the conditions (time and temperature) necessary for the pyrolysis of vapors of nitro compounds with the formation of NO_2.

4. Development of a procedure for preparing samples of nitro compounds

In this work, the following types of nitro compounds samples were used:
(a) – a sample of the solid substance dissolved by an acetone of special purity on an aluminum foil substrate 2.5 × 4 cm with a thickness of 0.9 μm;
(b) - a sample of the solid substance in a sterile container, dissolved by a fixed volume of acetone (0.5-10 ml), followed by precipitation of 100-200 μl of solution onto a substrate;
(c) – secondary dilution of a certain part of a solution of type (b) (100-500 μl) by a fixed volume of acetone (0.5-5 ml), followed by precipitation of 100-200 μl of solution onto a substrate;
(d) - secondary dilution of a solution of type (c);
(e) , etc - dilution of a solution of type (d): – next steps of the secondary dilutions by a fixed volume of acetone with precipitation of 100-200 μl of solution onto a substrate.

For example, to create a sample with a mass of about 10 μg it is necessary to take a dose of 100 μl from a solution of type (b) with a concentration 1 μg/μl (so 100 ± 3 μg of substance) and add 1 ml of pure acetone to it. In this way, we will obtain a new solution of type (c) with a concentration 91 ng/μl. Then, we can precipitate 100 μl of the solution on the foil, and after evaporation of acetone, we will get 9.1 ± 0.3 μg of the substance.

The relative error in the mass of the nitro compound in the sample was for a solution of the type (a) – 2 %, (b) – 3 %, (c) – 4 %, (d) – 5 %, (e) – 6 %; (f) – 8 %; (g) – 10 %; (h) – 12 %.

In the framework of this work, 2,4,6-trinitrotoluene was used as a sample of a nitro-containing substance. Dilution with acetone was carried out in sterile containers of glass and polypropylene. The shelf life of samples obtained by the multiple dilution methods with acetone at room conditions was no more than 7 days.

The substance was weighed using laboratory electronic scales CAUW-120D from manufacturer CAS with mean square deviation ≤ 0.02 mg. The dosage of acetone for the dissolution of the TNT samples was performed using single-channel pipette dispensers with a volume of 100 μl and 1-10 ml from manufacturer Thermo Fisher Scientific.

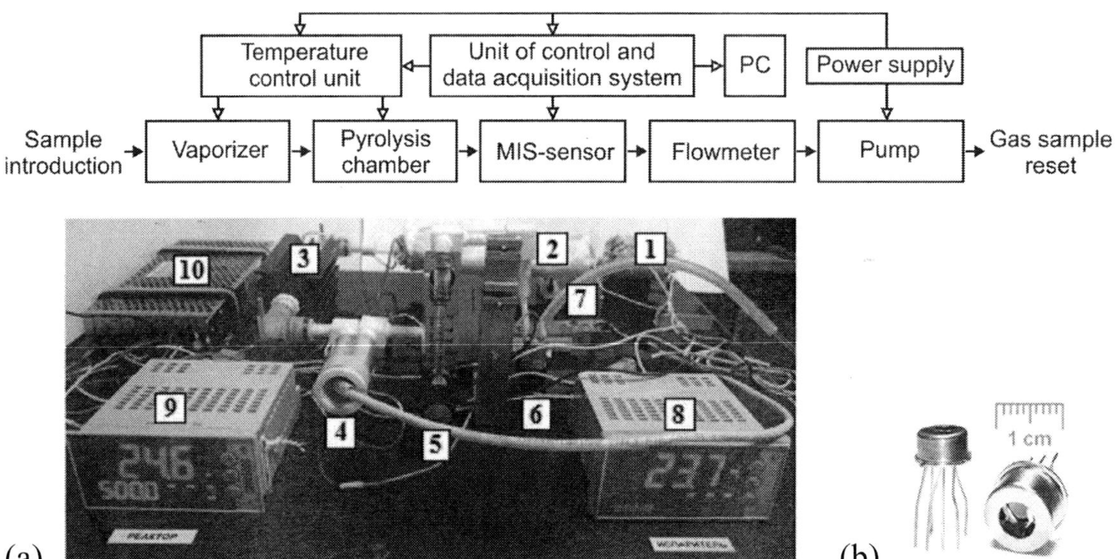

Figure 2. (a) Schematic diagram and photograph of the experimental testing workbench. Notation: 1 – the vaporizer; 2 – the pyrolysis chamber (or reactor); 3 – the device for cooling the gas mixture; 4 – MIS-sensor; 5 – the flowmeter; 6 – the pump; 7 - the electronic board; 8, 9 - thermoregulators, 10 - the power supply. (b) Photo image of the MIS-sensors.

5. Development and creation of an experimental testing workbench

Taking into account the above-described theoretical and experimental data, a test stand (Figure 2) was developed and constructed for a detailed study of the proposed method and determination of optimal operating modes.

The work on the test bench was carried out as follows.

A sample of the nitro compound on the foil was placed in an evaporator, where it was heated to 40-100 °C to increase the vapor pressure. The evaporator was made of a quartz glass tube 9 cm long with an inside diameter of 1.5 cm and a wall thickness of 2 mm. The nichrome filament was wound on the outer surface of this tube ($d = 0.5$ mm, $R = 12\ \Omega$).

Using a miniature diaphragm pump with a flow rate of 0.1-0.5 l/min, the carrier gas supplied to the evaporator inlet was withdrawn along with the TNT vapors and sent to the reactor. The reactor was also made of a quartz tube 25 cm long (heated working length 16 cm) with an internal diameter of 7 mm and a wall thickness of 1.5 mm. Nichrome wire was wound on the outer surface of the reactor tube ($d = 1$ mm, $R = 3\ \Omega$).

The nichrome heaters with the installed temperature sensors were fixed on the evaporator's and the reactor's quartz tubes by a heat-resistant electrically insulating compound and thermostated with glass fiber and heat-resistant glass cloth. Under the influence of high temperature in the reactor thermal decomposition of TNT vapors leads to NO_2 emission, which is registered by the MIS-sensor.

In this work, we used MIS-sensor of the Pd-SiO_2-Si-Pd type. The signal from the sensor was processed by an electronic circuit board and outputted using the appropriate software on the PC via the RS-485 interface.

Based on the geometric dimensions of the reactor and according to Figure 1, it was assessed of the required ratio "temperature of reactor - flow rate" (or what amounts to the same thing, time of the vapor residence in the reactor at a given temperature) in order to meet the conditions for the pyrolysis reaction by the radical C-NO_2 bond breaking mechanism.

The flow of sampling was regulated by a panel-type rotameter made of acrylic plastic with an adjusting valve. The temperature of the evaporator and the reactor was regulated by industrial thermoregulators. The power supply of all elements of the installation was carried out from a two-channel switching power supply.

6. Determination of optimal operating parameters of the experimental testing workbench

Before the experiments the MIS-sensor was calibrated by low NO_2 concentrations. The calibration results are shown in Figure 4.

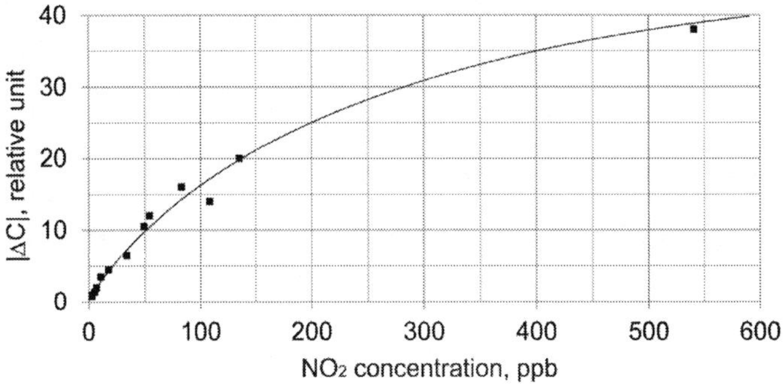

Figure 3. MIS-sensor calibration with respect to NO_2.

According to [7], MIS-sensors do not exhibit sensitivity to organic compounds, including nitro-containing ones. Therefore, assuming that at least one of the three nitro groups is detached from each TNT molecule, one can judge the amount of a substance successfully subjected to pyrolysis.

It was shown that the evaporation rate of TNT sample with a spot area $S \approx 5$ cm^2 (type (a)) from the foil surface at an evaporator temperature of 70 °C is 3.0 ± 0.5 μg/min (the influence of the flow in the range 0.1-0,5 l/min can be neglected).

The optimal temperature of the reactor was established with a fixed heating of the evaporator to 70 °C, a flow of 0.5 l/min, and the evaporation area of the TNT sample spot on the foil of 5 ± 1 cm^2. The time duration influence of the TNT vapor residence in the reactor on the response of the MIS-sensor ΔC_{MIS} was also studied. The results are shown in Table 1 and Table 2.

Table 1. Optimization of pyrolysis chamber temperature.

M_{TNT}, ±0,01 mg	$T_{reactor}$, °C	ΔC_{MIS}, rel.un.	$t_{rel.}$	$K_{rel.}$
1,60	700	12	370	0,1
1,16	600	27	25	0,4
1,29	550	51	5,3	1,6
1,44	500	37	0,8	0,9
1,76	400	1	0,01	0,01

Table 2. Influence of the flow rate or residence time of TNT vapors in the pyrolysis chamber.

M_{TNT}, ±0,01 mg	$T_{reactor}$, °C	Flow, l/min	ΔC_{MIS}, rel.un.	$t_{rel.}$	$K_{rel.}$
1,49	700	0,1	4	1850	→0
1,49	700	0,3	14	615	0,1
1,60	700	0,5	12	370	0,1
1,34	500	0,1	57	3,9	0,3
1,44	500	0,5	37	0,8	0,9

Notations: $t_{rel.}$ – the coefficient of the ratio of the duration of the actual and theoretically sufficient time of heating (see Figure 1) for pyrolysis of the TNT vapors; $K_{rel.}$ – the ratio of the measured NO_2 concentration (according to Figure 3) and the TNT concentration supposed.

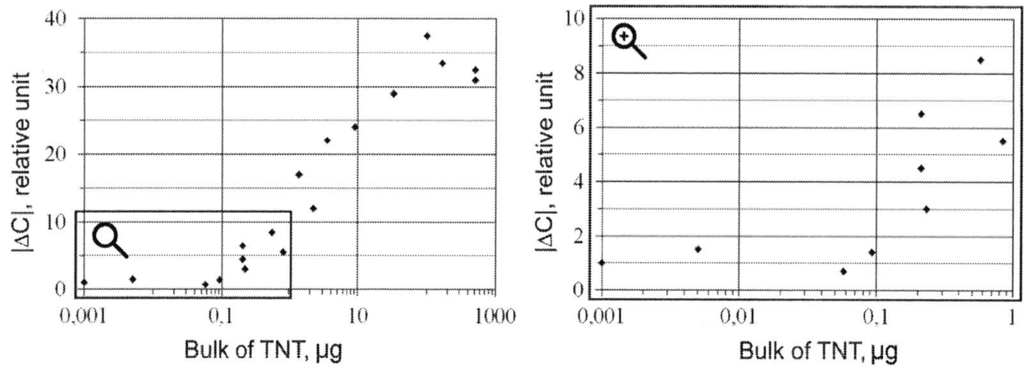

Figure 4. Calibration of the experimental stand by trace amounts of TNT.

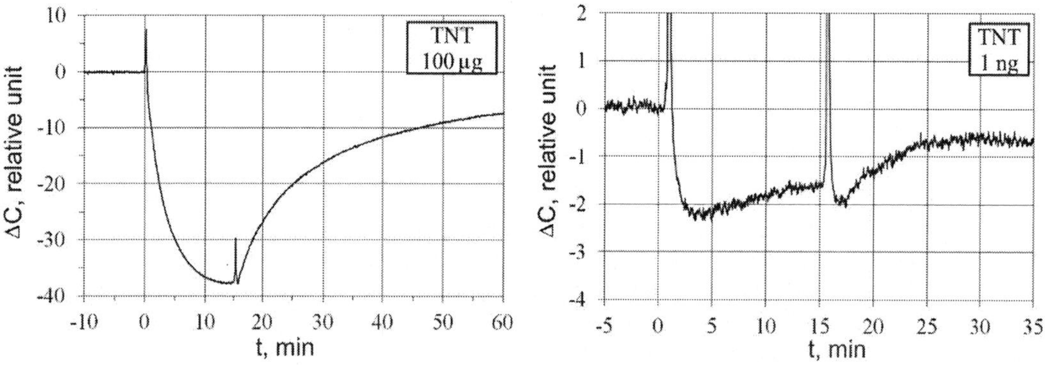

Figure 5. Example of the response of a MIS-sensor per 100 μg and 1 ng TNT.

If the sample productivity, the carrier gas flow rate and the duration of each sample supply (15 min) are known, one can estimate the concentration obtained and the total mass flow of TNT as:

$$3\,\frac{\mu g}{min}/0{,}5\frac{l}{min} = 6\frac{\mu g}{l}\ \text{(or 635 ppb under normal conditions)};$$

$$0{,}5\frac{l}{min}\cdot 15\ min = 7{,}5\ l\ \text{(total volume of gas sample)};\ 6\frac{\mu g}{l}\cdot 7{,}5\ l = 45\ \mu g.$$

The calculated data were confirmed by weighing results on the microbalance.

Analyzing the experimental data, we note:

1) at the optimum temperature of the reactor of about 500-550 °C the pyrolysis of TNT molecules occurs with the detachment of 1-2 nitro groups, judging by the expression $1 \leq K_{rel.} < 2$;

2) the optimal value of the mass flow of gas is selected in dependence on the geometric parameters and temperature of the reactor in such a way that the obtained regularity is correlated with the calculated one shown in Figure 1.

After determining the optimal operating parameters, the experimental stand was calibrated by TNT masses in the range from 500 μg to 1 ng (samples type (b)-(h)). The calibration results are shown in Figure 4. An example of the MDP sensor responses is in Figure 5.

7. Conclusion

We demonstrated efficiency of the method based on pyrolysis of TNT and following detection by a MIS-sensors of trace amounts of nitro-containing substances (including explosives). The method has a number of competitive advantages:

- a long service life of the sensor,
- an absence of consumables,
- the simplicity of operation and maintenance,
- stability of work and reproducibility of results,
- the possibility to create a miniature device (really achievable overall dimensions LWD of 80x50x50 mm and a mass of not more than 0.5 kg),
- an advantageous price-quality ratio,
- high sensitivity (detection threshold for TNT vapors is 10^{-12} g/cm^3 or 1 ng).

References

[1] Utochkin Yu A 1994 *Preprint MEPhI* **75** (in Russian)
[2] Bolodurin B A, et al. 2016 *Mendeleev Chemistry Journal* vol. **LX** 3 (in Russian)
[3] Nikolaev I N, Unichenko P O 2006 *Sensors and Systems* **10** (in Russian)
[4] Qxley C J, et al. 2009 *HMTD* **2** 1
[5] Andreev K K, Belyaev A F 1960 *Moscow: Oborongiz* 580 p. (in Russian)
[6] Nguyen V B 2014 *The Kazan national research Technological University* Ph.D thesis 165 (in Russian)
[7] Nikolaev I N, Onishenko P O 2006 *Sensors and systems* **3** 34 (in Russian)

IOP Conf. Series: Journal of Physics: Conf. Series **1238** (2019) 012013 doi:10.1088/1742-6596/1238/1/012013

Spectroscopic methods of investigation and the thermal stability of detonation nanodiamonds

V P Efremov and E I Zakatilova
Joint Institute for High Temperatures of the Russian Academy of Sciences, Izhorskaya 13 Bldg 2, Moscow 125412, Russia

E-mail: ei.zakatilova@mail.ru

Abstract. The behavior of detonation nano- and microdiamonds at increased temperatures is studied by synchronous thermal analysis, X-ray diffraction analysis, and scanning electron microscopy. The X-ray analysis of survived samples shows that a part of the nanodiamond is stable under heating to 1500°C. At the heating to 600°C, we see the decreasing of nanodiamond phase. Appearance of graphite phase isn't observed in the sample up to 1500°C. However, in the range Bragg angle 20-32° after heating to 1500°C, the x-ray amorphous graphite-like phase is observed. The analysis of microphotographs of samples before and after heating showed the influence of heating rate on the parameters of powder particles. A high thermal stability of detonation microdiamonds (above 1500°C) is established. Literature data on the radiation stability of diamond are analyzed. It is established that the character of the diamond damage depends on the type and dose of the irradiation.

1. Introduction

Carbon is the key material for the life on the Earth, it remains in the focus of various studies. Due to its special property to form various types of chemical bonds, carbon has many allotropic modifications: cubic and hexagonal diamond, graphite, carbine, and so on. Now, new nanocarbon materials are known: carbon nanotubes, fullerenes, graphenes, and detonation nanodiamonds. These substances possess unique properties. This allows one to use them in various fields.

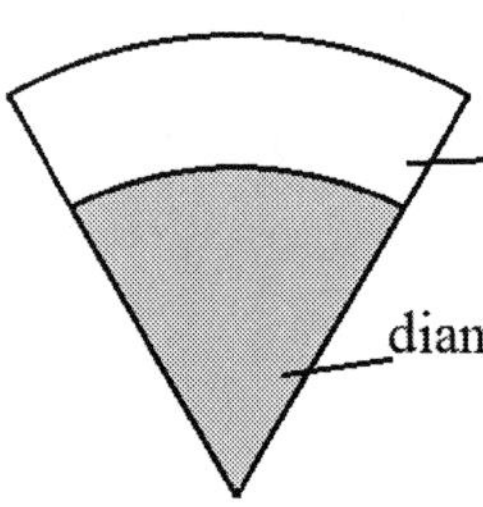

Figure 1. The model of detonation nanodiamond particle [1].

Detonation nanodiamonds (DND) are obtained by explosion of high explosives. A DND particle has heterogeneous structure. It consists of a diamond core and an impurity shell. The impurity shell consists of nondiamond carbon, adsorbed water, and C-O, C-N, C-H radical groups. The model of a DND particle is shown in figure 1.

With decreasing nanoparticle size, the surface fraction of atoms increases. The influence of the surface on the crystal lattice strengthens. The reason is that surface atoms have neighbors on one side only, and the center symmetry of the forces and masses is shifted. The interatomic distances change and the crystalline lattice is distorted [2]. Such structure modifies the properties of DND powder particles, as compared to macrocrystalline diamonds.

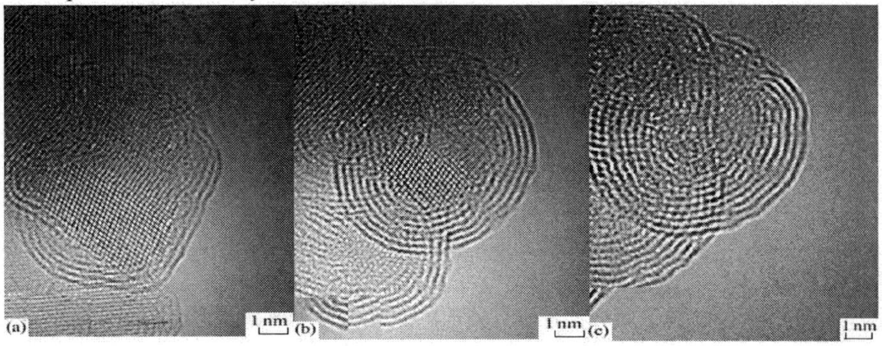

Figure 2. Layer-by-layer transformation of a diamond particle into an onion graphite [4].

The published data on the nanodiamond graphitization temperature vary in a wide temperature range from 670°C [3] to 980°C [4]. The graphitization mechanism is also unclear. Either graphitization occurs via the intermediate amorphous phase, or layer-by-layer transformation of a diamond particle into an onion graphite structure takes place. Upon heating, more and more diamond layers are transformed into an sp^2 structure. The photos of this transformation are shown in figure 2. Therefore, particles with a diamond core and a graphite shell can exist at certain temperatures. Graphitization, like a spherical wave, propagates into the particle. The degree of the graphitization depends on temperature. Thus, if the particle size and the temperature are known, it is possible to determine the amount of graphite and diamond phases in the particle.

The objective for this study is the experimental investigation of the thermal stability of detonation nanodiamond and the analysis of published data on radiation damage of diamond.

2. The thermal stability of detonation nano and microdiamond

In this study, the detonation nanodiamond powder is experimentally studied. The powder was obtained by explosion of the TNT/RDX (50/50) mixture [5]. Detonation microdiamond (DMD) is also studied. DMD was obtained by explosion of the TNT/RDX (50/50) mixture in water.

The study of elemental composition of the initial sample revealed the carbon purity of about 98%. The impurity content in the sample is small. Basically, the impurities include the following metals: Fe, Cr and Ti (from 0.3 to 0.2%). The particle sizes in the initial DND powder are from 4 to 7 nm. Detonation microdiamonds were studied in [6]. The typical microdiamond size is about 25 μm.

The DND and DMD heating was performed by synchronous thermal analysis (STA). STA includes thermogravimetry (TG) and differential scanning calorimetry (DSC). The heating was performed at Netzsch STA 409 PC. The samples were put in a corundum crucible (Al_2O_3) and the lid was closed. The heating was performed at ambient pressure in dynamic argon environment at temperature range from 30°C to 1500°C with constant heating rates of 2 and 10°C/min. The survived samples after the heating were studied by X-ray diffraction and scanning electron microscopy.

Detonation nanodiamonds were heated to temperatures of 600, 1000, and 1500°C for a heating rate of 10°C/min and to 1500°C for a heating rate of 2°C/min. The detonation microdiamonds were heated to 1500°C at heating rates of 2 and 10°C/min.

2.1. The thermal stability of detonation nanodiamond

The X-ray diffractograms of the studied samples are shown in figure 3. As seen, increase of treatment temperature leads to decrease of diamond sp^3 phase (figure 3, lines 1 – 4). The nanodiamond reduction begins at 600°C. A certain number of diamonds, however, can be found in the sample at a temperature of 1500°C. As the sp^3 peak from the base plane of the diamond (111) decreases, no graphite component appears. This may be connected with the fact that the nanodiamond structure is transformed to the amorphous phase upon heating. The amorphous phase is transformed into the graphite phase at further heating. The X ray structural analysis of the sample, heated to 1500°C, revealed a halo in the region of Bragg angles of 2θ=20-32°. This halo corresponds to the X-ray amorphous graphite-like phase. The interplanar spacing in this structure is larger than in graphite. This phase is not graphitized with further heating.

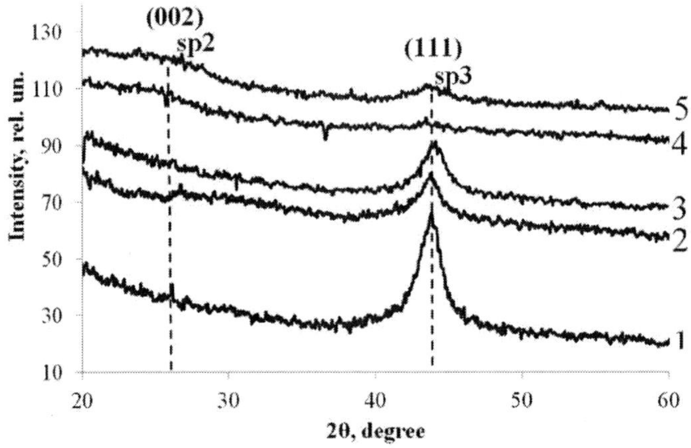

Figure 3. X-ray diffractograms of the samples of DND powder before and after the heat treatment: 1 – the initial powder, 2 – to 600 °C at a rate of 10 °C/min, 3 – to 1000 °C at a rate of 10 °C/min, 4 – to 1500 °C at a rate of 10 °C/min, 5 – to 1500 °C at a rate of 2°C/min. (002) – the base plane of graphite.

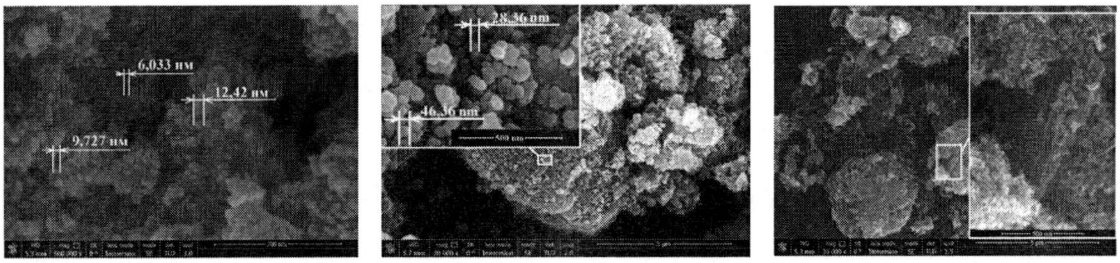

a) b) c)

Figure 4. Microstructure of powder nanodiamond: a) initial sample; b) after the heat treatment to 1500°C at v=10°C/min; c) to 1500°C at v=2°C/min.

The investigation of DND microstructure showed the influence of the heating rate on the powder particles parameters (figure 4). The heating at a rate of 10°C/min results in powder particles adhesion. As a result, spherical conglomerates with a size of 30 – 40 nm are observed in the sample (figure 4b). The powder heating at a rate of 2°C/min results in the formation of planar structures with a size of about 1 μm (figure 4c).

2.2. Thermal stability of detonation microdiamond

Heating of the detonation microdiamond powder was performed in a temperature range from 30 to 1500°C at rates of 2 and 10°C/min in dynamic argon environment. Under heating to a maximum temperature of 1500°C at rates of 2 and 10°C/min the mass loss of the sample was about 2%.

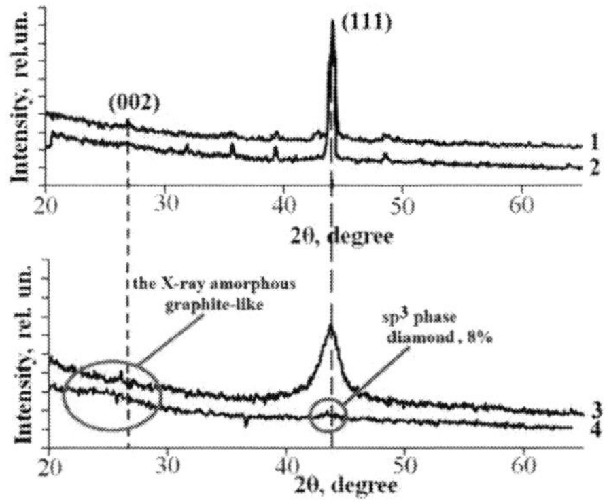

Figure 5. Microstructure of detonation micro and nanodiamond powder: 1 – initial microdiamond; 2 – microdiamond after treatment to 1500°C at a rate 10°C/min; 3 – initial nanodiamond; 4 – nanodiamond after treatment to 1500°C at a rate 10°C/min.

X-ray analysis of the DMD powder before and after the heating is shown in figure 5. As seen, the amount of the diamond phase has not changed even after the heating to a maximum temperature of 1500°C (figure 5, lines 1 and 2) [6]. No graphitization in the sample is observed upon heating. This indicates that the microdiamond is more stable to heating at atmospheric pressure than nanodiamond. The peak from the base plane of diamond (111) for nanodiamond is wider than that for microdiamond (figure 5, lines 1 and 3). The peak broadening results from the distortion of the crystalline lattice with decreasing particle size. The distortion of the crystalline lattice changes the substance properties. The X-ray peak from the base plane of diamond (111) for microdiamond is nearly the same as that for monodiamond. This indicates similarity of properties of the micro- macrocrystalline diamond.

At present, the graphitization mechanism of detonation nanodiamond is not completely clear. The graphitization temperature needs further study. For this purpose, it is necessary to study various impacts on the DND substance. The available published data were analyzed prior to begin the experimental study.

3. Radiation stability of diamond phase

Point defects, amorphous and graphite phases can be formed in the diamond, depending on irradiation type and dose. Irradiation changes the substance structure. Its properties are changed. Thus, it is possible to obtain a material with unique properties that do not exist in nature.

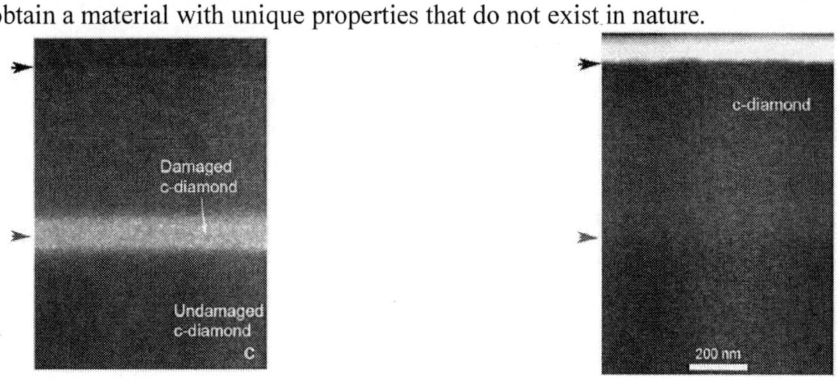

a) b)

Figure 6. The samples of irradiated single crystal diamond by ions Si^+ with 1MeV, at a dose $1*10^{15}$ $Si^+/см^2$ a) before heating, b) after heating to 1350°C during 24 hours. The black arrows show the surface; the gray arrows show damage bend; c-diamond – the bulk diamond. [7].

Thus, if single diamond is irradiated by 1 MeV Si ions at a dose of $1*10^{15}$ Si^+/cm^2 at the room temperature [7], point defect formation is observed (figure 6a). Annealing of the irradiated sample at 1350°C during 24 hours resulted in the recovery of defects to an ideal diamond lattice (figure 6b). If single crystal diamond is irradiated by 1 MeV Si ions at a dose of $7*10^{15}$ Si^+/cm^2, amorphous carbon is formed [7]. The annealing of the irradiated sample at 1350°C during 24 hours results in the formation of graphite from the amorphous phase.

The irradiation modeling of nanodiamond particles shows that the character of damage depends on the particles size [8]. Namely, 5 nm particles, being irradiated by heavy fast ions with an effective stopping energy of 17 keV/nm, are completely transformed into the amorphous phase. Particles with a size of 7 nm have an amorphous track at its center after irradiation. Therefore, nanodiamond with a size of 7 nm has two phases after irradiation: nanodiamond and amorphous diamond. For a particle with a size of 10 nm, point defects are only observed at its center. Thus, the particle size is the function of radiation damage.

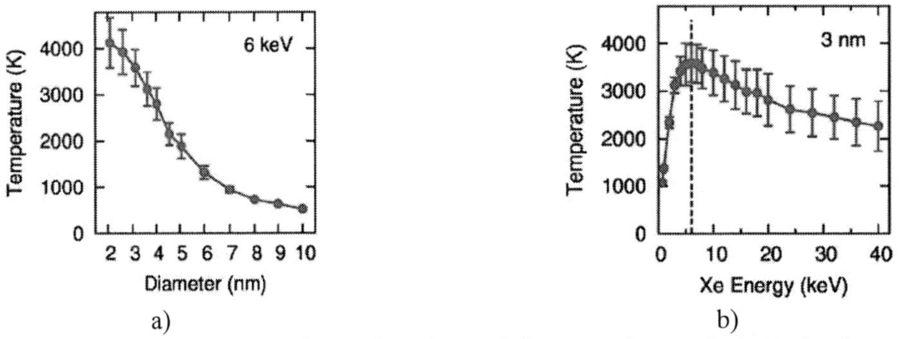

Figure 7. Peak temperature of ND grains as functions of diameter after a 6 keV Xe ion impact (a) and of incident Xe energy in a 3 nm ND (b) [9].

The irradiation of the nanodiamond powder (particle size from 2 to 40 nm) by Xe ions with a low energy from 6 to 40 keV showed [9] that 8 nm particles were completely destroyed by a small number of ions (with an energy of about 6 keV). It was found that the damage depends on the impact character. Central collisions completely destroy the particle. Non-central collisions result in lesser damage. Particles with a size larger than 8 nm are more stable to irradiation. The simulation of nanodiamond irradiation with the help of molecular dynamics showed (figure 7) that a 3 – 4 nm nanodiamond particle is heated to very high temperatures (over 3000K). The heating temperature decreases with increasing particle size. So, a particle with a diameter of about 10 nm, being irradiated by 6 keV Xe ions, is heated to less than 1000K.

4. Conclusions

It was found that the thermal stability of a part of the detonation nanodiamond is higher than 1500 °C. The analysis of the published data showed a wide spread of data on the nanodiamond powder graphitization temperature. According to our results, the basic process of graphitization of the nanodiamond powder occurs above 800 °C. However, the content of the diamond phase decreased after the sample heating. This is connected with the transformation of the diamond into the amorphous phase. The linear size of the powder particles increased after the heat treatment. The dependence of the limiting stability of nanodiamonds on the physical properties of particles requires further investigation. The graphitization temperature of the detonation microdiamond is above 1500°C. This value is higher than that for nanodiamonds. So, the properties of microdiamonds are close to those of macrocrystalline diamond. The solid-state transition sp^3 to sp^2 and the corresponding changes in the diffraction pattern is promising for experimental diagnostics. The combination of the diamond structure, the nanoscale size, and the porosity simplifies radiation release from the studied region. Such targets with the known thermophysical transition conditions can be used for WDM (Warm Dense Matter) diagnostics.

IOP Conf. Series: Journal of Physics: Conf. Series **1238** (2019) 012013 doi:10.1088/1742-6596/1238/1/012013

References

[1] Titov V M, Anisichkin V F, and Mal'kov I Yu 1989 *FGV.* **25** 117–25
[2] Petrunin V F 1991 *ZHurn. Vsesoyuz. him. obshchestva im. D.I. Mendeleeva* **36** 146–50
[3] Xu N S, Chen J, and Deng S Z 2002 *Diamond Relat. Mater.* **11** 249–56
[4] Popov V A, Egorov A V, Savilov S V, Lunin V V, et al. 2013 *J. Surf. Invest.: X-Ray, Synchrotron Neutron Tech.* 7 1034–43
[5] SHevchenko N V, Gorbachev V A, Ubej-Volk E Yu, and Danilenko V V 2004 *Konstr. Kompoz. Mater.* **3** 33 – 39
[6] Efremov V P, Zakatilova E I, Maklashova I V, and Shevchenko N V 2018 *J. Phys.: Conf. Ser.* **946** 012107
[7] Hickey D P, Jones K S, and Elliman R G 2009 *Diamond Relat. Mater* **18** 1353–59
[8] Valencia Felipe, Mella Jose´ D, Gonza´lez Rafael I, et al. 2015 *Carbon* 458–64
[9] Shiryaev A A, Hinks J A, Marks N A, et al. 2018 *Scientific reports* **8** 5099

IOP Conf. Series: Journal of Physics: Conf. Series **1238** (2019) 012014 doi:10.1088/1742-6596/1238/1/012014

Influence of the structure of the boundary layer of an adsorbed viscoelastic fluid on the dielectric properties of a metamaterial

Yu Agrafonov[1], I Petrushin[1], B Damdinov[2], Sh Tsydypov[2]

[1]Irkutsk State University, Karl Marx str., 1, 664003 Irkutsk, Russia
[2]Buryat State University, Smolina str., 24a, 670000, Ulan-Ude, Russia

E-mail: agrafonov@physdep.isu.ru

Abstract. We study the change in the shear elasticity of a viscous fluid near a solid surface of a metamaterial. The interest in studying the shear mechanical properties of these fluids is caused by their wide use in engineering. Surface forces in boundary layers or thin films of classic molecular systems should be considered when describing different phenomena in near-surface layers. We obtained the solution for two-particle distribution function at low densities. It is shown that the solution describes the transition from axial to spherical symmetry while each particle goes to infinity.

1. Introduction

Rheological measurements are a powerful method for studying the physical properties of liquids. Many methods for studying the mechanical properties of viscoelastic materials have been developed. Most of them are associated with acoustic measurements of various parameters of liquids.

To study very viscous (glassy) substances, melts of polymers, low-frequency vibrators are usually used, whereas for low-viscosity liquids and solutions, sound and ultrasonic vibrators are used. The frequency range of these methods extends from 10^{-4} to 10^{9} Hz, and the measured modules reach 10^{9} Pa. In recent years, much attention was given to viscoelastic liquids with a non-Newtonian behavior [1-3]. According to the existing concept of the nature of liquids, the shear elasticity of liquids should be observed under shear oscillations with the periods comparable with the time intervals of the particle jumps between the temporary equilibrium states. These intervals being identified with the relaxation time of the nonequilibrium state estimated as about 10^{-10}–10^{-12} s from the self-diffusion rate, and the calculations for low-viscosity. Hence, the shear elasticity of such liquids should be observed at the frequencies of shear oscillations of 10^{10} –10^{12} Hz.

However, as follows from experiments [4, 5] all liquids without exception demonstrate shear elasticity near frequency of 10^{5} Hz. Thus an unknown yet low-frequency viscoelastic relaxation process takes place in liquids. Presumably, it is caused by collective excitations of large groups of interacting molecules having a relaxation time which by many orders of magnitude exceeds the time of temporal stability of individual molecules. The characteristic feature of the shear mechanical properties of regular liquids is the small value of the mechanical loss tangent (< 1). According to the Maxwell rheological model, this means that a relaxation frequency of an investigated process is lower than the frequency used in the experiment.

2. Acoustical method

Our experiments were performed at the frequencies 40 and 74 kHz. The presence of the low-frequency shear elasticity was revealed by the acoustic resonance method [4, 5] described as follows. A piezoelectric quartz resonator made of 18.5° X-cut quartz crystal with zero Poisson's ratio oscillates at its fundamental resonance frequency. The horizontal surface of the crystal contacts with a liquid layer covered with a solid strap. Shear standing waves are excited in the layer experiencing shear strains. The parameters of the resonance curve of the system can vary in dependence on the film thickness. Equating the impedances of the piezoelectric quartz and the liquid, one can derive the complex shift of the resonance frequency of the oscillatory system [6]. These frequency shifts perform damped oscillations, and with the increase in the liquid layer thickness, they tend to some limiting values. When the liquid layer thickness is much less than the shear wavelength and the strap is practically at rest, we obtain the following expressions for the real shear modulus and the mechanical loss tangent:

$$G' = \frac{4\pi^2 M f_o \Delta f' H}{S} \tag{1}$$

$$\tan\theta = \frac{G''}{G'} = \frac{\Delta f''}{\Delta f'} \tag{2}$$

where $G^*=G'+iG''$ – complex shear modulus of liquid, $\Delta f^*=\Delta f'+\Delta f''$ – complex frequency shift, M – the piezoquartz mass, H – the liquid layer thickness, S – the cover-plate square, f_0 – the resonance frequency of the piezoquartz. The imaginary frequency shift is equal to the shift of the resonance half-width. When the liquid possesses a measurable shear modulus, according to (1) the real frequency shift must be inversely proportional to the liquid layer thickness. The latter was determined by the interference method. A positive shift of the resonance frequency indicates presence of the shear elasticity.

Several liquids presented in Table 1 were studied. The choice of these liquids is caused, first, by their wide use in engineering (studying the shear mechanical properties). Second, study of the homologous series of different liquids is of special interest, because a change in number of the chain links is accompanied by changes in the shape and size of molecules as well as in character of their interaction. All this affects the viscoelastic properties of liquids (i.e. structure) of the homologous series. For all studied liquids we obtained linear dependence of the resonance frequency shift on the inverse thickness of the liquid layer. According to (1), such a linearity indicates the presence of the bulk shear elasticity. Using (1) and (2), we calculated the values of G' and $\tan\theta$. The results obtained [6] at the frequencies 40 kHz and 74 kHz are presented in the Table 1.

Table 1. Viscoelastic properties of liquids at the frequencies 40 and 74 kHz

Liquid	40 kHz		74 kHz	
	G', 10^5 Pa	$\tan\theta$	G', 10^5 Pa	$\tan\theta$
Vaseline oil	1.13	0.60	1.36	0.50
Ethylene glycol	0.39	0.72	0.91	0.24
Diethylene glycol	0.51	0.44	1.22	0.31
Triethylene glycol	0.74	0.65	1.28	0.27
PES-1	1.26	0.18	9.63	0.11
PES-2	0.97	0.19	4.81	0.63
PES-3	0.85	0.21	3.45	0.87
Dibutyl phthalate	0.65	0.29	0.97	0.11
Oleic acid	0.76	0.23	1.63	0.91

As seen in Table, for all liquids the obtained values of the elastic modulus are smaller and the mechanical loss tangents are greater for 40 kHz than for 74 kHz. For example, for ethylene glycol, the real shear modulus is 0.39×10^5 Pa at the frequency 40 kHz and 0.91×10^5 Pa at 74 kHz. The mechanical loss tangent is 0.72 at 40 kHz and 0.24 at 74 kHz. In the simple Maxwell rheological model with a single relaxation time the mechanical loss tangent is equal to 1 at the relaxation frequency. Hence, the relaxation frequencies of all studied liquids are less than 40 kHz. Therefore, it is of interest to study the shear elasticity at lower frequencies. Thus, all liquids studied in the experiment possess measurable shear moduli at the given frequencies of shear oscillations. The presence of the low-frequency shear elasticity in regular and low-viscosity liquids means that the values of the shear stress relaxation time estimated on the basis of the time within which single liquid molecules remain in the temporal equilibrium states are wrong and confirms the hypothesis that the dominant role is played by the collective interactions of clusters.

We assume that any liquid represents a micro heterogeneous media which consisting of two dynamical components – arranged fields (clusters) disposed in nonarranged loose-packed matrix [7]. Clusters are rearranged by outer influence, which is expressed by diffusion change of fluctuation holes between those components. Getting off the atom from a cluster means formation of fluctuation hole, while its addition to cluster means hole shutting. Using data on temperature dependence of mechanical loss angle tangent of Vaseline oil at 40 and 74 kHz the activation energy of cluster disintegration process was evaluated. The obtained value, which is of the order of hydrogen bond energy (~5 kkal/mol), gives evidence that the low-frequency relaxation process in liquid is related to some low-energy process.

As a next step, we intend to cover the surface of the piezoquartz resonator with a hyperbolic metamaterial consisting of alternating dielectric and metal films and to measure the dependence of the dielectric constant of the nanocomposite on known fluid shear moduli [8]. In addition, the characteristics of the fluid are determined from the model of the liquid bordering with solid.

3. Structure and permittivity of metamaterial

In previously described experiments we studied boundary layers of liquids which thickness exceeds surface forces radius and, at the same time, is less than macroscopic scale (30–40 Å). In this case, the microstructure of boundary layer changes due to particle reordering near hard surface.

Structure properties define permittivity of liquid [9-11]. In particular, for bulk liquids static permittivity is evaluated via pair correlation function near hard surface. In the same way permittivity is evaluated for boundary layers.

Structure of hard surface defines potential field affecting boundary layers of liquids. For hyperbolic metamaterial which was studied in our experiments [8] it is necessary at first to calculate the potential field, produced by surface. Also we need to take into account real intermolecular potentials (like Lennard-Jones etc.)

However, we may quantitatively describe such system by model near the hard surface as already suggested in our papers [12-14]. In this work we obtained one- and two-particle distribution function for molecular system at low density. The results allow to obtain permittivity properties for molecular system near hyperbolic metamaterial.

4. Common equations

Statistics of the considered molecular system is based on Born-Green-Yvon-equations system (also known as BGY hierarchy) for l-particle functions and on $G_{l,\ldots,i}=(\mathbf{r}_l,\ldots,\mathbf{r}_i)$ for ensemble of identical particles. These particles interact with each other through potential $\Phi_{ij}(r_{ij})$, where $r_{ij}=|\mathbf{r_i}-\mathbf{r_j}|$ is the distance between centers of particles *i* and *j*. The diameter of each particle is σ.

The BGY-equations system may be transformed to equations for one- and two-particle distribution functions, which may be written as Ornstein-Zernike [11]:

$$\omega_1=n\int G_2C_{12}^{(1)}d(2)+\ln a\ ;\quad h_{12}=C_{12}^{(2)}+n\int C_{13}^{(2)}h_{23}d(3) \tag{3}$$

Here we integrating on coordinates of *i*-particle: $d(i)\equiv dr_i$, n – density, $G_1=\exp(-\Phi_i(kT)^{-1}+\omega_1)$ one-particle distribution function, which describes particle position; Φ_i - potential energy in external field; ω_1 - one-particle thermal potential; a – activity coefficient, which is defined by condition of passing to isotropic system. $h_{ij}=[\exp(-\Phi_{ij}(kT)^{-1}+\omega_{ij})-1]$ is the pair correlation function connected with two particle distribution function by expression: $G_{ij}=G_jG_i(1+h_{ij})$; ω_{ij} is two-particle thermal potential, which takes into account indirect interaction of two particles; $C_{\text{ij}}^{(\text{k})}$ – direct correlation functions:

$$C_{\text{ij}}^{(1)}=h_{ij}-\omega_{\text{ij}}-\frac{1}{2}h_{ij}\left(\omega_{\text{ij}}+M_{\text{ij}}^{(1)}\right);\qquad C_{\text{ij}}^{(2)}=h_{ij}-\omega_{\text{ij}}+M_{\text{ij}}^{(2)} \tag{4}$$

Functions $G_1(\mathbf{r}_1)$ and $G_{12}(\mathbf{r}_1,\mathbf{r}_2)$ are critical ones while they describe internal structure and let us obtain thermodynamic parameters of the system. Equations (3) and (4) are difficult to solve because $M_{\text{ij}}^{(1)}$ and $M_{\text{ij}}^{(2)}$ contains infinite series of distribution functions. To use the following equations in practice one should approximate these series by simple expressions (closures). Thus we can obtain approximated equations for high density systems. We use Martynov-Sarkisov approximation [15].

Space-heterogeneous systems (liquid near a hard surface) are described by one- and two-particle distribution functions: $G_1(\mathbf{r}_1)$ and $G_{12}(\mathbf{r}_1,\mathbf{r}_2)$. Boundary condition for these equations is a transition from a hard surface to a liquid.

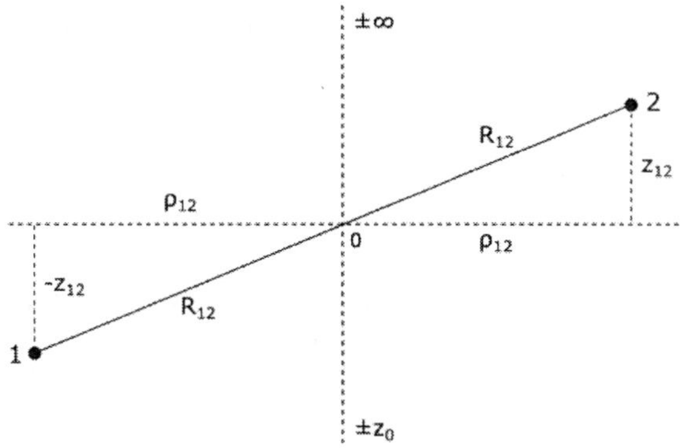

Figure 1. Axial symmetry of two-particle distribution function near hard surface

5. Molecular system of liquid near hard surface

Let's form the equations for molecular system near hard surface. When solving equation (3), the origin is located in the center of the particle contacting with hard surface. The *Z* axis is perpendicular to the surface, so the liquid is placed in the upper half-space ($z\geq 0$). The bottom half-space ($z<0$) is unavailable for the particles (Fig. 1). Such a system has axial symmetry,

$$G_1(\mathbf{r}_1)=G_1(z_1)=\exp(\omega_1(z_1))\quad G_{12}(\mathbf{r}_1,\mathbf{r}_2)=G_{12}(z_1,z_2,r_{12}) \tag{5}$$

where r_{12} is measured in particle diameter units, $z_i\geq 0$ – particle distance from the surface. Boundary conditions for $G_1(\mathbf{r}_1)$ and $G_{12}(\mathbf{r}_1,\mathbf{r}_2)$ are defined as follows

$$\omega_1(z_1) \to 0, \quad G_1(z_1) \to 1 \quad G_{12}^{(0)}(r_{12}) = \lim_{\substack{z_1 \to \infty \\ z_2 \to \infty \\ r_{12}=|\mathbf{r}_1-\mathbf{r}_2|=const}} G_{12}(z_1, z_2, r_{12}) \tag{6}$$
$$z_1 \to \infty \qquad z_1 \to \infty$$

6. Molecular system changes at low densities

Close order in a molecular system changes even at low densities. Let's take, for example, rarefied gas. Thermal potential is obtained through series expansion by density powers. At the first order we get:

$$\omega_1(z_1) = n\omega_1^{(1)}(z_1), \quad \omega_{12} = n\omega_{12}^{(1)}, \quad \mathrm{G}_{12} = G_{12}^{(0)} + nG_{12}^{(1)} \tag{7}$$

Substituting (9) into (7, 8) we obtain expression for factoring coefficients

$$\omega_1^{(1)} = \int f_{12} d(2) + \mu^{(1)} \tag{8}$$

$$\omega_{12}^{(1)} = \frac{1}{2}[\int_{-z_0}^{+z_0}\int_{-\infty}^{+\infty}\int_{-\infty}^{+\infty} \theta(1-r_{13}^2)\theta(1-r_{23}^2)dx_3dy_3dz_3 + \int_{-\infty}^{+\infty}\int_{-\infty}^{+\infty}\int_{-\infty}^{+\infty} \theta(1-r_{13}^2)\theta(1-r_{23}^2)dx_3dy_3dz_3] \tag{9}$$

Here f_{ij} – Mayer's function. Second integral in (9) is

$$\omega_{12}^{(10)}(R_{12}) = \frac{2\pi}{3}\left(R_{12}^3 - 3R_{12} + 2\right)\theta(1 - R_{12})$$

Integrating is carried out according to Fig. 1. Constant $\mu^{(1)}$ is derived from border condition at infinity. Interaction between particles and between particles and hard surface is set through hard-sphere potential. Evaluating the integral (9) we place the origin at the hard surface. The surface normal goes through the center of the particle. After integrating in cylindrical system over the upper half-space we get:

$$\omega_1^{(1)}(z_1) = \frac{\pi}{3}\left(z_1^3 - 3z_1 + 2\right)\theta(1 - z_1) = \frac{1}{2}\omega_{12}^{(10)}(z_1) \tag{10}$$

The origin is placed at center of mass when (9) is integrated. Through the centers of first and second particles we pass a plane normal to hard surface (Fig. 1).

$$f_{13}f_{23} = \theta\left(1 - r_{23}^2\right) = \theta\left(1 - \left((x_3 + \rho_{12})^2 + y_3^2 + \left(|z_{12}| + z_3\right)^2\right)\right) \tag{11}$$

where $\theta(\mathrm{x})$ – Heavyside function. The boundary condition at infinity corresponds to $r_{23}^2 = x_3{}^2 + y_3^2 + \left(R_{12} + z_3\right)^2$. Substituting this relation in (11), after integration we get final form for two-particle distribution function

$$\mathrm{G}_{12} = \theta(R_{12} - \frac{1}{2})\left\{1 + n\left[\omega_{12}^{(10)}(z_1) + \omega_{12}^{(10)}(z_2) + \frac{1}{2}\left(\omega_{12}^{(10)}(R_{12}) + \theta(z_0 + R_{12} - 1)\omega_{12}^{(10)}(R_{12})\right)\right]\right\} \tag{12}$$

The solution (12) describes transition from axial to spherical symmetry while each particle goes to infinity. For all other values of z_1, z_2 equation (12) describes close order change near surface. We should note that (12) is right only for low density molecular systems, but might be extrapolated for higher density field and get dielectric properties of liquid near surface of hyperbolic metamaterial.

Acknowledgements
We are grateful to prof. A. P. Menushenkov and Dr. A. V. Kuznetsov for their interest to our work. Following work is supported by Russian Foundation for Basic Research (grant № 18-02-00523a) and Ministry of Education and Science of Russia (grant № 16.8168.2017/8.9).

References
[1] Majeste J-C, Montfort J-P, Allal A and Marin G 1998 *Rheol. Acta* **37** 486
[2] Mankar R B, Saraf D N and Gupta S K 1998 *Rheol. Acta* **38** 84
[3] Phan-Thien N, Safari-Ardi M 1998 *J. Non-Newtonian Fluid Mech.* **74** 137
[4] Bazaron U B, Deryagin B V, Bulgadaev A V 1967 *Sov. Phys. JETP* **24** 645
[5] Derjaguin B V, Badmaev B B, Bazaron U B 1995 *Phys. Chem. Liq.* **29** 201
[6] Badmaev B B, Budaev O R, Dembelova T S, Damdinov B B 2006 *Ultrasonics* **44** 1491
[7] Badmaev B B, Damdinov B B and Sanditov D S 2004 *Acoust. Phys.* **50** 12
[8] Tumkur T, Zhu G, Black P, A. Barnakov Yu, Bonner C E and Noginov M A 2011 *Appl. Phys. Lett.* **99** 151115
[9] Bomont J M and Bretonnet J L 2003 *J. Chem. Phys.* **119** 2188
[10] Hafskjold B and Ikeshoji T 2002 *Phys. Rev. E* **66** 011203
[11] Tejero C F and Baus M 2003 *J. Chem. Phys.* **118** 892
[12] Agrafonov Yu V and Petrushin I S 2016 *J. Phys. Conf. Ser.* **747** 012024
[13] Agrafonov Yu V, Petrushin I S, Damdinov B and Tsydypov Sh 2015 *Proc. Mtgs. Acoust.* **24** 045002
[14] Agrafonov Yu V and Petrushin I S 2015 *Physics Procedia* **71** 364
[15] Martynov G A 2009 *J. Chem. Phys.* **129** 244509

Influence of antidots on the critical current density of HTSC in a magnetic field

A N Moroz, A N Maksimova, V A Kashurnikov and I A Rudnev

National Research Nuclear University MEPhI (Moscow Engineering Physics Institute), Kashirskoe sh. 31, 115409 Moscow, Russia

E-mail: moroz.anna@hotmail.com

Abstract. By using the Monte Carlo method, a numerical study has been conducted to calculate the current-voltage characteristics of a high-temperature superconductor with intrinsic pinning sites and the additional ones in the form of antidots in different magnetic fields. Varioustypes of antidot distribution among the sample such as a random distribution, triangular and square lattice and their conformal transformations, have been considered. Descending critical current density dependencies on the magnetic field for different samples have been acquired. Averaged vortex configurations of samples for different antidot distributions at the same magnetic field and transport current have been acquired. The most effective antidot distribution has been determined.

1. Introduction

Today high-temperature superconductors (HTSC) draw more and more attention to themselves. A vast amount of studies is devoted to finding the methods of increasing the critical current density J_c of HTSC. A great number of various methods is known today, and one of the newly discovered ones is the method of patterning superconductors with arrays of through or blind holes called the antidots [1–6]. Generally, those antidots are nanometer or submicrometer holes of different shapes distributed among the sample in a certain periodic way. Multiple research groups have shown that the presence of such antidots may significantly increase the transport characteristics of an HTSC [1–4]. The investigations of HTSCs with antidots in magnetic fields are especially relevant [5–6].

In this paper, we have conducted some numerical simulations of the influence of antidots and the type of their distribution among the sample on the critical current density of an HTSC in presence of a magnetic field.

2. Model for calculations

All calculations have been carried out within the limits of the model of a layered HTSC by using an algorithm based on the continual Monte Carlo method [7–10]. This model allows to carry out calculations of magnetic and transport characteristics of 2D and 3D layered HTSC and has already been repeatedly used, for example, in papers [9–10].

The principle of operation of the algorithm lies in the minimization of the Gibbs potential of a two-dimensional system with alternating number of vortices that can be described by (1). It consists of a sum of the vortex self-energies as well as the energy of their interactions with each other, the sample surface, external magnetic field, transport current and various pinning sites. For more detailed information see [7–8].

In our model, the antidots are cylindrical through holes about 300 nm in diameter. Each antidot is surrounded by a ring of columnar defects which appear in real samples when the antidots are fabricated. Thus, the interaction of vortices with the antidots consists of their interaction with the border and with the columnar defects.

$$G = \sum_{i<j} U(r_{ij}) + \sum_{i,j} U_{\mathrm{pn}}(r_{ij}) + \sum_{i,j} U_{\mathrm{surf}}(|r_i - r_j^{(\mathrm{image})}|) + \sum_i U_{\mathrm{m}}(x_i) + N\varepsilon \quad (1)$$

A BSCCO superconductor (λ = 180 nm, ξ = 2 nm, T_c = 84 K) of the size of 5 μm × 5 μm with the two-dimensional density of randomly distributed defects of approximately $4 \cdot 10^8$ cm^{-2} has been chosen as a sample for calculations. Various types of antidot distribution have been considered such as: a random distribution, triangular and square lattices and their conformal transformations (for the latter see [3–4]). In all types of distributions, the total amount of antidots is approximately the same and amounts to 50. All calculations have been carried out in different magnetic fields varying from 0 to 1000 Oe at helium temperatures. The critical current density J_c was determined by the standard electric field criterion of 1 μV/cm.

3. Results

Figure 1 shows series of current-voltage characteristics (CVC) of samples with different types of antidot distributions ((a) triangular lattice, (b) square lattice, (c) random distribution) and a sample with no antidots (d) in different magnetic fields. It is obvious from the picture that the critical current density J_c decreases with the increasing magnetic field. Also, when comparing the pictures for different distributions with one another one may notice that the extent of this decrease varies in different cases.

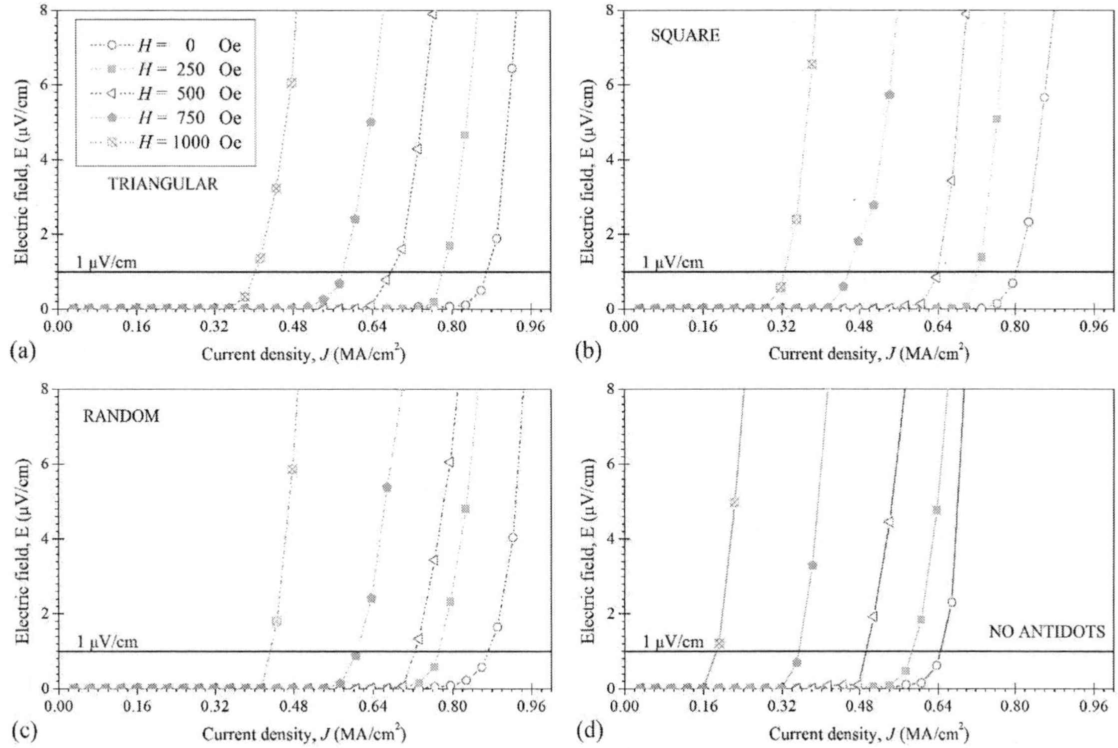

Figure 1. Series of current-voltage characteristics of samples with different types of antidot distribution: (a) triangular lattice; (b) square lattice; (c) random distribution; (d) no antidots, in different external magnetic fields.

This is clearly demonstrated in figure 2 where the acquired critical current density dependency on the external magnetic field for different types of antidot distribution including the sample with no antidots. It can be seen from the picture that the critical current density increases when antidots are added to the sample: it amounts up to approximately 15–20% for the conformal transformation of a triangular lattice of antidots. One may also notice that, though for weak magnetic fields different types of antidot distribution provide approximately the same result (in the left of figure 2), in strong magnetic fields the difference becomes quite significant. Thus, the conformal transformations of triangular and square lattices gave for the best critical current density resistance to the magnetic field (at $H = 1000$ Oe J_c dropped by approximately 25%) whereas the square lattice of antidots gave for the worst result (at the same conditions, J_c dropped by approximately 50%). Such a result is in qualitative correspondence with experimental data.

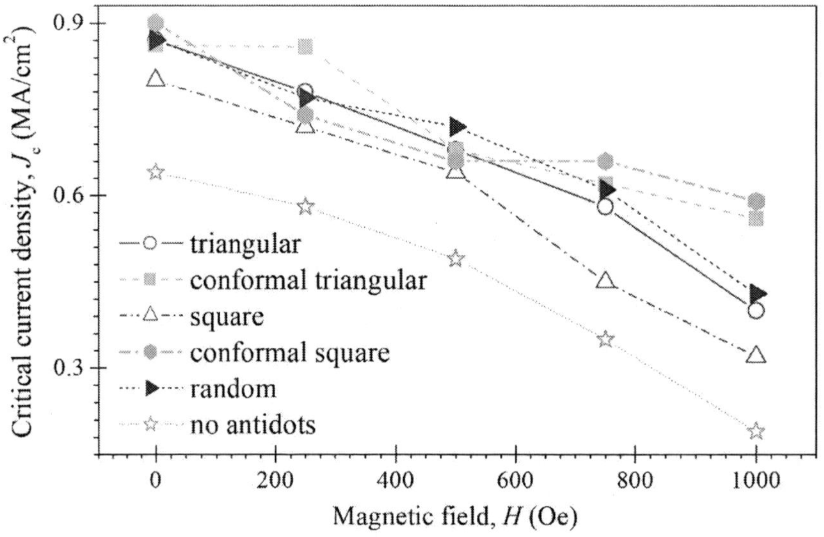

Figure 2. The calculated critical current density dependencies on the external magnetic field for different types of antidot distributions among the sample.

To demonstrate the influence of antidots and the type of their distribution on the vortex dynamics, averaged vortex configurations have been acquired for a magnetic field of 250 Oe and the transport current density of 0.74 MA/cm^2. They are demonstrated in figure 3. As was the case for figure 1, (a) is for the triangular antidot lattice, (b) is for the square lattice, (c) is for random distribution and (d) is for no antidots at all. It should be noted that in the cases of (a) and (b) the transport current density is lower than the critical value, in the case of (c) it is approximately equal to it and in the case of (d) it is much higher than J_c.

It can be seen from the picture that for the same current density the vortices enter the sample and move through it differently. In the sample with a triangular antidot lattice (a), the vortices got pinned by the antidots and defects near the sample border, and because of the repulsive interaction of vortices with one another, they prevent other vortices from entering the center of the sample. As a result, some of the antidots in the sample center were left unoccupied by vortices.

In the case of a square lattice (b) and a random distribution (c) of antidots, a similar situation is observed, however the distance between neighboring antidots is bigger than in the case of a triangular lattice (a), and the vortices are able to seep into the sample center between the occupied antidots. This is clearly demonstrated by thick lines (vortex tracks) in figures 3 (b) and (c). It should also be noted that in the cases of (b) and (c) almost all antidots are occupied by vortices.

In the case of the sample with no antidots (d) all the vortex pinning occurs on defects, and at the same current density the vortices move much more actively than in the case of a square antidot lattice (b). Thus, the antidots act as an additional effective pinning mechanism and increase the critical current density.

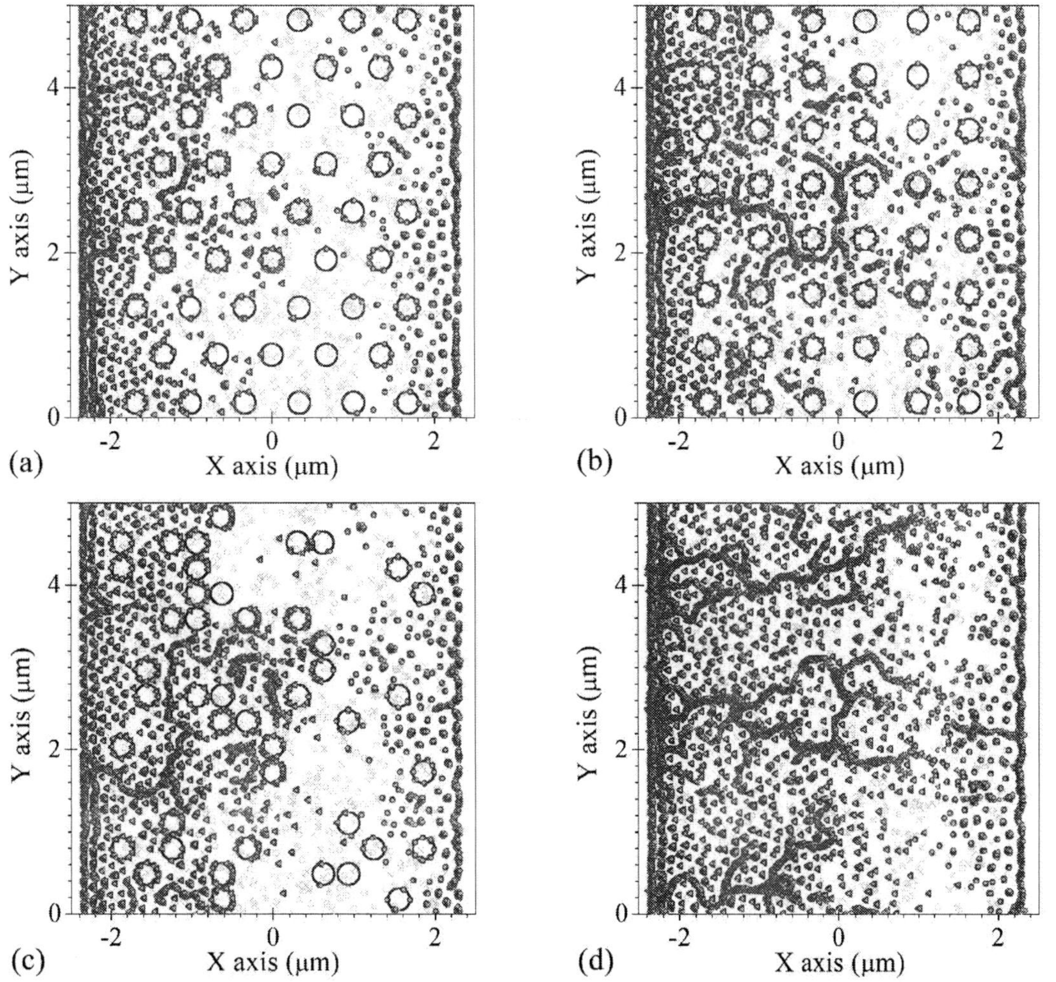

Figure 3. The averaged vortex configurations of samples with different antidot distributions in a magnetic field $H = 250$ Oe with the transport current density $J = 0.74$ MA/cm^2. (a) is for the triangular antidot lattice, (b) is for the square antidot lattice, (c) is for the random antidot distribution and (d) is for a sample with no antidots.

4. Conclusion

The vortex system of a layered high-temperature superconductor with intrinsic and artificial pinning centers in the form of antidots has been numerically studied in external magnetic fields and in the presence of a transport current. Different types of antidot distribution among the sample have been considered in order to determine the most optimal choice. The highest critical current density as well as its resistance to the external magnetic field was provided by the conformal transformations of a triangular and square antidot lattices.

It has been demonstrated that the vortex system behaves differently in samples with different antidot distributions in the presence of a magnetic field. The triangular lattice turned out to be more effective in terms of vortex pinning than the square lattice or a random distribution. When comparing the results with the ones for a sample with no antidots, it has been shown that the antidots contribute to the pinning potential of the superconductor thus increasing the critical current density.

Acknowledgements
The research was done under financial support of RFBR, grant no. 18-02-00278-a.

References

[1] Welling M, Wijngaarden R, Aegerter C, Wordenweber R and Lahl P 2004 *Physica* C **404** 410
[2] Laurent P, Mathieu J, Mattivi B, Fagnard J, Meslin S, Noudem J, Ausloos M, Cloots R and Vanderbemden P 2005 *Supercond. Sci. Technol.* **18** 1047
[3] Ray D, Olson Reichhardt C, Janko B and Reichhardt C 2013 *Phys. Rev. Lett.* **110** 267001
[4] Wang Y, Latimer M, Xiao Z, Divan R, Ocola L, Crabtree G and Kwok W 2013 *Phys. Rev.* B **87** 220501(R)
[5] Noudem J, Meslin S, Horvath D, Harnois C, Chateigner D, Eve S, Gomina M, Chaud X and Murakami M 2007 *Physica* C **463-465** 301
[6] Palau A, Monton C, Rouco V, Obradors X and Puig T 2012 *Phys. Rev.* B **85** 012502
[7] Kashurnikov V, Rudnev I and Zyubin M 2002 *JETP* **94** 377
[8] Rudnev I, Odintsov D and Kashurnikov V 2008 *Phys. Lett.* A **372** 3934
[9] Kashurnikov V, Maksimova A and Rudnev I 2013 *Phys. Sol. State* **56** 894
[10] Kashurnikov V, Maksimova A, Rudnev I and Odintsov D 2016 *Physica C* **528** 17

Critical current density in anisotropic superconductors containing columnar defects

V.A. Kashurnikov, A.N. Maksimova, I.A. Rudnev, D.S. Odintsov

National Research Nuclear University MEPhI (Moscow Engineering Physics Institute), Kashirskoe sh. 31, 115409 Moscow, Russia

E-mail: nastymaksimova@yandex.ru

Abstract. In order to reveal the influence of the tilt angle of columnar defects on angular dependence of critical current density the Monte Carlo simulation of vortex system in three-dimensional layered high-temperature superconductor (HTSC) was performed. For simulations, the three-dimensional model of layered HTSC generalized for the case of tilted magnetic field was used. The angular dependencies of the critical current were calculated at different splay angle of columnar defects and different anisotropy of HTSC. It was shown that simulation results correspond qualitatively to the experimental results for REBCO coated conductors.

1. Introduction

Nowadays, the improvement of critical current density in high-temperature superconductors is still required. It is also important to reduce the anisotropy of critical current since in most practical applications the superconductor is imposed in the magnetic fields of various orientations. Columnar pinning centers (columnar defects) have been shown to be considerably effective to prevent the motion of vortex lines and to increase the critical current density. Columnar defects can be either perpendicular to superconducting planes or tilted for an arbitrary angle. The angular dependencies of critical current on external magnetic field have been widely studied experimentally [1-5]. The maxima of experimental dependencies are normally associated with pinning of vortex lines on tilted columnar defects or superconducting planes [1,4], but despite the large number of experimental works the maxima do not have an explicit explanation.

The vortex lattice in layered HTSC in tilted magnetic field is widely studied theoretically [6]. In case of high anisotropy of HTSC (γ) it was shown that vortex lattice is composed of two interacting lattices of Abrikosov and Josephson vortices [6]. Abrikosov vortex lattice is produced by component of magnetic field perpendicular to superconducting layers and Josephson vortex lattice – by parallel component. The crossing lattices have been also observed experimentally [7].

Real HTSC are anisotropic materials with arbitrary configurations of pinning centers. To predict magnetic and transport properties of real HTSC-tapes it is necessary to take into account interaction of vortex lattice in tilted magnetic field with pinning centers at different values of anisotropy parameter. The aim of our work is numerical simulation of vortex lattice in tilted magnetic field and analysis of angular dependencies of critical current at different tilt angles of columnar defects.

2. Model of layered superconductor in tilted magnetic field

The simulations were performed in the framework of 3D-model of layered HTSC, by using Monte Carlo method. Gibbs energy of three-dimensional vortex system in this model can be given in the following form [8,9]:

$$G=\sum_z\left\{N_z\varepsilon+\sum_{i<j}U_{in-plane}(r_{ij})+\sum_{i,j}U_p(r_{ij})+\sum_{i,j}U_{surf}\left(r_{ij}^{(im)}\right)+U_m+\sum_i U_{\text{inter-plane}}\left(r_i^{z,z+1}\right)\right\} \quad (1)$$

where $\varepsilon=s\varepsilon_0(\ln[\lambda(T)/\xi(T)]+0.52)$ is a self-energy of vortex line, $\lambda(0)$, $\xi(0)$ - London penetration depth and coherence length at $T = 0$, respectively, N_z – is a number of vortices in z-plane, the second term describes a pairwise interaction between pancake vortices, third term is an interaction between vortex and pinning center, next two terms correspond to interaction of vortex with sample surface and meissner and transport current respectively, the last term in (1) – the inter-plane interaction between pancakes. $\varepsilon_0=\Phi_0^2/(4\pi\lambda)^2$, $\Phi_0=\pi\hbar c/e$ is a magnetic flux quant. The summation is carried out over superconducting layers. For a detailed description of the inter-plane interaction of vortices, see [9,10].

The last term describes the inter-plane interaction. For electromagnetic and Josephson interaction between pancakes in adjacent layers we used equations obtained in [11,12]:

$$U_{\text{inter-plane}}\left(r_i^{z,z+1}\right)=U_{em}\left(r_i^{z,z+1}\right)+U_{jos}\left(r_i^{z,z+1}\right) \quad (2)$$

$$U_{em}\left(r_i^{z,z+1}\right)=2s\varepsilon_0\left[C+\ln\left(r_i^{z,z+1}/2\lambda\right)+K_0\left(r_i^{z,z+1}/\lambda\right)\right] \quad (3)$$

$$U_{jos}^{z,z+1}\left(r_i^{z,z+1}\right)=\begin{cases}\varepsilon_0 d[1+\ln(\lambda/s)]0.25\left(r_i^{z,z+1}/r_g\right)^2\ln\left(9r_g/r_i^{z,z+1}\right), r_i^{z,z+1}\le 2r_g\\ \varepsilon_0 d[1+\ln(\lambda/s)]\left[\left(r_i^{z,z+1}/r_g\right)-0.5\right], r_i^{z,z+1}>2r_g,\end{cases}, \quad (4)$$

C=0.5772 is the Euler constant, $r_g=\gamma s$ is the Josephson length.

According to theoretical works [6,13], the magnetic field parallel to superconducting layers penetrate the superconductor in the form of Josephson vortices. Josephson vortices are located in the inter-layer spacing. In the absence of Abrikosov vortices Josephson vortices form a triangular lattice. We use the results [6,13] to obtain the energy of Josephson vortex system for Monte Carlo simulations.

The self-energy of Josephson vortex:

$$\varepsilon_1=\frac{\Phi_0}{4\pi}H_{c1}=\frac{\Phi_0^2}{16\pi^2\lambda_b(T)\lambda_c(T)}\left[\ln\frac{\lambda_b(T)}{s}+1.12\right],\ \lambda_b^2=\frac{s}{d_s}\lambda_s^2,\ \lambda_c^2=(\gamma s)^2+\frac{d_s}{s}\lambda_s^2,$$

The interaction energy of Josephson vortices (per unit length of Josephson vortex):

$$U_{1,2}^{jos}=\frac{\Phi_0^2}{8\pi^2\lambda_b\lambda_c}K_0\left(\sqrt{\frac{(y_2-y_1)^2}{\lambda_c^2}+\frac{s^2(n_2-n_1)^2}{\lambda_b^2}}\right),$$

For the interaction between Abrikosov and Josephson vortex

$$U_{1,2}^{v}=-\frac{\Phi_0^2}{8\pi^2\lambda_b\lambda_c}\frac{\frac{z}{\lambda_b}}{\sqrt{\left(\frac{y}{\lambda_c}\right)^2+\left(\frac{z}{\lambda_b}\right)^2}}\frac{x_0\pm d/2}{\lambda_b}K_1\left(\sqrt{\left(\frac{y}{\lambda_c}\right)^2+\left(\frac{z}{\lambda_b}\right)^2}\right).$$

s = 2.7 nm – the sum of inter-layer spacing and superconducting layer thickness d_s = 0.27 nm, γ is anisotropy parameter, d is the size of the superconductor in the direction perpendicular to the transport current.

The simulations were performed for the typical parameters of bismuth HTSC, for the sample size 5 μm × 3 μm. The algorithm takes into account the processes of production, annihilation and movement of Josephson vortices. The absolute value of magnetic field is $H = 320$ Gs.

3. Results: angular dependencies of critical current

Using the model described above we obtained angular dependencies of critical current for two different values of anisotropy parameter. We made the calculations for several different values of tilt angle of the columnar defects. The results are represented in fig. 1.

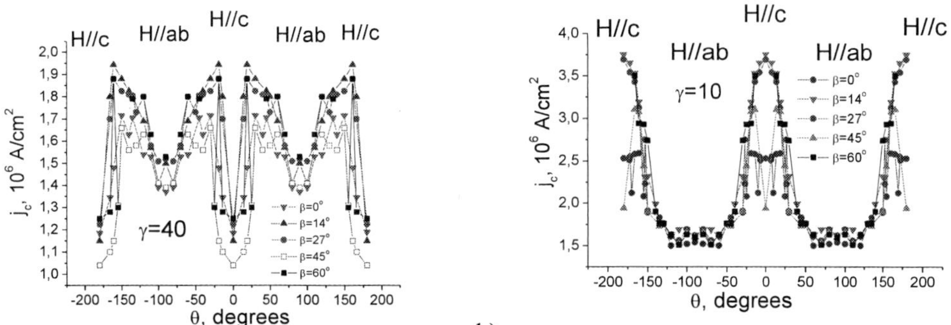

a) b)

Figure 1. The angular dependencies: a) γ=40, b) γ=10. β is the tilt angle of columnar defects.

The angular dependencies for γ=10 have pronounced minima at θ=90° and maxima at θ close to 0 and 180°. This result is in qualitative agreement with experimental results for REBCO superconductors in high magnetic field [1-5]. The $j_c(\theta)$ curves for γ=40 have a maximum at certain angle (θ_{max}) which value depends on tilt angle of columnar defects. In fig. 2 we plot the θ_{max} and $j_c(\theta_{max})$ as a function of β. These results show that θ_{max} is independent of tilt angle of the defects and maximal value of critical current decreases at high β when γ=40 and for γ=10 the critical current is about two times higher than for γ=40.

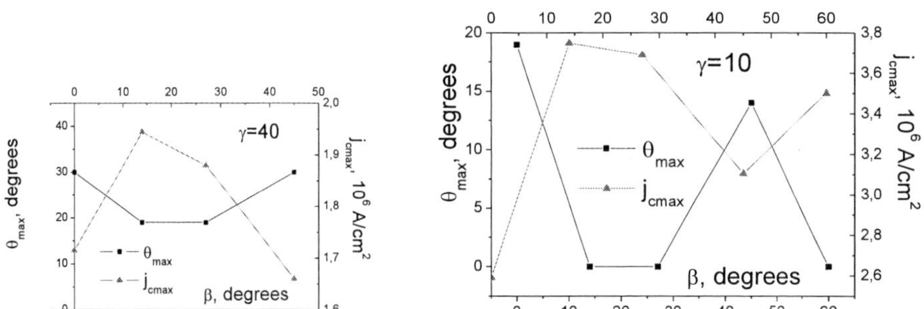

a) b)

Figure 2. θ_{max} and $j_c(\theta_{max})$ as a function of β.

The maximum of critical current at θ~30° has been observed experimentally for YBCO layer in [3]. For qualitative explanation of this result, we calculate a critical current in our model for γ=40 in magnetic field perpendicular to superconducting layers. The magnitude of the field equals to perpendicular component of tilted field. In fig. 3 we plot the critical current in perpendicular field on the angular dependence of the critical current at two different values of β.

The field dependencies for perpendicular magnetic field (fig. 3) have minima at $H = 320$ Gs (which corresponds to θ=0) and maxima at $H = 0$ (θ=90°). In the absence of parallel component of magnetic field Josephson vortices do not form. Hence, the maxima of $j_c(\theta)$ curves can be explained by interaction between Abrikosov and Josephson vortices and are not directly associated with the tilt angle of columnar defects.

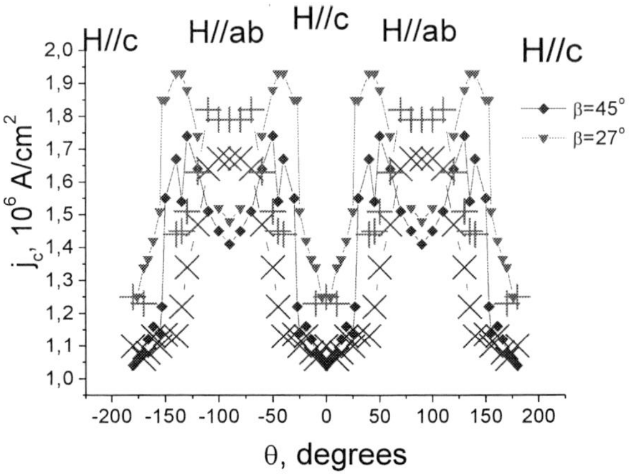

Figure 3. Values of critical current in perpendicular field are marked by crosses at the corresponding values of θ.

4. Conclusion

The 3D-model of layered HTSC was generalized for the case of layered superconductor in tilted magnetic field. The angular dependencies of critical current have been calculated at different tilt angle of columnar defects and at two values of γ. It was shown that $j_c(\theta)$ curves for γ=10 correspond qualitatively to experimental results for REBCO in high magnetic fields. The dependencies for γ=40 have additional maxima at the direction of magnetic field θ~30°. It was shown that the maxima are due to the interaction between Abrikosov and Josephson vortices and are not directly associated with tilt angle of columnar defects.

Acknowledgments

The research was done under financial support of RFBR, Grant № 18-02-00278

References

[1] Furuki Y, Sueyoshi T, Kai T, Iwanaga Y, Fujiyoshi T, Ishikawa N 2015 *Physica C* **518** 58–62
[2] Matsui Y, et. al. 2016 *Supercond. Sci. Technol.* **29** 065002
[3] Chudy M, Eisterer M, Weber H W 2010 *Physica C* **470** 1300–3
[4] Sueyoshi T, et. al. 2017 *IEEE Trans. Appl. Supercond.* **27(4)** 8001305
[5] Strickland N M, et. al. 2017 *Nucl. Instr. Meth. B* **409** 351-5
[6] Koshelev A E 2005 *Phys. Rev. B* **71** 174507
[7] Cole D, et. al. 2006 *Physica C* **437–438** 52–6
[8] Lawrence W E, Doniach S in *Proceedings of LT 12, Kyoto*, edited by E.Kanda (Keigaku, Tokyo, 1971), 361
[9] Rudnev I A, Odintsov D S, Kashurnikov V A 2008 *Phys. Lett. A* **372** 3934-6
[10] Kashurnikov V A, Maksimova A N, Rudnev I A, Odintsov D S 2016 *Physics of the Solid State* **58(8)** 1505-12
[11] Tyagi S, Goldschmidt Y Y 2004 *Phys. Rev. B* **70** 024501(14)
[12] Goldschmidt Y Y, Tyagi S 2005 *Phys. Rev. B* **71** 014503(9)
[13] Clem J R, Coffey M W 1990 *Phys. Rev. B* **42(10)** 6209

Electronic, phononic and superconducting properties of hydrides of yttrium and sulfur at high pressure

K S Grishakov, N N Degtyarenko, E A Mazur

National Research Nuclear University MEPhI (Moscow Engineering Physics Institute), Kashirskoe sh. 31, 115409 Moscow, Russia

E-mail: Eugen_mazur@mail.ru

Abstract. The electronic, phononic, and superconducting properties of double hydrides YH_6 and H_3S are analyzed and compared with that for triple hydride YS_4H_4 . A common physical and numerical approach was used to determine both the stability range and the critical superconducting temperature. The characteristics of the high-pressure phase of the triple hydride YS_4H_4 are obtained.

1. Introduction

At present, there is an urgent task to search for new high-temperature superconducting (HTSC) materials with high hydrogen content, in which the transition to the superconducting state occurs under close to normal conditions (ambient pressure and room temperature). In the review paper [1] the properties of 33 hydrides of the main groups of elements at high pressure are discussed and optimum values of pressure corresponding to maximum critical temperature T_c were found. For binary hydrides, superconductivity with T_c of about 100 K was predicted on the base of numerical simulation. Values of T_c above 200 K were predicted both in the covalent hydride H_3S and in the several polyhydrides having the YH_6 type structure.

Along with theoretical estimates of the superconducting properties of various hydrides, experimental results have also been obtained which have confirmed high values of superconducting temperature $T_c \sim 200$ K in sulfur hydrides under high pressure [2]. Thus, high temperature of the superconducting transition T_c is achieved in hydrides under high pressure. One of the main tasks is to search for compounds stable at lower pressures.

Using the USPEX program [3], we have found several structures for triple hydrides based on yttrium and sulfur. Yttrium sulphide (YS, Y_2S_3) exists under normal conditions. For the double hydrides at high pressures YH_6 and SH_3 the calculated values of T_c are sufficiently large. In this paper we present the results obtained for the high-pressure phases of YS_4H_4.

2.Method of calculation

The calculation of the properties of hydrides $Y_nX_mH_k$ was carried out within the framework of the density functional theory (DFT) in the plane waves basis, using the norm-preserving pseudopotential. The following sequence of calculations was used for each pressure value: 1) structural parameters, enthalpy of the equilibrium structure; 2) electronic Bloch spectra of the equilibrium structure; 3) phononic spectra for the equilibrium structure; 4) transition to the Wannier-functions basis; 5) calculation of electronic spectra, obtaining Eliashberg spectral function and the electron-phonon interaction con-

IOP Conf. Series: Journal of Physics: Conf. Series **1238** (2019) 012017 doi:10.1088/1742-6596/1238/1/012017

stant; 6) estimate of the superconducting transition temperature Tc, starting from the series of approximations of the solutions of the Eliashberg equations; 7) calculation of the normal state Wannier-functions of the system, taking into account the effect of the electron-phonon interaction on the electronic spectrum; 8) calculation of the superconducting properties taking into account an influence of the electron-phonon interaction on both T_c and the superconducting gap width.
QUANTUM ESPRESSO software package [4], proprietary software packages based on the results of work [5] and visualization algorithms were used calculations.

3. Results and discussion

A characteristic feature of the structure of the double hydrides YH_6 and H_3S is the high symmetry and the presence of a sublattice of homogeneous hydrogen in the H-H distances in the first coordination spheres. These structures are well studied and both calculated [6-8] and experimental data [2] were obtained for them. Variation of the basic elements of these hydrides leads to different magnitudes of the "chemical" pressure as well as to different ranges of their pressure stability. We recalculated these double hydrides and obtained results, in general, coincide with the results of previous studies [6-8].

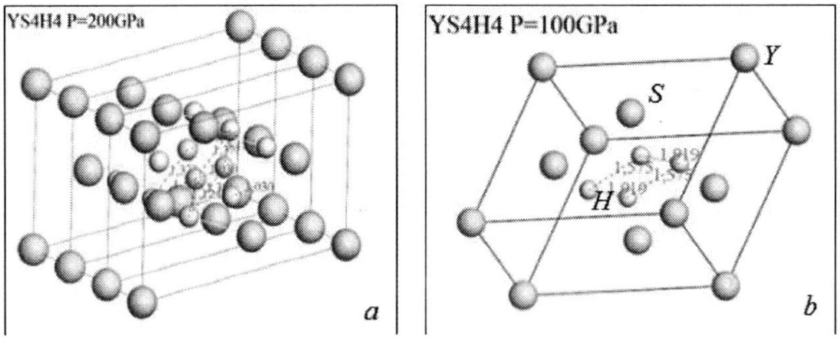

Figure 1 a. Cells YS_4H_4 with symmetry C2/M; b. primitive cell of the same structure.

The optimal structure of the triple hydride YS_4H_4, established by us, is shown in Fig 1. The primitive cell of this structure contains many-electron atoms Y and S. Their interaction keeps required structure symmetry at a lower pressure than the pressures needed to stabilize compounds YH_6 and H_3S. The hydrogen sublattice has the form of an "accordion" with the predominant quasi-chain spatial arrangement of atoms. The H-H interatomic distances have three different values depending on the pressure. Atoms of heavy elements form chains.

Oscillations in the interval 100-200 meV are associated with vibrations of the hydrogen sublattice of these hydrides. Almost the entire frequency range of oscillations is filled only for YH_6 and H_3S, while for YS_4H_4 empty frequency intervals are clearly pronounced. Continuity of the spectrum indicates connection of the sublattices in YH_6 and H_3S. Gaps in the YS_4H_4 spectrum are caused by different environments of different hydrogen atoms. We note that the density of the phonon modes number in the low-frequency region (100 meV for H atoms) is much larger than that for YH_6.

The phonon densities of the number of states for the YH_6, H_3S andYS_4H_4 compounds are shown in Fig.2. For YH_6 and H_3S the imaginary frequencies are absent in the spectrum for pressures larger than $P \approx 150$ and 175 GPa, respectively. The maximum phonon oscillation frequencies are about of 200 meV for YH_6 and H_3S, while for YS_4H_4 they are equal to 280 meV.

In these compounds, the electronic states density at the Fermi level is created by electrons of different types (s, p, d). The table shows the values of partial density of states PDOS at the Fermi level for a pressure $P = 200$ GPa for all three types of hydrides.

IOP Conf. Series: Journal of Physics: Conf. Series **1238** (2019) 012017 doi:10.1088/1742-6596/1238/1/012017

Table 1 Value of PDOS (E_F) P = 200 GPa

substance	s	p	d	Σ
H_3S	0.587	0.47	-	1.07
YH_6	0.294	0,039	0.424	0.762
YS_4H_4	0.133	0.108	0.566	0.807

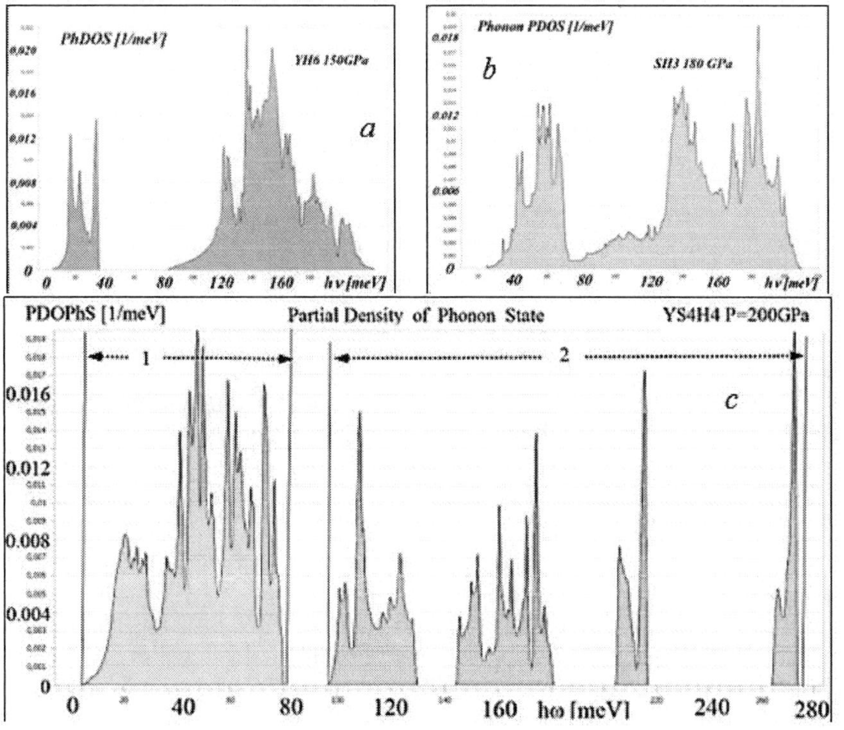

Figure 2. Partial densities of phononic oscillations of PhDOS for YH_6, SH_3 at characteristic pressures P = 150, 180 and 200 GPa, respectively. In Figure 2c for YS_4H_4, the vibration frequency band of hydrogen atoms is denoted as range 2, the energy range of the vibration modes of the heavy atoms Y and S is denoted by 1.

Variation of stoichiometry in triple hydrides allows to change the ratio between the partial contribution of different elements to the PDOS at the Fermi level. Such a variation leads also to changing the amplitude (s, p, d) of PDOS at the Fermi level and, therefore, the matrix elements of electron-phonon interaction.

The Eliashberg spectral functions preserve the main features of the phonon partial densities of the number of states PDOS. Calculations of the values of both the electron-phonon interaction constant λ and the critical temperature of the superconducting transition gave the following results: (SH_3 - P=180GPa - $\lambda \approx 2.3$ - $Tc \approx 200$ K); (YH_6 -P=125GPa - $\lambda \approx 3$ - $Tc \approx 165$ K); (YS_4H_4- P=200GPa - $\lambda \approx 1$ - $Tc \approx 20$ K).

Figure 3 shows the dependence of energy per atom on pressure: $\varepsilon=(E_{cell} - (mE_Y+nE_S+kE_H))/N_{cell}$, here E_{cell} and N_{cell} - total energy and number of atoms in the calculated cells of the systems; E_Y, E_S, E_H - calculated energies of solitary atoms.

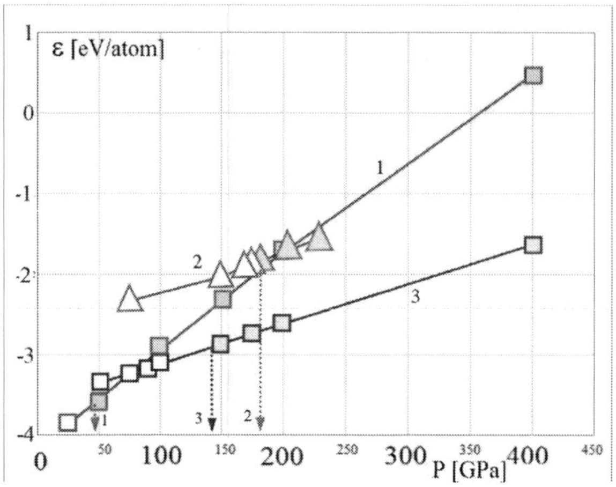

Figure 3. Dependences of the reduced specific energy per atom for the systems YH_6 (3), SH_3 (2), YS_4H_4 (1). The closed symbols show the minimum pressures at which the system is stable. Arrows indicate critical pressures

4. Conclusions

The obtained results indicate the complex character of the interrelationships in the structure, electronic and phononic spectra, types of internal bonds, and the charge state of the atoms of the compounds studied, especially taking into account changes in these properties with a change in pressure. The triple hydrides demonstrate a possibility to stabilize the sturcture at low pressures due to the "chemical" pressure of the parent matrix atoms.

Acknowledgments

The authors thank the program of competitiveness of the NRNU MEPhI for their support. This work has been carried out using computing resources of the federal collective usage center Complex for Simulation and Data Processing for Mega-science Facilities at NRC "Kurchatov Institute", http://ckp.nrcki.ru/.

References

[1] Duan D, Liu Y, Ma Y, Shao Z, Liu B and Cui T 2017 *National Science Review* **4** 121–35
[2] Drozdov A P, Eremets M I, Troyan I A, Ksenofontov V and Shylin S I 2015 *Nature* **525** 73–76
[3] Oganov A R et al. 2013 *Science* **342** 1502
[4] Giannozzi P et al. 2009 J. Phys. Cond. Matt. **21** 395502
[5] Grebenev V N, Mazur E A 1987 *Low temperature physics* **13** 479-85
[6] Errea I, Calandra M, Pickard C J, Nelson J, Needs R J, Li Y, Liu H, Zhang Y, Ma Y. and Mauri F 2015 *Phys. Rev. Lett.* **114** 157004
[7] Li Y, Hao J, Liu H, Tse J S, Wang Y and Ma Y 2015 *Scientific Reports* **5** 09948
[8] Liu H, Naumov I I, Hoffmann R, Ashcroft N W and Hemley R J 2017 *PNAS* **114** 6990-95
[9] Degtyarenko N N, Mazur E A 2015 *J. of Experimental and Theoretical Physics* **121** 5 1067-74.

IOP Conf. Series: Journal of Physics: Conf. Series **1238** (2019) 012018 doi:10.1088/1742-6596/1238/1/012018

Hydrodynamic processes in fused quartz under the action of laser radiation

V P Efremov, A D Kiverin I S Yakovenko

Joint Institute for High Temperatures of the Russian Academy of Sciences, Izhorskaya 13 Bldg 2, Moscow 125412, Russia

E-mail: alexeykiverin@gmail.com

Abstract. The paper analyzes numerically the hydrodynamic processes developing inside the fused quartz under the action of intense laser impulses. Depending on laser intensity two basic regimes are obtained and described in details. At low intensities, the slow regime driven by the heat transfer is observed. Herewith, the fracture took place in the heated region before the phase transition. At higher intensities, the high-speed propagation regime is established characterized by the fracture events exactly at the interfacial boundary between the hot plasma and condensed phase. The propagation of absorption wave coupled with the fracture wave is limited by the value of sound speed in the hot plasma, which determines the expansion of plasma into the spallation region ahead of the absorption front. The proposed model of the process agrees well with the recent experimental data, in particular with the characteristic velocity scales.

1. Introduction

An important issue related to the effective data transfer by optical fibers is the increase of optical breakdown probability with the power of laser radiation. Nowadays, one uses optical silica-based fibers as a carrier for data transfer with laser energy. Herewith, the alloyed fused quartz is a common material of the fiber core since the difference in refractive index is necessary for the effective energy transfer through the waveguide. The presence of impurities and micro defects in the bulk of fiberglass determines a possibility of localized optical breakdowns of the medium under the action of intense laser impulses. As a result, the plasma discharge arises together with the region of high concentration of free electrons. The laser radiation begins to be absorbed intensively by this region, and a stable absorption front starts propagating towards the laser beam. The temperature inside the absorption zone increases sharply while the corresponding expansion of the hot plasma causes compression in the cold medium. Eventually, the fracture of the material takes place and the fracture wave propagates together with the absorption front towards the energy source.

The experimental study [1] considers two basic regimes of the medium fracture under the action of intense laser impulses: (1) localized slow melting under the less intense laser action and (2) high-speed fracture wave formed under more intensive laser action. It should be noted that in most experimentally considered cases [1] the value of absorbed energy is relatively low and does not exceed the amount of 0.25–0.6 mJ. However, the focusing of the energy inside the optical fiber core of $\sim$ 7.0–9.5 μm diameter causes rather high intensities reaching the values of $\sim$ 4.0 GW/cm^2 [1]. Such an energy concentration on the micro scales determines specific physical mechanisms of absorption wave propagation.

The slow regime of melting front propagation associated with the absorption front is defined mainly by the mechanisms of heat transfer from the hot plasma region to the adjacent layers of condensed phase

[2]. Herewith, heat losses, radiant losses and multi-dimensional rarefaction should be taken into account. In this case the absorption wave propagates with a speed of ~ several meters per second [1, 2] similarly to the gas combustion under the action of laser impulse [3].

At laser radiation with higher intensity (of the order of GW/cm^2) one observes a regime of high-speed fracture wave propagation together with accompanying absorption front. In this case, the speed of absorption wave propagation is of the order of 2.6–3.2 km/s that is much higher than in the slow regime but almost two times lower than sound speed in the cold fused quartz (longitudinal sound speed in the fused quartz equals to 5.5 km/s). It is evident that such a high-speed regime is associated with high energy densities and the corresponding evolution of violent hydrodynamic processes. However, the detailed mechanisms and the sequence of processes evolution are still not clearly understood.

The paper considers numerically the hydrodynamic processes developed in the fused quartz under the action of intense laser impulse. The obtained sequence of hydrodynamic processes allows detailed description of the medium response to the energy impulse that is important for a clear understanding of the phenomenon of fused quartz destruction.

2. Problem setup and numerical method

Let us consider the following one-dimensional problem setup. The laser radiation propagates from right to the left through the planar layer of fused quartz. At a certain point, the medium starts to absorb the radiation according to the Beer–Lambert–Bouguer law. At the initial stage the absorption takes place in the cold solid quartz. As soon as the hot medium begins to expand heating the adjacent layers of solid quartz the absorption is assumed to proceed at the point with the temperature of 1000.0 K. It is assumed that the radiation is absorbed by heated plasma on the spatial scale of the order of wavelength presumed to be equal to 1.0 μm. Due to the lack of information on the absorption coefficient α the following form of density dependence is assumed $\alpha = (1/D)(\rho_0/\rho)^{1.5}$, where ρ is the local density of medium and $\rho_0 = 2.2$ g/cm^3 is the initial density of the cold fused quartz.

The hydrodynamics of the medium is described by the conventional system of partial differential equations [4] with account of heat conductivity and energy absorption at the interface between solid medium and hot plasma. The equation of state is built on the basis of Hugoniot experimental data for fused quartz, experimental data on heat capacity at conditions near normal ones and heat capacities of silica in disordered phase calculated using so-called chemical model of matter [5] and data from IVTANTHERMO database. The detailed description of the utilized semi-empiric equation of state is presented in [4]. The model for thermal conductivity is presented in [4] as well.

The formulated problem setup reflects all the peculiarities of the studied process. The absorbed laser energy is treated by the medium as the source of inner energy localized on the spatial scale of the order of radiation wavelength. This energy is re-distributed in space via heat conduction and hydrodynamic expansion of the heated region. As a result, the deformations arise in the bulk of medium. This leads to the formation of compression and rarefaction waves, medium spallation, fracture and related phenomena considered in details below.

A series of calculations was carried out with various energy intensities. As a result different regimes of absorption wave propagation and related mechanisms of medium fracture were analyzed. For detailed analysis two types of signal were used: sinusoidal ($I(x,t) \sim I_0 \sin(\omega t)$, I is intensity of laser radiation) as a model of experimental laser pulse and step-wise one ($I(x,t) \sim I_0$). In case of sinusoidal impulse the first half of 250 μs impulse was used.

3. Results and discussion

Let us first consider the sequence of processes developed in the region of energy absorption. The heating of the medium first causes its expansion relative to the initial state. As a result the heated region pushes the cold medium causing its compression. It should be noted that cold solid medium is almost incompressible and the discussed compression effect is negligible. Herewith, the expansion of the heated layer can be rather strong that causes localized spallation of the medium on the spatial scales of the order of heat layer

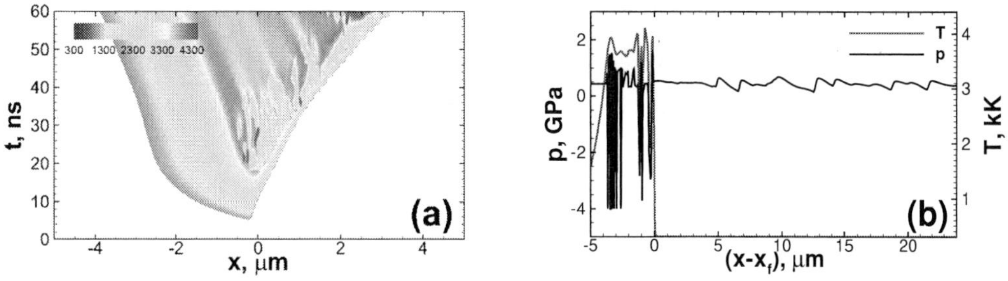

Figure 1. (a) Temperature field evolution in the region of expanding hot medium heated by the energy flux. (b) Characteristic flow pattern in the vicinity of the energy absorption zone ($x = x_f$), temperature and pressure profiles are presented. Sinusoidal signal, $I_0 = 10$ W/μm^2. The early stage of the process, before transition from the slow regime to the fast one, is shown. The laser radiation is transferred from right to left.

thickness. At the same time at sufficient energy release the heated medium undergoes phase transition to the hot gaseous plasma state. Hot plasma expands into the cold region much more intensively that causes more violent deformations of the solid medium at the interface. As a result one can observe spallation of the solid medium exactly at the interface. The hot plasma flows into the spallation region that results in the propagation of the interface towards the energy source. As well, the heat flux from the hot plasma transfers the energy to the cold layers that in turn can cause medium melting. In such a way two basic mechanisms determines the absorption wave propagation. The first one is related to heat transfer and subsequent melting. The second one is defined by hydrodynamic expansion and subsequent spallation. In turn, these two mechanisms define two basic regimes of absorption wave propagation. In the first case the slow regime is established while in the second case the high-speed regime arises. The characteristic velocity scales are respectively: speed of the heat wave in the solid medium (~ 20 m/s) and the sound speed in the hot plasma (~ 3 km/s).

Let us now consider in details each of the regimes of absorption wave propagation through the fused quartz. Figure 1(a) shows the $x - t$ diagram of the process of slow wave formation and subsequent propagation at laser energy flux of 1.0 GW/cm^2 (or 10.0 W/μm^2 in units used in [1]). At the early stage, one can observe the expansion of the heated region with local kernels of spallation at the background of the expansion. At the subsequent stages the heat is transferred to the adjacent layers of solid medium and according to the assumed problem setup the absorption front propagates together with the heat wave. As a result, the additional heating of this new layer takes place that leads to its expansion and spallation. After this, the process repeats itself for next layer etc. In such a way the slow regime of absorption wave propagation establishes. Herewith, this absorption wave is followed by the fracture wave representing the series of subsequent spallations. Figure 1(b) shows characteristic flow pattern in the vicinity of absorption front at the fixed time instant ($t = 50$ μs). At the background of temperature rise, one can observe pressure oscillations associated with the discrete manner of fracture.

Figure 2(a) shows the $x-t$ diagram of the process of high-speed absorption wave propagation under the action of laser pulse providing the energy flux of 4.0 GW/cm^2 (or 40.0 W/μm^2 in units used in [1]). The early stage of medium heating and expansion proceeds in the same manner as in the case of slow regime. Afterwards, as soon as the absorption wave is formed, the permanent spallation of the solid medium proceeds directly at the interface between expanding hot plasma and cold solid quartz. Corresponding flow pattern is shown in figure 2(b). The spallation region is characterized by the steep pressure drop up to the negative values. In such a regime the fracture wave propagates together with the absorption wave which speed is defined by the intensity of hot plasma expansion into the region of fractured solid quartz.

Figure 3 presents the diagram of speed regimes of absorption wave propagation through the fused

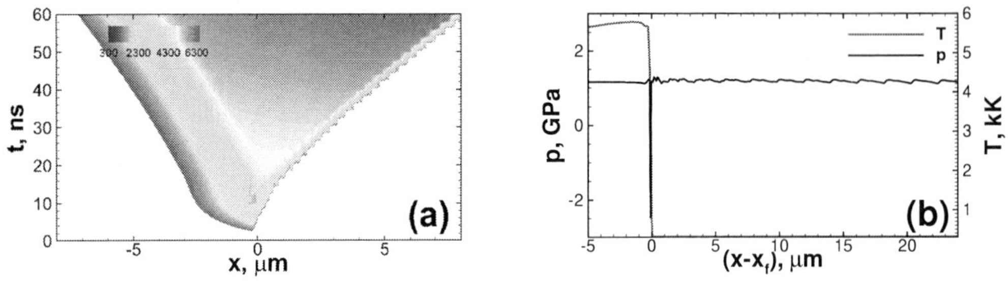

Figure 2. (a) Temperature field evolution in the region of expanding hot medium heated with the energy flux. (b) Characteristic flow pattern in the vicinity of the energy absorption zone ($x = x_f$), temperature and pressure profiles are presented. Sinusoidal signal, $I_0 = 10$ W/μm^2. The laser radiation is transferred from right to left.

quartz under the action of laser pulses of different intensity. Here it is important to note that the silica fibers are able to transpose without losses the radiation flux not higher than 10.0 GW/cm^2, even in pulse regime of laser operation. In view of this the data for higher values of radiation flux presented in figure 3 should be treated as qualitative for clearer understanding of the limiting scales determining different regimes. At energy flux in the range between $\sim$ 0.6–1.5 GW/cm^2 (or 6.0–15.0 W/μm^2 in units used in [1]) one can observe the transition from the slow regime to the high-speed one (zone II). Here it is interesting to note that in case of sinusoidal signal the transition from the slow regime to the fast one can be observed in this range of intensities. Thus, in case of 1.0 GW/cm^2 presented in figure 1 there is such a transition at time instant of 80 μs. In case of step-wise signal it is more convenient to define the margin between the slow and fast regime that is found to be equal to 0.6 GW/cm^2. Herewith, it is useful to note that experimental data obtained earlier correlates well with the obtained critical value [1]. In particular, in [1] the slow regime was observed at radiation fluxes up to 0.3 GW/cm^2 while the fast regime was observed in the range of radiation flux from 3.4 to 4.0 GW/cm^2. In [6] faster regimes were observed at radiation fluxes higher than 2.0 GW/cm^2. Moreover, as one can see from fig. 3 the maximum speed of the fast absorption wave obtained numerically is of the order of experimental value and equals to the calculated sound speed in the hot plasma. This experimental fact substantiates the formulated mechanism of absorption wave propagation in the high-speed regime. The velocity of hot plasma expansion is limited with the sound speed. Plasma cannot expand into the spallation region with velocity exceeding the sound speed value.

The main differences between the speed regimes observed experimentally and numerically are the following. The first distinction is that numerically reproduced slow regimes occur to be almost twice faster compared with the experimentally observed regimes which is most probably related to the one-dimensional approximation of the problem setup. The second distinction concerns the fast regimes speed values. The calculated values occur to be lower than the experimentally observed ones at the same intensities of laser radiation. On the one hand this can be also related to the multi-dimensional nature of the phenomenon. On the other hand it can be related to the fracture development in the cold quartz under the external impact on its boundary. Fused quartz represents a very brittle medium, so the spallation under the tension induced by the hot plasma expansion causes the development of deep cracks propagating into the bulk of solid medium. In such a case the hot plasma can expand into the volume of fracture with the maximum available speed. Due to this the establishment of highest possible propagation speed takes place at lower laser radiation intensities compared with the calculations predictions.

It is useful to note that although there are distinctions between the numerical and experimental data the basic physical mechanisms defining the development of fracture processes in the fused quartz under the action of intense laser energy pulses are reproduced well enough in the calculations. One more

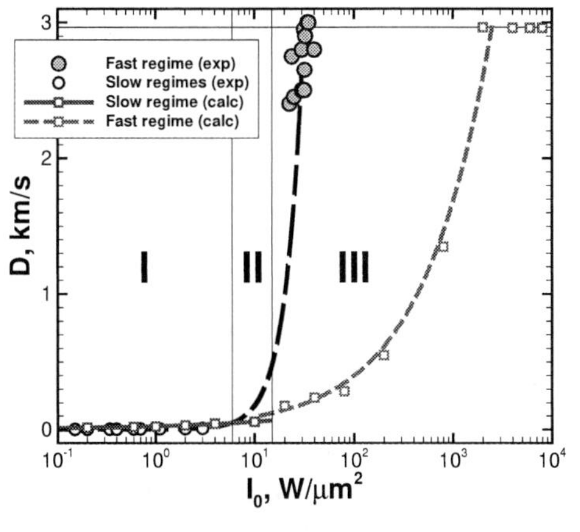

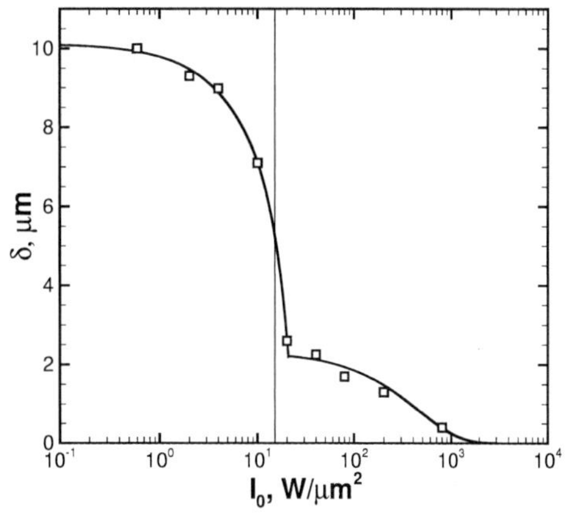

Figure 3. (left) Dependence of absorption wave speed on the intensity of energy flux (speeds are shown for the case of the sinusoidal signal). Blue line represents the data for slow regimes. Red dashed line — for fast regimes. For comparison, the experimental data from [1] and [6] are presented with the yellow signs (slow regimes) and green signs (fast regimes). Black dashed line represents the interpolation of experimental data. For convenience, we used the same units as in [1]. Vertical thin lines show the margins between three regimes: I — slow regime; II — transient regime (in case of the sinusoidal signal) or fast regime (in case of the step-wise signal); III — fast regime. A horizontal thin line shows the value of the maximal speed of absorption wave propagation at high intensities.

Figure 4. (right) Dependence of spatial scales related to the spallation of the medium on the intensity of energy flux. Vertical thin line shows the margin between regimes II and III for comparison with figure 3.

confirmation of this thesis is provided by the analysis of spallation scales obtained numerically and experimentally. Figure 4 shows the dependence of characteristic wavelengths arising in the process of quartz spallation associated with the propagation of absorption wave. The characteristic sizes of fragments measured experimentally in [6] for I_0 from the range between 2.0 and 4.0 $W/\mu m^2$ was found to be of the order of 0.1–1.0 μm that correlate well with the obtained numerical data for the fast regimes. In case of slow regime, the spatial scale of experimentally observed fractures was reported in [7] as 8.0 μm. Close scales were obtained experimentally in [1]. As one can clearly see, there is a steep change in the spatial size of fragments with the transition from the slow regime to the high-speed one. The spatial size decreases from the value of $\sim$ 10 μm to the value $\sim$ 2 μm and further decreases exponentially with the increase in radiation intensity.

4. Conclusion

As a result of detailed numerical analysis of the hydrodynamic processes associated with the absorption of intense laser pulses the basic mechanisms defining the solid medium spallation and absorption wave propagation are formulated. The slow regime is mainly determined by the heat transfer from the energy absorption region to the cold solid quartz. The localized heating caused by the heat transfer leads to the expansion of the heated medium and its subsequent spallation. Herewith, the characteristic spatial scale related to spallation is of the order of absorption length which correlates well with the laser wavelength. At higher intensity of laser radiation the hydrodynamic expansion of the hot plasma generated in the energy absorption zone causes significant tension and subsequent spallation of the cold solid quartz directly at the interface between the expanding plasma and cold medium. As a result the high-speed regime of absorption wave propagation is observed. The absorption front coincides with the interface between the expanding plasma and the cold medium. In the framework of the model considered here the front of fracture wave is driven by the plasma expansion. Therefore, spallation takes place exactly at the absorption front. Herewith, the maximum speed of the complex consisting of the absorption front and fracture front is limited with the sound speed in the hot plasma. In reality, the spallation at the interface of such a brittle medium as quartz causes the generation of magistrate cracks propagating with the near transversal sound speed [8]. In case of quartz, transversal sound speed equals to 3.4 km/s while the sound speed in the hot plasma equals to 3.0 km/s. Since the speed of magistrate crack occurs to be higher than the maximum speed of hot plasma expansion, the speed of wave propagation in a fast regime equals to the sound speed in the hot plasma. This fact was numerously observed experimentally [1,2,6,7], and the mechanisms formulated here explain well enough the high-speed regime of fracture wave propagation through the fused quartz under the action of intense laser impulse. The results are obtained in the one-dimensional approximation. However, despite in reality the dynamic processes develop in the three-dimensional geometry (including Gaussian distribution of the intensity in the cross section of core) the obtained results seem to be quite realistic in view of weak transversal rarefaction due to the fact that the energy is absorbed inside the narrow channel covered by a thick shell made of strong quartz.

Acknowledgments
The work was supported by Russian foundation for basic research (grant No. 17-08-00574). The development of computational codes for numerical simulations of dynamic processes in solid and liquid media is supported by the Program of Russian Academy of Sciences I.13 ''Condensed matter and plasma at high energy densities''.

References
[1] Dianov E M, Fortov V E, Bufetov I A, Efremov V P, Frolov A A, Schelev M Y and Lozovoĭ V I 2006 *JETP Lett.* **83** 75–78
[2] Gorbachenko V I, Dovzhenko A Y, Merzhanov A G, Rumanov É N, Fortov V E and Yachmeneva O E 2010 *Dokl. Phys.* **55** 384–387
[3] Raiser Y P 1974 *Laser Spark and Discharge Propagation* (Moscow: Nauka)
[4] Efremov V P, Ivanov M F, Kiverin A D and Yakovenko I S 2016 *J. Phys.: Conf. Ser.* **774** 012119
[5] Gryaznov V K, Fortov V E, Zhernokletov M V, Simakov G V, Trunin R F, Trusov L I and Iosilevski I L 1998 *J. Exp. Theor. Phys.* **87** 678–690
[6] Efremov V P, Fortov V E and Frolov A A 2015 *J. Phys.: Conf. Ser.* **653** 012013
[7] Duchateau G, Feit M D and Demos S G 2012 *J. Appl. Phys.* **111** 093106
[8] Kanel G I, Razorenov S V, Utkin A V and Fortov V E 1996 *Fizicheskie Velichiny* (Moscow: Yanus-K)

Temperature dependence of resistivity at the transition to a charge density wave state in rare-earth tritellurides

K K Kesharpu[1 †] **and P D Grigoriev**[1 2 3]

[1]National University of Science and Technology "MISiS", 119049 Moscow, Russia
[2]L. D. Landau Institute for Theoretical Physics, 142432 Chernogolovka, Russia
[3]P. N. Lebedev Physical Institute, RAS, 119991 Moscow, Russia

E-mail: [†]m1708377@edu.misis.ru

Abstract. About a half of the Fermi surface in rare-earth tritellurides $R\mathrm{Te}_3$ becomes gapped below the transition to a charge-density-wave (CDW) state, as revealed by ARPES data. However, the observed jump in resistivity during the CDW transition is less than 20%. Previously this phenomenon was explained by hypothesizing a very slow evolution of CDW energy gap below transition temperature in $R\mathrm{Te}_3$ compounds, which contradicts the X-ray measurements. Here we show that this weak change in resistivity can be explained in the framework of standard mean-field temperature dependence of the CDW energy gap in agreement with X-ray data. The change of resistivity caused by CDW is weak because the decrease in conducting electron density at the Fermi level is almost compensated by the decrease in their scattering rate. We calculate resistivity in $R\mathrm{Te}_3$ compounds using the Boltzmann transport equation and the mean-field description of the CDW state, and obtain a good agreement with experimental data.

1. Introduction

The charge-density-wave (CDW) ground state is rather common in strongly anisotropic metals [1, 2]. The typical precursors of the CDW transition are strong electron-phonon (EPC) or/and electron-electron (e-e) coupling and the nesting of Fermi surface (FS),[1] though the latter is not always a determining factor[3, 4]. In metals with (almost) ideal FS nesting the CDW creates an energy gap on the Fermi level and converts a metal to an insulator or semiconductor. In most CDW compounds the FS nesting is not perfect; then the FS is only partially gapped in the CDW state, and the compound retains its metallic properties even in the CDW state. An example of such partially gapped CDW compounds is the family of quasi-2D layered material, known as *rare earth tritelluride* $R\mathrm{Te}_3$ (R=Y, La, Ce, Nd, Sm, Gd, Tb, Ho, Dy, Er, Tm) [5]. These compounds are very convenient for the demonstration of various electronic properties in a partial CDW-metallic state, and their electronic structure was studied extensively and with rather high precision using Angle Resolved Photo Emission Spectroscopy (ARPES) [6, 7, 8] and through various other probing methods [9, 10, 11, 12, 13, 14].

$R\mathrm{Te}_3$ compounds have weakly orthorhombic layered crystal structure (Space Group: *Cmcm*) formed by sandwiching two alternate puckered R-Te layers in between two double Te planes. We will mainly concentrate on terbium tritelluride (TbTe_3), which has two incommensurate CDW: one at high temperature with $T_{c1} = 336$ K and the second below much lower temperature $T_{c2} = 41$ K. Above $T_{c2} = 41$ K it was shown to have unidirectional stripes as opposed to checkerboard pattern observed below $T_{c2} = 41$ K, which somewhat resembles the behavior of under-doped cuprates[15]. The FS in TbTe_3 at two different temperatures was measured by ARPES in Ref. [8], and the temperature dependence of resistivity along different directions has been investigated in Refs. [13, 14]. Experimentally it has been found that the resistivity in $TbTe_3$ shows a too small and very anisotropic jump at the transition temperature T_{c1} from metallic to CDW state (see figure 1 of Ref. [13]), which is in contrast to the expected much larger jump. This phenomenon has been explained in Ref. [13] only by assuming a very weak temperature dependence of the CDW energy $\Delta(T)$ just below T_{c1}, attributed to strong fluctuations and given by Eq. (9) below with $\alpha \approx 2$ instead of the mean-field-theory value $\alpha = 1/2$. However, the X-ray studies[11] suggest nearly the mean-field temperature evolution of the CDW order parameter below T_{c1} with $\alpha \approx 1/2$ (see Fig. 6 of Ref. [11]). In the present work we resolve this inconsistency and show that the observed T-dependence of resistivity can be explained in the framework of mean-field dependence $\Delta(T)$.

2. Theory

The anisotropic resistivity ($\rho_i = 1/\sigma_i$) is calculated by using the *Boltzmann transport equation* (see Eq. (3.16) of Ref. [16]), which gives

$$\sigma_i(T) = 2e^2 \sum_{\mathbf{k}} \tau v_i^2(\mathbf{k}) \left[-\partial f_0/\partial\varepsilon\right] = -e^2 \int v_i^2 \tau \frac{\partial f_0}{\partial \varepsilon} g(\varepsilon) d\varepsilon \frac{d\Omega}{4\pi}, \tag{1}$$

where $\sigma_i(T) = 1/\rho_i$ is the conductivity along main axes $i = x, y, z$, $v_i(k)$ is electron velocity, $k = \{k_x, k_y, k_z\}$ is the vector of electron momentum, $\partial f_0/\partial\varepsilon = -1/\{4T\cosh^2\left[(\varepsilon - E_F)/(2T)\right]\}$ is the derivative of the Fermi distribution function, which restricts the summation over momentum to the vicinity of FS, E_F is the Fermi energy, $g(\varepsilon)$ is the *density of states* (DOS), which depends on energy ε, and $d\Omega$ is the solid angle extended by an area in momentum space. τ is the mean scattering time. At $T \sim T_{c1} = 336K \gg T_D$, where $T_D \approx 180$ K is the Debye temperature, the electron scattering comes mainly from the short-wavelength phonons. This scattering is similar to the scattering by short-range impurites, because it also gives the momentum transfer of the order of Brillouin zone length, while the energy transfer is $\sim k_B T_D \ll T$. Hence, similarly to the scattering by short-range impurities, in the Born approximation or according to the golden Fermi rule, the scattering rate $1/\tau$ is proportional to the density of electron states $g(\varepsilon)$ [16], which has a jump at T_{c1} due to the opening of CDW energy gap. Also $1/\tau$ is proportional to the number of phonons, which grows linearly with T. Hence, the temperature dependence of scattering time is approximately given by $\tau(T) \propto [T\rho(\varepsilon, T)]^{-1}$. Substituting this to Eq. (1) we can hypothesize that, in the CDW state as the temperature decreases, the reduction of the density of electron states contributing to conductivity is compensated by the increase in their scattering time. Hence, one does not observe a large jump in resistivity during the transition between CDW and metallic state near T_{c1}. Below we substantiate this idea by direct calculations.

For materials with electron dispersion (ED) $\varepsilon(\mathbf{k}, T)$ the DOS is calculated as

$$g(\varepsilon) = \frac{2}{(2\pi)^3} \iiint dk_x dk_y dk_z\, \delta[E - \varepsilon(k_x, k_y, k_z, T)], \tag{2}$$

where the factor 2 is due to spin degeneracy of electrons. The integration in Eq. (2) is over the first Brillouin zone, but due to the δ-function it is restricted to the FS, which is found by solving the equation $\varepsilon(\mathbf{k}, T) = E_F$.

Because we are interested in layered materials, all the analysis is done for 2D case (although the same procedure can be applied to 3D materials). For 2D material the electron dispersion depends only on 2 momentum components (let it be k_x and k_y). Then Eq. (2) reduces to

$$g(\varepsilon) = \frac{2}{(2\pi)^2} \iint dk_x dk_y\, \delta[E - \varepsilon(k_x, k_y, T)]. \tag{3}$$

Below all functions in the text explicitly depends on k_x, k_y, T unless otherwise mentioned.

In the mean field approximation the electron dispersion in a CDW state is given by[1]

$$\varepsilon(\mathbf{k}) = \frac{\epsilon(\mathbf{k}) + \epsilon(\mathbf{k}-\mathbf{Q})}{2} \pm \sqrt{\frac{[\epsilon(\mathbf{k}) - \epsilon(\mathbf{k}-\mathbf{Q})]^2}{4} + \Delta_0^2(T)}\,, \tag{4}$$

where $\mathbf{Q}$ is the CDW wave vector and the energy gap Δ is momentum-independent. If the anti-nesting term $\epsilon(\mathbf{k}) + \epsilon(\mathbf{k}-\mathbf{Q}) < 2\Delta_0(T)$, the corresponding k-state acquires a gap at the Fermi level. The mean-field dispersion in Eq. (4) is often simplified to

$$\varepsilon(\mathbf{k}) \approx \sqrt{\frac{[\epsilon(\mathbf{k}) - \epsilon(\mathbf{k}-\mathbf{Q})]^2}{4} + \Delta^2(\mathbf{k}, T)}\,, \tag{5}$$

where the energy gap $\Delta(\mathbf{k}, T)$ now depends on electron momentum and reduces with the increase of the anti-nesting term $\epsilon(\mathbf{k}) + \epsilon(\mathbf{k}-\mathbf{Q})$, being non-zero only if $\epsilon(\mathbf{k}) + \epsilon(\mathbf{k}-\mathbf{Q}) < 2\Delta_0(T)$. Since for the well-nested parts of FS $[\epsilon(\mathbf{k}) - \epsilon(\mathbf{k}-\mathbf{Q})]/2 \approx \epsilon(\mathbf{k})$, Eq. (5) simplifies to

$$\varepsilon(\mathbf{k}) \approx \sqrt{\epsilon(\mathbf{k})^2 + \Delta(\mathbf{k}, T)^2}. \tag{6}$$

We will use this reduced ED for further calculations.

In RTe_3 materials in the CDW phase the FS is only partially gapped, i.e. for wave vector $|k_x| < k_{x0}$ the ED contains nonzero energy gap Δ, and for the remaining part $\Delta = 0$. Hence, the ED for $|k_x| < k_{x0}$ is given as $\varepsilon(\mathbf{k}) = \sqrt{\epsilon(\mathbf{k})^2 + \Delta(T)^2}$, and for the remaining part of FS it transforms to $\varepsilon(\mathbf{k}) = \epsilon(\mathbf{k})$. Due to this discontinuity in ED, the DOS $g(\varepsilon)$ for partially gapped FS is calculated piecewise. Hence, $g(\varepsilon)$ from Eq. (3) for the partially gapped FS is found as

$$\begin{aligned} g_{\text{tot}}(\varepsilon) &= g_{\text{gap}}(\varepsilon) + g_{\text{ungap}}(\varepsilon) \\ &= \frac{1}{\pi^2}\int_{-\pi/a}^{\pi/a} dk_y \left[\int_0^{k_{x0}} dk_x\, \delta(\varepsilon - \sqrt{\epsilon^2 + \Delta^2}) + \int_{k_{x0}}^{\pi/a} dk_x\, \delta(\varepsilon - \epsilon)\right], \end{aligned} \tag{7}$$

Where $g_{\text{tot}}(\varepsilon)$ is the total DOS in the CDW state, $g_{\text{gap}}(\varepsilon)$ is the DOS contribution from the gapped parts of FS, $g_{\text{ungap}}(\varepsilon)$ is DOS from the ungapped part of FS, a is the lattice constant. We should note that Eq. (7) is temperature dependent. Above T_{c1} the first term in Eq. (7) vanishes and total DOS is only given by ungapped part, which is the metallic DOS $g_{\text{met}}(\varepsilon)$.

For RTe_3 compounds ϵ is found in the tight-binding approximation (see Eq. (2) of Ref. [6]):

$$\epsilon = -2t_{\parallel}\cos[(k_x \pm k_y)a/2] - 2t_{\perp}\cos[(k_x \mp k_y)a/2] - E_F \quad , \tag{8}$$

where $t_{\parallel}$ and $t_{\perp}$ are the parallel- and perpendicular-to-chain electron hopping integrals. Momentum dependence of energy gap Δ is found by fitting the experimental data (see Fig. 13 of Ref. [6]):

$$\Delta = \Delta_0 \left(1 - \frac{T^2}{T_{c1}^2}\right)^{\alpha} \left(1 - \frac{k_x^2}{k_{x0}^2}\right), \tag{9}$$

In the mean-field approximation the temperature dependence of Δ is given by $\alpha = 1/2$, which we use in our calculations. If $t_{\parallel} \gg t_{\perp}$ the ED in Eq. (8) simplifies to a quasi-1D chain ED, given by Eq. (1.13) of Ref. [1].

The procedure of calculating the gapped part of DOS $g(\varepsilon)_{\text{gap}}$ taking into account CDW fluctuations has already been described in Ref. [17, 18, 19]. Here we briefly repeat the necessary steps of calculation for completeness of this article. First we find the temperature dependent coherence length ξ using the *Ginzburg-Landau functional* (GLF) and mean-field theory. In the next step we use ξ and the energy gap

IOP Conf. Series: Journal of Physics: Conf. Series **1238** (2019) 012019 doi:10.1088/1742-6596/1238/1/012019

Δ in the one particle Green's function. The major property of this Green's function is that it considers the effect of only near neighbor degenerate states. Using this Green's function we find the electronic properties of the material. The GLF is given as

$$F = F(0) + \int dx \left[a|\Delta|^2 + b|\Delta|^4 + c\left|\frac{d\Delta}{dx}\right|^2 \right],$$

$$\begin{aligned}
&\text{where}\\
&a = \frac{D_0(T - T_c)}{T_c} \quad ; \quad b = D_0\left[b_0 + (b_1 - b_0)\frac{T}{T_c}\right],\\
&c = D_0 + \xi^2(T) \quad , \quad \xi^2 = \frac{7\zeta(3)v_f^2}{16\pi^2k^2T^2},\\
&b_0 = 0.5/(1.76^2) \quad , \quad b_1 = \frac{7\zeta(3)}{16\pi^2},\\
&D_0 = \text{Density of states at } T = T_c \text{ at energy } E_f.
\end{aligned} \tag{10}$$

Here the values of a, b, c are found from the mean field calculation (see Sec. 5.1 of Ref. [1] for complete derivation of the values given in Eq. (10)). In next step we use ξ and Δ in one particle Green's functions and keep the terms with coupling between nearly degenerate states (See Eq. 5 of Ref. [17]), which is written as

$$\begin{aligned}
G^{-1}(\mathbf{k}, E) = E - \epsilon(\mathbf{k}) - \int d\mathbf{Q}\, S(\mathbf{Q})\langle\Delta^2\rangle\\
\times [E - \epsilon(k \pm \mathbf{Q}) + i\delta]^{-1}.
\end{aligned} \tag{11}$$

The structure factor $S(\mathbf{Q})$ is the Lorentzian with origin at $Q = 2k_f$ and of width $\xi^{-1}(T)$. Next we take the integral over $\mathbf{Q}$ in Eq. (11), which results in

$$\begin{aligned}
G^{-1}(\mathbf{k}, E) = E - \epsilon(\mathbf{k}) - \langle\Delta^2\rangle\\
\times [E - v_f(|k| - k_f) + \imath v_f\xi^{-1}(T)]^{-1}.
\end{aligned} \tag{12}$$

In the last step we integrate the imaginary part of Eq. (12) to get the DOS from the gapped part of FS as follows

$$\begin{aligned}
&\frac{g_{\rm gap}(E)}{g_{\rm met}(E)} = \frac{\alpha\sqrt{[2(y+x)]}}{\left[2(y+x) - \alpha^2\right] y},\\
&\text{where,}\\
&\tilde{\omega} = E/\Delta, \quad \alpha = v_f/(\xi^{-1}\Delta),\\
&x = 1 + 0.25\alpha^2 - \tilde{\omega}^2, \; y = \sqrt{(x^2 + \tilde{\omega}^2\alpha^2)}.
\end{aligned} \tag{13}$$

In Eq. (13) $g_{\rm met}(E)$ is DOS in the metallic state above CDW transition temperature.

The total DOS can be found as a sum of contributions $g_{\rm gap}(\varepsilon)$ from Eq. (13) and $g_{\rm ungap}(\varepsilon)$ from ungapped parts:

$$g_{\rm tot}(\varepsilon) = g_{\rm met}(\varepsilon)[(1-\eta) + \eta\, g_{\rm gap}(\varepsilon)/g_{\rm met}(\varepsilon)]\,, \tag{14}$$

where $\eta \approx k_{x0}a/\pi$ is the ratio between gapped and total FS length. Finally, from Eqs. 13, (14) we obtain

$$g_{\rm tot}(\varepsilon) = g_{\rm met}(\varepsilon)\left[\left(1 - \frac{ak_{x0}}{\pi}\right) + \left(\frac{ak_{x0}}{\pi}\frac{\alpha\sqrt{[2(y+x)]}}{\left[2(y+x) - \alpha^2\right] y}\right)\right]. \tag{15}$$

IOP Conf. Series: Journal of Physics: Conf. Series **1238** (2019) 012019 doi:10.1088/1742-6596/1238/1/012019

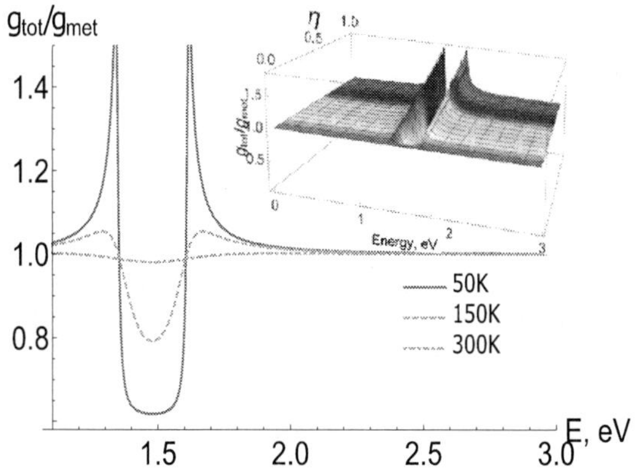

Figure 1. Plot of ratio of DOS under CDW state to DOS in normal metallic state calculated from Eq. (14) for $T = 50$ K (blue solid line), $T = 150$ K(orange dashed line), $T = 300$ K (green dot-dashed line). $E_F = 1.48$ eV, $\eta = 0.4$ has been used from Ref. [13]. Inset: The DOS dependence on η $(0 < \eta < 1)$ at $T = 50$ K.

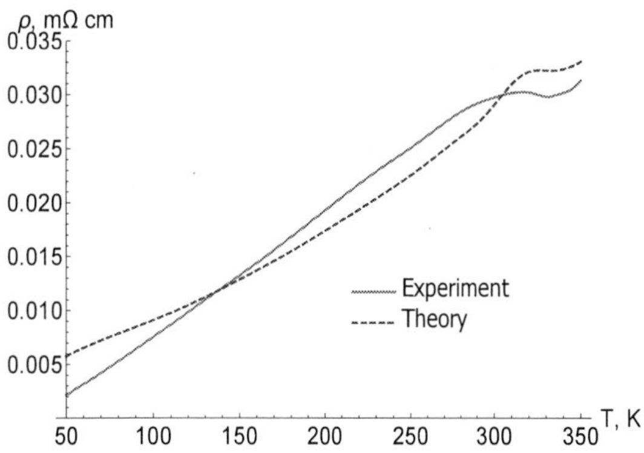

Figure 2. Comparison of the experimental resistivity in $TbTe_3$ (solid red curve), extracted from Fig. 1 of Ref. [13], and theoretical resistivity (blue dashed curve) calculated from Eq. (1). Experimental plot is found by taking average of ρ_a and ρ_c.

3. Comparison with experiment and discussion

The ratio of DOS in CDW and metallic phases in $TbTe_3$ for three different temperatures is plotted in figure 1. In this plot we used the experimental values for $\Delta_0 = 0.27$ eV, $k_{x0} = 0.29^{-1}$, $\eta = 0.4$, $t_{\|} = 2$ eV, $t_{\perp} = 0.37$ eV, $a = 4.4$ Å, $E_F = 1.48$ eV for $TbTe_3$ given in Ref. [6]. As expected, the DOS does not get to zero at the Fermi level for very low temperature due to contribution from the ungapped part of FS. In the inset figure 1 we show the evolution of DOS with the increase of fraction η of the gapped part of Fermi surface. We notice that as η goes towards unity, i.e the whole FS becomes gapped, DOS goes to zero at the Fermi level.

The conductivity is calculated using Eq. (1). The electron velocity v_i is found by differentiating the energy dispersion relation in Eqs. (6), (8): $v_i = \partial\epsilon/\partial k_i$. The temperature-dependent relaxation time decreases with the increase of temperature according to $\tau(T) \propto [T\rho(\varepsilon, T)]^{-1}$. Eq. 1 is evaluated numerically by integrating over the whole Fermi surface, and the resulting resistivity plot is shown in Fig. 2. As one can see from this figure, the jump of resistivity is very small at the transition temperature of 330K. This supports our hypothesis that although the number of charged quasiparticles contributing to conductivity below the CDW transition temperature decreases, it is compensated by the increase in their relaxation time τ.

The obtained theoretical curve for resistivity shows good agreement with experimental data. However, its calculation is done for the reduced electron dispersion in CDW state given by Eq. (6). It would be useful to perform similar calculation for the full mean-field dispersion in the presence of CDW with imperfect nesting, given by Eq. (4). Also it will be interesting to compare the results with and without the effect of fluctuations. We leave these and other relevant questions for future work.

Acknowledgments

The work was carried out with financial support from the Ministry of Education and Science of the Russian Federation in the framework of increase Competitiveness Program of NUST "MISIS", implemented by a governmental decree dated 16th of March 2013, No 211, and from Russian Science Foundation (project # 16-42-01100). P.G. also thanks RFBR grant # 17-52-150007.

References

[1] G. Gruner 1994 *Density Waves in Solids* Addison-Wesley Pub. Co.
[2] Monceau P., 2012 *Advances in Physics* **61**, 325
[3] Johannes M. D. and Mazin I. I. 2008 *Phys. Rev. B* 77, 165135
[4] Zhu X, Guo J, Zhang J and Plummer E W 2017 *Adv. Phys. X* **2** 622
[5] Norling B K and Steinfink H 1966 *Inorg. Chem.* **5** 1488
[6] Brouet V, Yang W L, Zhou X J, Hussain Z, Moore R G, He R, Lu D H, Shen Z X, Laverock J, Dugdale S B, Ru N and Fisher I R 2008 *Phys. Rev. B* **77** 235104
[7] Brouet V, Yang W L, Zhou X J, Hussain Z, Ru N, Shin K Y, Fisher I R and Shen Z X 2004 *Phys. Rev. Lett.* **93** 126405
[8] Schmitt F, Kirchmann P S, Bovensiepen U, Moore R G, Chu J H, Lu D H, Rettig L, Wolf M, Fisher I R and Shen Z X 2011 *New J. Phys.* **13** 063022
[9] Lavagnini M, Baldini M, Sacchetti A, Di Castro D, Delley B, Monnier R, Chu J H, Ru N, Fisher I R, Postorino P and Degiorgi L 2008 *Phys. Rev. B* **78** 201101
[10] Ru N, Chu J H and Fisher I R 2008 *Phys. Rev. B* **78** 012410
[11] N. Ru, C. L. Condron, G. Y. Margulis, K. Y. Shin, J. Laverock, S. B. Dugdale, M. F. Toney, and I. R. Fisher, Phys. Rev. B **77**, 035114 (2008).
[12] Sacchetti A, Degiorgi L, Giamarchi T, Ru N and Fisher I R 2006 *Phys. Rev. B* **74**
[13] Sinchenko A A, Grigoriev P D, Lejay P and Monceau P 2014 *Phys. Rev. Lett.* **112** 036601
[14] Sinchenko A, Lejay P, Leynaud O and Monceau P 2014 *Solid State Commun.* **188** 67
[15] LeBoeuf D, Kramer S, Hardy W N, Liang R, Bonn D A and Proust C 2013 *Nat. Phys.* **9** 79
[16] A. A. Abrikosov 1988 *Fundamentals of Theory of Metals* North-Holland
[17] Lee P A, Rice T M and Anderson P W 1973 *Phys. Rev. Lett.* **31** 462.
[18] Scalapino D J, Sears M, Ferrel R Al, Scalapino D J, Sears M and Ferrel R A 1972 *Phys. Rev. B* **6** 3409
[19] Suzumura Y 1987 *J. Phys. Soc. Japan* **56** 2494

Simulation of magnetization and heating processes in HTS tapes stacks

I V Anischenko, S V Pokrovskii, I A Rudnev

National Research Nuclear University MEPhI (Moscow Engineering Physics Institute), Kashirskoe sh. 31, 115409 Moscow, Russia

E-mail: mephizic@gmail.com

Abstract. In this work we present a computational model for a magnetic levitation system based on stacks of 12 mm x 12 mm second generation $GdBa_2Cu_3O_{7-x}$ high temperature superconducting (HTSC) tapes. In our model we have used the magnetic and transport characteristics of real industrial HTSC tapes. The thermal properties of each layer of high-temperature superconducting tape and the features of the layered structure of whole stack have also been taken into account. The numerical simulation was performed using the finite element method. Distributions of both magnetic field throughout the space and the current in every tape of the stack were calculated for two cases: (i) cryocooler cooling mode and (ii) cooling in the liquid nitrogen. The magnetization curves of the stacks in external field of a permanent NdFeB magnet and levitation force dependence on the gap between the magnet and the HTS tapes stack in these cooling modes were obtained. We have calculated heat transfer and temperature distribution in the system taking into account the effect of thermal properties of Hastelloy substrate, as well as Cu and Ag stabilization layers on the cooling process and dynamic magnetization. Simulation results were compared with the experimental data and a good agreement with numerical ones was shown.

1. Introduction

Currently, the main construction material for levitation bearings and suspensions are bulk high temperature superconducting samples. But bulk HTS has several drawbacks such as brittleness, the complexity of sample preparation, when samples of desired shape are needed. Also the critical current density (J_c) of the bulk samples is significantly less than J_c for the HTS tapes. Thus, an alternative way to construct levitation systems is to use HTS tapes stacks that has high strength, simplicity of thermal stabilization and the ability to vary the geometric parameters of the stack. The magnetic properties and levitation characteristics of the tapes stacks depend not only on critical characteristics of materials but also on its geometry (the number and shape of the tapes), conditions of magnetization and cooling. The present work is devoted to the development of a multiphysical numerical model that makes it possible to calculate the parameters of a magnetic system based on HTS tapes for two cooling modes: liquid nitrogen and cryocooler.

2. Computational model

The simulated magnetic-levitation system is a high-temperature superconducting tapes stack placed in a gradient magnetic field of a permanent magnet. The geometry of the system is shown in the Figure 1.

IOP Conf. Series: Journal of Physics: Conf. Series **1238** (2019) 012020 doi:10.1088/1742-6596/1238/1/012020

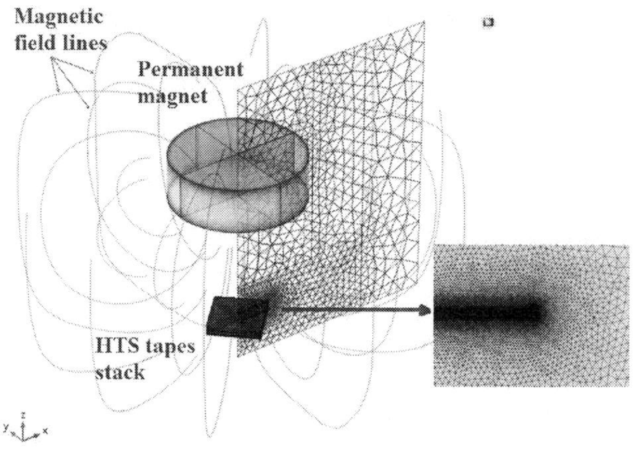

Figure 1. The geometry of the magnetic levitation system.

The measurements for the modeled stacks were performed on superconducting tapes produced by SuperOx (Russia) with the width of 12mm. Thickness of the superconducting layer $GdBa_2Cu_3O_{7-x}$ is 1 micron, and the tapes critical current is 300A (criterion 1µV / cm, 77K, self field). Each simulated tape contains a substrate, and also copper and silver thin layers, in addition to the superconducting layer.

To study the behavior of a HTS tapes stack in an external magnetic field a finite element method was used. The model was developed on the basis of the Partial Differential Equations (PDE) form of the Comsol Multiphysics software module.

In Comsol Multiphysics the general PDE standard equation is (1):

$$e_a \frac{\partial^2 \boldsymbol{u}}{\partial t^2} + d_a \frac{\partial u}{\partial t} + \nabla \times \Gamma = F \qquad (1)$$

Here the dependent variable, magnetic field $\boldsymbol{u} = \boldsymbol{H}(x,y,z)$, is chosen in the entire space. The parameters of equation (1) are chosen in such a way that equation (1) becomes Faraday's law (2):

$$\nabla \times \boldsymbol{E} + {d\boldsymbol{B}}/{dt} = \nabla \times \boldsymbol{E} + {d(\mu_0 \mu_r \boldsymbol{H})}/{dt} = 0 \ , \qquad (2)$$

where $\boldsymbol{E}$ - electric field, $\boldsymbol{B}$-magnetic induction, $\boldsymbol{H}$- magnetic field, μ_0 - relative permeability of free space, relative permeability for air and for superconducting elements $\mu_r = 1$. The specifications of the parameters in the equations (1) is given in [1].

The nonlinear resistance dependence on the current is given by the power law with the power 21, which is typical for the samples of HTS tapes. The current components are determined from the Amper's law. On the boundary of the computational domain, the Dirichlet boundary conditions were used. The critical current density J_{c0} is determined on the basis of the transport measurements of the used HTS tapes. Also, in this model it is taken into account the two-exponential dependence of the critical current on the magnetic field [2] and well-known temperature dependence of the critical current:

$$J_{c0} = \alpha \left(1 - \left(\frac{T}{T_c}\right)^2\right)^{1.5} , \qquad (3)$$

where T is the temperature, T_c is the critical temperature, α – a fitting parameter .

The description of the heat transfer physics (Heat Transfer in Solids module of the Comsol) is based on the expression:

$$\rho C_p \, {\partial T}/{\partial t} + \nabla \cdot (-k \nabla T) = Q + \rho C_p \boldsymbol{u} \nabla T \ , \qquad (4)$$

where C_p-heat capacity, ρ-density, k- the coefficient of thermal conductivity of the materials used in this work, which depend on temperature for each layer [3], $\boldsymbol{u}$ is the temperature velocity field, Q – all heat sources, including thermoelectric.

Local heat release in the system is calculated as:

$$Q = E \cdot J \tag{5}$$

The current distribution J of the superconductor is determined by the conditions of magnetization, and is calculated as the solution of the Maxwell equations, realized in the PDE form.

Within the model, the cryocooler cooling mode realized by a copper bulk plate whose external boundary is set as a cooling source with a certain temperature. The heat radiation from the outer cryostat walls, which are at room temperature, is taken into account. Calculations determine the required power for the cooling element, which should not exceed the power of the real cryocooler-cooling system [4]. Otherwise, there is a violation of the temperature regime, which leads to undesirable heating in the system.

Simulation of liquid nitrogen cooling reduces to setting the heat transfer coefficient for the space region completely encompassing the HTS tapes stack [5] as oppose to the cryocooler acting from the bottom of the stack. Actually, we specify the heat transfer coefficient for the convective boiling of nitrogen taking into account the possibility of boiling nucleation with the other heat transfer coefficient [6].

The interaction force of the HTS tapes stack with the permanent magnet (the levitation force) is calculated as follows:

$$\boldsymbol{F} = \int_V \boldsymbol{J} \times \boldsymbol{B}\, dV \ , \tag{6}$$

where $\boldsymbol{B}$ is the magnetic field induction of the permanent magnet. The currents distribution $\boldsymbol{J}$ in the superconducting tapes of stack was calculated as the result of the solutions of the PDE equations. Integration is performed over the entire volume of the superconductor.

Firstly, using the Rotating Machinery module of the Comsol Multiphysics software, the stationary study of the magnetic field distribution around the magnet was solved. Residual induction on the magnet surface must be in the range 0.30 - 0.35T, which corresponds to the experimental setup [7].

During the magnetization, the HTS stack slowly approaches the magnet (at a constant speed of 5mm/s), and then moves away from it in the same fashion (ZFC mode). To calculate the external magnetic field distribution during permanent magnet movement along the z axis we use the Moving Mesh module together with Rotating Machinery module. Obtained result can be applied as an external magnetic field on the next step – modeling the time dependent HTS tapes stack magnetization process.

It is notable that the last calculation step involves a simultaneous solution of all four modules used at the model. Equations (3) and (5) determine the interconnection of interfaces Heat Transfer in Solids and PDE. When currents appear in superconducting layers, heat release begins and the temperature of the tapes changes. This leads to changing in the critical current in accordance with equation (3). At the next solution time step, the new value of the critical current will be taken into account, so the simultaneous solution is more accurate than the segregated solution. However, the solution of such a simultaneous study requires large computing powers and requires much time. To reduce the calculation time, the functional dependencies on the number of tapes in the stacks were calculated in 2D mode. In 3D mode, some simplifications were also made: copper layers in each tape were specified as two-dimensional layers. Since these layers do not play a role in the magnetic fields physics, but have a great importance in the heat transfer in HTS tapes, their thickness was implicitly determined using Thin layer boundary conditions of Heat Transfer in Solids module.

Special attention should be given to the mesh construction (see fig.1). Using the multiscale structuring allows to thicken the mesh elements upon approaching to the sample, where the utmost accuracy of computation is required. In thin layers of the tape we have used swept tool for surface mesh. For the moving part of the model, the mesh of elements is rebuilt as soon as the quality of at least one of elements decreases below the value 0.2 on a scale from 0 to 1, where 0 is an element of absolutely unsatisfactory quality, whereas 1 is an ideal equilateral element.

3. Results and Discussion

The magnetic properties of the tapes stacks for various height were calculated in two cooling modes: the cryocooler cooling mode and the liquid nitrogen cooling mode. In the liquid nitrogen cooling mode, the remnant magnetizations and levitation forces were calculated for stacks of tapes containing from 1 to 100 layers (Figure 2). The calculations results are in good agreement with the experimental data (more about the experiment see [7]).

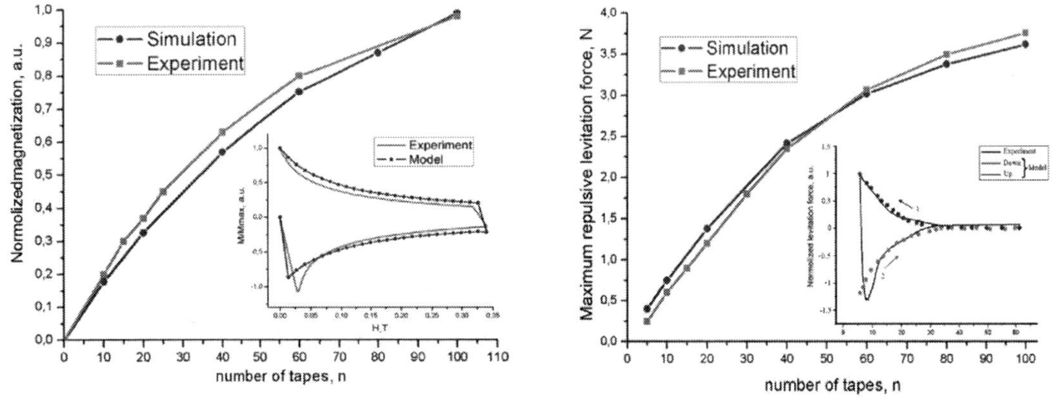

Figure 2. The normalized dependence of the magnetization (on the left) and the maximum repulsive force (on the right) on the number of tapes in a stack (inset: example of experimental and calculated curves for a stack of tapes containing 10 layers).

The dependences of the maximum repulsive force and remnant field strength on the thickness of the stack have a similar behavior. These dependences behave almost linearly at small stack heights, but then the dependences tend to saturation. This indicates an inefficiency of using large tapes stacks to achieve the largest trapped flux and the repulsive levitation force.

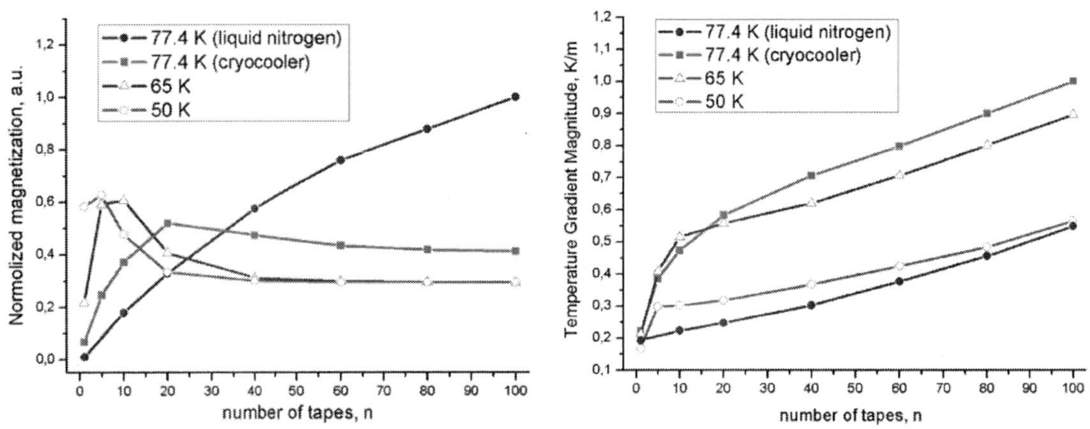

Figure 3. The normalized dependence on the number of tapes in a stack for liquid nitrogen and cryocooler cooling modes of trapped magnetic flux (on the left) maximum temperature gradient (on the right)

The upper tapes of the stack are closest to the magnet and they appear in a higher external magnetic field, than bottom layers. These layers make the main contribution to the trapped magnetic flux. For the given cooling parameters in the cryocooler-cooling mode, calculations show that the trapped flux increases only for stacks with small height. As a number of tapes in the stack further increases the trapped flux begins to decrease (Figure 3, on the left).

This behavior explained by the fact that cooling in a given mode is carried out only at the lower boundary of the stack, while the upper layers, which are most actively involved in the magnetic flux trapping, are at higher temperatures. From this point of view, nitrogen cooling for high stacks is more efficient. It should be noted that for a more powerful cryocooler-cooling (to temperatures of 65 K and 50 K), the magnetization drop with increasing number of tapes in the stack can be explained by the critical current increasing with decreasing temperature. In this case, a sufficiently weak permanent magnet magnetic field cannot penetrate into the tapes with higher critical current, and the tapes remain in the Meissner phase. Therefore, the maximum temperature gradient focuses on the edges of the stacks and increase with the grows of the stack height. (Figure 3, on the right). Despite the fact that the trapped flux is reduced substantially, the heat generation continues to rise. The magnetic field begins to enter the tapes from the edges, but it cannot enter into the tapes center which leads to more intensive heating at the edges of the stack. It is also worth noting that the significantly higher temperature gradient corresponds to a cryocooler cooling mode than to the liquid nitrogen cooling mode.

At the initial magnetization stages, the magnetic field H_z and the gradient temperature field permeate the sample practically at the same time. Then the resulting thermal spot begins to propagate through the sample due to the own thermal conductivity of the superconductor and the stabilizing layers. The main heat outflow is through the substrate. These data on the heat propagation in HTS tapes is also supported by other calculations [8]. The dynamics of the heat processes in the stack is somewhat different for the different cooling types. Figure 6 shows the dynamics of heat transfer processes in a stack of tapes containing 20 layers. The temperature change in such a stack is significant differ for the two considered cooling regimes (see Figure 3, on the right). Therefore, the data shown in Figure 4 is provided with different scales, which are convenient for illustrating the processes taking place in the tapes.

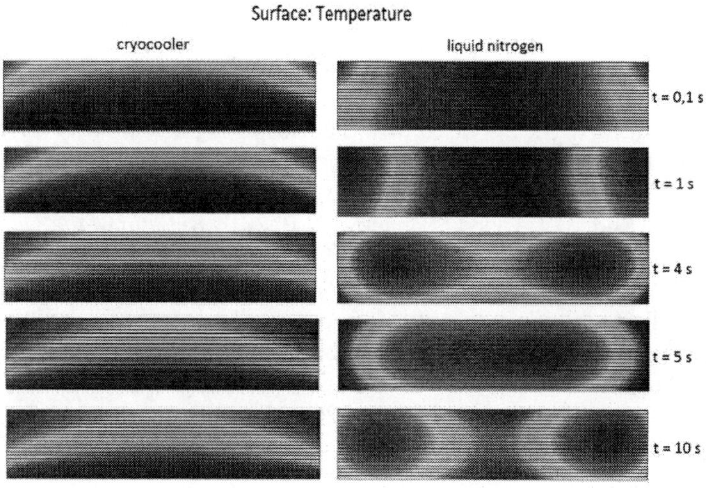

Figure 4. Dynamics of heat transfer in the HTS tapes stacks in the cryocooler cooling mode at the temperature 77.4 K (on the left) and in the liquid nitrogen cooling mode (on the right). Cross-section of the stack of 20 tapes.

At the initial moments of time, the magnetic field reaches a HTS tapes stack and heat generation begins at the edges of the upper tapes. The power of the cryocooler cooling system is sufficient to cool the layers closest to the cooling source. If the stack height is increased, the upper layers will be at higher temperatures, which negatively affects the amount of the trapped flux, hence it would be decreased. It can also be noted that the strongest heating for the cryocooler cooling mode occurs only at the edges of the stack. It can be assumed that for the efficiency of such cooling system it would be useful to cool not only at the lower boundary of the stack, but also along its lateral boundaries.

In the case of liquid nitrogen cooling, the stack of tapes, which is well cooled from all directions, turns out to be the most heated in its center. This turns out to be important for the stacks with only few tapes having a small height. For such stacks, cryocooling cooling mode is most effective, since it can maintain more uniform and constant shape of the temperature distribution throughout the stack during the magnetization. However, for the large stacks of tapes, nitrogen cooling is more promising. This can also be seen from the value of the trapped flux of HTS stacks of tapes (Figure 4)

4. Conclusions

By using the finite element method, the physical model of the magnetic levitation system element based on the second generation HTS tapes stack was developed. The model takes into account the real features of the tape's layered structure. In the frame of our model, the dependences of magnetization and the levitation force on the number of tapes in a stack in a gradient magnetic field of a permanent magnet were calculated. The dependence of the maximum repulsive force and the magnetization in ZFC mode on the thickness of the stack of tapes is non-linear. It also reaches a saturation for stacks of about 60-100 tapes. The comparison of the computational results and the experimental data shows consistent results.

The advantage of developed model is the possibility of changing all input parameters of the system, as well as properties of HTS samples, and also in the model we calculated real movement of the magnet, which makes the model applicable to the calculation and optimization of magnetic and levitation systems of various configurations. This may be useful in the fields of building new transport systems of various scales, electrical rotating machines, such as wind turbines, and other energy generators and energy storage.

Acknowledgments

The study was carried out at the expense of a grant from the Russian Science Foundation (Project No. 17-19-01527).

References

[1] Rodriguez-Zermeno V M, Sørensen M P, Pedersen N F, Mijatovica N, Abrahamsen A B, 2009 Fast 2D Simulation of Superconductors: a Multiscale Approach, *Proceedings of the COMSOL conference* (User presentations and proceedings CD) 1-6

[2] Podlivaev A I , Rudnev I A and Shabanova N P 2014 *Bulletin of the Lebedev Physics Institute* **41(12)** 351–4

[3] Zhang M, Matsuda K and Coombs T A 2012 *Journal of Applied Physics* **112** 043912

[4] [Online]. Available: http://www.cryomech.com/cryorefrigerators/pulse-tube/pt30/

[5] Fetisov S S, Vysotsky V S, Zubko V V 2011 *IEEE Trans. Appl. Supercond.* **21(3)** 2011

[6] Stephan K, Abdelsalam M 1980 *Int. J. Heat Mass Transfer.* **23** 73–87

[7] Pokrovskii S, Osipov M, Abin D, Rudnev I 2106 *IEEE Trans. Appl. Supercond.* **26(3)** 7416628

[8] Page G, Patel A, Baskys A, Hopkins S C, Kalitka V, Molodyk A and Glowackil B A 2014 *Supercond. Sci. Technol.* **28(8)** 2014

Structure and electronic properties of small gold clusters

U.N. Kurelchuk, O.S. Vasilyev, P.V. Borisyuk

National Research Nuclear University MEPhI, Kashirskoye sh., 31, Moskva, 115409

E-mail: unkurelchuk@mephi.ru

Abstract. Structure and electronic properties of small Au nanoclusters study was performed using density functional theory with pseudopotential in relativistic approximation. Density of states of the valence band, projected density of states was calculated, Bader charge density analysis is presented. The spatial behavior of the density of states was studied. Charge-momentum spatial anisotropy observed in symmetrical nanoclusters, electronic states on Fermi level of 1nm clusters founded to be caused mostly by $d_{5/2}$ surface states.

1. Introduction

Porous films and heterogenic structures, formed from metallic nanoparticles, can be considered as new perspective materials for high-effective thermoelectric devices. The efficiency of thermoelectric generation in thermoelectric material is expressed by the dimensionless function figure of merit $ZT=S^2\sigma T/k$. Here S is thermoelectric power (Seebeck coefficient), electrical conductivity σ, and thermal conductivity $k=k_{el}+k_l$ composed of electronic and phononic parts. Till nowadays it has been developed and used several ways to increase ZT of traditional thermoelectric materials (such as such as $CsBi_4Te_6$, Bi_2Te_3, Sb_2Te_3. Layering and alloying, doping with nanoparticles, using nanostructured semiconducters, was aimed at decreasing thermal conductivity k using disordering to diminish phononic transport, and increasing keeping the same value of the thermopower S (which appears quite high in several semiconductors). But using nanoparticles is limited by critically degeneration of electrical conductivity in semiconductor nanoparticles with decreasing their sizes [1-6]. At the same time, it was observed for metal nanoclusters, that they stay conductive down to ~1nm size and manifestation of semiconductor properties at lower sizes [7]. Also from experimental results for films composed from metallic clusters smaller than 1 nm^3 volume, the Seebeck coefficient S was indirectly measured and founded to be about 10 times higher than in corresponding bulk materials [8]. These encouraging phenomena lead us to suggest that one can reach an extremely increasing ZT of thermoelectric material by properly composing nanoclustered species from metallic clusters with tuned electronic properties. The thermoelectric properties generally depend on electronic properties in the complex law, which can be simply approximate by Mott's law for simple metals. It shows how the conductivity σ and thermopower S depends on the electronic density of states (DOS) behaviour near the Fermi energy E_f (the steeper the DOS the higher is S), states occupation at E_f. Band structure features and anisotropy effects on S and σ directly too. In this sense, d-metals with highly localized or correlated d-electrons, relativistic effects, and generally complex electronic structure seems to be good resource to reveal and use the features of their nanoparticles quantum properties, and construct from them the structures with especial electronic properties [9]. As it has been known, the quantum-size effects can reveal extraordinary fine effects of atom electronic structure, and size-geometry-quantum properties is interdependent. For example, it has been known that color of bulk gold caused by relativistic effects [10], spin-orbit interaction make the small $Pt_{n<5}$ nanoclusters flat [11,12]. Thus, modeling the small structures electronic properties one cannot neglect the relativistic effects that become essential. Moreover, an extraordinary electronic phenomena of such structures can lead to unexpected practical usage like spin Seebeck and invers Hall effect [13,14]. In the sense of thermoelectric efficiency, there is evidence that spin Seebeck effect can potentiate an electronic (charge) Seebeck effect [14] in structures combined from particles with special relativistic and magnetic properties. Another whish is the technological clearance in forming nanostructures, i.e. chemical inactivity and nonmagnetic, at least on laboratory conditions research, so the noble d-metals are appropriate to investigation and fine-engineering, particularly gold, because of highest spin-orbit splitting among them (to evidential exploring of the fine structure). Thus, research of dependency of the size, geometry, electronic and spin properties of metals, with initially complex electronic structure, can be essential to predict and tune the properties of nanostructures formed from this clusters.

In this work we study the size, spatial, charge, and electronic structure dependences for small Au nanoclusters.

2. DFT study of nanoclusters electronic properties

We define starting configurations of nanoparticals up to 1.1 nm size in a follow way. Configurations Au_{13} and Au_{55} were defined as icosahedrons, as has been showed by molecular dynamical simulations that structures is the most energetic-preferable and stable form of the d-metallic nanoparticles up to 1-2 nm [15,16]. Density functional theory (DFT) geometric optimization of structures was performed. DFT study of Au nanoclusters was performed in GGA Perdew-Burke-Ernzerhoff approximation of the exchange-correlation energy [17], uncluding spin-orbit interaction for the noncollinear start spin orientation and nonmagnetic initial state in plan wave basis set using periodic DFT code Quantum Espresso [18]. The interaction of valence electrons with the core is described with the full-relativistic ultrasoft Vanderbilt pseudopotential including semicore states in valence with core correction for atomic state Au(5d10 6s1 6p0) [19]. Relaxation was carried out until the forces reached 0.01 eV/A. In the periodic paradigm, single nanoclusters was placed at the center of a cubic pseudocell to provide a vacuum layer of 15-20 A between neighboring clusters. Calculations of wave functions and energy states for relaxed clusters, the total and projected density of electronic states, and Bader analysis of the converged charge density was carried out [20]. Then effective atomic charge, defined as the difference between calculated Bader charge N_e i.e. number of the valence electrons spatially associated with the selected atom, and N_{val} – number of electrons defined as valence in atomic pseudopotential, was calculated.

Icosahedral clusters has a shell spatially structure: Au_{13} consists of a central atom and 1st shell, formed from 12 atoms placed on icosahedral vertexes, in Au_{55} a 2nd shell of 42 atoms is added. In Fig.1 showed the scheme of clusters structure and diagrams of the effective charge distribution. In diagrams the shell effective charge as sum of the atomic effective charges of the shell atoms is plotted on the abscissa, the cluster shell numbers is plotted on the ordinates. It follows from the Bader analysis, that the neighboring shells has opposite charge. In Au_{13} found an even redistribution of the electron density between the center and the shell, with electron "leak" from center to surface, so the central atom in Au_{13} has an effective charge of -0.24 e, and shell +0.24 e. In Au_{55} founded that the charge is redistributed mostly between the surface shells till the central atom has negligible charge, and the extra electronic density about +0.55e founded to be located on surface.

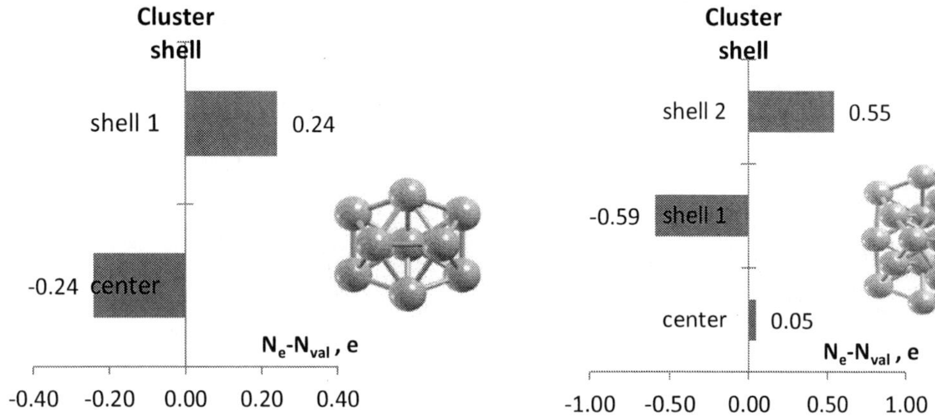

Au_{13} Au_{55}

Fig. 1. Nanoclusters Au_{13} and Au_{55} structure and effective atomic charge distribution. The abscise in diagram shows the "shell effective charge" as sum of the atomic effective charges of the shell atoms. The ordinates shows the cluster shell number. For Au_{13}: center, shell 1 - 12 icosahedral vertexes. For Au_{55}: center, shell 1 - 12 icosahedral vertexes, shell 2 - surface atoms.

Consider the density of electronic states DOS total and projected on d-band states where energy degeneracy of states with j=3/2 and 5/2 is removed by the spin-orbit interactions (See Fig.2). In Fig. 2 showed the spatial behavior of the DOS is significant depended on "shells" of the icosahedral clusters. In 55-clusters DOS on Fermi energy is provided mostly by $d_{5/2}$ states belonging to surface atoms (where is the extra-charge observed, according the Bader analysis). So one can suggest that 55-clusters of about 1 nm size can manifest conductive properties mostly due to $d_{5/2}$ states, which is principally in agreement with the experimental data for clusters 1-1.5 nm [1]. As the size of the clusters increases (comparing the investigated one) the spin-orbit energy splitting $d_{3/2}$-$d_{5/2}$ decreases. In the one given cluster splitting $d_{3/2}$-$d_{5/2}$ increases from center to surface.

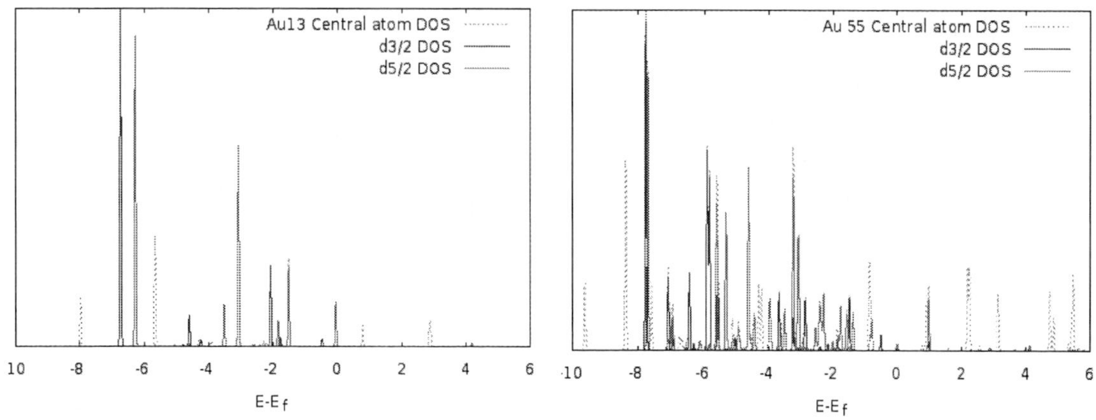

IOP Conf. Series: Journal of Physics: Conf. Series **1238** (2019) 012021 doi:10.1088/1742-6596/1238/1/012021

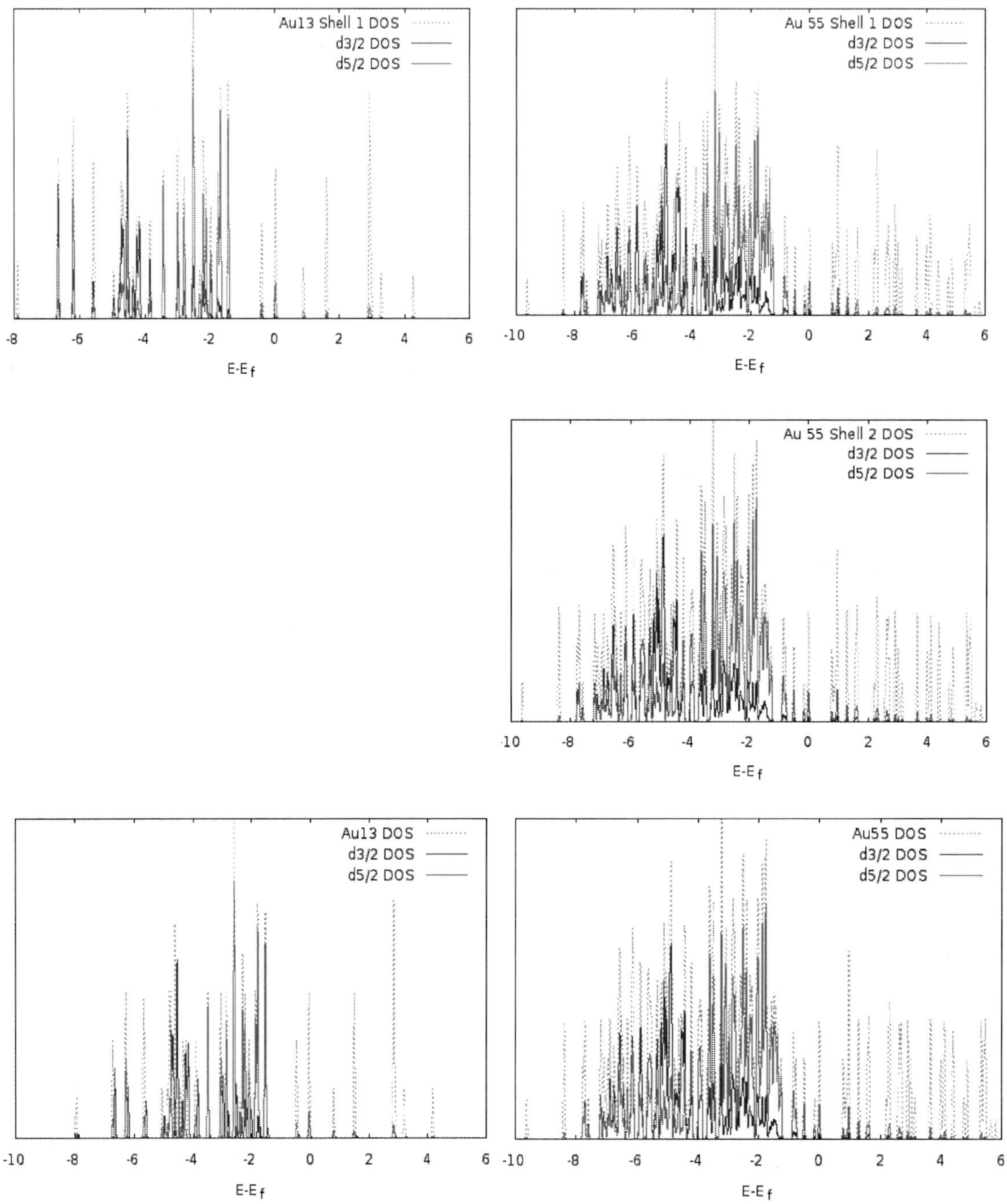

Fig. 2. Density of electronic states DOS projected on (l, j) and spatial atomic shells for Au_{13} and Au_{55} nanoclusters. Au_{13}: center, shell 1 - 12 icosahedral vertexes. Au_{55}: center, shell 1 - 12 icosahedral vertexes, shell 2 - surface atoms. Blue lines shows the DOS projected on $d_{3/2}$ states (l=2, j=1.5), red – $d_{5/2}$ states (l=2, j=2.5)

The structure and electronic properties of small Au nanoclusters was studied using density functional theory with pseudopotential in relativistic approximation. It was found that the spatial behavior of the electronic states density of the icosahedral clusters, is significant dependent on the "shell", in both cases the extra electronic density on surface was founded. Projected density on (l,j) states manifests the momentum anisotropy for the different "shells". 1nm clusters can manifest conductive properties, caused by the $d_{5/2}$ surface states localized near the Fermi energy. Summarizing this dependences, we can suggest that nanoclusters with a large proportion of the surface (in relate to volume) can be interesting for further investigation of their electronic properties thermoelectric properties of the materials formed from them. Although a larger surface-to-volume ratio usually means a smaller nanocluster size, or high roughness, in the first case one can also optimize the size with conductive properties and melting temperature. Knowledge about the evolution of DOS, interconnects the dimensional, geometry and relativistic effects, can provide to find the most effective way to combine small structures into macroscopic structures like porous nanoclustered films with high tunable properties.

Acknowledgements

This work was financial supported by the Russian Federation President Grant to support young scientists (№ 14.Y30.17.2948-MK)

References

1. Hilaal Alama, Seeram Ramakrishna. 2013 Nano energy, V 2, p. 190

2. Chung D.-Y, Hogan T., Brazis P., et al. 2000 Science, V 287, p. 1024

3. Jeffrey Snyder G., Toberner Eric S. 2008 Nature Materials, V.7, p. 105

4. Caillat T., Fleurial J.-P., Borshchevsky A. 1997 Journal of Physics and Chemistry of Solids, V. 58, p. 1119

5. H. Lee, D. Vashaee, D.Z. Wang, M.S. Dresselhaus, Z.F. Ren, G. Chen, Effects of nanoscale porosity on thermoelectric properties of SiGe, J. Appl. Phys. 107 (9) (2010) 1–7

6. Ferry D.K., Goodnick S.M. and Bird J. 2009. Transport in Nanostructures, Cambridge University Press.

7. Borman V.D. et al. 2007 JETP Lett. V.86, I.6, P.393

8. Borisyuk P.V., Krasavin A.V., Troyan V.I., et al. 2015 Appl. Surf. Sci. V. 336. P. 359

9. Pei Y., Wang H., Snyder G.J. Adv.Mater. 2012. V.24. P. 6125

10. Pyykko, P. 1988. Chem. Rev. 88, 563–594

11. Huda M. N., Niranjan Manish K., Sahu B. R., and Kleinman L. 2006. Phys. Rev. A 73, 053201

12. Sebecti A. 2009 Phys. Chem. Chem. Phys., 11, 921-925

13. Zhang S., 2000 Phys. Rev. Lett. 85, 393

14. Salitoh E. 2006 Appl. Phys. Lett. 88, 182509

15. Zhang M. and Fournier R. 2009 Phys.Rev. A, 79, 04320

16. Wenping Zeng, Jian Tang, Pu Wanga, Yong Pei. 2016 RSC Adv., 6, 55867

IOP Conf. Series: Journal of Physics: Conf. Series **1238** (2019) 012021 doi:10.1088/1742-6596/1238/1/012021

17. Perdew J.P., Burke K., and Ernzerhof M. Phys. Rev. Lett. 77, 3865
18. Giannozzi P. et al. 2009 Journal of Physics: Condensed Matter. 21 (39): 395502
19. Dal Corso A. 2014 Computational Material Science. V. 95 P. 337
20. Tang W., Sanville E., Henkelman G. 2009 J. Phys.: Condens. Matter 21, 084204

IOP Conf. Series: Journal of Physics: Conf. Series **1238** (2019) 012022 doi:10.1088/1742-6596/1238/1/012022

Thermal degradation of metamorphic HEMT InAlAs/InGaAs/InAlAs grown on GaAs substrates

A N Vinichenko[1,2], S S Ermakov[1], N I Kargin[1], and I S Vasil'evskii[1]

[1]National Research Nuclear University MEPhI (Moscow Engineering Physics Institute), Kashirskoe shosse 31, 115409 Moscow, Russian Federation
[2]Immanuel Kant Baltic Federal University, A. Nevskogo 14, 236016 Kaliningrad, Russian Federation

Corresponding author's e-mail address: ivasilevskii@mail.ru

Abstract. termal annealing effect on electron transport properties of MHEMT nanostructures is studied for inhert and atmospheric conditions. Iti is revealed that surface modification is mainly responsible for changes in electron properties rather than dislocation sliding and threading into active layers.

1. Introduction

Quantum well heterostructures with high electron mobility (HEMT) is most widely used material for microwave compound semiconductor electronics. Metamorphic technology provides an additional degree of freedom for heterostructure design because of flexibility of the semiconductor compoud alloy composition that determines the lattice parameter. Gradual relaxation of the epitaxially grown metamorphic layers' mechanical stress allow to fit the desirable lattice parameter and thus InAs chemical composition in InAlGaAs heterosystem in metamorphic HEMTs (MHEMTs) [1,2]. The resulting advantages of MHEMT are are increase of quantum well depth, electron velocity and density, accompanied with high enough breakdown electric field and good manufacturability. Drawbacks of metamorphic HEMTs are the presence of ripple heterointerface and surface relief because of misfit dislocation array forming during the metamorphic growth [3,4]. However rather small average roughness (RMS less than 2 nm) can be achieved for that cross hutched relief by appropriate control of substrate temperature and flux ratio during the metamorphic buffer growth [5].

The best performance of the monolithic microwave integrated circuits devices based on MHEMT technology are shown for Ku, Ka and V bands for both low noise [6,7] and power applications [8,9]. Usually the best compromise between the MHEMT operation frequency and power performance leads to the coise of InAs composition ~ 35-40% in the active layers. Typical electron transport properties of MHEMTs are electron concentration in the range (2.5÷3.5) 10^{12} cm^{-2} and electron mobility in the range 8000÷9000 $cm^2/(V{\cdot}s)$ at room temperature for the InAs composition 35-40% [1-7]. It is well established that dislocation scattering can play important role in MHEMTs, especially in the case when threading dislocation of considerable density attains the quantum well where electrical conductivity occures [10,11]. As the transport properties are highly sensitive to formation conditions during Molecular beam epitaxy, the comparable degradation due thermal stress in HEMT, especially power, can take place during they duti cycle. The accelerated thermal test procedures mainly involves the statistical couts of device failure [12], where the criteria of functionality fault is clear. In case of material properties degradation it isn't threshold character of degradation, so continuous dependencies

IOP Conf. Series: Journal of Physics: Conf. Series **1238** (2019) 012022 doi:10.1088/1742-6596/1238/1/012022

of electron transport properties on the annealing temperature and time, as well as the certain environmental conditions of annealing are important. The only known report [13] unfortunately takes into account only long-therm storage degradation of MHEMT at room temperature.

For the HEMT applications the thermal stability and performance degradation upon the thermal stress are of major importance, because they can affect the mean time to failure. Despite that wide metodology elaborated and used for the serial MMIC products, the issue of HEMT material degradation effect on MHEMT performance is still open. For the physical reasons it is very important to investigate the temperature treatment effects on the fundamental electron transport properties changes.

In this paper we have investigated the MHEMT quantum well structure electron properties changes due to thermal annealing in the different regimes.

2. Sample growth and preparation

Metamorphic $In_{0.2}Ga_{0.8}As$ quantum well structure with linear $In_{0.15}Al_{0.85}As \rightarrow In_{0.40}Al_{0.60}As$ metamorphic buffer with one side delta-Si doping was grown by molecular-beam epitaxy on semi-insulating (001) GaAs substrates without misorientation of (001) plane. Layer structure of the sample shown in Figure 1. Quantum well L_{QW} has 14.5 nm width. The sample has delta doping by Si from the upper side through 5.3 nm spacer with concentration of $3.4 \cdot 10^{12}$ cm^{-2}. Growth temperature was 420 °C during metamorphic buffer growth, after that high temperature smoothing layer was grown at the temperature of 460 °C, accordingly to our previous study [14]. Inverse step introduced without growth interruption for the cancellation of residual strain at the top of metamorphic buffer.

For the electron transport measurements samples were processed by photolithography and mesa etching followed by the AuGe/Ni contact metallization and without an annealing. Mesa structure has van-der Pauw geometry for the simple material characterization, providing both sheet resistance and electron Hall concentration and mobility extraction without interference of specific contact resistance. I-V characteristic of all contacts showed linear performance within the range up to 20 mA. Cental mesa has width of 1 mm. The sample view is presented in Figure 2.

i- $In_{0.37}$ $Ga_{0.63}As$ (cap layer)	8 nm
i- $In_{0.37}Al_{0.63}As$ (barrier)	11 nm
δ-Si dopant	
i- $In_{0.37}Al_{0.63}As$ (spacer)	5.3 nm
$In_{0.37}$ $Ga_{0.63}As$ (quantum well)	14.5 nm
$In_{0.37}Al_{0.63}As$ (smoothing layer)	180 nm
$In_{0.40}Al_{0.60}As \rightarrow In_{0.37}Al_{0.63}As$ (inverse step)	48 nm
$In_{0.15}Al_{0.85}As \rightarrow In_{0.40}Al_{0.60}As$ (metamorphic buffer with linear compositional profile)	1180 nm
5 x {$Al_{0.2}Ga_{0.8}As$/GaAs} (Superlattice)	{5 nm/ 3 nm}
GaAs (buffer)	34 nm
GaAs (100) undoped substrate	

Figure 1. Metamorphic HEMT sample layer structure design (InAs content 37% in active layers).

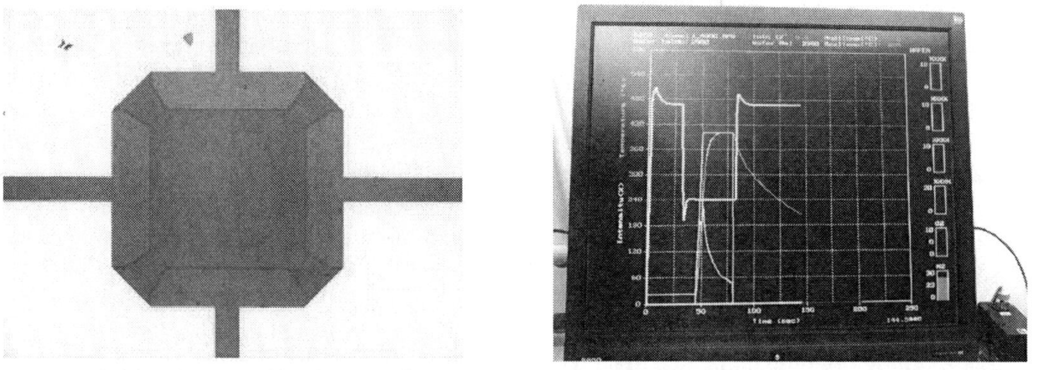

Figure 2. Fabricated mesa chip for Van der Pauw measurements. Central mesa area is 1x1 mm^2 (left). Termal processing temporal profiles example (right panel).

3. Experimental

For the temperature processing several techniques have been involved. Temporal degradation for the the moderate heating have been studied with in-situ data acquisition in atmospheric environment. Sample with the 4-contact probes on the Teflone fixture was placed on the ceramic heater with temperature stabilisation by loop controller within the accuracy of ± 1 K. Magnetic field for the Hall effect measurement have been applied by automatically retracted permanent magnet. Measurement system with the commutation for the current reversal allows to exclude a thermopower voltage due to the existing temperature gradients. This setup allows to measure long therm variation of electron transport properties within the temperature range of 300 K - 550 K and durations up to 10^2 hours.

For the heating in the inhert gase standart Modular RTP system have been adopted with permanent 4N Nitrogen blowing in the chamber during the thermal processing within the temperature range of T_{ann}=300 K - 550 K. For the latter techique the electron transport parameters have been measured after the certain time of annealing. For the comparison, temperature dependencies of electron concentration and mobility have been investigated in both thermal treated as-grown samples in the lattice temperature range 77 K - 300 K.

Several identical sample chips have been processed by RTP during 8 minute treatment in the temperature range of T_{ann}=300°C÷450°C and measured after that.

4. Results and discussion

Long term medium temperature annealling at 250°C at ambient atmosphere gives slow resistanse rize, accompanied with both mobilily and concentration decrease. There is no pronounced square root law dependence, typical for diffusion processes. Sample sheet resistance change shown in Figure 3 (left). Short term high temperature annealing in contrary gives significant electron concentration decrease and respective resistance increase. Overall resistance change not exceed 0.7% within 24 hour experiment at 250°C. Providing that our mesa structures haven't additional passivation by SiO_2 or Si_3N_4 this result shows good stability for ambient condition accelerated oxidation. For the so small variation of resistance we can't also detect any changes in electron concentration and mobility by Hall effect.

High temperature annealing showed the changes in both electron concentration and resistance. Electron concentration drop follow square-root like character with time of annealing as shown in Figure 3 (right). Surprizingly, there weren't any detectable degradation of electron mobility. Electron transport properties were ex-sity measured at the different temperatures in the range 77 K to 300 K. All samples demonstrate the metallic behavior of the sheet resistance. Electron mobility rises while the temperature is decreased. Electron concentration measured by Hall effect, showed different behaviour - in the native sample it slightly decreased with temperature lowering, in the annealed samples it becomes nonmonotonic. Substantial lowering of electron concentration was registered. Hall concentration is presented in Figure 4 (left) and Hall mobility (right). Thus electron concentration variation showed decrease by 10% for T_{ann}=400°C.

Accordingly to model [11], electron scattering by dislocation gives mobility inversely proportional to dislocation density. If dislocations is reason for electron capture, the degradation would accompanied by commensurate drop in electron mobility, but it is not the case. Another responsible mechanizms of electron concentration depletion in quantum well is change in doiant activation (self-compensation) or changes in spatial potential boundary conditions. Evolution of threading dislocation not affects narrow band channel but could give rize to the conduction band upward shift at the buffer resulting in quantum well depletion. Another possibility is change of structure surface potential due to arsenic loss in atomic scale, but it is of minor probability due to high temperature stability of InGaAs compounds.

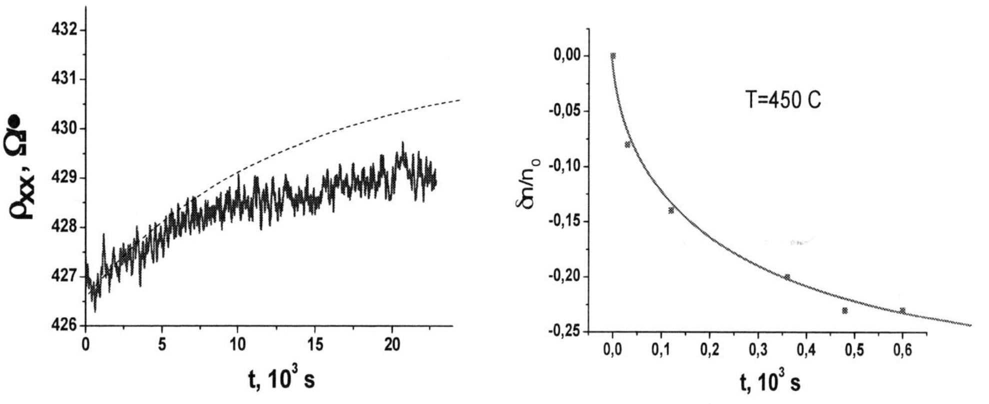

Figure 3. Temporal in-situ measurement of electron transport properties upon annealing temperature resistivity at T_{ann}=250 °C (left) and concentration at T_{ann}=450 °C (right).

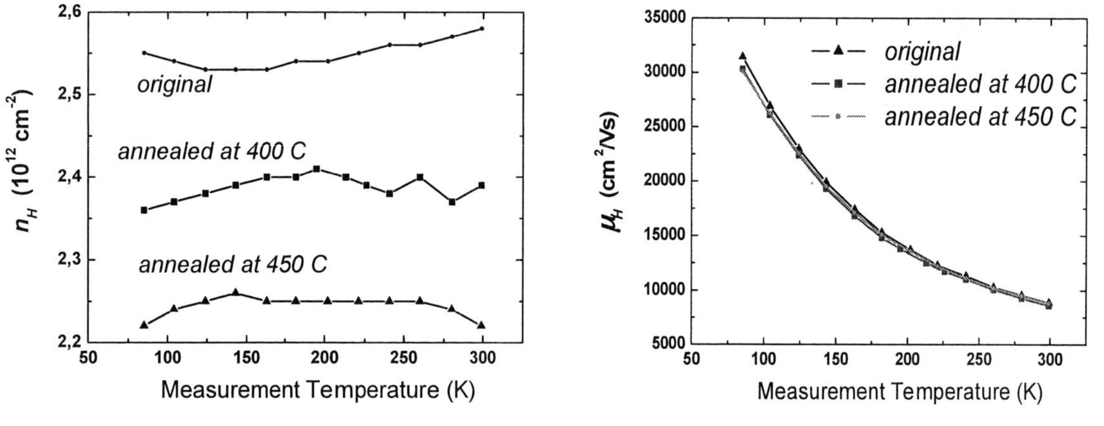

Figure 4. Temperature dependence of electron Hall concentration (left) and mobility (right) of metamorphic HEMT before and after rapid thermal annealing at T_{ann}=400 °C and 450 °C.

5. Conclusion

The effect of low temperature long term and rapid high temperature annealing on the electron transport properties of metamorphic high electron mobility quantum well with 37% InAs content is presented. The structure is stable in the long term treatment at relatively small temperatures up to 250°C. The short time annealing upon Nitrogen environment at temperatures 400°C and 450 °C gives rize to specific resistance of MHEMT, mainly associated with electron concentration drop. Temporal behavior of concentration capture follow diffusion-like dependence. However, there is not accompanied drop in electron mobility, so the effect isn't connected to dislocation threading into channel. Possible mechanism can be conduction band potential pinning shift at the metamorphic buffer side. The results will be useful for conditions estimation for MHEMT transistors fabrication, especially for rapid thermal annealing of ohmic contacts.

This work was supported by the Competitiveness Program of National Research Nuclear University "MEPhI".

References

[1] S. Bollaert, Y. Cordier, M. Zaknoune, H. Happy, V. Hoel,S. Lepilliet, D. Theron, A. Cappy. Sol. St. Electron., 44, 1021 (2000)

[2] Yvon Cordier, Sylvain Bollaert, Mohammed Zaknoune, Jean Dipersio, Denise Ferre. Jpn. J. Appl. Phys., 38, pt 1, N 2B, 1164 (1999)

[3] O. Yastrubchak, T. Wosinski, T. Figielski, E. Lusakowska, B. Peez, A.L. Toth. Physica E, 17, 561 (2001)

[4] Jia-Lin Shieh, Mao-Nan Chang, Yung-Shih Cheng, and Jen-Inn Chyi J. Appl. Phys. 82, 210 (1997)

[5] Y. Cordier, P. Lorenzini, J.-V. Chauveau, D. Ferre, Y. Androussi, J. Di Persio, D. Vignaud, J.-L. Codron. J. Cryst. Growth, 251, 822 (2003)

[6] A. Leuther, A. Tessmann, M. Dammann, C. Schwörer, M. Schlechtweg, M. Mikulla, R. Lösch, G. Weimann, Proc. IPRM 2007, pp. 24-27, 2007.

[7] A.Tessmann, IEEE J. Solid-State Circuits, vol. 40, no.10, pp. 2070-2076, 2005

[8] W.E. Hoke, T.D. Kennedy, A. Torabi et al. Journal of Crystal Growth 251(1-4):804-810

[9] Shu-Jenn Yu, Wei-Chou Hsu et al. Solid-State Electronics 50 (2006) 291–296

[10] D. Lubyshev, W. K. Liu, T. R. Stewart, A. B. Cornfeld et al. J. Vac. Sci. Technol. B 19 4, 1510 (2001)

[11] U Penner, H Rücker and I N Yassievich. Semicond. Sci. Technol., 13, 709 (1998)

[12] Temperature Acceleration Models in Reliability Predictions: Justification & Improvements Franck Bayle, 2010 Reliability and Maintainability Symposium, San Jose, CA, USA, January 25-28, 2010.

[13] E. V. Nikitina A. A. Lazarenko E. V. Pirogov M. S. Sobolev T. N. Berezovskaya Technical Physics Letters September 2017, Volume 43, Issue 9, pp 863–865 |

[14] G. B. Galiev, I. S. Vasil'evskii, E. A. Klimov, S.S. Pushkarev, et al. Journal of Crystal Growth, Vol. 392, p. 11-19 (2014)

IOP Conf. Series: Journal of Physics: Conf. Series **1238** (2019) 012023 doi:10.1088/1742-6596/1238/1/012023

Thermoelectric properties of the metallic nanoclustered model material.

U.N. Kurelchuk, O.S. Vasilyev, P.V. Borisyuk

National Research Nuclear University MEPhI, Kashirskoye sh., 31, Moscow, 115409

E-mail: unkurelchuk@mephi.ru

Abstract. In this work the Seebeck coefficient of simple model of nanoclustered material with respect to bulk form was investigated. The first-principle calculation of Seebeck coefficient of the bulk and model nanoclustered form of Ta was performed via semiclassical Boltzmann transport approximation with constant relaxation time, using first-principal bandstructure calculated in density functional theory. 5-15-times increasing of Seebeck coefficient was estimated for simple model of nanoclustered material.

1. Introduction

Nowadays challenges of modern energetics demands the high effective thermoelectric materials. The efficiency of thermoelectric generation in thermoelectric material is expressed by the dimensionless value figure of merit $ZT=S^2\sigma T/k$. Here S is the thermoelectric power (Seebeck coefficient), σ is the electrical conductivity, and $k = k_{el} + k_l$ is the thermal conductivity, composed of electronic and phononic parts. One can see from ZT expression, that it can be increased mostly by increasing S, then σ, and decreasing κ. Improving the ZT of traditional high-S thermoelectric materials (such as $CsBi_4Te_6$, Bi_2Te_3, Sb_2Te_3) by doping, alloying, layer composing have been successful till recent time, but effectiveness of this way is limited [1-4]. Another way is to decrease thermal conductivity k by the disordering or clusterisation: it is expected that phoninic transport will be restricted in structure made from disordered localized nanoparticles but electronic conductivity will be provided by percolation effects [1]. But for semiconductors further increasing ZT by diminish the particle size is limited due to the fact of critical degradation of electric conductivity of semiconductor nanoparticles with decreasing their size [5,6]. On the other hand, it was observed for metal nanoclusters, that it stay conductive down to ~1nm size and manifestation of semiconductor properties at lower sizes [7]. Moreover in experiment [8] it was estimated that S of d-metallic nanoclusters with volume decreasing lower than $1nm^3$, appears 10 times higher than S of corresponding bulk sample. Summarizing these facts we can presume, that disordered, porous structures composed of metallic nanoclusters, can be promising materials for high-efficiency thermoelectric elements. The thermoelectric properties generally depend on electronic properties in the complex law, which can be simply approximate by Mott's law for simple metals. It shows how the conductivity σ and thermopower S depends on the electronic density of states (DOS) features near the Fermi energy E_f (the steeper the DOS the higher is S), occupation at E_f, band structure features and anisotropy effects on S and σ directly too. So, information about electronic structure helps to predict the thermoelectric properties. On the other hand we can construct the structures of nanoclusters with especial electronic propertie, because its size, geometry and electronic structure are interdependent.

In this work we investigate how the "clusterization" or slight disordering effects on Seebeck coefficient with respect to bulk form. We consider the simplest model of metallic nanoclustered material: simple cubic lattice formed by small nanoclusters. For the reasons mentioned above, d-metals have proper electronic structure, with highly localized d band, spin-orbit splitting, and mostly high conductivity. But for the complex electronic structure of the d-metals the simple analytical electronic-thermoelectric approximations is inappropriate. For first-principle thermoelectric coefficients calculation we use the semiclassical upproach Boltzmans equations in constant relaxation time approximation with tensor coefficients derived directly from electronic band structure E_i ($\mathbf{k}_j$) (bolzttrap algorithm) [9]. Bands data is obtained via density functional theory calculations on large set of **k**. Within this method obtain Seebeck coefficient S(T, μ), conductivity σ/τ(T, μ) where μ is the chemical potential, τ is the uncertain constant relaxation time. As soon as Seebeck coefficient is independent of τ value and $ZT \sim S^2$, we consider temperature dependence S(T, $\mu \sim E_{fermi}$) at several temperatures. Generally, nanoclusters melting temperature is lower than bulk [10,11], so we consider refractory d-metal Ta to safely consider temperatures up to 500 K, that corresponds to the range of thermoelectric elements usage. The calculated S(T) for bulk and nanoclustered model of Ta is compared.

2. Calculation methodic and results

Study of the electronic structure of bulk Ta was provided using density functional theory (DFT) method with pseudopotential in plane-wave basis set implemented in Quantum Espresso code [12]. DFT calculations was performed in generally-gradient approximation of the exchange-correlation energy (GGA PBE) including the spin-orbit interaction for the noncollinear spin orientation and nonmagnetic state. For the valence configuration of Ta (5s2 5p6 6s2 5d3) fully-relativistic ultrasoft Vanderbild pseudopotential with core correction uncluding semicore states in valence was used [13]. Self-consistent energy calculation was performed on 12x12x12 k-mesh with 750 eV plan wave kinetic energy cutoff. Electronic energy bands and density of states (DOS) calculations was performed on dense 48x48x48 Monkhorst-Pack k-mesh.

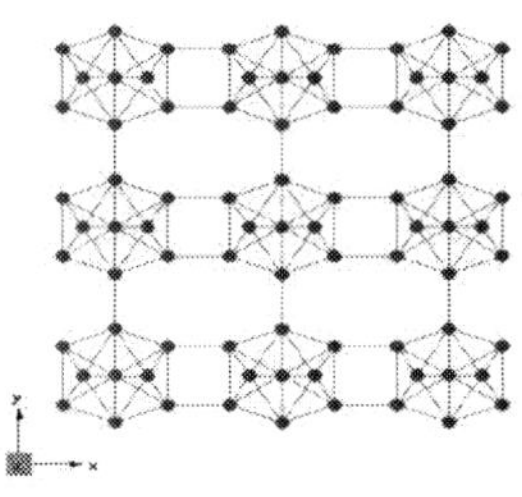

Fig. 1. Lattice of nanoclusters Ta_{13} model scheme.

It was investigated by molecular dynamic simulations that icosahedral Ta_{13} structure is the most energetic-preferable and stable form of the <1nm nanoparticles [14]. Let's consider the simple cubic superlattice composed of small ~0.5nm Ta_{13} nanoclusters (see Fig.1). Structure relaxation was performed with embedded atom model potential [15] and force threshold of 0.05 eV/A. Superlattice parameter or intercluster distance was defined by this procedure in order to save ability of appearance new bends, but to avoid total disordering by interatomic forces and save icosahedral units. It was provided purposely to investigate structure with controlled (parametrized) disordering, as first step of disordering simulations. Thus the global symmetry became decreased only two times with respect to the ordinary bulk lattice (24vs48 translation operators). Then DFT self-consistent and band structure calculation was performed using the former parameters.

The semiclassical Seebeck coefficient S(T, μ=Ef) was obtained from DFT bandstructure $E_i(\mathbf{k}_j)$ for T = 100-500K for bulk and model nanoclustered Ta. S(T) results compared with experimental is shown in Fig.1.

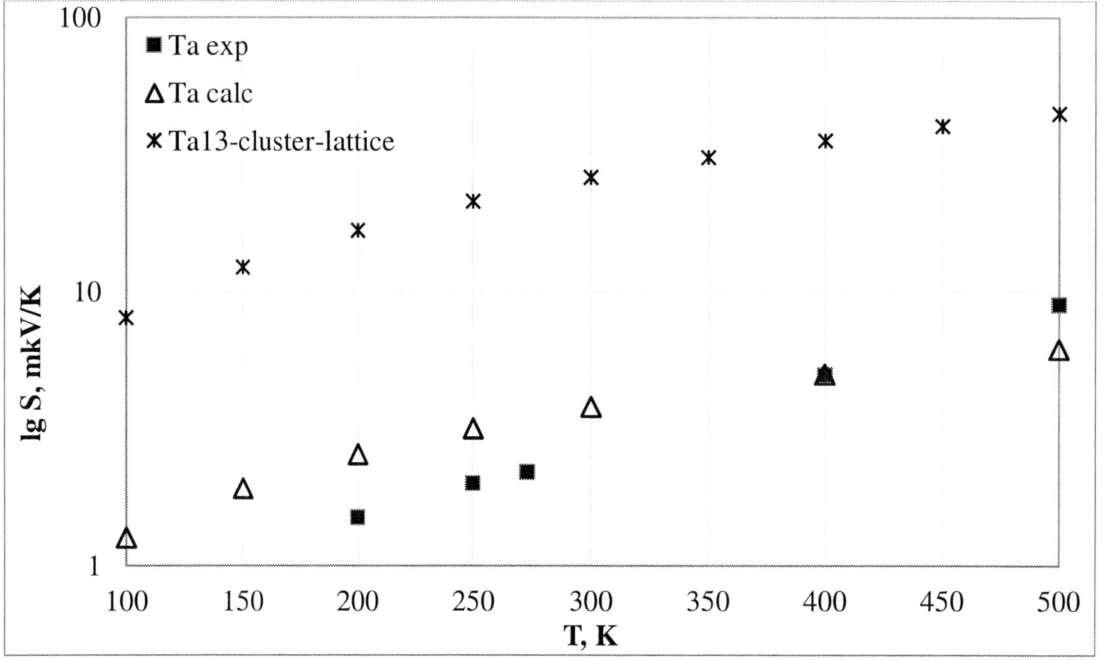

Fig. 2. Seebeck coefficient logarithmic dependence of temperature for Ta structures. Black squares marks experimental values of Seebeck coefficient S, white triangles – S calculated for bulk Ta. Stars marks calculated S for model lattice formed from Ta_{13} nanoclusters.

Fig.2. showed results of the S(T) calculation for bulk and model nanoclustered Ta material. The bulk S calculations show the best accuracy in T=250-400K range, the minimal declination with experimental one is 1% at 400K. One can see that in simple model of nanostructured material constructed by 0.5 nm Ta13 clusters, S value can grow by 5÷15 times with respect to bulk, at 300 K it appears 26 mkV/K. In the resent experimental research [16] the thermopower of the film, formed from 1.3 nm Ta clusters, was indirectly measured and it found to be about 15 mkV/K at room temperature. That proves the results of current study.

Thus ab initio calculations of the thermopower of nanoclustered metal, presented by simple model of sulerlattice formed from small nanoclusters Ta_{13}, showed that Seebeck coefficient appear to be 5~15 times higher than bulk metal. We suggest that more disordered structure, formed from nanoparticles of larger and different sizes, can be more thermoelectrical effective, having higher S and lower k for the reasons recently studied.

Acknowledgements

This work was financial supported by the Russian Federation President Grant to support young scientists (№ 14.Y30.17.2948-MK)

References

1. Hilaal Alama, Seeram Ramakrishna. 2013 Nano energy, V 2, p. 190

2. Chung D.-Y, Hogan T., Brazis P., et al. 2000 Science, V 287, p. 1024

3. Jeffrey Snyder G., Toberner Eric S. 2008 Nature Materials, V.7, p. 105

4. Caillat T., Fleurial J.-P., Borshchevsky A. 1997 Journal of Physics and Chemistry of Solids, V. 58, p. 1119

5. H. Lee, D. Vashaee, D.Z. Wang, M.S. Dresselhaus, Z.F. Ren, G. Chen, Effects of nanoscale porosity on thermoelectric properties of SiGe, J. Appl. Phys. 107 (9) (2010) 1–7

6. Ferry D.K., Goodnick S.M. and Bird J. 2009. Transport in Nanostructures, Cambridge University Press.

7. Borman V.D. et al. 2007 JETP Lett. V.86, I.6, P.393

8. Borisyuk P.V., Krasavin A.V., Troyan V.I., et al. 2015 Appl. Surf. Sci. V. 336. P. 359

9. Madsen G. K.H., Singh D. J. 2009 Computer Physics Communications. V.175. I.1. P. 67

10. Borisyuk P.V. et al. 2010 Measurement Techniques. V. 53, I. 2, pp 128–133][Gao F., Gu Z. (2015)

11. Melting Temperature of Metallic Nanoparticles. In: Aliofkhazraei M. (eds) Handbook of Nanoparticles. Springer, Cham

12. Giannozzi, P. et al. 2009 Journal of Physics: Condensed Matter. 21 (39): 395502

13. A. Dal Corso. 2014 Computational Material Science. V. 95 P. 337

14. Min Zhang and René Fournier. 2009 Phys.Rev. A, 79, 043203

15. NIST. Interatomic Potentials Repository Project. https:www.ctcms.nist.gov/potentials

16. Borisyuk P.V. et al. 2018 Surface & Coatings Technology 343, 69–74

ANALYSIS OF INFLUENCE OF SEISMIC OSCILLATIONS ON PRECISION LASER PLANTS' WORK

K Gubskiy[1], S Sorokin[1], M Reshetova[2] and V Reshetov[3]

[1] National Research Nuclear University MEPhI, Russia
[2] Lyceum 1523, Moscow, Russia
[3] FSBI TISNCM, Moscow, Russia

Abstract. The seismic situation in a number of rooms of the NRNU MEPhI, including the Laboratory Building, where the construction of the multifunctional laser complex ELF is planned, is studied. Estimations of seismic and vibrational noise at the main and isolated parts of the basement of the room that hosts the laser complex are obtained.

Introduction

In the modern world, which is saturated with technological equipment, we are constantly confronted with an increased level of vibration effects on the measuring systems. Examples of the spectral density of typical seismic noise in the territory of a large city (Warsaw) and in a deserted area (Hanover region) are shown in Figure 1. The black dotted lines in these graphs represent the averaged values of the minimum and maximum seismic noise typical for our planet [1, 2].

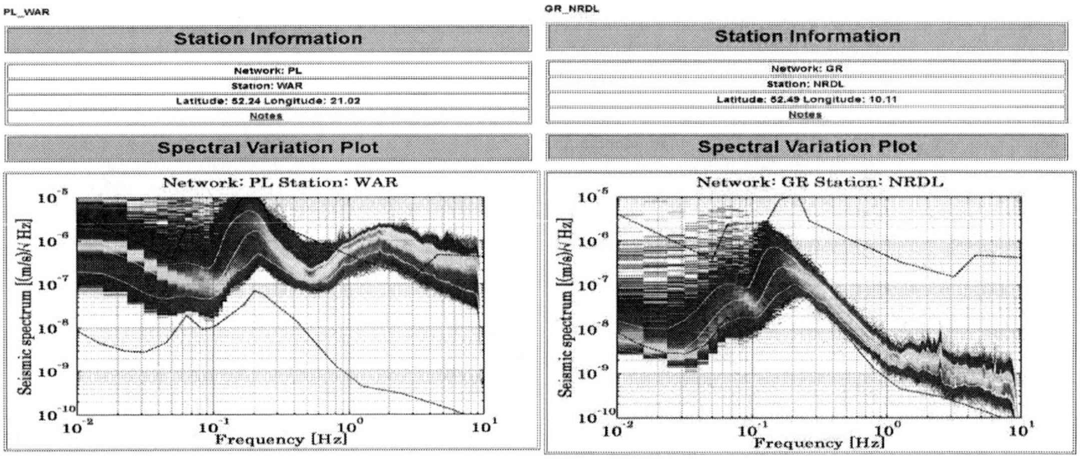

Figure 1. Seismic noise in Warsaw and near Hanover [1,2]

The integrated level of vibrational noise in a seismically and technologically quiet region of the Earth can be roughly estimated as 1 mkm/s. Moreover, because of the dependence of the spectral density of the background noise on the frequency, the integral estimation for displacement and acceleration will also be 1 mkm and 1 mkm/c^2. Due to the high variability of seismic noise, specific measurements should be made exactly in the place where the multifunctional laser complex is planned to be located.

The study of the level of vibration interference

The research presented in this paper were carried out using a digital three-component seismic recorder ZET 048 with a working frequency band (0.3-400) Hz and a sensitivity of 0.5 V/(m/s^2) [3]. Daily diurnal variations of the levels of vibrational acceleration in three buildings of the National Research Nuclear University MEPhI were measured. The main long-term measurements were taken in the Building “K” on the 10th floor in the K1010 audience. Short-term observations were carried out in the basements of the building K and the Main Building. In the Laboratory building, in addition to vibration control in the ELF placement hall, the mapping of the noise level over the working area was carried out and the degree of mutual isolation of the independent part of the foundation, intended for placement of the main ELF elements, and the rest of the working surface of the room was estimated.

The most complete picture of seismic noise was obtained for the 10th floor of the Building K. Measurements were carried out in different weather, day and night, during classes and on weekends. Examples of seismograms and spectra of the signals registered in the Building K are shown in Fig. 2. A characteristic feature of the recorded signals is the presence of low-frequency picks in the region of 1 Hz, related to the fact that high-rise buildings are swinging at their resonant frequencies under the influence of wind and seismic loads. The second feature of the noise spectrum is related to the operation of various types of electric motors, the rotational speed, which are subharmonics of the power frequency. The wide peak in the region (60-120) Hz is due to traffic on the Kashirskoye highway. The characteristic fractional frequency components caused by the movement of the railway transport and the subway were not found.

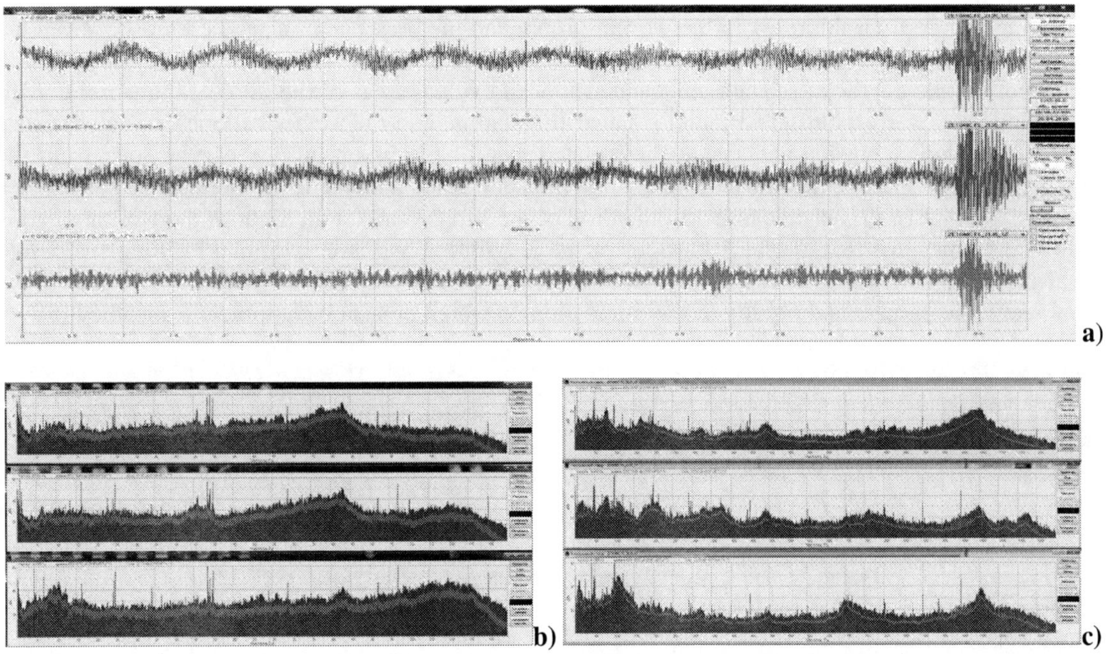

Figure 2. Seismograms in windy weather (a), and variants of seismic noise spectra in three axes in windless weather in the K1010 case (b, c), blue denotes instantaneous noise spectrum, green averaged over the last 10 samples

The amplitude of low-frequency oscillations of the case K at frequencies of 0.98 Hz and 1.05 Hz with the force of the wind in gusts up to 20 m/s was (2-4) mm/s^2, which, in terms of the displacement, gives 0.1 mm. This value is three orders of magnitude higher than the typical value of the seismic noise in this frequency band on the ground in the MEPhI region. This circumstance in a compartment with a huge number of narrow-band vibrations caused by technological equipment (each with an individual level of ~ 1 mm /s^2) excludes the possibility of placing large laser systems on the upper floors of multi-storey buildings. The root-mean-square level of the seismic noise in the frame K did not exceed 2 mm/s^2. In this case, the peak values of vibrations associated with the movement of people, the fall of chairs and the slamming of the door were 2 orders of magnitude higher than the average level and reached 0.5 m/s^2. Such a level of seismic noise is unacceptable for the normal functioning of precision laser installations.

Measurements of the background level of the seismic noise in the Laboratory building demonstrated the absence of low-frequency components in the noise spectrum, the weak effect of vehicles and (20-40) dB a lower level of noise power spectral density in Fig. 4a. An important feature of the noise spectrum was the absence of powerful discrete components in the spectrum. Such behavior for the background level of seismic activity in the absence of industrial interference [4, 5].

To study the degree of vibration isolation of the foundation ELF from the bearing structures of the building, a pulsed broadband source of seismic noise was used. It was a solid rubber ball weighing 0.5 kg. When it fell to the floor from a height of 2 m and a jump after an impact of 1 m into the energy of the vibrations of the building's structures, the kinetic energy of the rubber ball passed to 5J. This energy was sufficient to excite the oscillations of the concrete overlap in the housing K with a peak acceleration value of ~ 5m/s^2, while in the basement location of the Main Building the vibrational acceleration caused by the drop of the rubber ball was of the same order of ~ 5 m/s^2, which indicates the presence of free space under the slab forming the floor of the basement floor. In the Laboratory building, the impact acceleration dropped to 0.3 m/s2, both on a general and isolated foundation, indicating a large floor thickness in the working room and the presence of a sand cushion beneath the entire floor surface.

To control the degree of seismic independence of individual areas of the floor of the working hall, the seismic receiver was installed on an isolated foundation, and the rubber balloon fell onto various parts of the floor - outside and inside the isolated part of the basement. As can be seen from the presented data in Fig.3 b. the impact levels of the impact outside and inside the isolated part differ by a factor of 10, that is, the seismic effect of the drop of the rubber ball on the floor in the working hall is a thousand times less than on the 10th floor of the K building and is only 5mm/s^2 while it is tenfold greater than the level of background noise in the Laboratory building. These data indicate the significant role of mass and rigidity of the foundation on which precision equipment is installed [6].

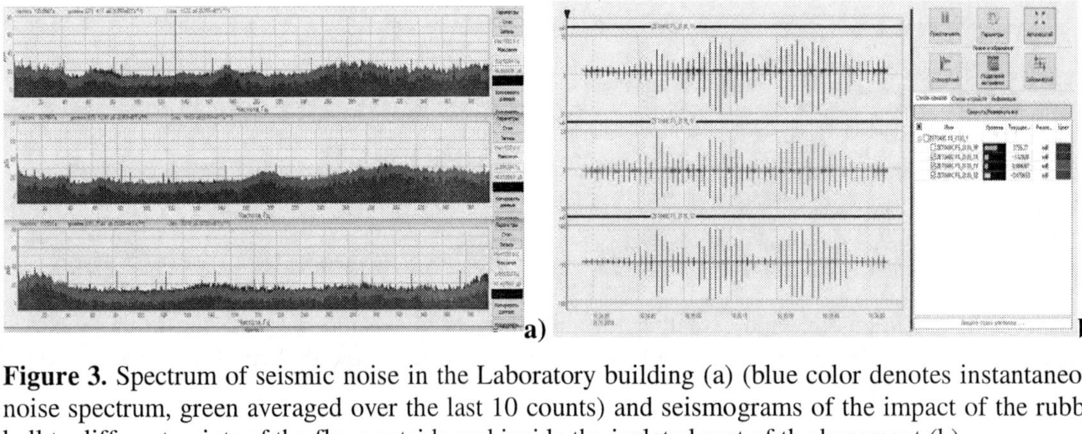

Figure 3. Spectrum of seismic noise in the Laboratory building (a) (blue color denotes instantaneous noise spectrum, green averaged over the last 10 counts) and seismograms of the impact of the rubber ball to different points of the floor outside and inside the isolated part of the basement (b)

Conclusion

The measured values of the seismic noise in the Laboratory building indicate the optimal choice of this particular location on the territory of the NNIU MEPhI for the construction of the multi-functional laser complex ELF. The conducted research shows that there is no significant influence of the subway and railway transport on the level of seismic noise and the dominant influence of seismic disturbances caused by vital activity inside the Laboratory Building. Based on the obtained spectral data and design documentation at the ELF, the vibration behavior of the laser complex will be simulated in the future and the question of the necessity and nature of the vibration isolation equipment included in ELF will be solved. We consider it advisable to introduce a seismic monitoring system into the ELF, excluding the carrying out of laser experiments during the increased seismic and vibrational activity on the territory of MEPhI and in the premises of the Laboratory building.

Acknowledgments

This work was supported by the Ministry of Education and Science of the Russian Federation (project ID RFMEFI59317X0007; the agreement No. 14.593.21.0007); the work was done using the Shared research facilities “Research of Nanostructured, Carbon and Superhard Materials” FSBI TISNCM.

References

[1] https://www.google.com/earth/download-earth.html
[2] http://www.ligo.caltech.edu/~jharms/data/GoogleEarth/
https://labcit.ligo.caltech.edu/~jharms/data/GoogleEarth/Freq_Europe.kmz
[3] https://zetlab.com/en/shop/seismic-stations/zet-7152-n-ver-3-seimograph/
[4] https://en.wikipedia.org/wiki/Seismology
[5] https://en.wikipedia.org/wiki/Earthquake
[6] V Gusev Vibraciya oborudovaniya inzhenernyh sistem i sposoby zashchity ot nee, *J. AVOC* [VHC: Ventilation, heating, conditioning], 2010, no. 5.

Two-channel interferometer system of direct optical heterodining for destruction by shock waves processes studying

M A Gorbashova, K L Gubskiy, R G Glukhov, D V Dyagileva, A P Kuznetsov

National Research Nuclear University MEPhI (Moscow Engineering Physics Institute), Moscow, Russia

E-mail: gorbashovama@gmail.com

Abstract. The article describes a multichannel modular system for velocity measuremens in shock physics. The two-channel time compression system is based on PDV method. It was tested in lead dusting experiments.

1. Introduction

Research of shock wave processes under strong pulse loadings of targets is one of the actual problems of high-energy-density physics [1]. Particle velocity is one of the few parameters available for direct measurement. Due to the extremes of the experimental conditions, contactless methods of its diagnostics are preferable. The only contactless and time continuous method of velocity measurement is laser interferometry.

One of the most common and actively developing systems for measuring particle velocity is VISAR (Velocity Interferometer System for Any Reflector). Developed in the early 70s, VISAR system undergone many modifications that significantly increased its accuracy and reliability. VISAR technique allow to measure velocities of various studied objects surfaces in shock-wave experiments in a large range of speeds (50-10000 m/s) [2]. However, such interferometers also have certain disadvantages. This technique does not allow to measure the speed of several objects localized at one point. In addition, this type of system has a time resolution limit, which restrict the range of possible measurements.

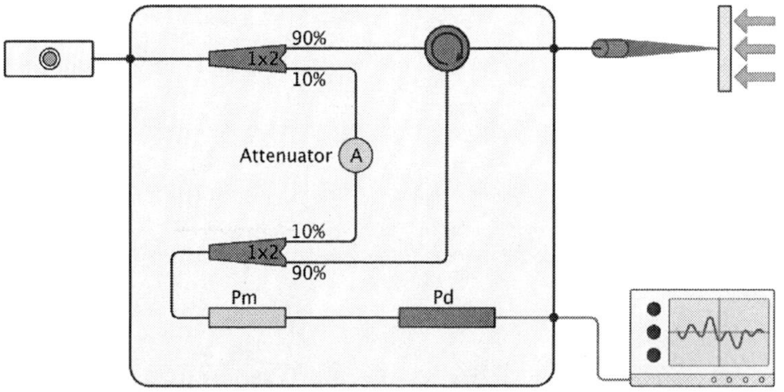

Figure 1. Scheme of the basic diagnostic system using the PDV technique.

In order to research fragmentation and dusting of objects (i.e. registration of several objects simultaneously) PIV (Particle Image Velocimetry) technique is used. However, PIV allows to measure only the average velocities of the studied particles. Continuous velocity measurements can be conducted using PDV (Photonic Doppler Velocimetry) methods. Scheme of the basic diagnostic system using the PDV technique is shown in Fig. 1. Similar measurement system was used in research of the processes of chondritic targets destruction under high-power laser radiation on SATURN facility [3].

The principle of operation of this system is following: beam of a single-frequency continuous laser is transported to the target via a fiber optic cable. After passing through the circulator, the beam is divided into two parts. First beam hits the target, reflects from its moving surface and returns into the GRIN lens. After passing through the circulator, the light reflected from the object is added to the second beam that has passed through the attenuator. As a result of mixing of these two signals, shifted in frequency due to the Doppler effect, the photodetector generates beats, which subsequently registered by an oscilloscope. The speed is obtained by recalculating using the following equation:

$$V = {}^{\lambda}/_{2} f_b \qquad (1)$$

where λ – probing radiation wavelength, f_b– beats frequency.

At a speed of 1 km/s, the beat frequency f_b=1.29 GHz. To measure such speeds, it is required that the detection system has a bandwidth of the order of several GHz.

In this manner, it is possible to measure the velocities of all reflecting particles. This technique is immune to external influences, accurate, easy to implement [4,5].

2. System description

Each shock physics experiment is unique in its own way and has a number of hardly reproducible parameters. Increasing the number of registration channels allows to increase not only the information content of the experiment, but also the reliability of the obtained data. Latter is performed by controlling the experiment parameters responsible for the adequacy of the assumptions used in the calculations. An increase of the detected points amount allows to study the particle velocity behavior of several objects simultaneously (or several points of the same object), and also to measure the shock wave parameters before its interaction with the target along with estimation of the wave front flatness.

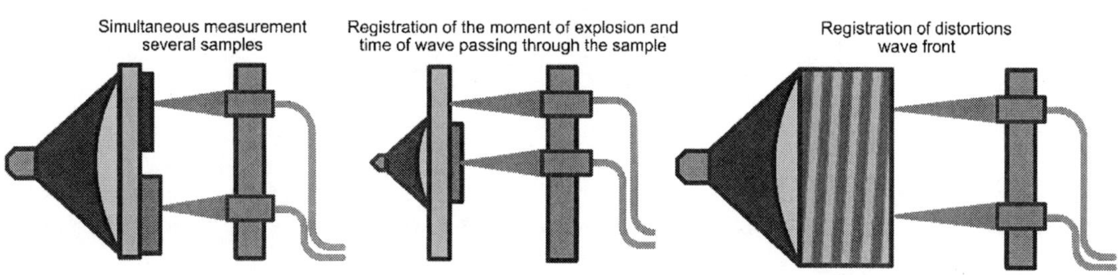

Figure 2. Schemes for the possible use of dual-channel PDV.

Increasing the number of registration channels for PDV based devices is complicated by the high cost of the wide-band (capable to record signals in the several GHz band) recording systems. However, application of the time compression technique provides the possibility of several optical channels registration without increasing the amount of the photoelectron conversion channels and analog-digital conversion channels [6].

The time compression implies a more efficient use of the time base of the oscilloscope, due to which it becomes possible to record signals corresponding to two different points of the object. The implementation of the technique is based on the two optical recording channels forming by the splitter with the subsequent delay of light in one of them by the delay line (Fig. 3). When the light from the second channel passes the delay line, the signal of the first one is recorded. After that the fast fiber switch changes its state and sends light from the second channel to the photodetector.

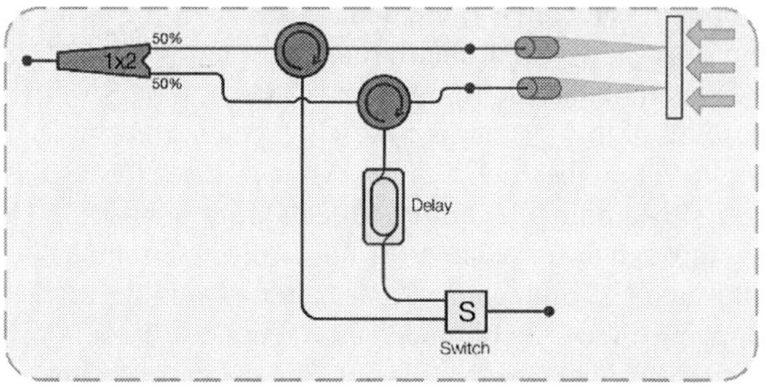

Figure 3. Scheme for the implementation of a time compression.

The memory capacity of the oscilloscope allows to record several signals in succession, with these signals displayed sequentially on the chart.

3. Experimental setup

For the research of the shock-wave loading processes under explosion load two-channel diagnostic system was developed, its scheme is shown in Fig. 4. This scheme was obtained by upgrading the original basic one (Fig. 1) by adding time delay unit in measuring arm (Fig. 3). The length of the delay line corresponds to 25 μs signal delay, the response time of the switch is ~ 300 ns.

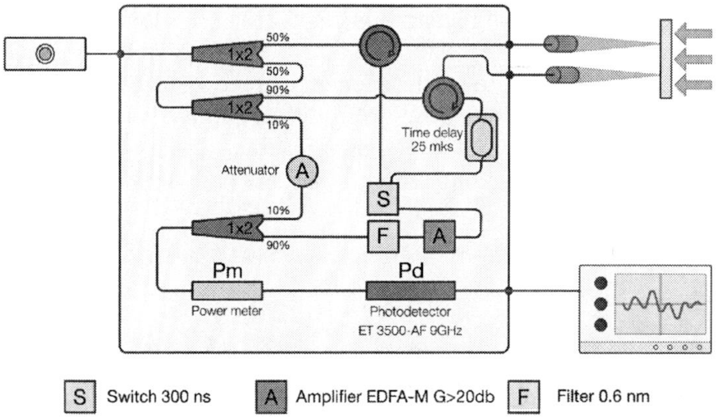

Figure 4. Scheme of a two-channel fiber interferometer with direct conversion of the optical signal with a time compression.

To increase the reflected signal power, in front of the mixer an AEDFA–PA–35 fiber amplifier was built into the scheme. Also, to cut off its noise a filter with a central wavelength of 1550.86 nm and a bandwidth of 0.5 nm, was used. Fiber laser radiation tunable in wavelength from 1550.5 nm to 1550.9 nm was used as a probe radiation. The signal was recorded using an ET–3500AF photo sensor (9 GHz band) and a LeCroy WaveRunner 640 Zi oscilloscope (4 GHz band).

The developed measurement system is a combination of optical and electronic elements assembled into several interconnected blocks. Diagram of the system components is shown in the Fig. 5. One optical unit allows to measure the speed at two points of the object. The modular configuration of the system assumes that with an increase of the optical blocks number to four, it will be possible to measure the speed at eight independent points. The 2W laser was estimated as sufficient to work with 4 optical units simultaneously. The last element of the complete assembly is four-channel oscilloscope.

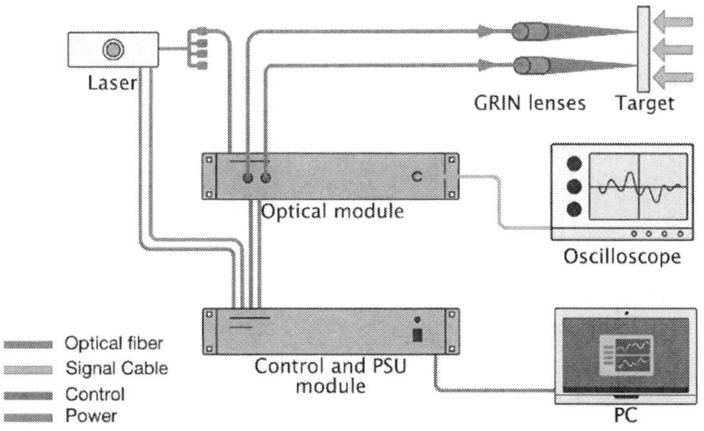

Figure 5. Communication scheme of optical and electrical units.

Developed two-channel PDV system was tested in the study of dusting in the xenon atmosphere during shock wave loading of solid targets (Fig. 6). Lead samples were used as targets. Irregularities with parameters ≈45 µm, ≈300 µm were applied to the surface by the pressurized matrix in order to ease cumulative effects occurrence.

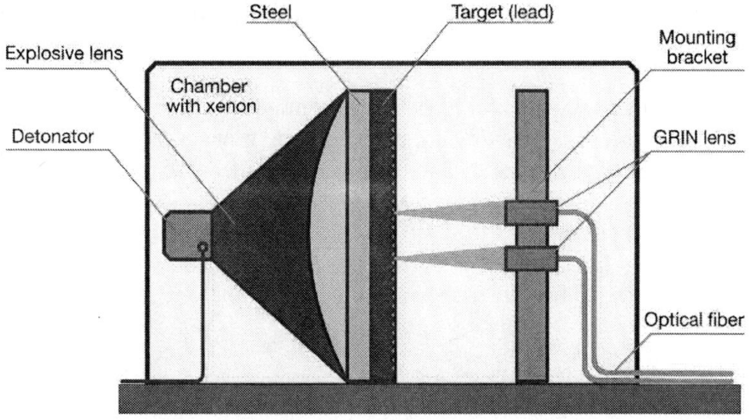

Figure 6. Scheme of the experiment on the study of lead dusting.

In order to process experimentally obtained interferograms software based on short-time Fourier transform was developed in MATLAB computing environment [7]. During processing the signal is multiplied by a window function of a fixed length. Fourier transform of resulting dependency is taken, which is assigned to a given point in time. The result of processing is a set of spectrograms corresponding to different points in time. Calculation according to the following equation must be made to obtain the dynamics of the object velocity from the spectrum (1). The time and velocity resolutions are determined by the processing parameters such as window function size.

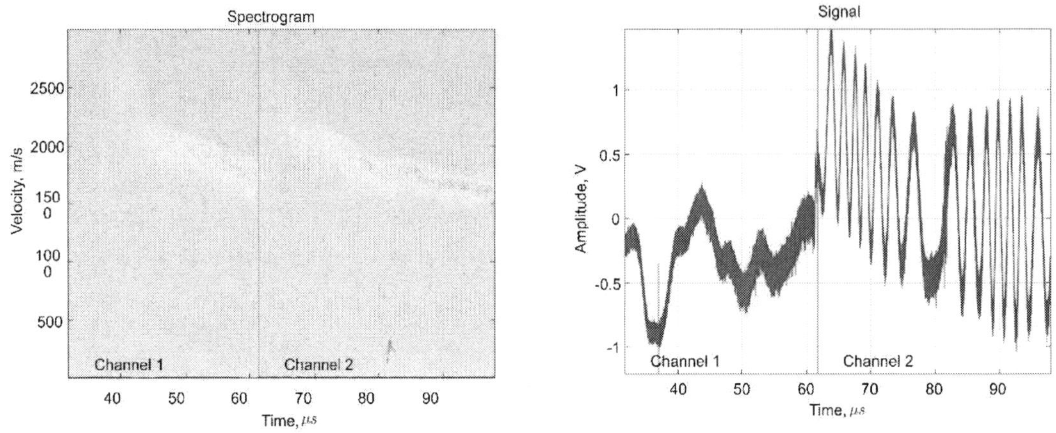

Figure 7. The signal recorded in the study of lead dusting in the atmosphere of xenon and its spectrogram.

Fig. 7 shows an example of a signal recorded during an experiment and its spectrogram. During the process of target velocity measurement dusting of the bottom surface of the sample was observed, which is displayed on the spectrogram. In this case, the signal from the first GRIN lens corresponds to the time interval 1. In interval 2, the system recorded the signal from the second GRIN lens.

4. Conclusion

Measurement system based on a two-channel PDV interferometer for the study of processes of dusting and the destruction of solid targets of was developed. The system has a modular composition and can be expanded to up to 8 channel applications. Using the developed measurement system test experiments on dusting of lead under shock loads were conducted.

Acknowledgments

Funding: this work was supported by the Russian Science Foundation (RSF) (grant number 17-72-20164).

References

[1] Kanel G I, Fortov V E and Razorenov S V 2007 Shock waves in condensed-state physics *Physics-Uspekhi* **50** 771–791
[2] Barker L M and Hollenbach R E 1972 Laser interferometer for measuring high velocities of any reflecting surface *Journal of Applied Physics* **43(11)** 4669-4675
[3] Shock wave generation in the high-power laser interaction with polycrystalline targetsI.N. Burdonsky, A.Yu. Golsov, A.G. Leonov, K.N. Makarov, I.S. Timofeev, V.N. Yufa *VANT Series: Nuclear Fusion 2013 33 Vol. 2* ISSN 0202—3822
[4] Strand O, Berzins L, Goosman D, Kuhlow W, Sargis P and Whitworth T 2004 *Proc. SPIE* **5580** 593
[5] Strand O, Goosman D Martinez C and Whitworth T 2006 Compact system for high speed velocimetry using heterodyne techniques *Rev. Sci. Instum.* 77 083108
[6] Moro E 2014 New developments in photon Doppler velocimetry *Journal of Physics: Conference Series* **500** 142023
[7] Bergland G A guided tour of the fast Fourier transform 1969 *IEEE Spectrum* **6(7)** 41-52

The engraftment dynamics evaluation of skin grafts via aluminium phthalocyanine nanoparticles using spectroscopic methods

D S Farrakhova[1,2], V I Makarov[2], V B Loschenov[1,2]

[1] National Research Nuclear University MEPhI (Moscow Engineering Physics Institute), Moscow, Russia
[2]Prokhorov General Physics Institute of Russian Academy of Science, Moscow, Russia

E-mail: farrakhova.dina@mail.ru

Abstract. The research and comparison of spectroscopic methods for assessment of skin grafts engraftment/rejection after the back cross skin transplantation of laboratory mice were carried out. The spectral analysis was performed via inflammatory reaction evaluation of the studied area. The inflammatory reaction intensity correlates with the fluorescence intensity of the aluminum phthalocyanine nanoparticles. The photodynamic therapy was carried out to improve the skin engraftment and reduce the inflammatory reaction.

1. Introduction

The main problem of transplantology is inflammation of transplanted tissue that leads to serious complications of septic (microbial) or aseptic (autoimmune) nature. Today there are no non-invasive methods for assessing engraftment of grafts that are not time-consuming. The most well-known methods for vascular imaging used in clinic are MRI and CT. Unfortunately, these methods require contrast agents injection, have low resolution (0.5 mm), long scanning time and high cost [1]. One of the perspective method is laser-induced fluorescent diagnostics [2]. Usually wounds and burns are accompanied by inflammatory processes that affect to the fluorescent properties of tissue [3]. These properties depend on the fluorescence of normal tissues, microbial products in tissue and exogenous fluorophores that are used to accelerate wound healing [4].

One of the most perspective classes of nanophotosensitizers for inhibiting the inflammation are photosensitizers based on porphyrin [5] and nanoparticles of aluminium phthalocyanine (nAlPc), which have been studied intensively by various research groups [6-10]. We used nAlPc, the molecular nanocrystals capable of forming stable colloidal suspensions.

The main reason for the nAlPc use is the extremely low or nonexistent fluorescence of the nano form compared to high fluorescence intensity of the soluble form of AlPc (figure 1b). This fact is explained by the fluorescence quenching of nAlPc crystal structure [11]. But nAlPc begin to show fluorescent properties when entering the inflamed biological tissue. According to the authors [11] this is due to the presence of a large number of macrophages and pathogenic microflora in the inflamed tissue. nAlPc have several competing processes, such as dissolution and/or change of the particle surface structure while interacting with this microenvirenment. Now, the physics of this process are being studied, but the authors [12] have proposed the most likely hypothesis about the transition of nanoparticle surface molecules from para-position to ortho-position with nanoparticles becoming fluorescent and photodynamically active.

The fluorescence lifetime is one of the important characteristics of fluorescence. According to the literature, the number of components and the fluorescence lifetime changes in the interaction with cells and depending on the pH. Thus, the presence of two fluorescence lifetimes were recorded at pH=2: 5 ns which corresponds to the molecular form in the solution and 1.5 ns which corresponds to the bound state of the phthalocyanine molecule. The study of the fluorescence lifetime of nAlPc in macrophages showed the presence of two components at 9 ns and 4.5 ns [13]. During our study, a change in the fluorescence decay kinetics was recorded, expressed in the appearance of additional fluorescence decay components.

2. Methodology

2.1. Nanophotosensitizer

nAlPc were used for the spectral analysis of skin healing grafts. The nAlPc colloidal solution was obtained by ultrasonic dispersion of large-dispersed crystals of aluminum phthalocyanine [14]. Thus, nAlPc with a hydrodynamic radius from 30 nm to 250 nm are possible to obtain via this method. Nanoparticles are plates with a diameter of 240±35 nm and a thickness of less than 50 nm (figure 1c).

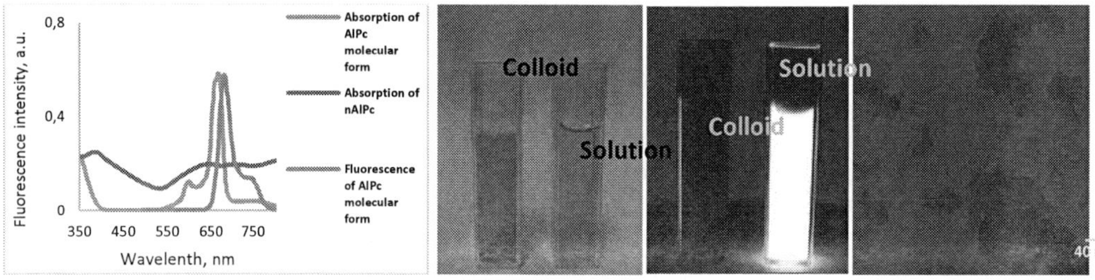

(a) (b) (c)

Figure 1. (a) absorption spectra of nAlPc colloidal solution and sulfated AlPc solution and fluorescence spectra of sulfated AlPc solution, (b) AlPc molecular solution and nAlPc colloidal solution in visible light and fluorescent image, (c) transmission electron microscopy image of nAlPc.

2.2. Experimental setup

Changes in the interstitial distribution of macrophages and other immune cells were evaluated via a streak-camera with picosecond time resolution (15 ps) by measuring the fluorescence lifetime of nAlPc. The streak-camera is paired to a fiber-optic spectrometer with 637 nm wavelength excitation and 65 ps pulse duration (figure 2).

The luminescence excited in the sample by the light from the laser source is collected via an optical fiber. A laser pulse is delivered to the sample through the central fiber and 6 peripheral fibers collect the backscattered light as well as the fluorescence. The central-symmetrical position of the fibers on the distal probe tip makes it possible to detect fluorescence regardless of the probe location to the sample (figure 2b). The luminescence and backscattered light are decomposed into the spectrum which then falls onto the photocathode at the streak scope input. The emitted electrons are accelerated by an electrical field parallel to the optical axis inside the streak scope. A perpendicular field is applied to the electrons for the duration of each scan. The electrons fall on the MCP (microchannel plate – analogue of the photoelectron multiplier with a 2d resolution), which amplifies the electron flow as a result of secondary electron emission (figure 2a) thus creating a flash on the phosphorescent screen. This flash is registered by a CCD camera and converted to a single photon event by the software. The mathematical processing of these events allows obtaining information on the fluorescence lifetime and the distribution of the number of photons between the components of the fluorescence decay registered by the streak-camera.

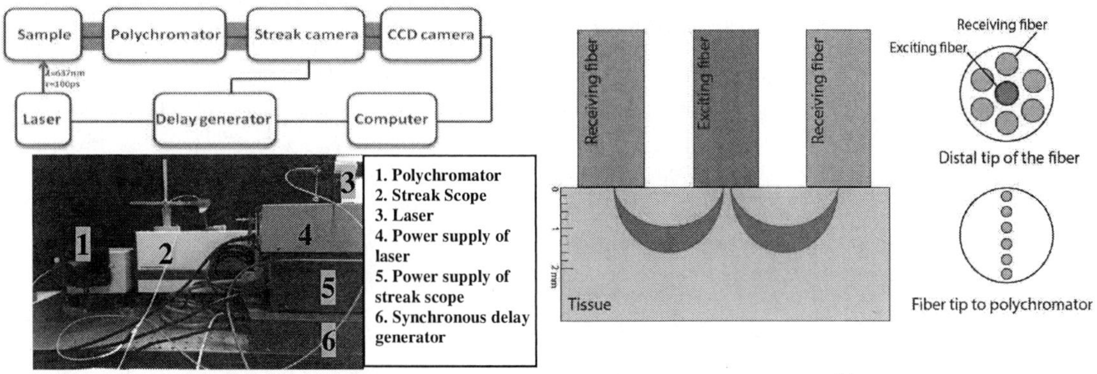

(a) (b)

Figure 2. (a) The flow chart of the experimental setup and the image of the experimental system, (b) the working principle of the fiber-optic probe consisting of 7 optical fibers.

2.3. Animal experimental design

The studies were carried out on the laboratory mice. Cross skin transplantation on the mouse back was performed similar to the study [4]. 10 μl of nAlPc with a 100 mg/kg concentration were added under the right skin graft (G500) while the left skin graft was used as control until the skin tissue was fully healed. In addition, skin tissues from mouse ears were used as "thin" skin grafts (G250). The thickness of the grafts were 500±50 μm for G500 and 250±30 μm for G250.

The photodynamic therapy was performed on day 18 after cross skin transplantation with 195 J/cm^2 dose with 500 mW/cm^2 power density and a laser source excitation wavelength of 668±10 nm in order to reduce the degree of inflammation in the area of the tissue to be healed. Fluorescence spectra were registered by the method of diffuse backscattered light from biological tissue using 633 nm laser source providing information about the state of the skin at depth. Measurements of nAlPc fluorescence lifetime and registration of nAlPc spectra were carried out on days 1;4;7;10;16;18;22;29 after cross skin transplantation.

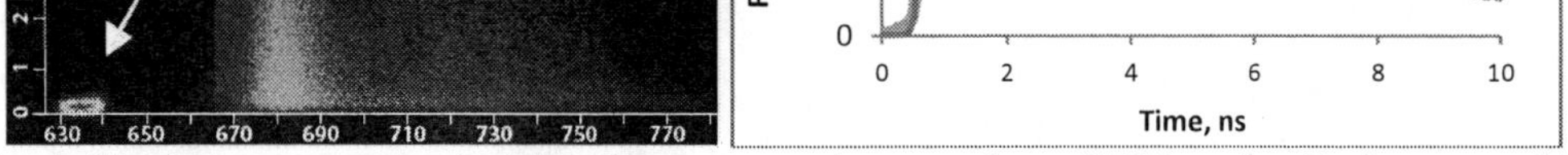

Figure 3. (a) the spectrum of nAlPc fluorescence decay under a G500 on day 7 after cross skin. transplantation, (b) the fluorescence decay after picosecond light pulse excitation.

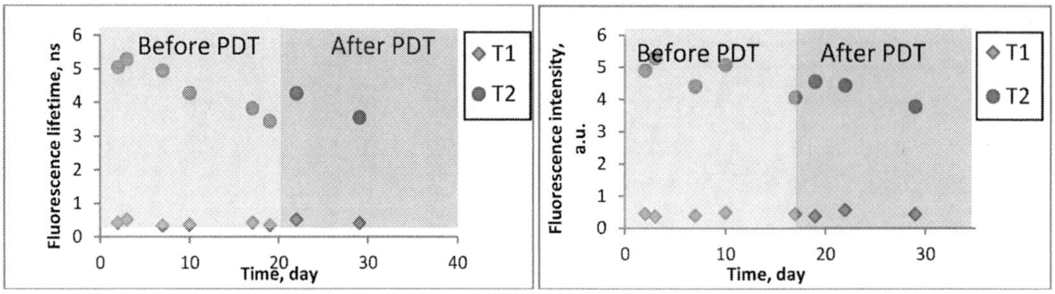

(a) (b)

Figure 4. The dependence of nAlPc fluorescence lifetime on time (a) G500 (b) G250.

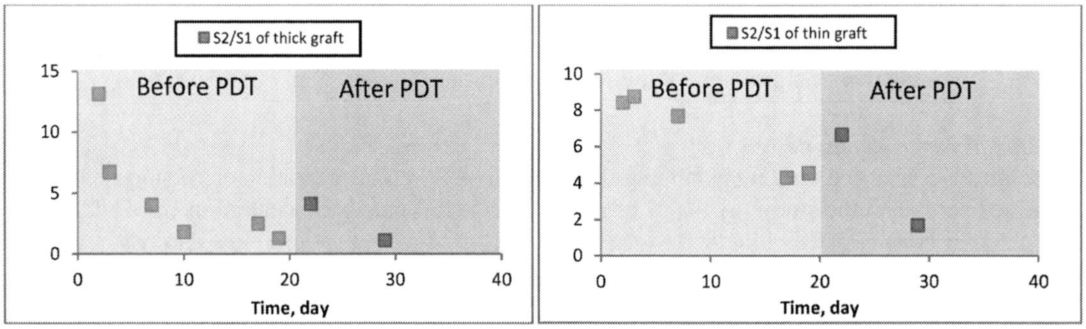

(a) (b)

Figure 5. The dependence of the long (S_2) and short (S_1) components photon number ratio of the nAlPc fluorescence lifetime on time (a) G500 (b) G250.

3. Results and discussions

In this work, the dynamics of the nAlPc fluorescence intensity under G500 and G250 grafts was obtained.

Two components of the nAlPc fluorescence lifetime under both grafts were recorded during the experiment. The nAlPc fluorescence lifetime itself varies slightly, but the contribution of each component to the total fluorescence varies significantly (figure 5). Particularly, the intensity ratio between long and short components changed from 13:1 to 1.5:1 during the experiment for G500 and from 9:1 to 2:1 for G250.

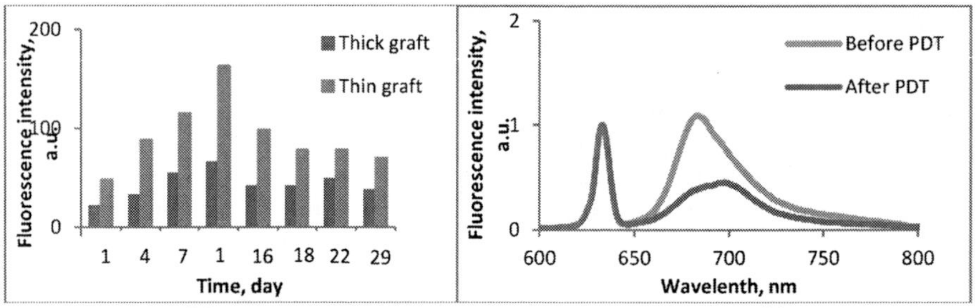

(a) (b)

Figure 6. (a) Histogram of the fluorescence intensity of G500 and G250 grafts with nAlPc, (b) fluorescence intensity of G500 before and after PDT.

Figure 6a shows the fluorescence index calculated as a ratio between the nAlPc fluorescence and the backscattered laser light for G500 and G250 according to the equation:

$$k_{fl} = \frac{\int_{660}^{725} I(\lambda)\mathrm{d}\lambda}{\int_{623}^{640} I(\lambda)\mathrm{d}\,\lambda} \qquad (1)$$

The nAlPc fluorescence intensity increases within 10 days after cross skin transplantation of both grafts. Furthermore, the fluorescence intensity in the graft tissue decreases which indicates the successful graft engraftment. On average, the fluorescence intensity of G250 with nAlPc is lower due to the absorption of laser radiation by blood cells. Consequently, the thickness of the graft tissue greatly affects the obtained results in this measurement geometry. Meanwhile the changes of the nAlPc fluorescence intensity are similar in both grafts. PDT was performed on day 18 after cross skin transplantation with the aim to improve the engraftment speed of the skin graft and nAlPc photobleaching (figure 6b). There is a noticeable decrease of the nAlPc fluorescence intensity in G250 after PDT. The spectral curve change is due to a different environment inside the graft after PDT, which affected the change in the nAlPc fluorescence lifetime.

The successful engraftment of skin grafts took place in the experiment, and while there is a strong purulent inflammation under the control graft, there was no pus under the graft with nAlPc.

4. Conclusion

The engraftment of skin grafts was evaluated via nAlPc during the experiment. By changing the fluorescence intensity, which correlates with the inflammatory reaction, the degree of graft engraftment can be assessed.

An interesting new effect was demonstrated that the ratio of the long to short components of the nAlPc fluorescence decay is significantly different after PDT. Presumably, this characterizes the changes in nAlPc local microenvironment, which is determined primarily by the type of immune-competent cells involved in the inflammation. This method can be useful for assessment of graft engraftment and allows preventing the rejection.

Acknowledgements
This work was supported by a grant RFBR # 15-29-04869-ofi-m.

References

[1] Makarov V I, Akhlyustina E V, Farrakhova D S, Pominova D V, Ryabova A V and Loschenov V B 2016 *Biomedical Photonics* **5** 30-40
[2] Loschenov V B, Konov V I and Prokhorov A M 2000 *Laser Physics* 10 1188-207
[3] Etra J W, Raimondi G and Brandacher G 2018 *Current Opinion in Organ Transplantation* 23 28-33
[4] Makarov V I, Pominova D V, Ryabova A V, Saveleva T A, Ignateva I V, Reshetov I V and Loschenov V B 2018 *Unconventional Optical Imaging* **10677** 1067734
[5] Meerovich G A, Akhlyustina E V, Tiganova I G, Panov V A, Tyukova V S, Tolordava E R, Alekseeva N V, Linkov K G, Romanova Y M, Grin M A, Mironov A F , Loshchenov V B, Kaprin A D and Filonenko E V 2017 *Biomedical Photonics* **6** 16-32
[6] De Paula C S, Tedesco A C, Primo F L, Vilela J M C, Andrade M S, Mosqueira V C F 2013 *European Journal of Pharmaceutical Sciences*, **49** 371-81
[7] Asem H, Abd El-Fattah A, Nafee N, Zhao Y, Khalil L, Muhammed M, Hassan M and Kandil S 2016 *Photodiagnosis and Photodynamic Therapy* **12** 48-57
[8] Calori I R and Tedesco A C 2016 *Journal of Photochemistry and Photobiology B: Biology* **160** 240-7
[9] Vasilchenko S Y, Volkova A I, Ryabova A V, Loschenov V B, Konov V I, Mamedov A A, Kuzmin S G and Lukyanets E A 2010 *Journal of Biophotonics* **3** 336-46
[10] Steiner R, Breymayer J, Ruck A, Loshchenov V and Ryabova A. 2015 *Optical Methods for Tumor Treatment and Detection: Mechanisms and Techniques in Photodynamic Therapy XXIV* **9308** 93080R
[11] Kuznetsova J O and Makarov 2016 V I *Journal of Physics: Conference Series* **737**
[12] Bystrov F G, Makarov V I, Pominova D V, Ryabova A V and Loschenov V B 2016 *Biomedical Photonics* **5** 3-8
[13] Maklygina Yu S, Sharova A S, Kundu B, Balla V K, Steiner R and Loschenov V B 2016 *Biomedical photonics* **5** 4–12

IOP Conf. Series: Journal of Physics: Conf. Series **1238** (2019) 012027 doi:10.1088/1742-6596/1238/1/012027

Multichannel laser radiation collection and transport system for particle velocity measurements in shock wave experiments

I Yu Tishenko, K L Gubskiy, V A Pirog, T V Kazieva, K S Lukyanov, A A Morozov, A P Kuznetsov

National Research Nuclear University MEPhI (Moscow Engineering Physics Institute), Moscow, Russia

E-mail: kgubskiy@mail.ru

Abstract. VISAR type measuring laser systems are used for the determination of surface velocities in high-speed processes. Their operational characteristics mostly depends on the efficiency of collection of scattered radiation. In this paper we present results of the comparative analysis of several radiation collection schemes along with experimentally obtained data on the scheme parameters effect on collection efficiency and recording length.

1. Introduction

One of the most popular methods for measuring shock wave velocity is based on VISAR systems usage [1]. VISAR is a reliable measuring instrument, its various modifications have been successfully used in experimental studies on fast and transient processes. Modern measuring systems allow to perform multichannel high-precision velocity measurements of the investigated moving surface Moreover, utilization of the nonius scheme allows to eliminate the velocity ambiguity due to its step increase upon shock wave reaching surface of the sample. [2].

Operating principle of VISAR type systems based on Doppler shift of light reflected off of the surface of a moving object. During the experiment, the surface of the object often rotates, so that usage of specular reflective surfaces is unreliable. This allows operation with rough unpolished surfaces, or with objects without reflective coatings. On the other hand, the reflection of rough surfaces is close to Lambert reflectance, so that efficiency of light collection is low.

Experimentally measured light collection efficiency $\mathcal{E}$ is of the order of 10^{-5} (10 μW of radiation was collected per 1 W of radiation delivered):

$$\mathcal{E} = \frac{P}{P_0} \tag{1}$$

where P is the power of the radiation collected from the surface of the investigated object, when the object is illuminated by laser radiation with the power P_0.

Lasers with a power of 2–10 W CW can be used as sources of probing radiation, photomultipliers are used for recording. With that said, powerful local heating is unacceptable for the unstable and sensitive to heating objects.

Possibilities of operation on samples with scattering surfaces without an additional reflective ones are also limited by the obtainable efficiency of radiation collection. The signal-to-noise ratio depends on the amount of collected radiation, which directly affects the accuracy of the results obtained.

2. Light collection and transport systems

In this regard, usage of the appropriate light collection and transport system (LCTS) is essential in order to obtain meaningful experimental results. Adequate LCTS should provide high collection efficiency along with long recording length (the distance beyond which the signal level drops to the noise level). Three LCTS were used as optical sensors setups: LCTS on a free beam, fiber LCTS without collimating optics, and LCTS fiber with the fiber optic sensor.

2.1. LCTS on a free beam

The most simple LCTS scheme is a mirror with a hole and collecting lens located at the focal distance from the object (Fig. 1). Laser radiation passes through the hole and focuses on the investigated object. The scattered light returns through the collecting lens and forms a parallel beam, which is reflected from the mirror to the interferometer.

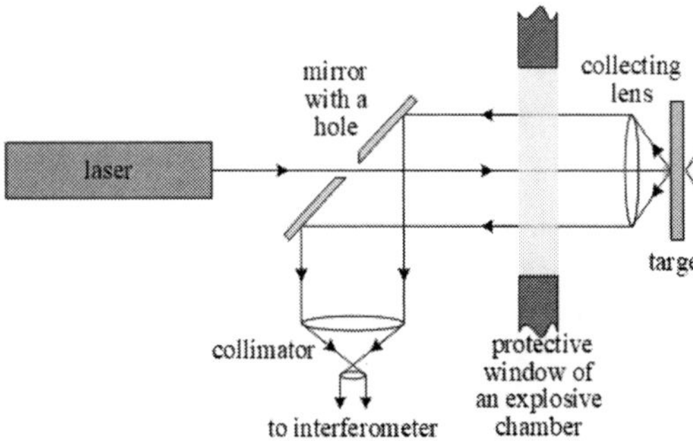

Figure 1. The scheme of transportation and collection of radiation using a mirror with a hole.

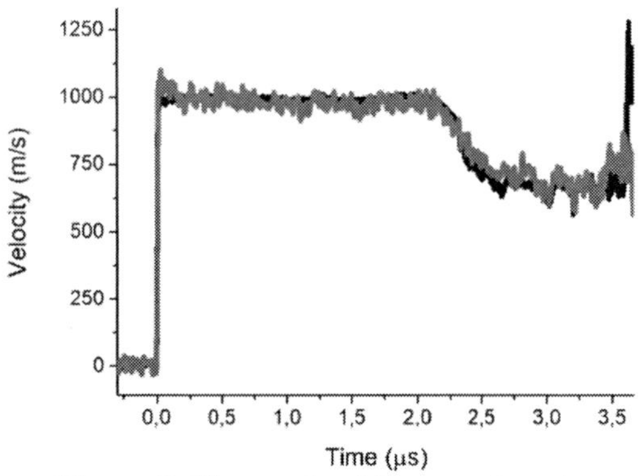

Figure 2. The result of the experiment on shock-wave loading of aluminum plate using the free beam scheme. The black line is the data of the interferometer with a constant of 280 m/s/band, the red line is the data of the interferometer with a constant of 1280 m/s/band.

The movement of the object results in its surface leaving the focal plane of the lens. In this case, the scattered radiation collected by the lens stops forming a parallel beam and begins to diverge. This leads to the fact that the power of the collected radiation decreases rapidly as the studied object moves. This imposes restrictions on the effective length of the signal recording.

For the analysis purposes scheme with following parameters was tested: lens focal length 15 cm and distance between the collecting lens and the mirror with a hole is 4 m. The effective length of the signal recording for such scheme was calculated to be 3.5 mm.

Operational characteristics of the LCTS on a free beam were determined in test experiment on the shock-wave loading of the aluminum plate. The result of the experiment is shown in Fig. 2. After 3.5 μs the signal is lost and the time dependence of the velocity becomes unreliable. Thus the length of the effective signal recording assessed as 3.4 mm.

Multipoint measurement setup using a free beam scheme is extremely complicated, as it requires bulky optical system consisting of a large number of optical elements. However multipoint measurements could be significantly simplified by the use of optical fiber systems for the optical signal transport.

2.2. Fiber LCTS without collimating optics

LCTS without collimating optics is the system of fiber-optic transport of radiation from a laser source to a test object and an interferometer.

IOP Conf. Series: Journal of Physics: Conf. Series **1238** (2019) 012027 doi:10.1088/1742-6596/1238/1/012027

In this case, the polished ends of the optical fibers are located at the distance of 2-5 mm from the surface of the test object (Fig. 3). Radiation from the laser source is transported to the surface of the test object using one optical fiber, another fiber (or few others) are used for the collection of scattered radiation and it transport to the interferometer. In the most simple case fibers can be fastened in rigid plane plate made of plastic or even plywood. In that case collection efficiency assessed to be no more than $2 * 10^{-4}$.

For such similar test experiment on the shock-wave loading of an aluminum plate was conducted (Fig. 4). The length of the effective signal recording was 0.82 mm.

Described scheme is easy to use, cheap, allows multipoint experiments, but at the same time it has low effective signal recording length and a large measurement area, as it is determined by the fiber numerical aperture (Measurement area is 1.44 mm^2 for NA = 0.22 and the 3 mm distance between the fiber and the object).

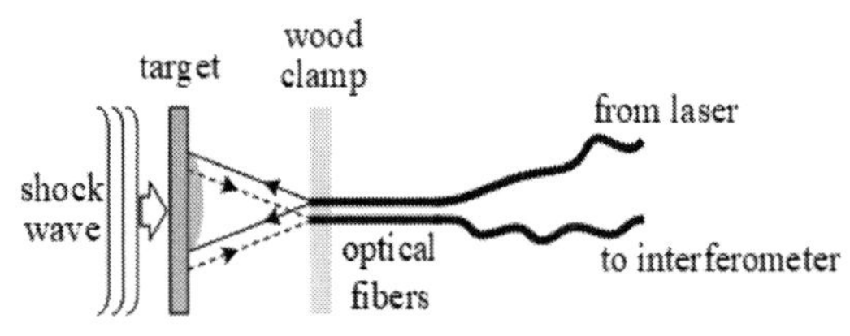

Figure 3. Fiber LCTS scheme without collimating optics

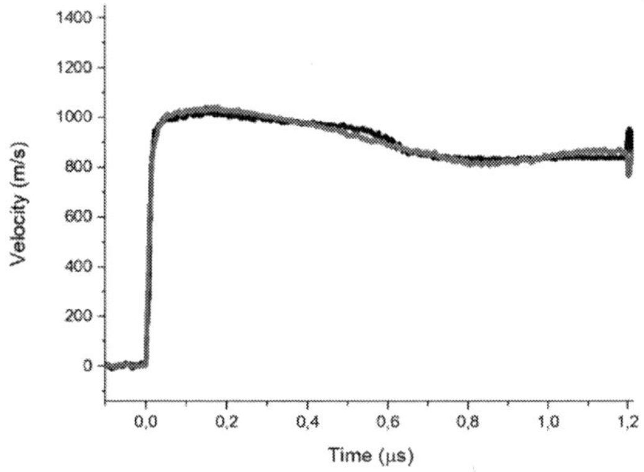

Figure 4. The result of a test experiment on shock-wave loading of an aluminum plate using a non-collimating scheme. Black and red lines show surface velocities at two different points, 280 m/s/ band.

2.3. Fiber LCTS with a fiber-optic sensor

Convenience of fiber transport of radiation with the collecting lens systems efficiency can be combined with the applications of probes, i.e. fiber based optomechanical systems (Fig. 5). Disposable fiber-optic probes are built on a base of the metal case with a through hole for radiation transport and mounting holes. Probe is set on a special table at some distance from the surfaces of the test samples parallel to them.

Lens with a 10 mm diameter and a 10 mm focal length is glued in the case for radiation collection, the other side of probe contains thread, which holds hollow plastic adjustment screw. In the screw through hole assembly of the optical fibers is fixed.

The ends of optical fibers in the probe, are aligned along the optical axis of the lens; fibers axes are equally distanced from it and grouped together.

The probe design allows to adjust the distance between the faces of the optical fibers and the lens, as well as to adjust the angle sensor's axis and surface of the researched object.

The choice of fibers for the sensor is dictated by the requirement of maximum collection efficiency and acceptable collimability of radiation at the fiber output for radiation processing in the interferometer.

A fiber with a larger diameter collects radiation more efficiently because of the larger cross-sectional area of the core, but using larger diameter fibers leads to increased radiation losses at the interferometer input. Modelling in Zemax software have shown that the use of fibers with a core diameter of 200 μm is optimal. Additionally, studies have been conducted on the possibility of using standard telecommunication fibers with a core diameter of 62.5 microns as they are significantly cheaper and more technologically advanced.

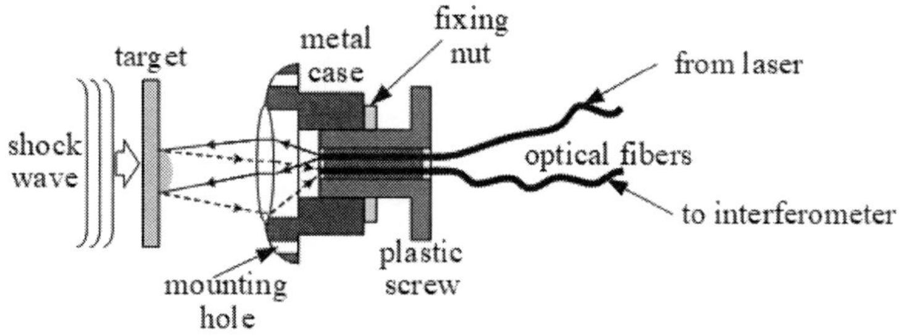

Figure 5. Scheme of a fiber optic sensor with single lens.

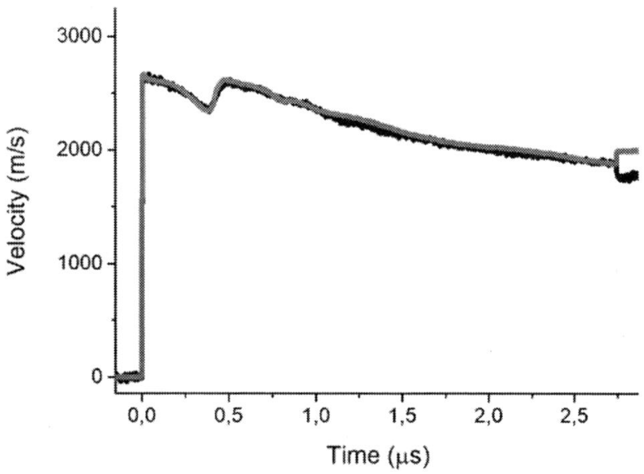

Figure 6. The result of a test experiment on shock-wave loading of an aluminum plate using a single lens probe. The black line is the data from interferometer with a constant of 1280 m/s/band, the red line is the data from interferometer with a constant of 280 m/s/band.

Characteristics of single lens probes with different fibers were determined by the measurements of dependencies of the radiation collection efficiencies on the distance between the sensor's collecting lens and the test surface on different positions of the ends of optical fibers relative to the collecting lens. Various combinations of transferring and collecting optical fibers were used: transferring and collecting fibers with 62.5/125/250 μm diameters (core/sheath/protective coating) (Fig. 7, a); transferring fiber with 62.5/125/250 μm diameters and collecting fibers with 205/230/340 μm diameters (Fig. 7, b); transferring and collecting fibers with 205/230/340 μm diameters (Fig. 7, c). Matte polished aluminum plate surface was used as reflective test surface.

The highest radiation collection efficiency is provided by probes using 205/230/340 μm optical fibers for both input and output and 13 mm distance between faces of the optical fibers and collecting lens with 30 mm distance between the target and the lens. With this configuration, the collection efficiency was $5.6 * 10^{-4}$.

Test experiment was conducted on the shock-wave loading of an aluminum plate using single lens sensor. In the experiment the sensor was set so that the distance between the collecting lens and the surface of the researched sample was 25 mm. The distance between the ends of the optical fibers and the collecting lens is 13 mm. Optical fibers with 205/230/340 microns diameters were used for transferring supply and collecting of the radiation. The results of the experiment are presented in Fig. 6. the effective length of the signal recording was 6.0 mm.

Such configuration, according to measurements of the efficiency collection, allows achieving an effective recording length of up to 25 mm. However, destruction of the target during the experiment sets practical limit to the recording length.

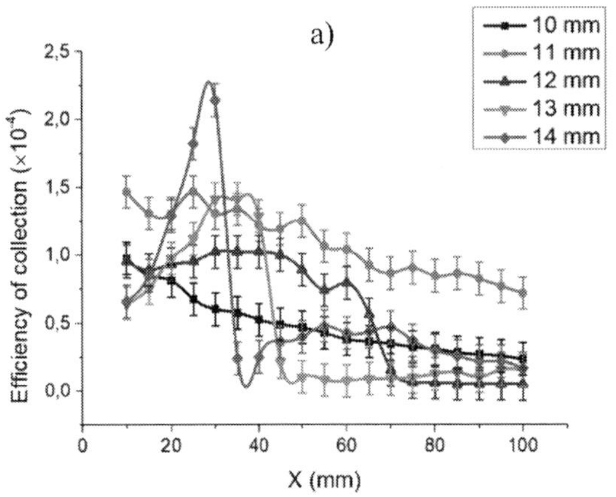

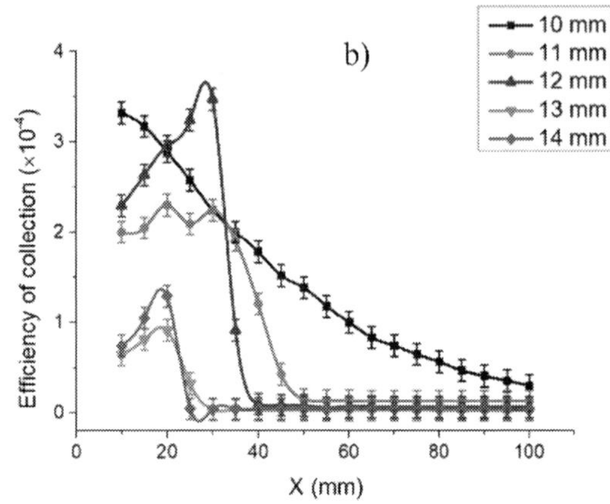

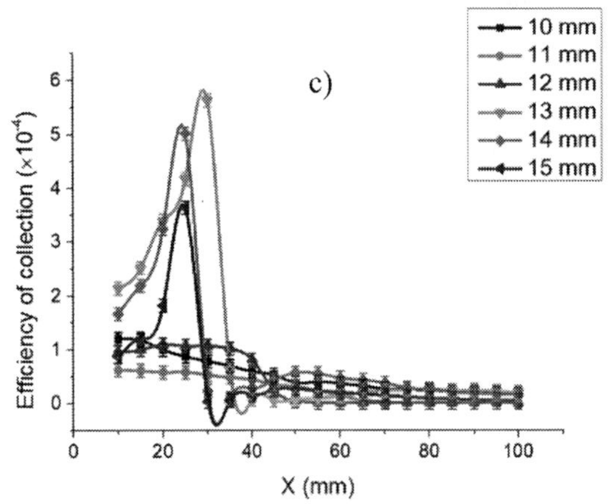

Figure 7. Efficiency of collection of radiation vs the distance between the collecting lens and the surface of the target on different positions of the ends of the optical fibers relative to the lens (a) for a probe with 62.5/125/250 µm both transferring and collecting fibers, b) for a probe with a 62.5/125/250 µm transferring fiber and 205/230/340 µm collecting fibers, c) for a probe with 205/230/340 µm both transferring and collecting fibers)

3. Conclusion

Three types of LCTS for VISAR interferometric systems were studied.

LCTS based on the single lens fiber-optic probe has the longest effective recording length. Changing the distance between the ends of the optical fibers and the collecting lens allows to achieve an effective signal recording length of the tens of millimeters order.

The highest radiation collection efficiency achieved by using single lens probe was $5.6 * 10^{-4}$. At the same time, the effective recording length was 16 mm with, the distance between the ends of the optical fibers and the collecting lens 13 mm.

Conducted test experiments confirmed the effectiveness of disposable single lens fiber-optic probes for particle velocity measurements in multipoint experiments on shock-wave loading.

Acknowledgments

Funding: this work was supported by the Russian Science Foundation (RSF) (grant number 17-72-20164).

References

[1] Barker L M and Hollenbach R E 1972 Laser interferometer for measuring high velocities of any reflecting surface *Journal of Applied Physics* **43(11)** 4669-4675

[2] Gubskii K L, Koshkin D S, Mikhaylyuk A M, Korolev A M, Pirog V A. and Kuznetsov A P 2015 Multipoint vernier VISAR Interferometer system for measuring mass velocity in shock wave experiments *Physics Procedia* **71** 212-216

IOP Conf. Series: Journal of Physics: Conf. Series **1238** (2019) 012028 doi:10.1088/1742-6596/1238/1/012028

Laser Milling Ceramics Dioxide Zirconium and Disilicate of Lithium

D V Panov, V N Petrovskiy, D V Ushakov, A V Osintsev, P S Dzhumaev, V I Polskiy

National Research Nuclear University MEPhI (Moscow Engineering Physics Institute), Moscow, Russia

E-mail: panovdv96@yandex.ru

Abstract. The paper considers the possibility of using laser milling technology for precision processing of ceramic samples from pre-sintered and sintered ceramics zirconia and lithium disilicate. To find the best radiation source were compared different lasers. To find highest removal rate with acceptable accuracy was studied influence laser and scanning system parameters on ceramics. Accuracy problem of laser milling was considered. Examples of surfaces of prostheses were made.

1. Introduction

Laser processing was established as one of the precision methods to obtain three-dimensional structures. The widest distribution laser processing gets in micromachining. Nowadays transition from micromachining to full-scale milling of objects starts. For "big" objects laser milling may compete with tradition mechanical milling only with comparable material removal rate and range of milling materials.

Lately a few studies [1,2] where authors considered laser milling as new technology for creating dentures were published. These works show possibility of laser milling, but low removal rate shows the lack of prospects for laser processing.

The purpose of this work was to increase material removal rate up to comparable with mechanical milling with saving of accurate of the surface.

2. Materials and methods

2.1 Materials

The experiments were performed on commercially available blocks of pre-sintered and sintered blocks of Y-TZP (yttria-stabilized tetragonal zirconia polycrystal) ceramics and unsintered blocks of lithium disilicate.

2.2 Radiation source

To choose optimal laser we used range of radiation sources with different parameters. The basic parameters of radiation sources are listed in Table 1.

2.3 Methods

Measurement of different parameters was made with area processing. After that depth and roughness of the surface was measured. By this data was calculated material removal rate and other parameters. Surface analysis and depth was made with mechanical profilometer Dektak-150.

Table 1. The parameters of the lasers

Parameter	Solid-state, 3d harmonic	Solid-state, 2d harmonic	Fiber	Fiber
Wavelength, nm	355	532	1064	1030
Average power, Wt	up to 2	up to 10	up to 20	up to 30
Pulse frequency, kHz	10-100	20-200	20-100	200-2000
Pulse width	2-6 ns	20-50 ns	100 ns	5-20 ps

3. Results and discussions

3.1 Radiation source

Y-TZP ceramic is a typical dielectric with band gap 5.2-6 eV [3], also lithium disilicate is a dielectric. This type of material has good interaction with CO_2 laser [4], but minimal focused spot diameter is larger than it needs for the precession laser milling. So solid-states and fiber pulsed lasers from UV to IR was chosen as main radiation source.

3.1.1 Wavelength

An estimate of the effectiveness of the radiation source can be obtained from measurement of absorption. The measurements are done at low energy, so it does not take into account high energy processes, but helps assess the prospects of radiation sources. Figure 2 (a) shows absorption of ZrO_2 doped Y_2O_3 and figure 2 (b) shows absorption of lithium disilicate. As can be seen from graphs the highest absorption of the both materials are in UV part area the graphs. With lithium disilicate it proves in experiment. Ablation of lithium disilicate with nanosecond laser was obtained only on wavelength 355 nm and less. Material removal rate amounted up to 0.1 cm^3/h with roughness Ra 3.4 μm.

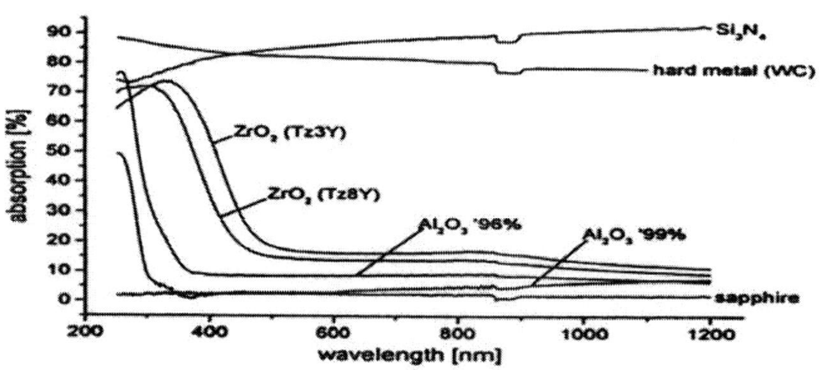

(a)

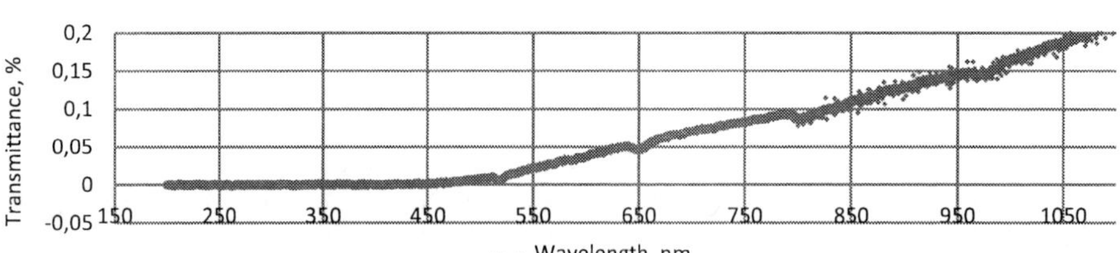

(b)

Figure 1. Absorption of ZrO_2 (a) [5] and Transmittance of Lithium disilicate (b)

3.1.2 Pulse width

Laser processing occurs by the mechanism of ablation. Ablation starts when fluence reached threshold fluence. Depends between pulse width and threshold fluence considered in [6]. In connection with it was used picosecond laser. This laser has lower threshold fluence and lower thermal affect on surface.

3.2 Laser and scanning parameters

There are a lot of different laser and scanning system parameters which have great influence on milling process: scan speed, hatch distance, scheme of scanning, frequency, average power, fluence, focus depth, spot diameter. The main parameters are scan speed and fluence.

3.2.1 Scan speed

Scan speed has affects on a thermal distribution and laser spot overlap (LSO). Thermal distribution is particularly important for ablation with nanosecond pulses. LSO determines roughness of surface. Dependence between scan speed and MRR for pre-sintered Y-TZP and lithium disilicate is on figure 2.

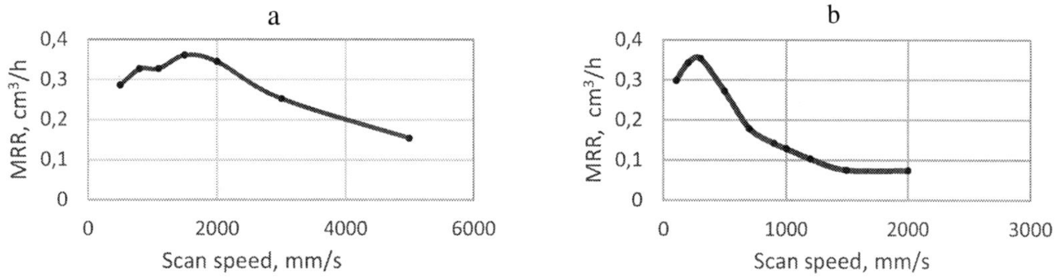

Figure 2. Dependence between MRR and Scan speed for lithium disilicate (a) and pre-sintered Y-TZP (b)

Graphs show that with increasing of scanning speed, MRR decrease. However, after some speed, reducing of scan speed is not increased MRR. It caused increasing of absorption of radiation by plasma. It caused by incubation effect during multi-shot laser ablation. Other materials have the same dependence.

Also scan speed related to LSO. Our experiments show that LSO is the main factor of roughness. Dependence with high MRR is similar for different materials (fig. 3). Roughness decrease up to “0%” LSO and after do not change much. For low MRR the lowest Roughness was obtained on “50%” LSO.

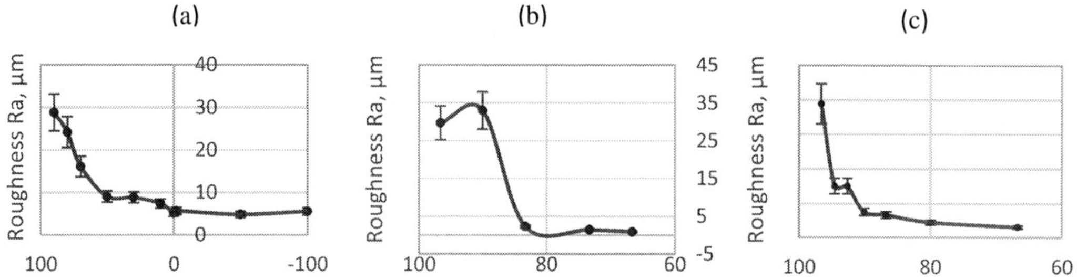

Figure 3. Dependence between Roughness Ra and LSO for pre-sintered Y-TZP on nanosecond laser (a), pre-sintered Y-TZP (b) lithium disilicate (c) on picosecond laser

3.2.2 Laser parameters

As the main characteristic for laser radiation was chosen fluence. With changing only average power pronounced linear dependence can be seen (fig. 5).

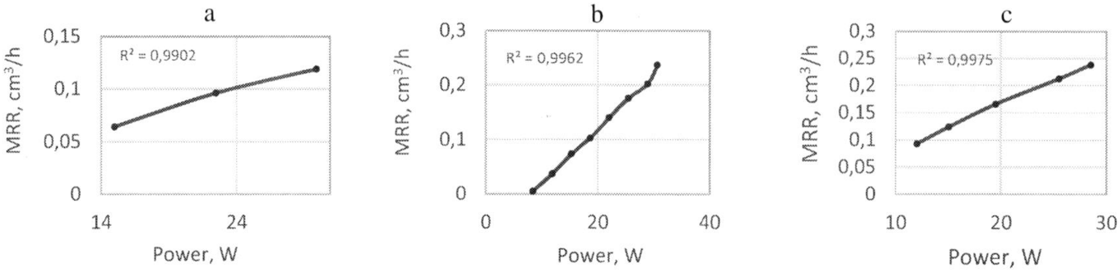

Figure 4. Dependence between MRR and Average power for pre-sintered Y-TZP on nanosecond laser (a), pre-sintered Y-TZP (b) lithium disilicate (c) on picosecond laser

That's result let to chose average power as the main parameter for managing of the milling. With increasing of the average power this dependence would not be saved because of appeared plasma absorption [7]. High power lasers wouldn't give high increment of removal rate.

3.3 Material removal rate

For each laser, the highest MRR was obtained with a given roughness. The results are in table 2.

Table 2. Material removal rate and roughness of the samples

Laser	Material	Pre-sintered Y-TZP	Sintered Y-TZP	Lithium disilicate
Fiber picosecond	MRR, cm³/h	up to 1.2	up to 0.41	up to 0.44
	Ra, µm	2.1	1.76	2.5
Fiber nanosecond	MRR, cm³/h	up to 0.44	up to 0,11	no ablation
	Ra, µm	4	3.7	
Solid-state, 2d harmonic	MRR, cm³/h	up to 0,17	up to 0,053	no ablation
	Ra, µm	1.7	1.2	
Solid-state, 3d harmonic	MRR, cm³/h	-*	-*	up to 0.1
	Ra, µm			3.4

*- no data

3.4 Prosthesis' surfaces

As result we get different prosthesis surfaces from pre-sintered Y-TZP with fiber nanosecond laser (fig. 5). That's laser was chosen be because it has Z axis. Scanning system has repeatability up to 10 mkm, so flat sizes of prosthesis was precise. But Z sizes are different from given. But their differences were predicted and based on different processing area.

IOP Conf. Series: Journal of Physics: Conf. Series **1238** (2019) 012028 doi:10.1088/1742-6596/1238/1/012028

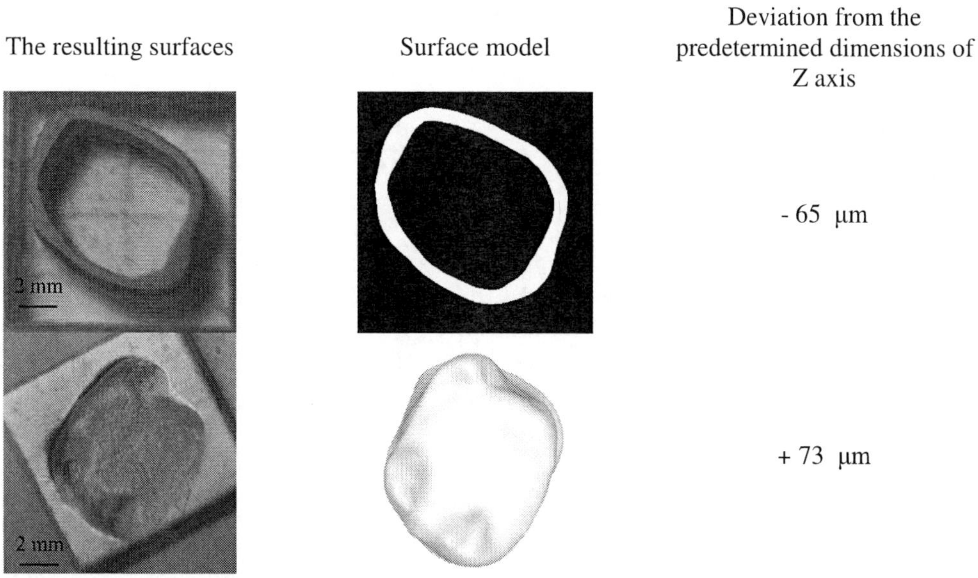

Figure 5. Comparison between prothesis surface and its model with deviation from the predetermined dimensions of Z axis

4. Conclusions

In this research the following results was obtained:

- The different radiation sources were compared.
- Dependence from main parameters was considered.
- Prosthesis surfaces were made. The time to made one was from 20 to 30 min.

The laser milling technology can be used in dental restorations manufacturing because of its benefits. There are some difficulties with precession and removal rate. In next researches these difficulties are being solved.

References

[1] Kazama-Koide, Miku and Ohkuma, Kazuo and Ogura, Hideo and Miyagawa, Yukio 2014 A new method for fabricating zirconia copings using a Nd:YVO4 nanosecond laser *Dental materials journal* **33** 422

[2] Andreev A, S Kosenko, V N Petrovskiy 2015 The use of laser milling for prototyping of dental structures *Bulletin of the Lebedev Physics Institute* **42** 255-259

[3] French RH, Glass SJ, Ohuchi FS, Xu YN, Ching WY Experimental and theoretical determination of the electronic-structure and optical-properties of 3 phases of ZrO_2 1994 *Phys Rev B Condens Matter* **49** 5133

[4] A P Menushenkov, V N Nevolin, V N Petrovsky 2010 *Physical principles of laser technology* (Moscow: NRNU MEPhI) p 34

[5] Dirk Hellrung, Li-Ya Yeh, Frank Depiereux, Arnold Gillner, Reinhart Poprawe High-accuracy micromachining of ceramics by frequency-tripled Nd:YAG lasers 1999 *Proc. SPIE 3618, Laser Applications in Microelectronic and Optoelectronic Manufacturing IV*

[6] Stuart B C, Feit M D, Herman S, Rubenchik A M, Shore B W Nanosecond-to-femtosecond laser-induced breakdown in dielectrics 1996 *Phys Rev B Condens Matter* **53** 1749

[7] Anisimov Sergei & Luk'yanchuk Boris 2007 Selected problems of laser ablation theory. *Physics-Uspekhi* **45** 293

IOP Conf. Series: Journal of Physics: Conf. Series **1238** (2019) 012029 doi:10.1088/1742-6596/1238/1/012029

Microstructure and mechanical properties of stainless steel 316L obtained by Direct Metal Laser Deposition

D P Bykovskiy, E D Ishkinyaev, V N Petrovskiy, A V Osintsev, P S Dzhumaev, V I Polskiy, K L Sergeev, A S Shchekin
National Research Nuclear University MEPhI (Moscow Engineering Physics Institute), Moscow, Russia
E-mail: ishkinyaev.emil@gmail.com

Abstract. The microstructure of 316L stainless steel obtained by layer-by-layer direct metal laser deposition is reviewed. Mechanical tests of the samples were performed in accordance with GOST 1497-84. Studies show that changes in power of laser radiation to grow parts lead to changes in their mechanical properties. The research shows dependencies between the strength characteristics of materials and the power of laser radiation. Causes of the forementioned changes are studied through the analysis of the microstructure. Nanosized inclusions of spherical shape were found in the process of studying the microstructure of materials. A study in the nature of the formation of these inclusions and their effect on the properties of the obtained material was performed.

1. Introduction

Addictive technologies not only let one obtain details of complex geometric shapes which can't be obtained by traditional methods, but also give an opportunity to give additional properties to the materials. By choosing various technological parameters of cladding process we can impact the inner structure of materials and obtain details with various required properties.

In this research, direct metal laser deposition (DMLD) technology is used. This method is based on layered jet printing, when the powder is dispensed through the coaxial laser-beam nozzle. This technology is frequently used both for damaged surface recovering and for obtaining details from a scratch.

The research is made towards the analysis of properties and microstructures of the details obtained with DMLD technology. These objects have improved qualities compared to fully identical details obtained by the traditional means of molding without thermal treatment. Possible reasons behind detail hardening compared to molded details and their property changes under various laser radiation capacities in the process of DMLD are observed in this research.

A cyclic impact of laser radiation is made on the obtaining object in the process of layered jet printing. The number of cycles equals the number of layers. A multiple heat treatment in the process of DMLD makes a great impact on the alteration of its mechanical properties.

Besides that, oxides of metals compounded with the powder appear in the obtained details. The reasons of their appearance may be both primary oxidation of powders and interaction with air in the process of cladding, which can't be entirely avoided using this printing technology. These inclusions can alter the properties of materials for the better, since they can be centers of material defect dislocation.

IOP Conf. Series: Journal of Physics: Conf. Series **1238** (2019) 012029 doi:10.1088/1742-6596/1238/1/012029

2. Sample growing

Stainless steel mark 316L powder was taken to perform this experiment. The content of chemical elements in this material is shown in the Table 1.

Table 1. Chemical element content. Here, the nominal values are given, without taking into account air oxidation.

Mark powder	Fe	C	Cr	Mo	Si	Mn	Ni
316L	66.17	0.03	17	2.5	0.8	1.5	12

Samples made under various radiation power were obtained using HUFFMAN HC-205 installation equipped with fiber laser capable of delivering 3.5 kilowatts. The radiation power was altered as the most influential parameter for the properties of materials [1]. Technological parameters of laser in the process of detail growing are shown in Table 2. For the purpose of residual tension removement, the layers were cross-cladded [2, 3]. Further on, electroerosional cut was used to obtain blades for rupture testing.

Table 2. Technological parameters of the cladding process. Laser beam diameter on the surface is 300 mcm. The scanning was made in 0.5 mm step and the rollers are cladded tight without any zones unfilled by metal.

Sample number	Power, W	Scanning speed, mm/min	Powder consumption gr/min
1	200	1000	1.2
2	250	1000	1.2
3	300	1000	1.2
4	350	1000	1.2

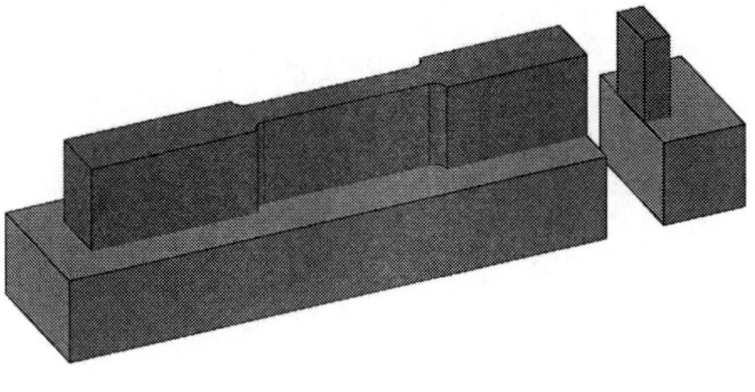

Figure 1. 3D model of the grown sample. The lines show the cutting planes for obtaining the plates for the tensile tests and the sections for analyzing the microstructure.

3. Mechanical properties

Sample stretching using INSTRON 5966 machine showed the following characteristics:

- Tensile strength decreases with the rise of radiation power amount, but the plasticity increases.
- Yield strength increases with power increase.

The Figure 2 shows the results of rupture testing of samples cut from the same areas, so they are matched in the order up from the substrate and are processed through the same number of heating cycles.

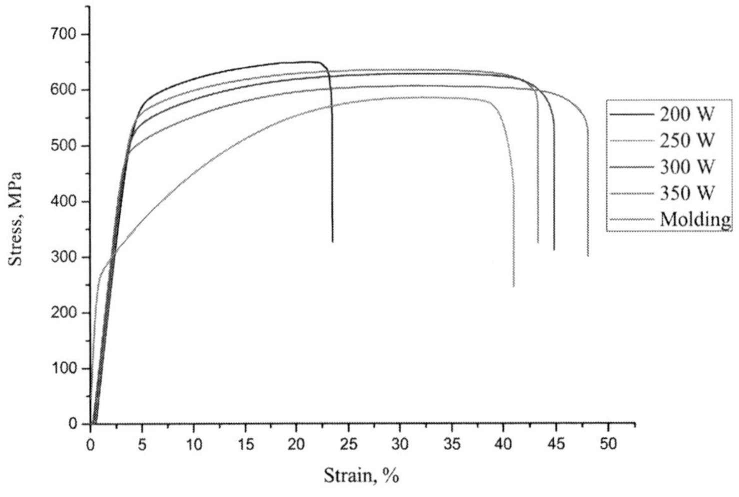

Figure 2. Stress-strain diagram of the samples grown at different laser radiation powers.

While Figure 3 shows the results of rupture testing of all samples on all layers. The diagrams show inevitable heterogeneity in samples properties depended on the height, because every layer is processed through different impact of laser radiation during the cladding.

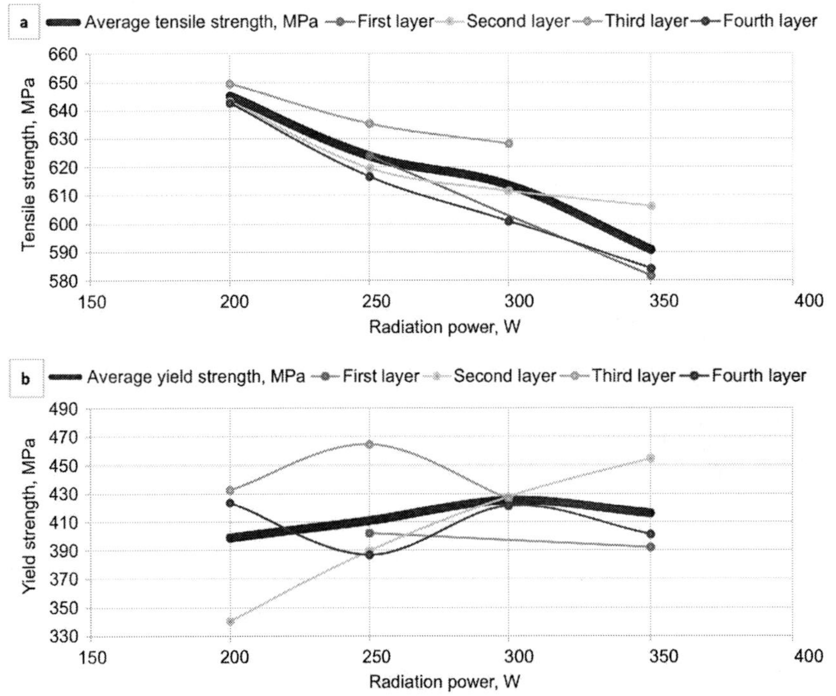

Figure 3. Dependency of tensile (a) and yield (b) strengths on laser radiation intensity. A layer is meant a plate 2 mm thick, located at a certain height from the substrate, the numbering of the layers is from the upper surface of the samples.

Obtained microstructures and mechanical properties of addictive technology materials are not well-studied because of uniqueness and complexity of thermal cycles during the machining. Heat treatment hugely impacts the materials and, due to it, objects obtained through addictive technology methods can have better strength characteristics than the usual ones [4, 5].

4. Microstructure

Structure of the obtained samples is studied using the structural method. Structural method is based on direct observation of material's structure and includes both microscopic and macroscopic analysis. Surfaces of the cladded samples were studied using Carl Zeiss EVO 50 microscope (Made in Germany).

Figure 4 shows that more dark areas appear with radiation power increase. The contrast depends on the crystallographic orientation of the structure's elements. The crystals grow towards the heat sink. It is clearly seen that the orientation of the cells is more chaotic, and that's the reason why these samples have lower tensile strength.

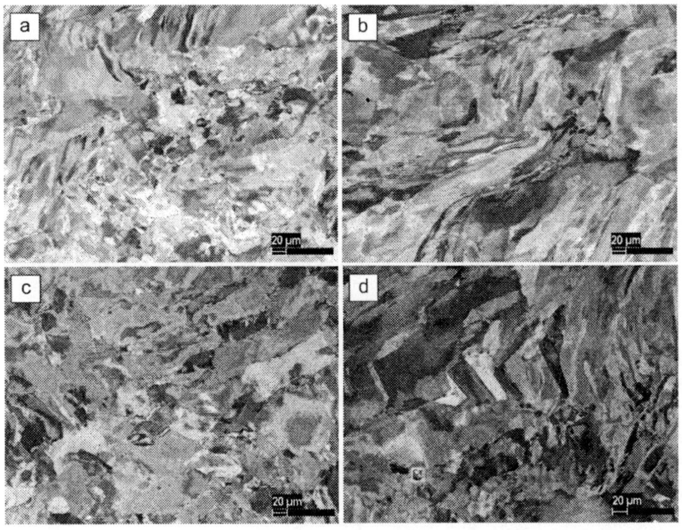

Figure 4. Microstructure (×250) of grown samples in 3 mm from the edge (2nd layer) with different radiation power: a) 200 W, b) 250 W, c) 300 W, d) 350 W.

An even meshed structure resulting in higher tensile strength can be observed in the samples shown in Figure 5. This effect is observed in medium rate of cooling. Elements of such structure are ground with the increase of cooling rate. With higher cooling rate branchings, indicating the start of dendrite crystallization, appear instead of cells. Smaller cells result in higher tensile strength [6].

Figure 5. Microstructure (×1000) of grown sample in 3mm from the edge (2nd layer) with radiation power – 300 W (other areas are the same).

IOP Conf. Series: Journal of Physics: Conf. Series **1238** (2019) 012029 doi:10.1088/1742-6596/1238/1/012029

Microstructure on the Figure 6 shows minor changes with the height of the sample. Those changes derive from different number of thermal cycles of the layers and are much less significant than the previous ones depending on the power of the laser radiation.

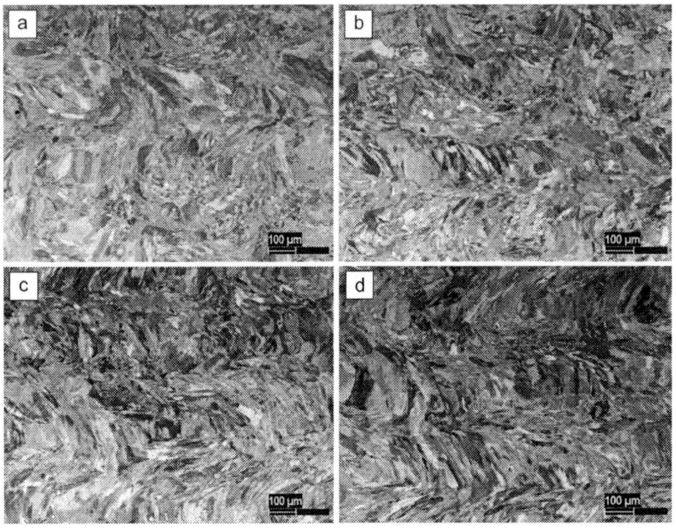

Figure 6. Microstructure (×100) of samples grown at one radiation power (300 W), but at different heights from the upper edge: a) 1 mm, b) 3 mm, c) 5 mm, d) 7 mm.

The analysis of samples structure in the areas of spalling shows presence of spherical particles what look like unmelted powder, though their size is smaller than the size of powder particles - around 100nm, which is 2-3 times smaller than unmelted powder particles (Figure 7).

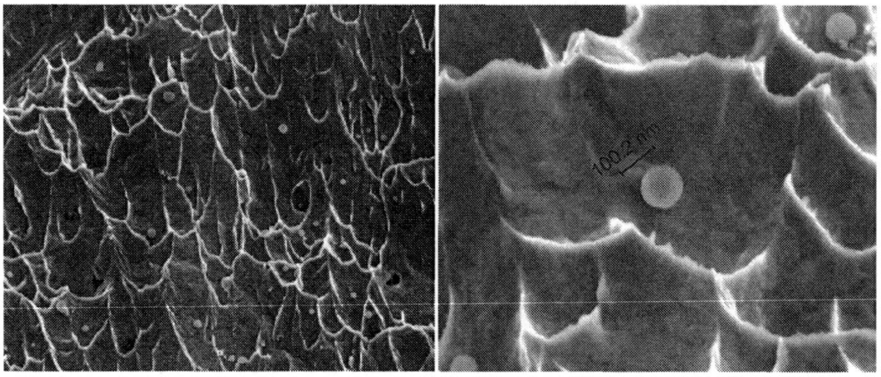

Figure 7. Topography of spalled surface (microscopic view).

Chemical analysis of these particles shows them as being oxides of metals presented in material matrix. They are formed at interaction with air as a result of fast heating-cooling processes. In the process of cladding, defects will necessarily form [7]. This defect areas can form evenly around these particles, resulting in tensile strength improvement.

These oxides appear to be black in the photographs made with electron microscope (Figure 5). They are evenly distributed in on the edges of grains and their concentration and size have no correlation with radiation power during growing of samples and their placement in sample's volume.

5. Results

The influence of the radiation power on the mechanical properties of the grown samples is presented in this research. With increase of radiation power, the ultimate strength decreases, but the plasticity properties increases. The most common explanation of these results is the fact that at a lower radiation power, a more ordered crystallographic structure is observed. All samples have higher performance of these characteristics than conventional parts of the same material, obtained by traditional methods of molding. Apparently, this is associated with simultaneous thermal treatment of the material during growing. The research shows no heterogeneity of properties over the volume of the material, due to the fact that each layer passes through different processing cycles. The heterogeneity of the obtained values of the strength characteristics are not very noticeable in comparison with the previous one and can subsequently be reduced by changing the laser radiation power during the cladding process.

Based on the results of the microanalysis, spherical particles were found in the volume of the samples, which are the oxides of the matrix metals. Their size and concentration do not depend of the radiation power and the location of the height of the part. They are formed due to the primary oxidation of the powder or the interaction of the melt bath with air. Despite the fact that surfacing occurs with the supply of an inert gas, it is impossible to completely avoid oxidation. There are suggestions that this phenomenon can affect the properties positively, since these particles can collect defects around themselves, evenly distributing them throughout volume.

References

[1] Grigoryants A G, Shiganov I N and Misyurov A I 2008 *Technological processes of laser processing* 2nd ed (Moscow: Bauman Moscow State Technical University)
[2] Kudzal A, McWilliams B, Hofmeister C, Kellogg F, Yu J, Taggart-Scarff J and Liang J 2017 *Materials and Design* **133** 205–215
[3] Wang D, Yu C, Ma J, Liu W and Shen Z 2017 *Materials and Design* **129** 44–52
[4] Popovich V A, Borisov E V, Popovich A A, Sufiiarov V S, Masaylo D V and Alzina L 2017 *Materials and Design* **131** 12–22
[5] Hengsbach F, Koppa P, Holzweissig M J, Burns M, Nellesen J, Tillmann W, Tröster T, Hover K and Schaper M 2017 *Materials and Design* **133** 136–142
[6] Bunin K P and Baranov A A 1970 *Metallography* (Moscow: Metalurgiya)
[7] Darvish K, Chen Z W and Pasang T 2016 *Materials and Design* **112** 357–366

IOP Conf. Series: Journal of Physics: Conf. Series **1238** (2019) 012030 doi:10.1088/1742-6596/1238/1/012030

Nano- and microstructured plasmonic substrates for laser and spectral applications

T T H Nguyen[1,3,4], S I Kudryashov[1,2], P A Danilov[1], A A Ionin[1], R A Khmelnitskii[1], A A Rudenko[1] and I N Saraeva[1]

[1]Lebedev Physical Institute, Leninskiy prospect 53, 119991 Moscow, Russia
[2]ITMO University, 197101 St. Peterburg, Russia, Russia
[3]Moscow Institute of Physics and Technology, 141701 Dolgoprudny, Moscow Region, Russia
[4]Ha Tinh University, precinct Dai Nai, city Ha Tinh, province Ha Tinh, Viet Nam

E-mail: trang.nguyenthihuyen@htu.edu.vn

Abstract. Nano- and micro-structures of thin plasmonic films by focused femtosecond laser pulses was demonstrated as chemo- and bio-sensing substrates for surface-enhanced spectroscopies. As an example, large regular arrays of microholes (mm-sized laboratory-scale) of millions holes with variable periods were produced in thin films of diverse plasmonic metals or alloys, demonstrating surface plasmon-polariton enhanced extraordinary transmission.

1. Introduction

Plasmonic and all-dielectric functional nano- and micro-elements and their regular and irregular arrays as sensors, circuits and metasurfaces are invoking the growing attention in nanophotonics, photovoltaics and photo-catalysis, ultra-sensitive chemo- and bio-sensing, during the last decade [1-6]. Intricate morphology of advanced nano- and microstructures implies their versatile state-of-art, but elaborate, expensive, time-consuming and non-flexible fabrication procedures – electron, ion- and photolithography [7, 8]. In this lieu, the rapidly developing laser patterning technologies present a promising viable alternative for the current well-developed lithographic nano- and micro-fabrication routines in terms of fabrication costs, throughput, modalities, ecological shortcomings and flexibility. Furthermore, such laser patterning technologies are currently or potentially compatible with ultra-precise laser-induced additive – multi-photon polymerization, surface chemistry, displacive – thin-film dewetting, communicative – (sub)micro- scale bulk inscription and drilling in supporting transparent substrate materials, and many other functional laser-based and non-laser post-processing subtractive procedures – chemical etching, ion-beam polishing [9, 10].

Nano- and micro-structured plasmonic and dielectric films make use of nano- and micro-elements in the form of shaped nano- and micro-holes, which are the structures with their optical/plasmonic responses complementary to separate, "free-standing" nano- and micro-elements (or their arrays) in terms of transmittance, reflectance or absorbance [11, 12] (the "complementarity" principle in optics [6]). Owing to small – nano- or (sub)micro-scale dimensions, localized plasmon-polaritonic or pure polaritonic modes in metallic or dielectric structures, respectively, become dominating in their electromagnetic response irrespectively their "closed" (a nanoshaped piece matter) or "open" (a nanoshaped hole in metallic or dielectric film) morphology [13]. Moreover, comparing to nanoshaped elements made of plasmonic materials with considerable dissipative losses, considerably increased optical strength is provided for "hollow" nano-elements because of the efficient heat sink to the surrounding reservoir metallic films, which is currently the crucial issue in plasmonic and all-dielectric nanophotonic sensing. Additionally, enhanced surface adhesion, chemical stability and mechanical robustness are characteristic of "hollow"-element arrays in thin non- and metallic films in comparison to those of shaped nano-elements [14].

In this work, we implicate thin films of diverse plasmonic metals (silver, copper, aluminum, gold-palladium alloy) as a starting point for fabrication of elaborate and multi-scale nano/micro-patterns through diffractive optical elements (DOE)-mediated femtosecond laser nano- and micro-scale ablative and displacive processes. The advanced patterning technology was also employed in ultrafast fabrication of mm-sized laboratory-scale of surface enhanced infrared absorption (SEIRA) sensors, comprised by regular arrays of ~1 million micro-holes in thin CaF_2-supported films with variable micro-hole period. These sensors allowed enlightening experimental IR-spectral studies of extraordinary transmission (EOT) effect in these micro-hole arrays and its many-fold surface plasmon-polaritonic (SPP) enhancement. Finally, label- and luminescence-free sensing of a mono-layer of highly luminescent Rhodamine 6G dye with many-fold (a few hundred times) analytical enhancements of its IR-absorption signals and the internal calibration of the deposited analyte concentration via IR-absorption measurements in the overtone range were demonstrated.

2. Extraordinary transmission in arrays of nano- and micro-holes

EOT is mediated by SPP and localized plasmonic effects on the metal surfaces (in thin films – both front and rear ones [6]) and on hole edges, respectively, effectively increasing the actual hole size available for transmission and revealing, cfg., a few resonant transmission bands for different directions in square arrays of holes as 2D photonic crystals [15]:

$$\nu_{SP}(i,j) = \frac{\sqrt{i^2+j^2}}{n_{SP}P} = \frac{\sqrt{i^2+j^2}}{P}\sqrt{\frac{\varepsilon_M+\varepsilon_D}{\varepsilon_M\varepsilon_D}} \quad (1)$$

where i,j are the integer indexes for different directions (i,j) in the array (grating), n_{SP} is the effective refractive index for surface plasmon-polaritons, ε_M and ε_D are the dielectric functions of the metallic and the dielectric media [15].

In this study, we harnessed the advantage of large-scale, high-throughput laser patterning of thin plasmonic films to comprehensively characterize IR-range transmission in large (~10^5-10^6 holes per array) arrays of micro-holes of variable periods in thin films with thickness 50 nm from diverse plasmonic materials (silver Ag, copper Cu, aluminum Al and gold-palladium (Au-Pd, 80%/20%) alloy.

Equation (1) based on the SPP-mediated transmission model for hole arrays in metal films [6, 15] enables to assign the main peaks in the transmittance maximum in Fig.1 to different directions in the gratings as 2D photonic crystals. In Fig.1 such normalized transmittance spectra are presented for different materials (Ag, Al, Cu, Au-Pd alloy) and different grating periods P. In agreement with this equation, the corresponding (1,0), (1,1) and (2,0)-peaks blue-shift monotonously versus decreasing P with their particular experimental spectral positions in the quantitative agreement with the predicted ones, considering the metal-dielectric (CaF_2, IR dielectric constant $\varepsilon_{CaF2} \approx 2$), rather than metal-air (IR dielectric constant $\varepsilon_{air} \approx 1$), interface of the gratings. Previously, such air-metal interface of nano-hole gratings was also demonstrated as predominating in their transmission [15].

Moreover, since both localized surface plasmons on micro-hole edges/walls and SPP propagating on the metal-air and metal-CaF_2 film interfaces are considered to be involved in the EOT phenomenon, it is instructive to evaluate their relative contributions from our experimental data for the thin metallic films, supporting waveguide-like transmission. Similarly, in Figure 1 the fixed 4-μm hole diameter (perimeter, square) is favorable for increasing amplitudes (contrast) of the (1,1)- and (2,0)-

peaks at increasing P, as the corresponding surface density of hole decreases versus P, decreasing the surface corrugation.

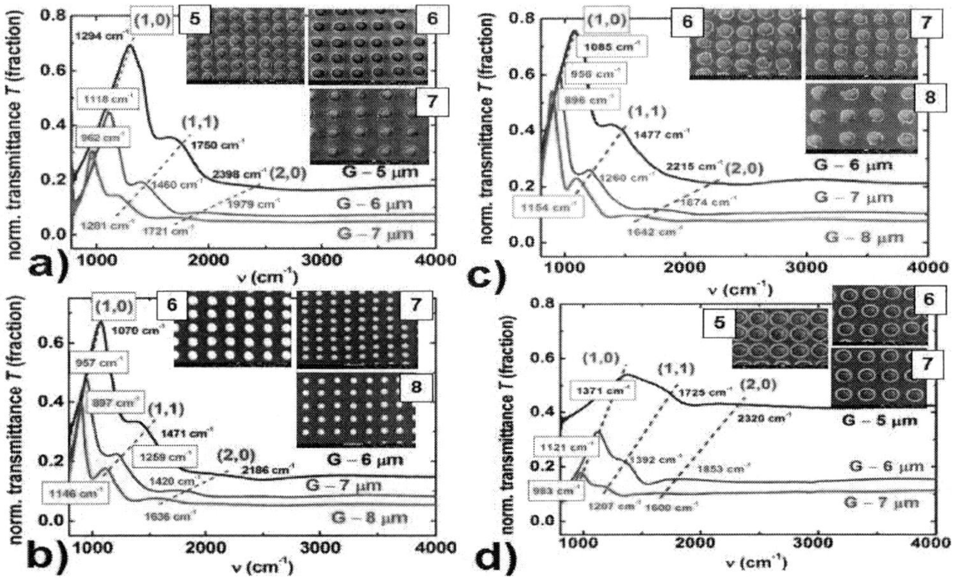

Figure 1. Normalized transmittance spectra of micro-hole gratings with the diameter D ≈ 4 μm on the 50 nm thick Ag (a), Al (b), Cu (c) and Au-Pd alloy (d) films on the CaF_2 substrates with the variable periods (shown by the same colors as the corresponding spectra), the colored numbers showing the spectral positions of their (1,0), (1,1) and (2,0)-peaks and the red dashed lines showing their evolution versus P. Insets: (a) top-view SEM images of the gratings with the diameters shown in the frames (scale bars can vary).

Finally, the abovementioned SPP enhancement of the IR-transmittance appears differently for the micro-hole gratings with the fixed 4-μm hole diameter and different similar periods on the films of the diverse metals used in this study. Specifically, the micro-hole gratings on the Au-Pd alloy film demonstrate considerably – by almost 50% – lower (1,0)-peak transmittance, comparing, e.g., to the similar gratings on the Ag-film (Fig.1a). Likewise, in the whole series of gratings the 4-μm hole diameter with similar periods of 6 and 7 μm in Figure 1, the ranking of metals according to (1,0)-peak amplitude is 1) Cu (0.75/0.6), 2) Al (0.7/0.5), 3) Ag (0.45/0.3) and 4) Au-Pd (0.3/0.2), indicating the best plasmonic response for copper and aluminum, the medium one for Ag and the worst one for the Au-Pd alloy (80%/20%), which has indeed main transport characteristics strongly – five-fold – diminished due to the high Pd-impurity concentration. In the case of silver films, their grain-like (island) structure and minor surface oxidation during low-vacuum (10^{-2} bar) magnetron sputtering can hamper the usual outstanding plasmonic properties of this metal.

3. Advanced chemo-sensing with mm-sized laboratory samples of micro-structured IR-sensors

Chemo-sensing of the produced 4×4 mm^2 gratings of micro-holes (approx. half-million of micro-holes) on the 30-nm thick silver film on the IR-transparent CaF_2 substrate (Fig.2a,b) was investigated in terms of their resonant EOT-enhanced IR-transmission as surface-enhanced IR-absorption (SEIRA) [3]. FT-IR transmission spectra of the pristine CaF_2 substrate, as-deposited Ag-film and the micro-hole grating (D ≈ 4 μm, P ≈ 6 μm) with an analyte – Rhodamine 6G molecules, was measured in vacuum in the near-mid IR range of spectral wave-numbers ν = 400-5000 cm^{-1}, using a FT-IR spectrometer V-70 (Bruker). These spectra normalized to the CaF_2 transmittance near the cut-off wavenumber ~ 1000 cm^{-1} are presented in Fig.2c,d, indicating the very minute characteristic fingerprint dips in the spectra in the ranges ν = 900-1300 cm^{-1} and 2800-3000 cm^{-1}, as R6G analytical signals. The extra-large micro-hole array with its SPP-enhanced EOT provide strong and robust transmission signals, enabling enlightening studies of the R6G distribution (i.e., potential pre-concentration or dis-concentration effects) and spectral characteristics on the array surface owing to different ultrafine physi- and chemi-sorption interaction with the structured Ag surface and the CaF_2 substrate.

The large-hole sensor with monolayer R6G demonstrates a few much stronger absorption lines for ν = 900 – 1300 cm^{-1} (the strongest one at 1261 cm^{-1}) (Fig.2c). In comparison with this sensor, similar surface density of R6G molecules on the bare CaF_2 substrate indicates much stronger absorption in the range of 2800 – 3000 cm^{-1}, while, surprisingly, much less absorption in the mid-IR range of 900 – 1300 cm^{-1}. Comparison of the mid-IR absorbance for the sensor- and CaF_2-supported R6G molecules demonstrates the analytical SEIRA enhancement – 40 ± 11, 47 ± 19 and 455 ± 230 times for the mid-IR spectral bands at 1024, 1097 and 1261 cm^{-1}, respectively (Fig.2c). However, being normalized to the six-fold enhancement of the absorption in the near-IR range (see above), the spectrally sensitive physical SEIRA enhancement appears to be more moderate –7 ± 2, 8 ± 3 and 76 ± 38 times at 1024, 1097 and 1261 cm^{-1}, respectively. Furthermore, the limit of R6G detection for the sensor, taken for the most SEIRA-enhanced band at 1261 cm^{-1}, at the 3σ-level is equal to LOD = Δ/A1261 ≈ 10^{-2} monolayer.

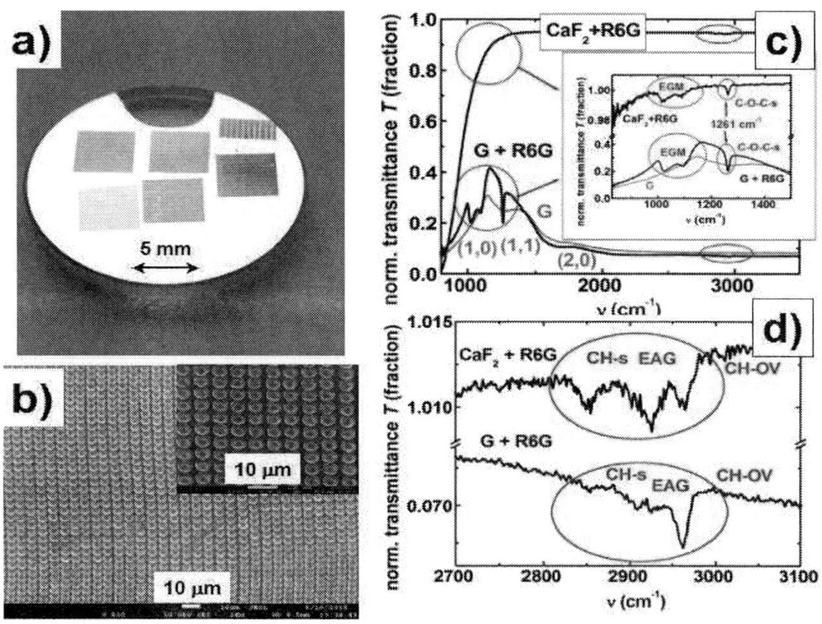

Figure 2. (a) Optical image (general view) of the 11-mm wide CaF_2-slab with the top 30-nm thick Ag-film and a number of micro-hole gratings (typical square – 4×4 mm^2). b) Top-view SEM image of the grating with D = 4 μm and P = 6 μm (inset: its magnified view). c) Normalized FT-IR transmission spectra of the grating with D = 4 μm and P = 6 μm (G, (1,0), (1,1) and (2,0) – resonance transmission peaks) and of the grating with the R6G monolayer atop (G+R6G); FT-IR transmission spectrum of the CaF_2 substrate with the R6G monolayer is given for comparison. Inset: the magnified view of their normalized low-ν transmittance with the assignment of the R6G absorption bands on the CaF_2 substrate and the grating (see the different top and bottom scales) discussed in details in the text. d) Magnified view of their normalized high-ν transmittance in the overtone region with the assignment of the composition R6G absorption bands.

Such internal calibration was performed, considering the waveguide-like transmission through the array in the mid-IR range of 2800-3000 cm^{-1} at the transmittance level ≈0.08 (8%) in the clean array and ≈0.069 (6.9%) in the array with the R6G monolayer. The corresponding transmittance dips of R6G bands are 0.11% (≈2850 cm^{-1}), 0.35% (≈2920 cm^{-1}) and 0.2% (≈2960 cm^{-1}) on the CaF_2 substrate, representing combination CH-stretching modes of the main xanthene ring (XR), phenyl, ethyl and methyl groups with the lower-wavenumber external group modes (EGM), but with the overall wave-numbers less than those of C-H stretching overtones (Fig. 2d). On the array, these R6G absorption bands demonstrated, on average, much – almost 7-fold – lower signals of 0.04% (≈2850 cm^{-1}), 0.03% (≈2920 cm^{-1}), while the R6G absorbance at 2960 cm^{-1} unexpectedly appears even two-fold higher (0.4%). Comparing to the 13-fold stronger IR-transmission through the CaF_2 substrate – 92% vs 7% (this array), such 7-fold lower R6G absorbance on the array effectively means its 2-fold, on average, enhancement of R6G absorbance, with the corresponding 26-fold enhancement of the absorption band at 2960 cm^{-1}. Generally, such enhancement could occur due to plasmonic (spectrally homogeneous over the narrow range of 2800-3000 cm^{-1}) or chemical (R6G-Ag-comlexing, spectrally-selective for the different R6G bonds) factors, which will be discussed below during the analysis of the characteristic absorption bands. Meanwhile, taking the minimal enhancement conditions at ≈2920 cm^{-1}, 13-fold difference in the R6G absorbance on the CaF_2 and the grating could be expected over the entire IR spectrum at the absence of the relevant plasmonic and chemical enhancement factors.

4. Conclusion

DOE+MHz fs-laser through easy, green and low-cost high-throughput fabrication of complementary nano- and micro-structures – large arrays of plasmonic (or even dielectric) nano- and micro-elements. Enable advanced studies of topology- and material-dependent physical phenomena in metasurfaces by common laboratory spectroscopic tools, but also label- and luminescence-free laboratory-scale, robust and high-sensitivity sampling of chemicals and biosamples via plasmonic and chemical contributions, uneven and structurally-sensitive regarding different functional groups of the analyte molecules and band structure of the plasmonic metal.

Acknowledgments

This work was supported by the grant of the RAS Presidium program, the Russian Foundation for Basic Research (grant nos. 15-02-08400, 16-32-00880), as well as by the Government of the Russian Federation (Grant 074-U01) for ITMO University.

References

[1] Baikova T V et al 2017 *AIP Conference Proceedings* **1874** 040052
[2] Baikova T V et al 2016 *Las. Phys. Lett.* **13 (7)** 070602
[3] Danilov P A et al 2016 *Laser Phys. Lett.* **13** 055602
[4] Emelyanov V I 2014 *JETP Letters* **99(9)** 518
[5] Zayarny D A et al 2014 *JETP Lett.* **100** 295
[6] Klimov V 2014 *Nanoplasmonics* (Boca Raton, FL: CRC Press)
[7] Im H et al 2010 *ACS Nano* **4** 947
[8] Qiuming Yu et al 2008 *Nano Lett.* **8 (7)** 1923
[9] Danilov P A et al 2014 *Quantum Electron* **44** 540
[10] Kulchin Yu N, Vitrik O B, Kuchmizhak A A, Nepomnyashchii A V, Savchuk A G, Ionin A A, Kudryashov S I and Makarov S V 2013 *Opt. Lett.* **38** 1452
[11] Masatoshi Osawa, Masahiko Ikeda 1991 *J. Phys. Chem.* **95(24)** 9914
[12] Hartstein A, Kirtley J R and Tsang J C 1980 *Phys. Rev. Lett.* **45** 201
[13] Kuchmizhak A A, Ionin A A, Kudryashov S I, Makarov S V, Rudenko A A, Kulchin Yu N, Vitrik O B and Efimov T V 2015 *Opt. Lett.* **40** 1687
[14] Ionin A A, Kudryashov S I, Makarov S V, Levchenko A O, Rudenko A A, Saraeva I N, Zayarny D A, Nathala C R and Husinsky W 2015 *Laser Phys. Lett.* **13** 025603
[15] Ghaemi H F, Thio, Tineke, Grupp D E, Ebbesen T W, and Lezec H J 1998 *Phys. Rev. B.* **58(11)** 6779

IOP Conf. Series: Journal of Physics: Conf. Series **1238** (2019) 012031 doi:10.1088/1742-6596/1238/1/012031

Experimental study of the filamentation phenomenon and measurement of filaments parameters

M A Tarasova, K S Khorkov, D A Kochuev,V G Prokoshev and A V Ivaschenko

Vladimir State University named after A. G. and N. G. Stoletovs, Vladimir, Russia

E-mail: trsvmargarita@gmail.com

Abstract. The article experimentally investigated the phenomenon of filamentation in the propagation of femtosecond laser radiation in transparent solid media. The experimental scheme for measuring the parameters of filaments using spherical and cylindrical optics was developed and assembled. A series of experiments on the spatial ordering of filaments was carried out. The sample was structured in the filamentation mode with a predetermined distribution of laser radiation intensity. The spatial redistribution of laser radiation passing through the formed structures in the quartz glass sample is also investigated.

1. Introduction

The phenomenon of filamentation of femtosecond laser radiation consists in the spatiotemporal localization of energy in a high-power laser femtosecond pulse under the action of self-focusing in a medium and nonlinearity in a self-induced laser plasma [1]. Unique possibilities of filamentation open new possibilities for using femtosecond laser technologies in micro-optics, material modification, atmospheric optics and other applications [2]. The necessary condition for the development of filamentation is the excess of the critical power of self-focusing. In the process of filamentation, the intensity increase in the self-focusing of femtosecond radiation is limited by the nonstationary defocusing of radiation in a self-induced laser plasma, which occurs when the intensity of the photoionization threshold of the medium is exceeded [3]. The change in the refractive index of the medium associated with the induced plasma can be regarded as the appearance of a defocusing lens [4]. With increasing intensity, multiphoton absorption begins to play an increasing role. Multiphoton absorption is a nonlinear process in which a material simultaneously absorbs several photons to excite an electron. The result of ionization of the medium passing through filamentation with laser radiation is the formation of a narrow plasma channel, the filament. Pulsed radiation, the peak power of which is tens or more times higher than the critical power of self-focusing, forms a multitude of filaments.

2. Description of the experimental scheme

The ytterbium femtosecond laser TETA-10 was used as a source of laser radiation. The laser has the following parameters: the wavelength λ = 1033 nm, the radiation pulse duration τ = 280 fs, the pulse repetition rate f = 10 kHz, the pulse energy ε_{max} = 150 μJ. Polarization of laser radiation is linear.

As a sample, KU–1 quartz glass with geometric dimensions of 5×5×20 mm was used. KU - 1 is the optical quartz glass that is transparent in the ultraviolet and visible regions of the spectrum, without absorption bands in the wavelength range of 170-250 nm, with absorption bands in the wavelength ranges of 2100-2300 nm and 2600-2800 nm, non-luminescent, radiation-optically stable [5].

A series of experiments was carried out to measure the filaments parameters, namely the length and intensity distribution in the transverse profile of the beam [6]. The scheme of the experiment is shown in Figure 1.

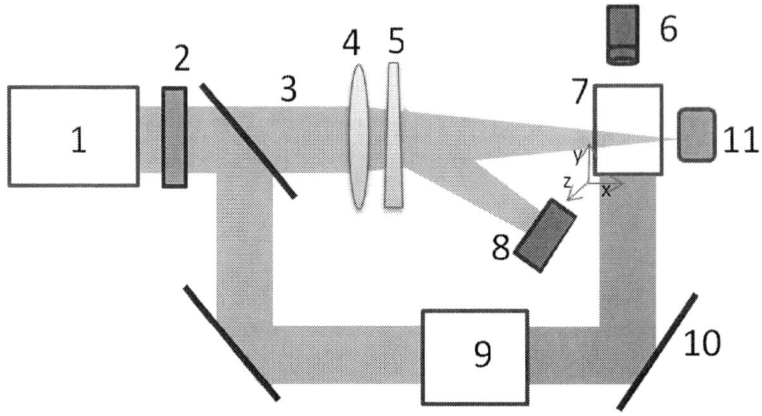

Figure 1. Scheme for measuring the parameters of filaments: 1 - laser system; 2 - polarization attenuator; 3 - laser radiation (LI); 4 - spherical lens; 5 - optical wedge; 6 - camera and micro lens; 7 - sample; 8 - photodiode sensor, 9 - frequency converter unit (514 nm, 257 nm); 10 - the mirror; 11 - measuring device of the transverse profile of the laser radiation intensity distribution.

A spherical lens with a focal length of 150 mm was used to focus the laser radiation.the Sample was fixed on a precision moving table x-y-z, which allowed to move it relative to the focus. In our case, the focus was located behind the back of the sample to prevent its destruction. Photo-registration of the filamentation zone was carried out with the help of a camera and a micro lens. They were placed on a three-coordinate precision table above the sample.

3. Measurement of filament parameters

The parameters of filaments were measured, namely: the length of the filament, the dependence of the length on the power, the critical value of the power of the beginning of filamentation, as well as the power necessary for the beginning of the multiple filamentation effect. Images of the filamentation zone at different power values are shown in Figure 2. As the radiation power values increase, the length of the filament increases.

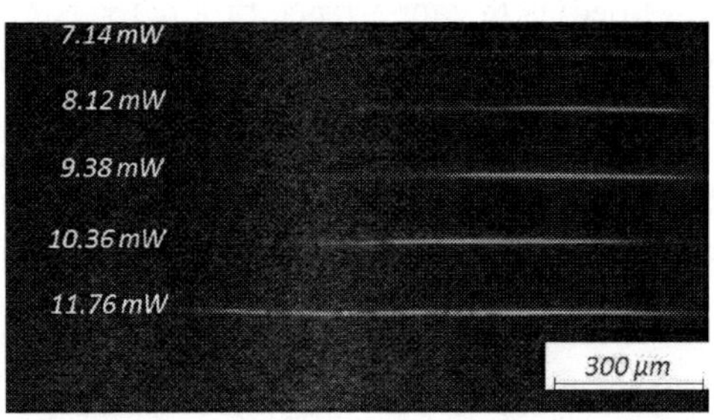

Figure 2. Filamentation in fused quartz KU-1.

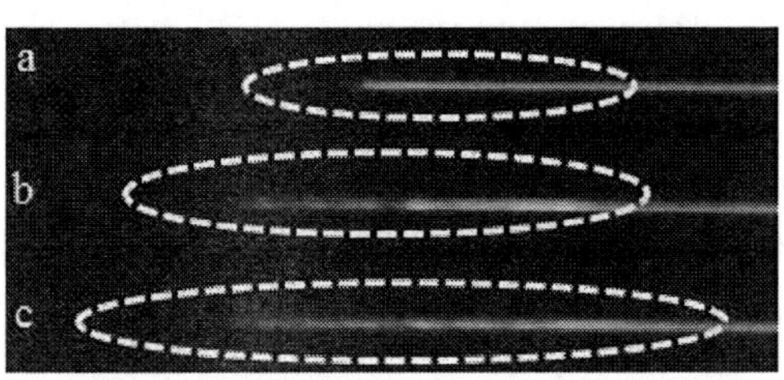

Figure 3. Growth of the filament: (a) at the beginning; (b) after 5 minutes; (c) after 10 minutes.

The critical power value was determined experimentally, which was 5.18 mW. When the value of 38.5 mW was reached, the effect of the formation of multiple filamentation was observed.

Also in the course of the experiment, luminescence of the impact area was observed, which indicates structural changes in the environment. The growth of the filament has been observed over time, which is clearly shown in Figure 3.

During filamentation, the intensity is redistributed in the transverse profile of the beam. When high-intensity radiation passes through the sample, areas with a changed refractive index are formed in the places of self-localization of radiation. And in the future, the formation of filaments occurs on modified areas. Figure 4 shows the intensity distribution in the transverse profile of the beams at different values of the radiation power.

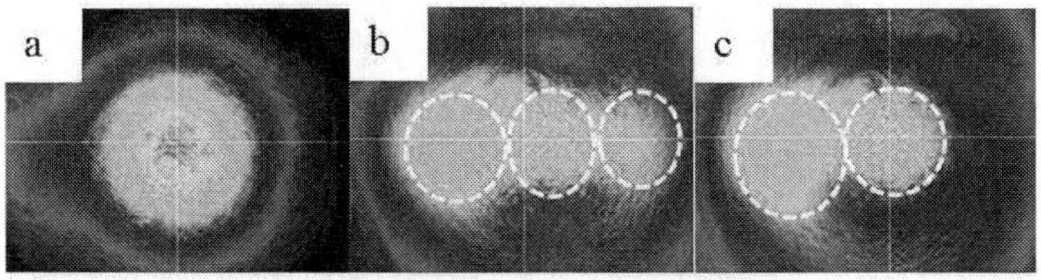

Figure 4. Distribution of the transverse intensity profile of the laser beam: (a) initial beam; (b) power 4.2 mW; (c) power 9 mW.

Thus, using the phenomenon of filamentation, it is possible to obtain the redistribution of intensity, forming several local maxima.

4. Control of the spatial arrangement of filaments

Multiple filamentation of powerful femtosecond laser pulses is inevitable and stochastic. It is therefore of interest to examine the question of control of filamentation. The use of focusing, for example, cylindrical or axial optics, is one of the simplest ways to control the parameters of laser filaments and their plasma channels [7].

A series of experiments using cylindrical optics was carried out [8]. The use of cylindrical optics in experiments allows the formation of filaments arrays propagating in one plane [9]. In this case, the intensity redistribution in the transverse profile of the beam, shown in Figure 5.

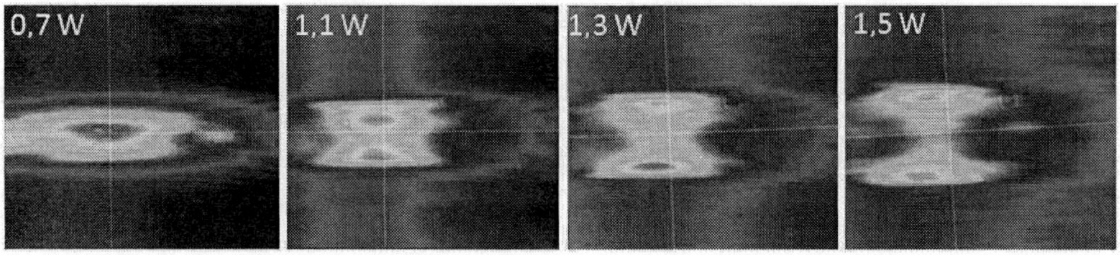

Figure 5. Intensity distribution in the transverse beam profile at different laser radiation power values using cylindrical optics.

Figure 6. The development of filamentation in the volume of medium at a power of 1.5 W

Figure 6 shows the distribution of filaments in the sample at a laser radiation power of 1.5 W. The distribution of filaments occurs in one plane.

An experiment was carried out to modify the sample using the phenomenon of multiple filamentation, when the cylindrical lens was rotated several times around the radiation axis. When the

IOP Conf. Series: Journal of Physics: Conf. Series **1238** (2019) 012031 doi:10.1088/1742-6596/1238/1/012031

cylindrical lens rotated around the radiation axis, the spatial position of the focus changed. To visualize the structures formed, the sample was illuminated by ultraviolet radiation with a wavelength of 257 nm, according to the scheme shown in Figure 1. Ultraviolet radiation, getting into the sample of quartz glass, caused a pronounced luminescence of the formed structures.

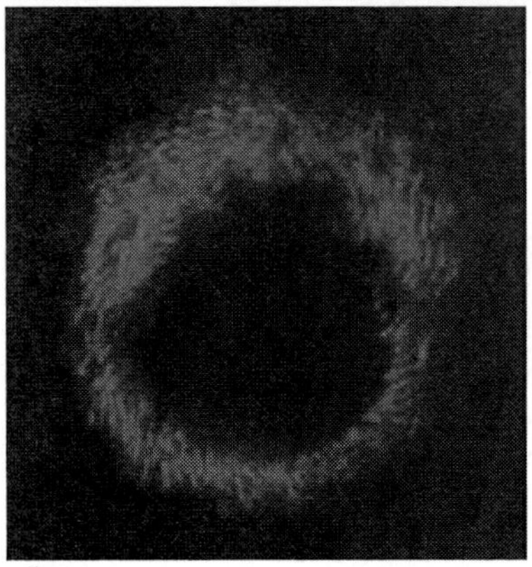

Figure 7. The structure formed in the process of filamentation.

Figure 7 shows the structure recorded in the sample during filamentation using cylindrical optics. Formation of the ring structure achieved the rotation of the cylindrical lens, in this experiment, it rotated 10 times around the axis of the laser radiation.

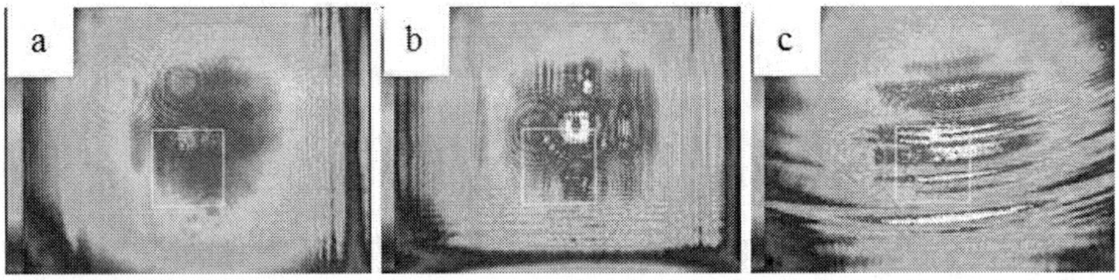

Figure 8. The intensity distribution in the beam: (a) when passing through a clean sample (b) when passing through a modified sample located along the laser radiation (c) as it passes through a modified sample perpendicular to the laser radiation.

Figure 8 shows the intensity distribution of the laser radiation passing through the modified sample (Figures 8(b), 8(c)). It can be seen that structures have formed, and there is also a redistribution of intensity. Laser radiation is propagated by the modified area.

To create more equal structures, the laser power must be stable and the sample must be rotated, not the lens. This requires special technical equipment and a computer program that will carry out the automated movement of the sample along a specified trajectory.

5. Conclusion

Thus, the parameters of the filaments were measured. The processes occurring during the interaction of femtosecond laser radiation with a dielectric transparent medium are studied. A series of experiments using cylindrical optics was carried out, which makes it possible to obtain an ordered arrangement of filaments in one of the planes. The control of the spatial location of the filaments is of great interest, since it can be used for the modification and structuring of the surface of materials.

In the course of the experiments, the formation of structural changes in the medium in the places of laser radiation passage was recorded, which directly affects the further spatial distribution of filaments. Thus, controlling the spatial arrangement of the filaments, we obtain the intensity distribution with a given geometry. The sample was modified using the phenomenon of filamentation and cylindrical optics. The transverse profile of laser radiation distribution was measured. This phenomenon can have a wide range of practical applications, for example, in the precision processing of materials and various laser processing operations.

Further optimization of the experimental scheme for sample structuring and obtaining a specified laser beam intensity distribution is planned.

Acknowledgements
This study was performed as part of the state task VlSU 1106/17 and grants of the Russian Foundation for Basic Research 16-48-330651, 16-08-01226.

References

[1] Chekalin S V, Kompanets V O et al. 2010 *Quantum Electronics* **43** (4) 326-331
[2] Kandidov V P, Shlenov S A, Silaeva E P, Dergachev A A, 2010 *Optics of the atmosphere and ocean* **23** (10) 873-884
[3] Chekalin S V, Kandidov V P, 2013 *Successes of physical sciences* 183 (2) 133- 152.
[4] Kandidov V P, Fedorov V Yu, Tverskoi O V, Kosareva O G, Chin S L, 2011 *Quantum Electronics* **41** (4) 382–386
[5] Zverev V A, Krivopustova E V, Tochilin T V, 2013 *Optical materials. Part 2.* (St. Petersburg ITMO) p 248
[6] Tarasova M A, Kochuev D A, Khorkov K S, Prokoshev V G, Ivashchenko A V, 2018 Experimental study of the filamentation phenomenon and measurement of filaments parameters *IV Int. Conf. "Laser, plasma studies and technology » LaPlaz-2018* 518-519
[7] Theberge F, Liu W, Simard P.Tr. et al, 2006 *Physics Review E.* **74** 036406
[8] Tarasova M A, Khorkov K S, Kochuev D A, Ivashchenko A V, Chkalov R V, 2018 Experimental study of the filamentation phenomenon of femtosecond laser radiation XXV *Int. conf. of students, graduate students and young scientists on fundamental sciences "Lomonosov-2018". Section "Physics" (Moscow)* 618-619.
[9] Bérubé J–Ph, Vallée R et. al, 2010 *Optics Express* 18 (3) 1801–1819

Modelling of temperature profiles and stress in CW laser diode bars

V V Bezotosnyi[1], **V P Gordeev**[1,2], **V A Oleshchenko**[1]

[1] P.N. Lebedev Physics Institute, Russian Academy of Sciences, Moscow, Russia
[2] National Research Nuclear University MEPhI (Moscow Engineering Physics Institute), Moscow, Russia

E-mail: vs.gordeev@yandex.ru

Abstract. We present the results of 3D thermal regimes modelling of CW laser diode arrays (LDA) assembled on the CS-mount type heat sinks. The dependences of the LDA temperature profiles on the fill factor (FF) of the LDA aperture at different levels of thermal load in different sections on LDA, including those inaccessible for direct experimental measurement calculated and discussed.

1. Introduction

Semiconductor LDA have found a wide range of practical applications, ranging from medical devices in surgery and laser hair removal, to materials processing and laser inertial thermonuclear fusion. LDA's are compact, efficient, the output parameters of LDA continue to improve, thus expanding applications at the forefront of scientific and technical progress. Unfortunately, in many cases of practical applications, the conditions of LDA's operation do not allow to achieve the most effective heat flux removal from semiconductor crystal. This is a main limiting factor for the output power, especially in high-power CW operation. To increase the output power and reliability it is necessary to optimize the distribution of thermal energy released in the emitting clusters of the array. Not uniform distribution of heat and the mechanical stresses accompanying them in LDA leads to a loss in the quality of radiation, which also significantly limits their applications efficiency, particularly in systems of incoherent spectral summation of LDA radiation in order to achieve high brightness and high power.

Thus seems useful to study the temperature profiles and mechanical stress in the LDA under different operational conditions and their dependences on the main constructive parameters: FF, cluster width, pitch, optical resonator length, etc. in order to optimize these parameters for the increase of the output power, improvement of power uniformity from cluster – to cluster as well as to improve the spectral characteristics. This optimization is also valuable for guaranteed service lifetime.

The Comsol Multiphysics software environment was used for three-dimensional modeling of the thermal regime of CW LDA assembled at CS-mount type heat sinks by the finite element method. The dependences of temperature profiles on filling of the LDA aperture by radiating clusters (FF) and the number of clusters in the LDA at different levels of thermal load were studied. LDA width was 10 mm, FF varied in a commonly used range of 30% - 70%, modeled number of clusters was 20, 47 and 67 and resonator length 1.5 and 2 mm. Conditions of operation - CW pumping, conductive cooling, boundary condition - a constant temperature of the down face of the CS-mount equal to 20 ° C corresponding to an experimental condition of water cooling with a chiller. The model contains the following assumptions - uniform heat transfer from the active area of LDA to the bottom of the heat-conducting element and the planar geometry of the heat transfer boundary. The range of parameters variations were determined by the characteristics of experimental samples of LDA produced and studied in [1-4].

2. Simulation result

The simulation was conducted with a 10% step variation of the fill factor, depending on the power, thermal load and the number of clusters. Figure 1 shows a General view of a three-dimensional model of deformations in LDA assembled at a CS-mount heat sink type, operating at a typical CW thermal load 60 W, corresponding to the experimental conditions for devices operated in the reliable regime.

For a more convenient understanding of the qualitative picture of arising thermo-elastic stress, the deformation values presented at figure 1, were reinforced by 40 times. This picture clearly shows the nature of LDA strain. The central part of the upper surface of C-S mount and its front face are deformed upwards and outwards respectively, relative to their position in the "cold" LDA (in the absence of thermal load) as a result of non-uniform heating of the LDA crystal. That non-uniform profile is formed by more temperature rise in central part of LDA in comparison with peripheral parts.

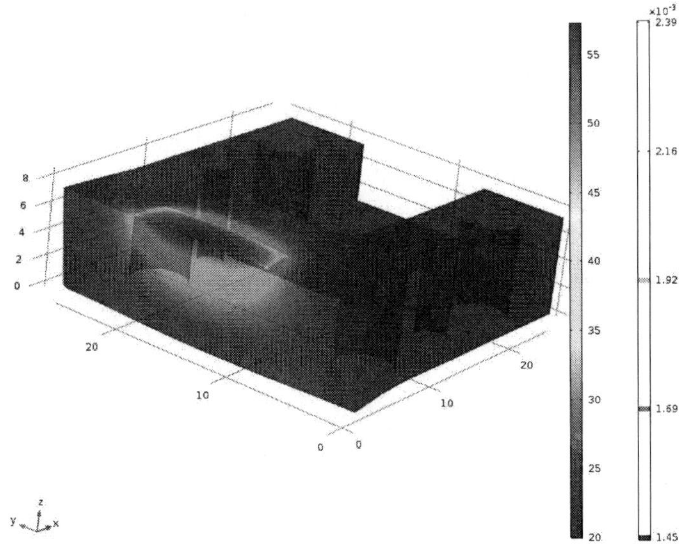

Figure 1. Model of laser diode bar

Figure 2 shows the typical temperature profile of the LDA depending on the fill factor at a fixed thermal load power of 60 W and 47 clusters. The temperature values are given for output mirror plane at the position of the active layer. The obtained distribution of the temperature profile line indicates the inhomogeneous character of the thermal field associated with the width and spacing of the emitting clusters. This graph also shows that the temperature in the central part of the LDA is much more than in the peripheral part. The temperature difference is about 10 ° C. Temperature steps in the central part of LDA between emitting clusters and passive areas are about 2.5 ° C. With an increase in FF from 0.3 to 0.7, the temperature steps amplitude is reduced by about 2 times. The absolute temperature of the clusters also decreases, at the same time the width of the cluster temperature peaks increases, exhibiting the smoothing of the temperature distribution.

IOP Conf. Series: Journal of Physics: Conf. Series **1238** (2019) 012032 doi:10.1088/1742-6596/1238/1/012032

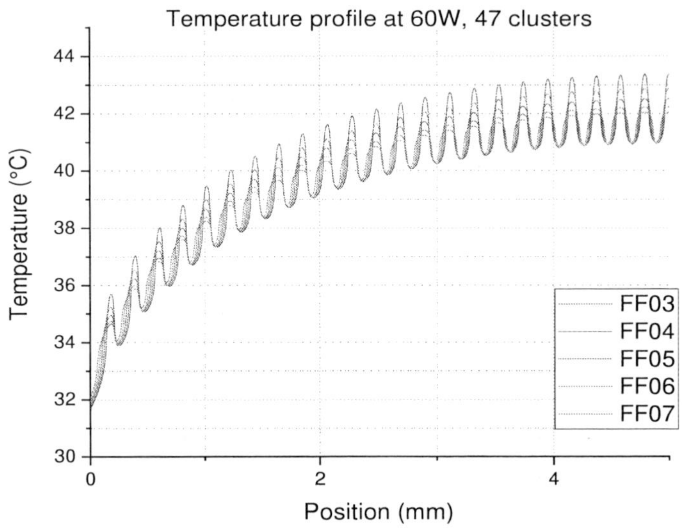

Figure 2. Temperature profile at 60W thermal load for 47 cluster LDA

Such changes in temperature values for different FF under fixed thermal load can be explained by the difference in the heat flux density because the conditions of the heat transfer are improved with the FF increase due to the increase of the emitting clusters surface area.

An important parameter for understanding the thermal processes in LDA is the temperature difference between the central and peripheral radiating clusters. This difference shows the degree of temperature non-homogeneity of the LDA in operation, which, in turn, affects the value of mechanical stresses, leading to a loss of radiation spectrum quality and change in other important parameters, in particular, reliability and service life.

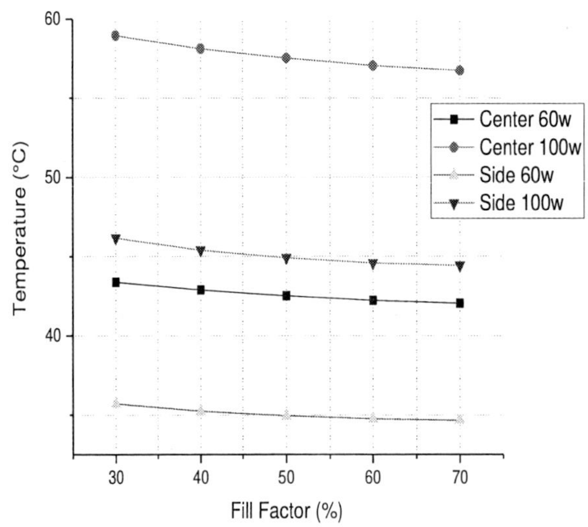

Figure 3. Temperature of the central and side emitter for 47 clusters LDAat 60 W and 100 W thermal load in FF range 0.3-0.7

At the base of the calculated data, the dependence of the temperatures of the central and peripheral clusters of the LDA on the filling factor were obtained. The graph of this dependence presented in figure 3 for a fixed thermal load power of 60 W and 100 W and 47 radiating clusters in LDA. As can be seen from the graph, according to an increase in the filling factor, the temperature tends to decrease. At FF = 70 % it is approximately 42 °C for the central cluster at 60 W of thermal load, while the temperature at the same thermal load is about 35 °C for the peripheral cluster.

According to the obtained data, the dependence of the temperature difference between the central and peripheral clusters on the number of clusters in the bar was built, the graph of this dependence at a fixed thermal load of 60 W is shown at figure 4.

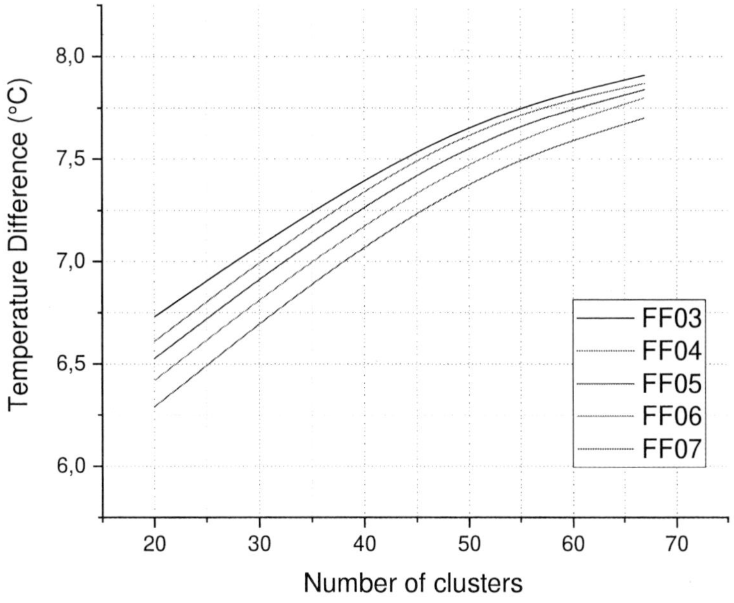

Figure 4. Temperature difference between central and side clusters in cluster number range 20-67 and FF range 0.3-0.7

According to the graph, with the increase in the number of clusters, the temperature difference tends to "saturation", the nature of that dependence is nonlinear. Such dependence character can be explained by the effective smoothing of the temperature field as the heat transfer area is filled with emitting clusters corresponding to the increase in number and density of heat sources. The total temperature difference in the whole range of the emitting clusters number 20-67 is not more than 1.5 °C. With increasing FF in the range of 0.3-0.7, the temperature difference decreases from 0.5 °C to 0.25 °C for 20 and 67 clusters.

3. Conclusion

3D simulation of temperature profiles and induced thermoelastic stress in LDA, assembled on CS-mount type heat sink in the range of CW thermal load power 20 – 100 W, was performed using finite element method. Temperature distribution in LDA in the range of FF 0.3 - 0.7 and emitting clusters number in the range 20-67 was obtained. At a fixed thermal load, increasing FF leads to a decrease in the cluster maximum temperature, equalization of the temperature field and induced thermo-elastic stress. The temperature field in the LDA laser crystal exhibit effective smoothing with the increase of the number of emitting clusters due to a more uniform filling of the heat spreading area by heat sources.

References

[1] Bezotosnyi V V, Kozyrev, A A, Kondakova, N S, et al 60-W continuous wave laser diode arrays emitting at a wavelength of 808 nm. Bulletin of the Lebedev Physics Institute 2016 V43 I12 p369-370

[2] Bezotosnyi V V, Kozyrev A A, Kondakova N S, et al Quasi-CW 808-nm 300-W laser diode arrays. Quantum Electronics 2017 V47 I1 p5-6

[3] Bezotosnyi V V, Gordeev V P, Krokhin O N, Mikaelyan G T, Oleshchenko V A, Pevtsov V F, Popov Yu M, Cheshev E A, Modelling and experimental study of temperature profiles in cw laser diode bars. Quantum Electronics 48(2) 115–118 (2018)

[4] Bezotosnyi V V, Gordeev V P, Oleshchenko V A Pumping level impact on optical power and spectra uniformity of CW laser diode bars. Quantum Electronics 48(6) 502-505 (2018)

IOP Conf. Series: Journal of Physics: Conf. Series **1238** (2019) 012033 doi:10.1088/1742-6596/1238/1/012033

Interaction of femtosecond laser radiation with titanium in a liquid hydrocarbon medium

A V Ivashchenko, D A Kochuev, K S Khorkov, V G Prokoshev and M A Tarasova

Vladimir State University named after A. G. and N. G. Stoletovs, 87 Gorky str., Vladimir, Russia

E-mail: alenaenergie@gmail.com

Abstract. A method for the synthesis of microspheres coated with titanium carbide is described. The processes occurring in the region of the action of laser radiation are considered. The formation of microspheres was carried out by the action of femtosecond laser radiation on the surface of titanium located in the reaction medium - the ultimate hydrocarbon. The hydrocarbon source was the medium in which the treatment was carried out. The obtained microspheres have an average size of 1-3 μm and can be used in applications of additive engineering, powder metallurgy as the main raw material, or as an alloying additive.

1. Introduction

Metal powders serve as the primary raw materials in the technologies of additive engineering. Products made on the basis of titanium alloys have high corrosion resistance, chemical inertness, heat resistance, low specific gravity and high strength. Powder materials have found their application in the technologies of aircraft building, in medicine in the manufacture of implants for dentistry and orthopedics.

At present, there are many ways and approaches for obtaining powders of various materials: by mechanical grinding, chemical precipitation, and magnetron sputtering. Each of these methods has its advantages and disadvantages, which determine the field of application of the obtained powder material. In spite of the very limited distribution of laser technological installations, the method of laser synthesis of powder material is of high interest due to the exceptional properties of laser radiation, unattainable by other sources of concentrated energy. Thus, it is possible to obtain powders from virtually any material. By selecting the processing mode it is possible to synthesize granules of different fractional and chemical composition. The peculiarities of laser action make it possible to achieve a change in the structural and phase states, to control the introduction of foreign impurities into it, which makes it possible to obtain pure powder materials in comparison with other synthesis methods. When introducing the necessary impurities, it is possible to obtain complex alloys, composites, and cold solutions in the required aggregate state.

In the field of obtaining powdered materials, the process of material destruction due to the action of laser radiation is of great interest, in particular, it is required to obtain not a melt, not steam, but directly "sprays", which in the process of rapid cooling acquire a spherical shape. By creating certain conditions it is possible to obtain particles with a narrow particle size distribution. For such purposes, pulsed laser systems capable of effectively transferring laser radiation energy for the development of the laser ablation process are used. The use of additional means, such as processing media, various sources of physical impact, can influence the dynamics and result of this synthesis process. In recent years, the processes that occur during laser ablation in a liquid have been extensively studied [1-4]. Laser ablation in a liquid medium allows rapid cooling of ablation products, which promotes the formation of a "highly spherical" surface. The liquid medium used can exhibit both inert properties during processing and reaction.

2. Equipment and samples

In experiments we used a Ti:Sapphire laser system with the following parameters: pulse duration 50 fs, average power 450 mV, wavelength 800 nm, repetition rate 1 kHz. A laser beam with a diameter of 30 μm was focused on the surface of a titanium sample located in the reaction medium. A titanium plate of grade VT1-0 was used as a sample. Figure 1 shows the scheme of the experiment. The cuvette was a container isolated from atmospheric air, having a point of attachment of a sample, a transparent window for introducing laser radiation. The sample was located under a layer of liquid hydrocarbon. The laser radiation power was controlled by removing a part of the radiation from the beam splitter plate. Scanning with a laser beam was carried out with the galvanometer scanner with a speed of 0.1-200 mm/s.

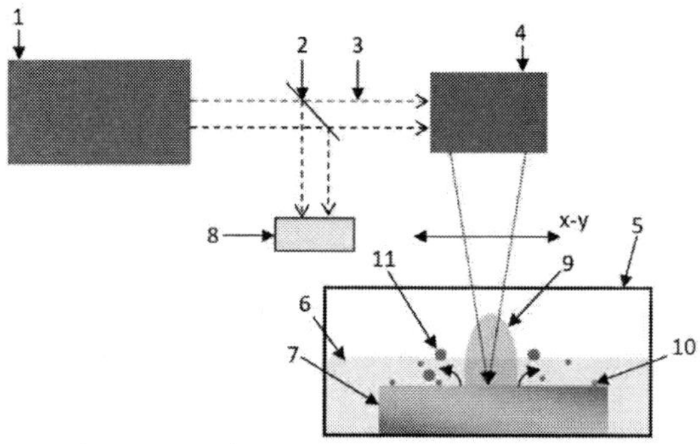

Figure 1. Scheme of the experiment. 1 - femtosecond laser system; 2 - light-emitting plate; 3 - laser radiation; 4 - optical head with galvanometer scanner (X-Y); 5 - vessel isolated from the atmosphere with laser input window; 6 - reaction medium; 7 – sample; 8 - optical powermeter; 9 - a plasma torch; 10- an aerosol of a liquid medium; 11 - a powder granule.

3. Conducting an experiment and discussing the results

A titanium sample of the brand VT1-0 was placed in a cuvette with the reaction liquid with the limiting hydrocarbon-hexane. The depth of immersion of the sample was from 2 to 4 mm. Laser radiation with an optical system focused on the sample, resulting in the formation of a vapor-gas cloud together with the ejection of material particles.

When laser radiation interacts with the sample surface located in a liquid medium, the laser radiation passes through the liquid layer, is absorbed by the surface of the sample. With a small energy contribution, the morphology of the surface changes, including the formation of surface periodic structures. The use of various media can lead to changes in the chemical composition of the treated surface, conditions for the interaction of titanium with the liquid medium are formed, unattainable when processing in gases in view of the greater density of the liquid medium. A schematic representation of the ongoing processes is shown in Figure 2a, the characteristic surface formed as a result of the processing is shown in Figure 2b.

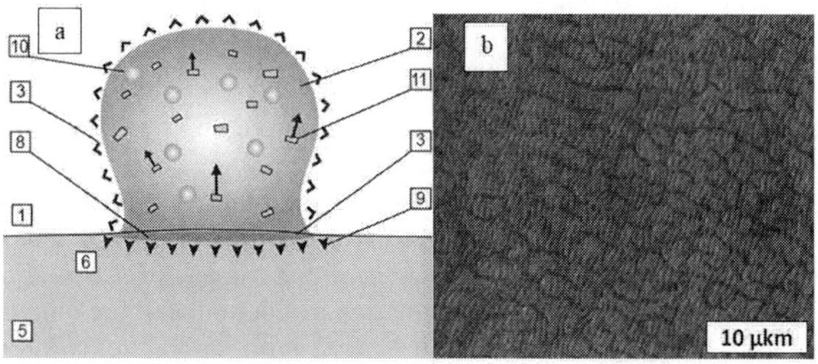

Figure 2. (a) 1 - liquid medium; 2 - steam and gas area; 3 - near-surface layer; 4 - formed micro-bodies; 5 - a sample; 6 - the area of heat removal into the volume of the material, 7 - subsurface overheating region; 8 - front of spreading / rising pressure; 9 - front of temperature propagation due to thermal conductivity; 10 - gas; 11 - the direction of dispersion of the material of the surface layer; (b) the surface formed as a result of the treatment.

An increase in the power of the laser radiation leads to the formation of a near-surface plasma, an intensive expansion of the ablation products into the liquid volume occurs, is schematically shown in Figure 3a. The ablation products rapidly cooled as a result of contact with the liquid and settled on the bottom and walls of the vessel. Rapid cooling of the ablated material promotes the formation of particles with a spatial shape corresponding to the body with the smallest surface area - the sphere.

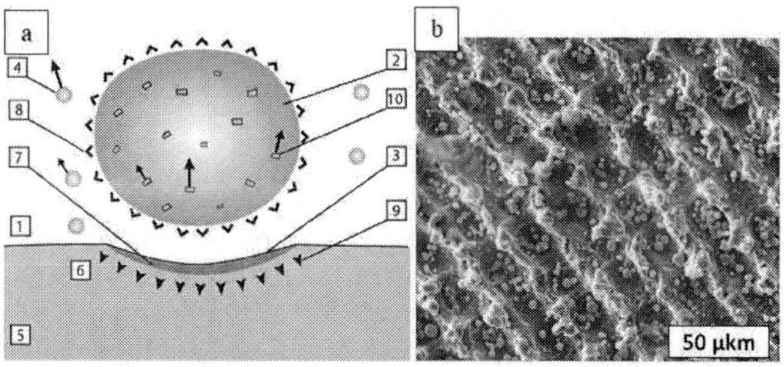

Figure 3. (a) 1 - liquid medium; 2 - plasma torch; 3 - area of action of laser radiation; 4 - formed microgranules; 5 - a sample; 6 - area of heat removal into the volume of material, 7 - zone-intensive reflow; 8 - front of temperature and pressure; 9 - front of temperature propagation due to thermal conductivity; 10 - particles of material forming a plasma torch; (b) the surface formed as a result of the treatment.

During the interaction of laser radiation with the surface of a titanium sample, dynamic erosion of the material was achieved by moving the laser beam along the processing plane, the SEM image is shown in Figure 3b. With prolonged exposure to laser radiation, a significant decrease in the intensity of the ejection of the processed material occurs over a localized area of the material over time. This effect is observed in view of the significant screening properties of the plasma. Laser radiation is partially absorbed by the plasma, partially reflected, the effect of laser radiation on the surface of the material is practically not realized. The structure of the formed surface is not typical for the treatment regime by ultrashort laser pulses. On the surface there are traces of reflow and ejection of the liquid phase of the material. The dominant role is played by the thermal contribution from the plasma torch, whose lifetime is estimated by hundreds of nanoseconds. The properties of the formed plasma flame largely depend on the pulse repetition frequency, wavelength, and laser radiation intensity. Under the experimental conditions created, the lifetime of the plasma torch was much greater than the gap between the laser pulses, which led to intense melting. The liquid in the processing area, when in contact with ablation products, was intensely sprayed and evaporated, and a fog formed. The formed fog in the caustic region promoted the formation of a plasma flare due to the absorption of laser radiation.

The treated surface was investigated by Raman spectroscopy. Figure 4a shows the Raman spectrum of the treated titanium surface, as shown in Figure 2a, in Figure 4b, the Raman spectrum of the micro granule surface corresponding to Figure 5 is shown. The pronounced peaks corresponding to the titanium carbide peaks are indicated by white arrows with a black outline at positions 135, 249, 431, 601 cm^{-1}. The detected peaks of titanium carbide both on the treated surface and on the surface of the microspheres are characterized by intensity and width, which can speak of conditions conducive to the growth of crystallites of titanium carbide. The formation of titanium carbide on the sample surface can be formed as a result of the contact of the heated metal with carbon containing media, by sorbing carbon with a metal surface [5]. Thus, due to the dissociation of hydrocarbon molecules, carbon sorption takes place on the surface of the sample to form titanium carbide.

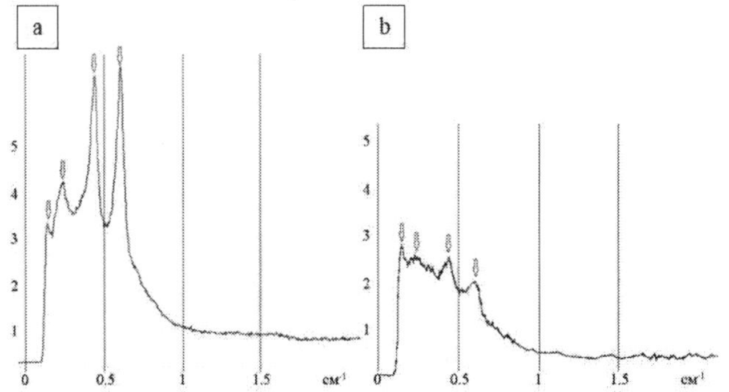

Figure 4. (a) Raman spectra of the treated surface; (b) Raman spectra of the microgranule surface.

As a result of the studies, a direct relationship between the duration of exposure of laser radiation in the treatment area was established. When scanning a surface with a high speed of conditions for dynamic erosion does not occur, the material ejected as a result of laser ablation leaves the area of action. The adhered particles do not feed the plasma torch, which significantly reduces the heat load on the surface being treated, thus creating conditions for the formation of surface periodic structures. Reducing the speed of scanning laser radiation leads to an increase in the thermal contribution to the processing area due to the formation of a plasma flame, which is an intense source of thermal energy. As a result, the sample surface acquires a relief containing reflow zones, liquid phase emissions and grad. This exposure mode can be used in the synthesis of micro granules of powder materials. As a result of the study of ablation products, particles of a spherical shape with a size of the order of 1-3 µm were found with a sufficiently long and intensive action - Figure 5.

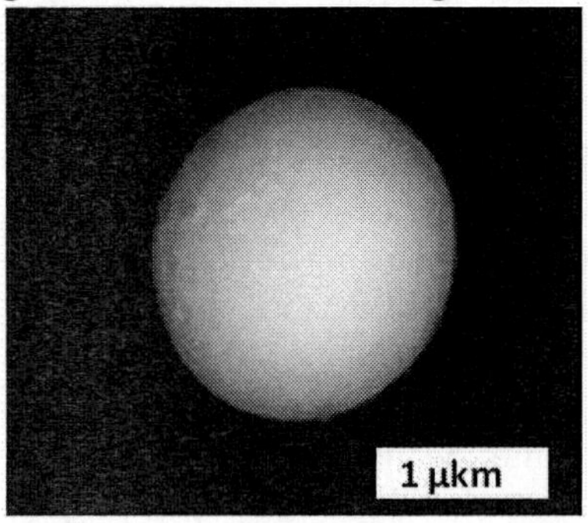

Figure 5. SEM image of the obtained microgranules.

4. Conclusion

The results obtained in this paper can be used to produce micro-structured surfaces of titanium alloys, to obtain coatings on the surface of materials using reaction media, as well as to obtain titanium microgranules of a highly spherical shape, a narrow granulometric composition. The use of this kind of powder materials, for the cultivation of various medical instruments and parts, is a promising and developing direction due to the production of products of high dimensional accuracy. These technologies open up new opportunities for creating new materials with a wide range of properties, in some cases unique, not achievable by other methods. The process of laser processing of materials in a liquid has great potential, both from the point of view of research activity, and from the point of view of technological applications of this kind of processing.

This study was performed as part of the state task VlSU 1106/17 and grants of the Russian Foundation for Basic Research 16-48-330651, 16-08-01226.

References

[1] Kazakevich VS, Kazakevich PV, Kamynina DA and Yaresko PS, 2018 *Brief communications on physics FIAN* (3) 15-20
[2] Barmina E V, et al. 2010 *Quantum Electronics* **40** (11) 1012-1020
[3] Kostishko B B, Svetukhin V V and Yavtushenko I O, 2016 *News of higher educational institutions, Electronics* **21** (6) 510-514
[4] Golubenko U V, Savkin A N, Sidorodnina T U and Timoshenko V A, 2016 *Proceedings of the Russian Academy of Sciences, Physical Series* **80** (8) 1078-1079
[5] Simakin AV, Voronov VV and Shafeev GA, 2014 *Russian Academy of Sciences* **60** 83-107

Electron spectroscopy of various nanostructured carbon films

O Yu Nishchak, N F Savchenko, O A Streletsky, E Yu Zykova and V V Khvostov

Faculty of Physics, Lomonosov Moscow State University, Leninskie Gory, 1-2, Moscow, 119991, Russia

e-mail: nischak@physics.msu.ru

Abstract. X-ray photoelectron spectroscopy technique have been used to study carbon films obtained by means of ion-plasma deposition in the plasma of different chemical composition. The detailed analysis of electron spectroscopy data made it possible to determine the structure of the films and the influence of nitrogen and hydrogen admixture in plasma on the structural properties. Diamond-like films are obtained in pure argon plasma, graphite-like and carbon chain structures correspond to nitrogen and hydrogen addition respectively.

1. Introduction

Thin carbon films are promising objects for application in various heterostructures for micro- and nanoelectronics due to their unique physical, chemical and mechanical properties [1].

At the present time, nanocarbon films are obtained by different techniques including both high and low temperature deposition. The mean size of their structural element (cluster) has typical order of nanometers. At these nanocluster sizes the quantity of surface atoms is almost equal to the quantity of bulk ones, which significantly affects on the properties of the structure. All the three types of carbon bonds hybridization (sp^1, sp^2, sp^3) are present inside the clusters, on their surface and between them.

To study electrophysical properties of complex nanocarbon structures, it is necessary to use analytical techniques, which can give us information on both atomic and electronic structure of a sample. The most informative methods are electron spectroscopy techniques, which include such analysis techniques as X-ray Photoelectron Spectroscopy (XPS), Auger Electron Spectroscopy (AES) and Electron Energy Loss Spectroscopy (EELS). However, there is a lack of modern works related to investigation of carbon materials by electron spectroscopy.

Electron spectroscopy provides qualitative and quantitative information on chemical composition of matherial. More sophisticated application of the method can yield information about electronic structure of the studied object. For example, in work [2] Auger spectra were interpreted assuming that specific features observed in spectrum depend on the electron structure of carbon material. However, this interpretation of spectra is not straightforward, and accurate theoretical modelling is required.

In this work we present the results of experimental study of thin nanocarbon films by means of EELS and AES using the spectra interpretation technique described in paper [3].

2. Experiment

The samples were deposited by ion-plasma pulse arc sputtering of graphite (discharge power 1 kW, pulse duration ~ 1 ms) assisted with ion beam with the energy of 400 eV in an ultra-high vacuum chamber with an ambient pressure of 10^{-6} Torr [4]. The temperature of substrates did not exceed 30°C. The thickness of the deposited films was about 100 nm.

Before the deposition, silicon plates used as the substrates were cleaned by 1 keV Ar ion etching. Carbon films deposition was carried out at the working pressure of 8.5×10^{-4} Torr in various gas atmospheres: in argon, in argon and nitrogen mixture (70/30) and in argon and hydrogen mixture (70/30).

3. Results and discussion

3.1. X-ray photoelectron spectroscopy

Photoelectron spectra provide information about the electronic structure of both surface and bulk phase of a sample. In the latter case, the information can be obtained by analyzing the spectrum of inelastic electron energy losses. Therefore, the consideration of x-ray photoelectron spectra was divided into two steps: analysis of carbon 1s line shapes and interpretation of electron energy loss spectra. In the spectra of each sample we observed a peak shifted by ~ 4eV relatively to carbon 1s peak. This peak can be attributed to either carbon oxygen bonds presence on the surface or plasmon excitation. According to electron spectroscopy for chemical analysis data, all the three sets of the samples have the same oxygen concentration, but their loss spectra are greatly different. So we suppose this peak to arise due to plasmon excitation.

Figure 1 shows the electron energy loss spectra after subtraction of 1s electron elastic peak. The experimental spectra are shifted in such a way that the zero energy corresponds to 1s carbon peak position. For carbon materials, the energy range up to 10 eV is associated with excitation of weakly bound π- electrons, whereas the energy losses in the range of 20 -30 eV are due to excitation of the whole electronic system (π+σ). It is known that the energies of collective electronic excitation are 27 eV in crystalline graphite and 35 eV in diamond. In contrast to diamond, there are non-hybridized π-orbitals in graphite, which correspond to plasmon energy 7 eV. In carbon materials having non-hybridized orbitals the values of π-plasmon energy are from 3.7 to 7 eV depending on the material structure. Low energies of the range correspond to quasi-one-dimensional carbon (polyacetylene and polydiacetylene) [5, 6]. π-plasmon excitation energy has a momentum-dependens, which is determined by the angle of electron exit relative to the carbon structure [5].

As can be seen from the figure, the spectra of the films obtained in different conditions differ from each other by the presence of π-plasmon peak and its energy. The data give evidence of existence of different carbon structures in the samples. The spectrum of the film deposited in pure argon plasma has no π-plasmon peak; hence, this film is a diamond-like one. In the spectrum of the film deposited in the Ar + N_2 plasma, we can observe the peak at the energy of 6 eV, position of which is close to the energy of π-plasmon in crystalline graphite. The 4 eV peak is a characteristic feature of electron energy loss spectrum of the carbon film obtained in hydrogen atmosphere.

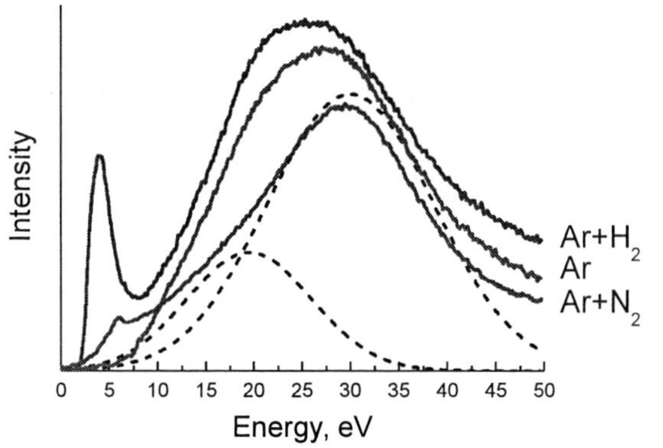

Figure 1. Electron energy loss spectra after subtracting C1s line

Energy of π-plasmon exitation in the investigated film (6 eV) has lower values than in high-oriented graphite under the same electron exit angles. This can be explained with the existence of arbitrary oriented graphite clusters in the film structure. The peak corresponding to energy loss 4 eV unambiguously indicates to the presence of carbon chains in the film [6, 7].

High energy range of the spectra (> 10 eV) is associated with excitation of the whole electronic subsystem and determined by electron concentration and, accordingly, material density. As mentioned above, (π+σ) plasmon energies in graphite and diamond are respectively 27 and 35 eV. In our case, the energy of (π+σ) plasmons does not exceed 30 eV.

Such a low value of π + σ plasmon energy can be explained by the size effect. For example, it was shown in [8] that electronic properties of the core and of the periphery of a diamond nanocrystal are significantly different. The energy loss spectrum of such a nanodiamond film has two maxima at energies 23 and 33 eV, corresponding to excitation of surface and bulk plasmons, respectively. In our case, the spectrum of the diamond-like film also has two components, shown in a dash line. The energy ratio of these components is approximately the same as one for surface and bulk plasmons. However, in contrary to an ideal diamond, in our case the maximum energy is 30 eV instead of 35 eV, which can be due to the low density of the diamond-like structure.

Generally, the loss function (spectrum) can be expressed as following:

$$S(E) = -\mathrm{Im}\left(\frac{1}{\varepsilon}\right) \tag{1}$$

where $\varepsilon = \varepsilon_1 + i\varepsilon_2$ is a complex dielectric function of a sample, ε_1 corresponds to collective electron excitations; ε_2 corresponds to one-electron interband transitions between valence and conductive bands and is a convolution of electron state density or joint density of states (DOS).

Therefore, electron energy loss spectra can give us information about electronic structure of a sample. In order to obtain this information, it is necessary to determine one-particle excitations using the experimental spectrum with Kramers-Kronig's relation:

$$S'(E) = -\mathrm{Re}\left(\frac{1}{\varepsilon}\right) = 1 - \frac{2}{\pi}\mathrm{p}\int_0^{\infty}\frac{E'S(E')\mathrm{d}E'}{(E')^2 - E^2} \tag{2}$$

where p means the principal value of the integral.

The calculated real ε_1 (dash line) and imaginary ε_2 (solid line) parts of the dielectric function of the films are shown in figure 2. The spectra were normalized using the measured value of optical refractive index n=2.2 ($\varepsilon_1 = n^2$), the value of which was determined by elipsometry. The curve representing joint density of states for diamond-like film demonstrates the presence of band gap (zero values of density in the region of 0-2 eV). The exact width of the band gap cannot be obtained from this spectrum because of the high noise in this energy range. The spectrum of joint DOS has a maximum at the energy of 12.8 eV, a plateau centred at 7 eV and a feature at 10.5 eV. Each of them represent an electron transition and such a picture is well consistent with the energies theoretically calculated on the base of a convolution of valence and conduction bands of diamond-like films (α-C:H) [9]. The presence of π-states at 7 eV can be explained by nanodiamond film structure with a layer having deformed sp^{3-x} hybridized bonds being formed on the surfaces of nanoclusters. These bonds have binding energy lower then sp^3 bonds, and the corresponding electronic states lie near the upper edge of the band gap.

Joint DOS of graphite-like film is similar to graphite one [10] and has specific interband transitions denoted π+π^* and π+σ^* in the figure (dot-dash line). In [10] the spectra are recorded from the highly oriented graphite for different exit angles. For comparison, the spectrum of electrons exiting normally to the base plane of graphite is also shown. The spectra are shown in the same scale and without normalization. In this case, we can speak of the existence of a nonzero band gap, which is in agreement with our results for the film. The displacement of π electron excitation peak to higher energies can be explained by deformation of graphite bonds and formation of sp^{2+x} orbitals, which means a shift toward diamond sp^3 bonds.

Joint DOS of the film deposited in the Ar+H_2 plasma corresponds well to joint DOS of one-dimensional or quasi-one-dimensional systems, which have maxima at the edges of both valence and conduction bands. The insert in figure 4 represents the band structure of an infinite carbon chain calculated by the LCAO method [11]. Thus, in the spectrum there should be the maxima corresponding to the two types of transitions marked by arrows. The first maximum is due to the transition from the top of valence band to the bottom of conduction band (through the forbidden band), the second one represents the transition between the band edges. As can be seen from figure, the experimental curve is in a good agreement with the calculated data. The second transition on the experimental curve is broadened, since the width of the valence band is much larger than the width of the conduction band.

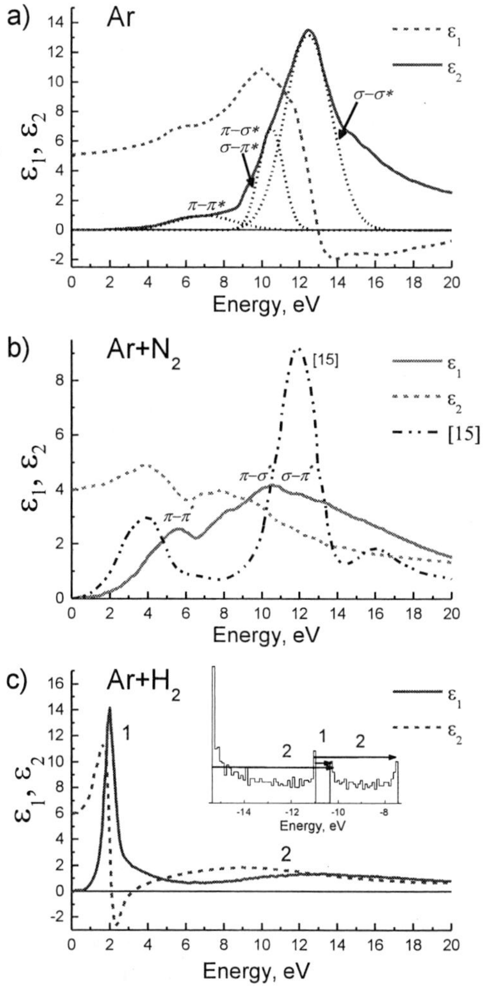

Figure 2. Real ε_1 (dash line) and imaginary ε_2 (solid line) parts of the dielectric function, calculated using Kramers-Kronig's relation

Thus, a diamond-like carbon film with a small number of defected sp^{3-x} bonds in a band gap can be obtained in pure argon atmosphere. The spectrum of the film deposited in nitrogen plasma is very similar to the spectrum of graphite single crystal, which does not have a band gap. Adding hydrogen into the plasma leads to formation of chain structures characterized by specific position of the π-plasmon peak.

3.2. Auger spectra analysis

The shape of carbon KVV peak heavily depends on the structure of carbon materials.

Figure 3 depicts Auger C KVV spectra of the investigated samples. Here, the energies are measured from the Fermi level of carbon. As can be seen from the figure, the position of the main maximum is in the energy region of 20 - 25 eV for all three samples. A wide maximum at 40-50 eV arises from the Auger electrons, which lose their energy for plasmon excitations (20-30 eV). The

IOP Conf. Series: Journal of Physics: Conf. Series **1238** (2019) 012034 doi:10.1088/1742-6596/1238/1/012034

analysis of presented spectra is not possible because of the significant background of inelastically scattered electrons and plasma losses.

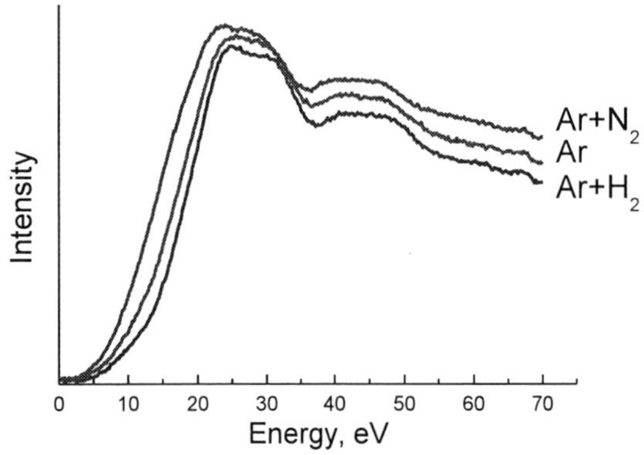

Figure 3. Auger spectra of deposited carbon films

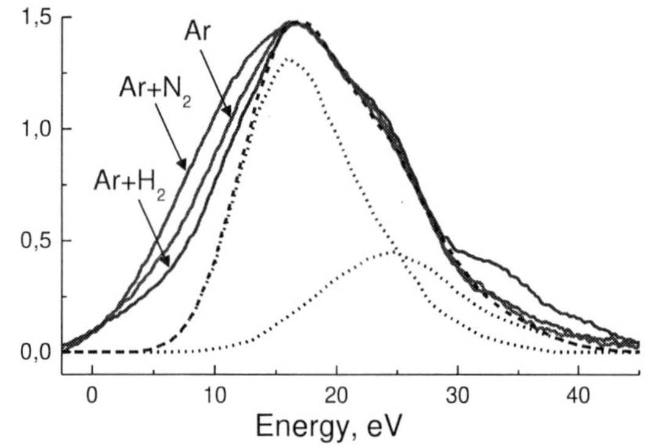

Figure 4. Experimental Auger spectra after inelastic and plasma losses background subtraction

Figure 4 shows the experimental Auger spectra after inelastic and plasma losses background subtraction. For comparison, theoretical spectrum is also presented in the figure. As it can been seen, the spectra of the samples differ from each other in the energy range from Fermi energy to the energy of spectrum maximum. Besides, there is an additional maximum at the energy of ~35 eV in the spectrum of the film deposited in argon and hydrogen mixture.

To explain the features of C KVV Auger peak of the investigated samples we used the one-electron approach in which the Auger lineshape of solids is determined by self-convolution of the density of valence electron states.

The valence band of graphite is made up of σ- and π-subbands, whereas only σ-electrons form the diamond valence band. Thus, the shape of KVV Auger peak can be derived from following expressions:

for graphite
$$I(E) = (\sigma + \pi) * (\sigma + \pi) = \sigma * \sigma + 2\pi * \sigma + \pi * \pi \quad (3)$$

and for diamond
$$I(E) = \sigma * \sigma \quad (4)$$

where * denotes a self-fold procedure.

Because of sp^3-hybridization of σ-orbitals in diamond, self-convolution of its σ-band can be decomposed as

$$\sigma * \sigma = s * s + s * p + p * p \quad (5)$$

The partial density of states was calculated within the framework of density functional theory (DFT) [12] using the plane wave basis set. As π-subband is closer to the Fermi level than σ-subband, the low energy region of Auger spectrum is determined mainly by the π*π term. Figure 4 depicts the theoretical spectrum for diamond shown as the dashed line. The components of the spectrum are shown with dotted lines.

As can be seen from the figure, there is a large difference between the experimental and theoretical spectra in the low energy region which corresponds to π-subband. To analyse this difference we subtracted the theoretical spectrum from the experimental ones. The curves obtained are presented in figure 5.

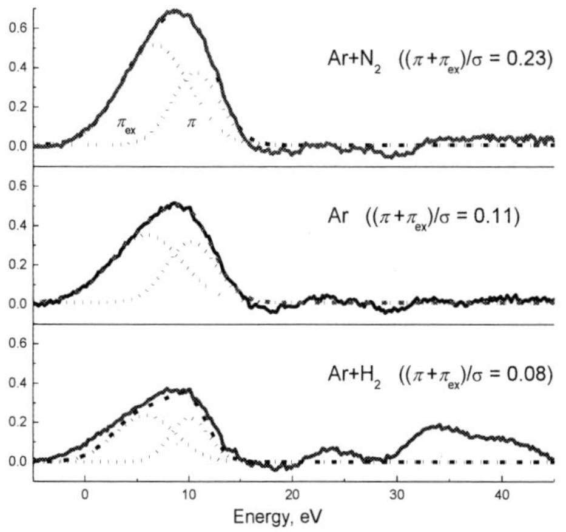

Figure 5. The curves obtained by subtracting theoretical spectrum of diamond from the experimental spectra of carbon films

First, note a wide and asymmetric peak on self-convolution curve of π-subband. Its asymmetry can be explained by the contribution of multi-particle effects accompanying Auger process. The ionization of 1s core level increases the effective nuclear charge experienced by valence electrons. This causes relaxation of the valence band represented by excitation of a valence electron into an unoccupied orbital above the Fermi level (formation of an exciton associated with long-lived holes in the valence band) [13]. As the result, we observe some broadening of π-subband, which can be decomposed into two Gaussian lines (figure 5, π and π_{ex}).

4. Conclusion

Electron spectroscopy methods are sufficiently sensitive to local structure of amorphous nanocarbon. The results showed that theoretical analysis of both photoelectron and Auger spectroscopy data makes it possible to identify sp, sp^2 and sp^3 hybridised bonds in carbon nanocrystals. Absence π-plasmon peak approves that diamond-like carbon films are obtained under pure argon atmosphere. Nitrogen addition into plasma leads to dramatic increase of graphite-like phase having π-plasmon with the same energy as the graphite one. With hydrogen addition, the film structure corresponds to carbon chains; it is indicated by both π-plasmon position and joint density of states shape.

References

[1] Avjya K E, Matevosyan L A, Ohanyan K S and Petrosyan L G 2016 *Instrum. Exp. Tech.* **59** 60
[2] Shikin A M, Prudnikova G V and Adamchuk V K 1994 *J. Elect. Spec. Rel. Phen.* **68** 413
[3] Ramaker D E 1991 *Crit. Rev. Solid State Mater. Sci.* **17** 211–76
[4] Streletskii O A, Ivanenko I P, Khvostov VV, Savchenko N F, Nishchak O Yu and Aleksandrov A F 2016 *Phys. Solid State* **58** 2119–25
[5] Ritsko J J, Mele E J, Heeger A J, MacDiarmid A G and Ozaki M 1980 *Phys. Rev. Lett.* **44** 1351
[6] Ritsko J J, Crecelius G and Fink J 1983 *Phys. Rev*. B. **27** 4902
[7] Kudryavtsev Yu P, Evsyukov S E, Babaev V G, Guseva M B, Khvostov V V and Krechko L M 1992 *Carbon* **30** 213
[8] Baitinger E M, Belenkov E A, Brzhezinskaya M M and Greshnyakov V A 2012 *Phys. Solid State* **54** 1715–22
[9] Zajičkova L, Franta D and Nečas D 2011 *Thin Solid Films* **519** 4299–308
[10] Stephan O, Taverna D and Kociak M. 2002 *Phys. Rev*. B. **66** 155422
[11] Babaev V G, Guseva M B, Novikov N D and Khvostov V V 2006 *Polyynes: Synthesis, Properties, and Applications* ed. F Cataldo (CRC Press) pp.219–52
[12] Kohn W 1999 *Rev. Mod. Phys.* **71** 1253
[13] Perfetto E, Cini M, Ugenti S, Castrucci P, Scarselli M, De Crescenzi M, Rosei F and El Khakani M A 2007 *Phys. Rev*. B. **76** 233408

Formation and investigation of plasma nanostructured hafnia coatings

S V Savushkina[1,3], M N Polyansky[1], E A Vysotina[1], A A Ashmarin[1,2], G V Panasova[1], and L E Agureev[1]

[1]Keldysh Research Center, 8 Onezhskaya st., Moscow, Russia
[2]Institute of Metallurgy and Materials Science A.A. Baikova of the Russian Academy of Sciences, 49 Leninsky prospect, Moscow, Russia

sveta_049@mail.ru

Abstract. Nanostructured coatings HfO_2-9wt%Y_2O_3 were obtained by plasma spraying in vacuum using supersonic nozzle creating Prandtl-Mayer expansion fan. The coatings were investigated by scanning electron microscopy, X-ray diffraction analysis, and synchronous thermal analysis (differential scanning calorimetry, thermogravimetric analysis, and differential thermogravimetric analysis) in the temperature range of 40÷1600°C. The coatings are predominantly characterized by the nanostructured globular structure. Thermal analysis showed a peak corresponding to the exothermic reaction in the temperature range of 1400÷1600 °C, accompanied by a mass loss. *c*-HfO_2 phase remained in the coating after thermal analysis with a decrease in the lattice parameter *a* from 5.18 Å to 5.13 Å.

1. Introduction

Hafnia is a perspective material for the thermal barrier layer of the thermal barrier coating due to its similarity with zirconia and its elevated temperatures of phase transitions. Hafnia has a lower thermal conductivity than zirconia and shows lower sintering rates due to lower partial oxygen pressures for transition of the ionic conductivity to electronic conductivity compared to zirconia [1-5]. In [6] it was shown that the thermal stability of the HfO_2-7.5wt%Y_2O_3 coating increases by at least 100°C compared to the coating of ZrO_2-8wt%Y_2O_3. However, the disadvantage of hafnia use is a lower coefficient of thermal expansion in comparison with zirconia.

Nanostructured coatings are characterized by more homogeneous distribution of pores therefore their influence on the coating structure is less destructive [7, 8]. In addition, the presence of nanopores can contribute to an effective damping of the volume changes in the coatings when the operating temperature changes. Also, the nanostructured coating can have better thermal insulation properties due to the increase in the number of boundaries that promote the scattering of lattice phonon waves.

One of the methods to obtain a nanostructured coating by plasma spraying is the formation of rarefaction waves in Prandtl-Maier expansion fan in vacuum. An abrupt expansion of the vapor phase in Prandtl-Maier expansion fan contributes to condensation of nanoparticles of the deposited material [10, 11]. The Prandtl-Mayer expansion fan is a supersonic gas flow with a continuous increase in speed [9]. It can be obtained when the freely expanding gas jet flows around a convex corner into a region of reduced pressure. In this case, in the strongly underexpanded gas jet a shock wave (so-called "first barrel") appears. Inside barrel-shaped shock wave the flow coincides with the flow in the jet outgoing into vacuum. . When the supersonic flow is turned in the Prandtl-Mayer flow at angles of the order of several tens of degrees, the local flow parameters (static pressure and temperature) drop sharply. In the vicinity of the corner of the turn the cooling of the spraying material vapor phase occurs faster. The static pressure also falls faster than inside the nozzle.

In this work a nanostructured hafnia coatings formed by plasma spraying in vacuum using a supersonic nozzle creating a rarefaction waves in a Prandtl-Mayer expansion fan were investigated by scanning electron microscopy and thermal analysis.

2. Experimental setup and characterization techniques

Hafnia coatings with the thickness up to 60 μm were obtained on samples of bronze alloy (1wt%Cr) with a diameter of 30 mm and a thickness of 2 mm. The coatings were formed by DC plasma torch with a power up to 10 kW in VS-2 setup of Keldysh Research Center under conditions of dynamic vacuum (the pressure ~ 120 Pa) in the chamber. Plasma spraying was carried out using the powder of HfO_2-9wt%Y_2O_3 with a particle size of 10÷40 μm. A supersonic nozzle of plasma torch was equipped with two turning sections creating Prandtl-Maier expansion fan: the first by 6 and the second by 27 degrees. Such parameters of the nozzle lead to the condensation of hafnia nanoparticles as a result of cooling the vapor phase of the sprayed material in the vicinity of the turning sections. The consumption of the plasma-forming gas (N_2) was 0.87 g/s, the powder consumption was 0.23 g/s. The arc current was 120 A, and the voltage was 60 V. The spraying process was carried out by reciprocating displacement of the plasma torch along the rotating holder with the samples.

The coatings were investigated using a Quanta 600 scanning electron microscope (SEM), X-ray diffractometer (Empyrean PANalytical, CuK_{α} radiation), and synchronous thermal analysis (NETZSCH STA449F 1 Jupiter). The coating was studied by thermal analyzer in a corundum crucible in the temperature range of 40÷1600°C at a rate of 5 °/min. The protective and purge gases were argon with the flow of 50 ml/min and air with a flow of 50 ml/min.

3. Results and discussion

The study of hafnia coatings surface showed the presence of several types of particles (Figure 1 (a)): deformed planar particles with size up to 30 μm, smaller fragments of particles formed in a collision with the surface of superheated powder particles, the particles in the form of disks with hole in the centre obtained as a result of a collision with the surface of the particles with a molten shell and a solid centre. Between big particles there are regions containing particles with size of 2 ÷ 5 μm that mainly have the form of flat discs with uneven edges, and submicron particles (Figure 1 (b)). Submicron particles prevailing in the structure of the coating are represented in the form of globules with a size of 0.2 ÷ 0.4 μm (Figure 1 (c,d)), which are formed mainly of nanoparticles (Figure 1 (d)).

It can be assumed that the presence of nanostructured globules in the structure of the coating is due to formation of rarefaction waves in the Prandtl-Mayer expansion fan in the nozzle of the plasma torch. In this case, the central part of the nozzle contributes predominantly to the formation of the deformed particles but the peripheral part of the nozzle with turning sections contributes to the formation of predominantly nanoparticles in the coating structure.

The hafnia coating was investigated using synchronous thermal analysis including differential scanning calorimetry (DSC), thermogravimetric analysis (TG), and differential thermogravimetric analysis (DTG) . In figure 7 the graphs of DSC, TG and DTG on temperature are presented. The sample for analysis was obtained by scraping down the coating from the substrate.

Synchronous thermal analysis did not show significant mass changes in the temperature range from 30 to 1200 °C that indicates the stability of the chemical compound in this temperature range, see figure 2 (1). In DTG curve, a small peak corresponding to a decrease in mass can be noted at a temperature of 210 °C, see figure 2 (2). It can be associated with the release of gases from the coating (water vapor, CO_2). Small peaks on the curves of DSC, TG, and DTG corresponding to a slight increase in mass and exothermic reaction at the temperature of about 700 °C can be associated with oxidation of residual carbon on the surface of the coating. At a temperature of more than 1300 °C, the mass decreases by 2%. In this temperature region there is also a peak in the DSC graph corresponding to the exothermic reaction completed by 1600 °C, see figure 2 (3). The temperature corresponding to the maximum of this peak, according to the DTG graph, is ~ 1412 °C. This peak can be associated with changes in the crystal structure; for example, redistribution of the stabilizing oxide (yttria) and partial release of it from the hafnia crystal lattice, because the sprayed powders often have an uneven distribution of the stabilizer, which is then partially retained in the coating structure.

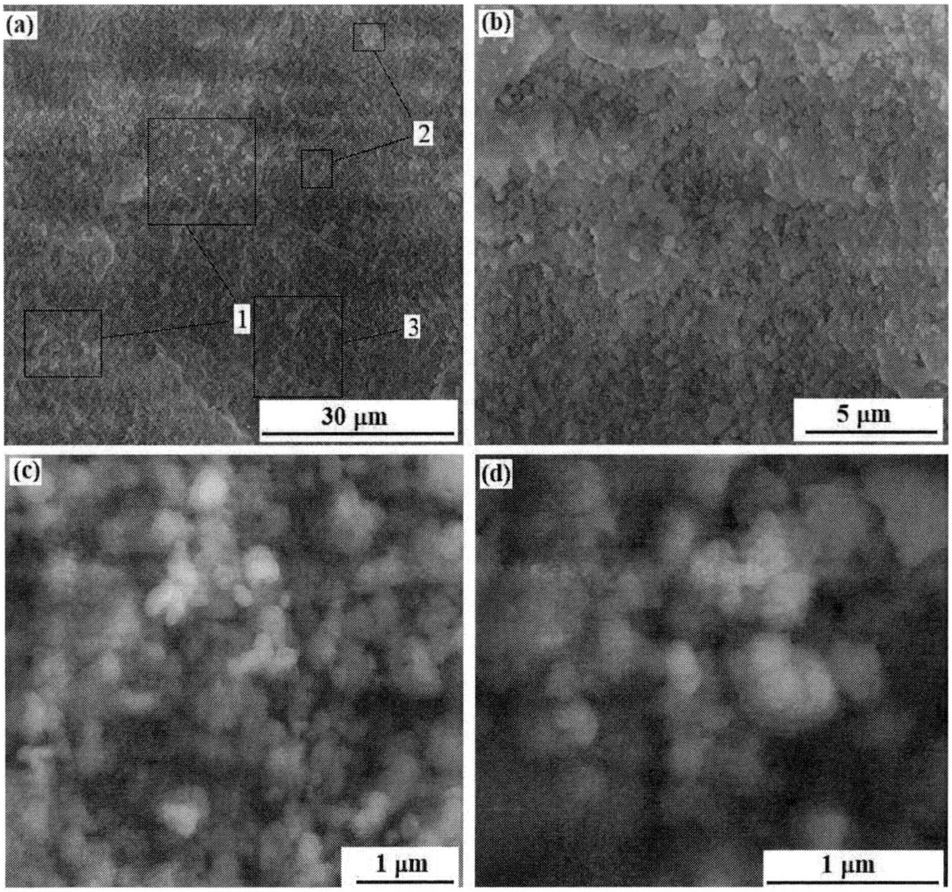

Figure 1. SEM images of HfO_2-9wt%Y_2O_3 coating surface: a) The morphology of the coating, b) the structure of coatings in the regions between the big particles, c) the globular structure at higher magnification, and d) nanostructured particles.
1 - Deformed particles in the form of disks, 2 - fragments of molten particles in the structure, 3 - particle in the form of discs with a hole in the center.

X-ray diffraction analysis of the coating after thermal analysis showed that the cubic modification of hafnia remained in the coating with the crystal lattice parameter $a \approx 5.14$ Å. The parameter a was 5.18 Å before the thermal analysis. A decrease in the parameter a may indicate partial losses of yttria from the crystal lattice of hafnia. It may also be due to a decrease of imperfections and internal stresses in the coating caused by a strong deformation of the hafnia particles sprayed at supersonic speeds. The parameter $a \approx 5.18$ Å corresponds to a 9wt% yttria content, which confirms the distortion of the crystal lattice in the coating prior to thermal analysis [12].

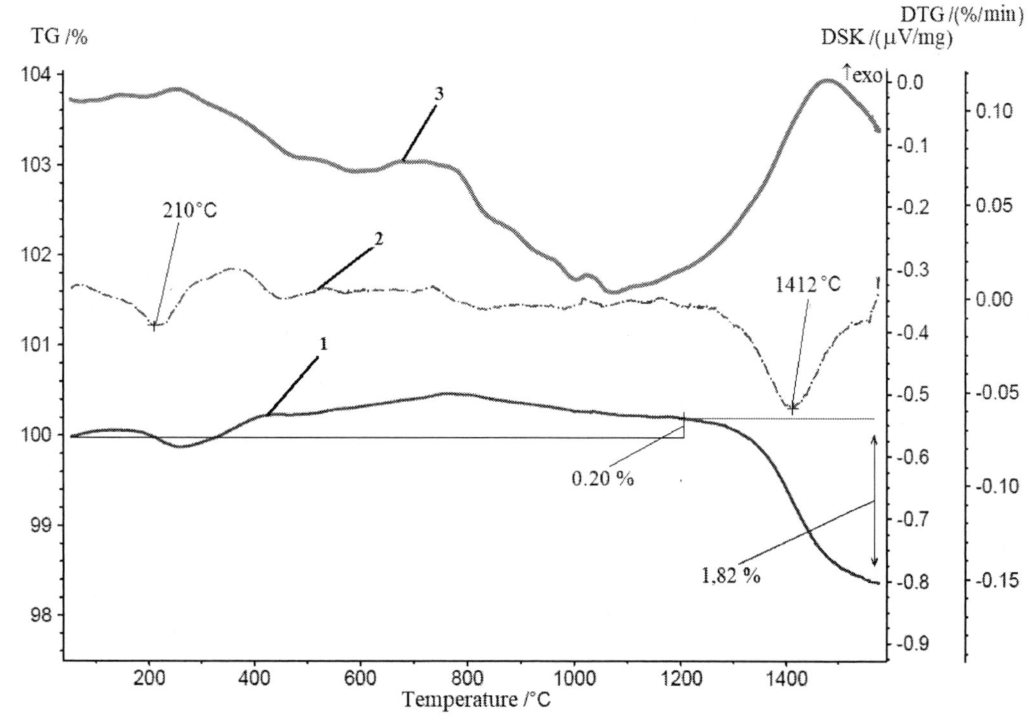

Figure 2. Graphs of thermogravimetric analysis (TG) (1), differential thermogravimetric analysis (DTG) (2) and differential scanning calorimetry (DSC) (3) on temperature.

The peak on the DSC graph in the temperature range from 1400 to 1600 °C can also be related to the crystallization of the amorphous component in the coating, the presence of which can be caused by high rates of spraying and cooling.

4. Conclusions

Nanostructured coatings of HfO_2-9wt%Y_2O_3 with a thickness of up to 60 μm were obtained by plasma spraying in vacuum using supersonic nozzle creating rarefaction waves in the Prandtl-Mayer expansion fan. The coatings are predominantly characterized by a globular structure that consists of nanoparticles. Thermal analysis of the coatings showed the presence of a peak corresponding to the exothermic reaction in the temperature range 1400÷1600 ° C accompanied by a mass loss that may be due to the partial release of the stabilizing oxide, its redistribution, and reduction of defects and internal stresses in the coating. X-ray diffractometry showed the stability of the cubic structure of hafnia after thermal analysis with a decrease in the lattice parameter a from 5.18 Å to 5.13 Å.

Acknowledgments
The work was supported by Russian Science Foundation, project 17-79-10309.

References

[1] Zhu D, Miller R A *Surf. Coat. Technol.* 1998 **108–109** 114–20
[2] Winter M R, Clarke D R *Acta Mater.* 2006 **54** 5051–59
[3] Singh J, Wolfe D E, Miller R A, Eldridge J I, Zhu D M *J. Mater. Sci.* 2004 **39** 1975–85
[4] Stöver D, Pracht G, Lehmann H, Dietrich M, Döring J E, Vaßen R *J. Therm. Spray Technol.* 2004 **13(1)** 76–83
[5] Ibigazne H, Alpirine S, Diot C *J. Mater. Sci.* 1995 **3** 938–51.

[6] Matsumoto K, Itoh Y, Kameda T *Sci. Technol. Adv. Mater.* 2003 **4** 153–158
[7] Wang Y, Bai Y, Yuan T, Chen H Y, Kang Y X, Shi W J, Song X L, Li B Q *Surf. Coat. Technol.* 2017 **319** 95-103
[8] Liu C, Zhang Z, Jiang X, Liu M, Zhu Z *Trans. Nonferrous Met. Soc. China* 2009 **19** 99–107
[9] Schetz J A, and Fuhs A E 1999 *Fundamentals of Fluid Mechanics. 3rd Edition* (John Wiley & Sons)
[10] Polyanskiy M N, Savushkina S V J. *Surf. Invest.: X-Ray, Synchrotron Neutron Tech.* 2014 **8** 144–48
[11] Borisov A M, Vostrikov V G, Polyansky M N, Romanovsky E A, Savushkina S V, Suminov I V, Tkachenko N V J. *Surf. Invest.: X-Ray, Synchrotron Neutron Tech.* 2015 **9** 248–52
[12] Stacy D W, Wilder D R *J. Am. Ceram. Soc.* 1975 **58** 285–88

The Influence of the Presence of Viscosity on Evolution of Perturbations in the System of Colliding Plates

Denis Ilnitsky[1,2], Kirill Gorodnichev[1], Alexey Serezhkin[1], Sergey Kuratov[1], Nail Inogamov[1,2]

[1] Dukhov Research Institute of Automatics (VNIIA), 22, ul. Sushchevskaya, Moscow 127055, Russia

[2] L.D. Landau Institute for Theoretical Physics, 1a, ul. Akademika Semenova, Chernogolovka, Moscow Region, 142432, Russia

E-mail: cyrgo85@gmail.com

Abstract. The results of research on the effect of the presence of viscosity on evolution of perturbations in the system of colliding plates are presented in this work. The range of amplitudes of the initial perturbations wavelengths, at which the effect of viscosity may be neglected, are determined. Comparison of analytical and numerical results are shown.

1. Introduction

The interaction of planar shock waves with various perturbation fields in a medium has been studied in the last half century in numerous theoretical and experimental works(see, e.g. [1–12]). In particular, in the linear approximation, the interaction of a shock wave with acoustic waves was considered in [2, 11], with an isotropic turbulent vorticity fields in [8], and with the perturbation of the density in [9, 12]. In all this works hydrodynamic approach has been used for the research. It is known, that the presence of viscosity provides a stabilized effect on the evolution of perturbations [13–15].

The interaction of the shock wave with the anisotropic density perturbation field was considered in the linear approximation in [16]. The evolution of perturbation field in the system of colliding plates, one of which has initial density perturbation field, was investigated by means of the hydrodynamic approach. Here, in this work, the results of consideration of analogous problem with account for the presence of viscosity are presented. The range of amplitudes of the perturbation wavelengths, when viscosity plays a significant role on perturbation evolution, are obtained.

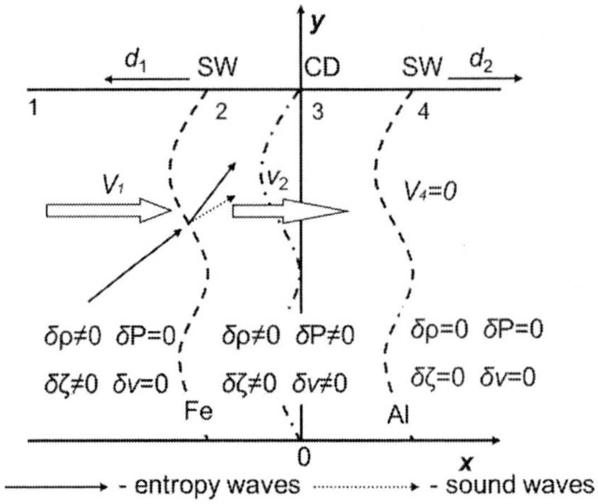

Figure 1: Illustration of the problem.

2. Definition of the problem

We consider the collision of two semi-infinite plates one of which is at rest (target)(see Fig. 1). A spatially anisotropic density perturbation field initially exist in the incident plate(impactor). Collision results in the appearance of shock waves diverging in both plates from the contact discontinuity (see Fig. 1). It is assumed that the velocity of the impactor is quite high (about 5 km/s). For this reason, the elastoplastic properties of the colliding plates can be neglected and the interaction of the shock wave with the density perturbation field in the impactor can be described by the Navier-Stokes equation system

$$\rho\left[\frac{\partial \mathbf{v}}{\partial t}+(\mathbf{v}\nabla)\,\mathbf{v}\right]=-\nabla P+\eta\,\Delta\,\mathbf{v}+\left(\nu+\frac{\eta}{3}\right)\,\mathrm{grad\,div}\,\mathbf{v}, \tag{1}$$

$$\frac{\partial \rho}{\partial t}+\mathrm{div}\rho\mathbf{v}=0, \tag{2}$$

$$\frac{\partial}{\partial t}\left(\frac{\rho v^2}{2}+\rho\varepsilon\right)=-\mathrm{div}\left[\rho\mathbf{v}\left(\frac{v^2}{2}+\varepsilon+\frac{P}{\rho}\right)-(\mathbf{v}\sigma')\right], \tag{3}$$

where

$$\sigma'_{ik}=\eta\left(\frac{\partial v_i}{\partial x_k}+\frac{\partial v_k}{\partial x_i}\right)-\frac{2}{3}\delta_{ik}\frac{\partial v_l}{\partial x_l}.$$

In following calculations we set the second viscosity coefficient to zero $\nu=0$. The system of Eqs.(1)-(3) should be supplemented by the Rankine-Hugoniot conditions at the shock front [17]:

$$v_{1\tau}=v_{2\tau}, \qquad v_{1n}-v_{2n}=\sqrt{(P_2-P_1)\left(\frac{1}{\rho_1}-\frac{1}{\rho_2}\right)}, \tag{4}$$

$$j^2=\rho_1\rho_2\frac{P_2-P_1}{\rho_2-\rho_1}, \qquad \varepsilon_1-\varepsilon_2=\frac{1}{2}\left(\frac{1}{\rho_2}-\frac{1}{\rho_1}\right)(P_1+P_2), \tag{5}$$

where $v_{1\tau}(v_{1n})$ and $v_{2\tau}(\ v_{2n})$ are the tangential (normal) components of the velocity ahead of and behind the front, respectively; $j=v_1\rho_1=v_2\rho_2$ is the mass flux; and ε_1 and ε_2 are specific internal energies on both sides of the shock wave.

The equation of state of the medium is used in the form of the Mie-Gruneeisen equation [18]:

$$P = \frac{\rho_0 c_0^2}{n}\left(\left(\frac{\rho}{\rho_0}\right)^n - 1\right) + \Gamma\rho\zeta, \tag{6}$$

where ρ_0 is the density of the uncompressed ("cold") medium, c_0 is the speed of sound in this medium; ρ is the current density; Γ is the Gruneisen coefficient, and n is a constant; and ζ is the specific thermal energy.

3. Propagation of the perturbations in the impactor

Let the initial density perturbation field in the impactor be described by the expression $\delta\rho_1 = A_{\delta\rho_1}\exp(ik_{0x}x + ik_{0y}y)$. The axis of anisotropy for such perturbation is directed along the vector $\mathbf{k_0}$ (Fig. 1). The shock wave moves from the right to the perturbation at the velocity $(\mathbf{d_1} - \mathbf{v_1})$ in the the negative direction of the x axis ($d_1 - v_1 < 0$). The angle between the vector $\mathbf{k_0}$ and $0x$ axis is θ_0 ($\mathrm{tg}\theta_0 = k_{0y}/k_{0x}$). At the interaction of the density perturbation field with the shock wave, acoustic and entropy-vortex waves are formed behind the shock front [17].

Perturbations of the hydrodynamic values are the functions of spatial coordinates and time $f(x,y,t)$. Taking into account the expression for the initial perturbation field we will find the solution for acoustic waves in the form $f(x,y,t) = A_f^{(s)}\exp(ik_{0y}y + ik_{2x}x - i\omega_2 t)$, and for entropy-vortex waves - $f(x,y,t) = A_f^{(ev)}\exp(ik_{0y}y + i\kappa_{2x}x - i\Omega_2 t)$. From linearized system of the equations (1) – (3) we obtain:

for acoustic waves

$$\left(\omega_2 - v_{2x}k_{2x} + \frac{i\eta}{\rho_2}k_2^2 + \frac{i\eta}{3\rho_2}k_{2x}^2\right)\delta v_{2x}^{(s)} = \frac{k_{2x}}{\rho_2}\delta P_2 - \frac{i\eta}{3\rho_2}k_{2x}k_{0y}\delta v_{2y}^{(s)}, \tag{7}$$

$$\left(\omega_2 - v_{2x}k_{2x} + \frac{i\eta}{\rho_2}k_2^2 + \frac{i\eta}{3\rho_2}k_{0y}^2\right)\delta v_{2y}^{(s)} = \frac{k_{0y}}{\rho_2}\delta P_2 - \frac{i\eta}{3\rho_2}k_{2x}k_{0y}\delta v_{2x}^{(s)}, \tag{8}$$

$$(\omega_2 - v_{2x}k_{2x})\,\delta P_2 = \rho_2 c_2^2\left(k_{2x}\delta v_{2x}^{(s)} + k_{0y}\delta v_{2y}\right), \tag{9}$$

for entropy-vortex waves

$$(\Omega_2 - v_{2x}\kappa_{2x})\,\delta v_{2x}^{(ev)} = -\frac{i\eta}{\rho_2}\kappa_2^2\delta v_{2x}^{(ev)} - \frac{i\eta}{3\rho_2}\left(\kappa_{2x}^2\delta v_{2x}^{(ev)} + k_{0y}\kappa_{2x}\delta v_{2y}^{(ev)}\right), \tag{10}$$

$$(\Omega_2 - v_{2x}\kappa_{2x})\,\delta v_{2y}^{(ev)} = -\frac{i\eta}{\rho_2}\kappa_2^2\delta v_{2y}^{(ev)} - \frac{i\eta}{3\rho_2}\left(k_{0y}^2\delta v_{2y}^{(ev)} + k_{0y}\kappa_{2x}\delta v_{2x}^{(ev)}\right), \tag{11}$$

$$\kappa_{2x}\delta v_{2x}^{(ev)} + k_{0y}\delta v_{2y}^{(ev)} = 0, \tag{12}$$

where $k_2 = \sqrt{k_{2x}^2 + k_{0y}^2}$, $\kappa_2 = \sqrt{\kappa_{2x}^2 + k_{0y}^2}$.
From (4) – (5) it results

$$k_{0x}(v_1 - d_1) = \omega_2 - k_{2x}d_1 = \Omega_2 - \kappa_{2x}d_1. \tag{13}$$

It follows from Eqs. (7) – (9) and (13) that the x - projection of the wave vector of acoustic waves behind the shock front k_{2x} is given by the expression k_{2x}

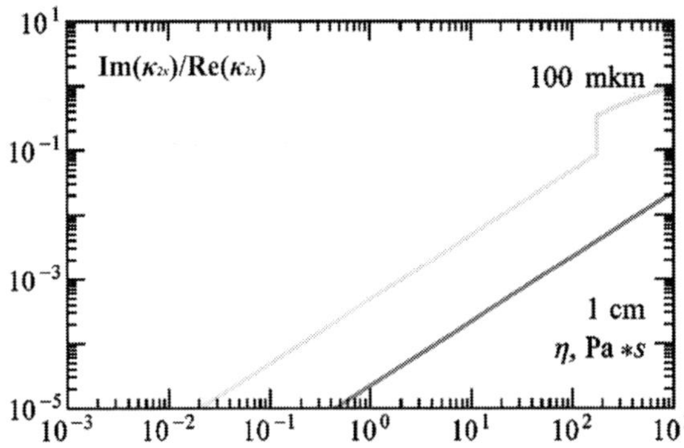

Figure 2: The dependence of imaginary part to real part ratio $\mathrm{Im}\left(k_{2x}\right)/\mathrm{Re}\left(k_{2x}\right)$ for the x - projection of the wave vector of acoustic waves behind the shock front on the first coefficient of viscosity η.

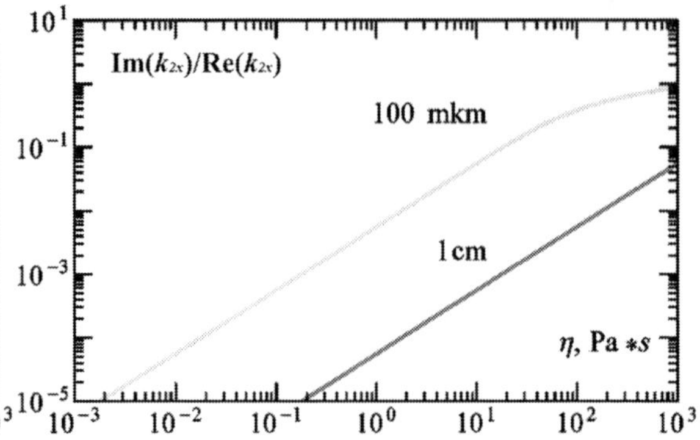

Figure 3: The dependence of imaginary part to real part ratio $\mathrm{Im}\left(k_{2x}\right)/\mathrm{Re}\left(k_{2x}\right)$ for the x - projection of the wave vector of entropy-vortex waves behind the shock front on the first coefficient of viscosity η.

$$d_1\left(k_{0x}-k_{2x}\right)-k_{0x}v_1+k_{2x}v_{2x}=$$
$$=\frac{3\rho_2 c_2^2 k_2^2}{-4i\eta k_2^2+3d_1k_{0x}\rho_2-3d_1k_{2x}\rho_2-3k_{0x}\rho_2v_1+3k_{2x}\rho_2v_2}.$$

From Eqs. (10) – (12) and (13) we receive the expression for x - projection of the wave vector of entropy-vortex waves behind the shock front κ_{2x}:

$$\kappa_{2x}=-\frac{i}{2\eta}\left[d_1\rho_2+\rho_2v_2-\sqrt{\left(d_1\rho_2+\rho_2v_2\right)^2-4i\eta\left(i\eta k_{0y}^2+\left(v_1+d_1\right)k_{0x}\rho_2\right)}\right].$$

The frequencies for both types of waves are obtained from (13).

In order to receive values of the perturbations amplitudes the system of Eqs.(10) – (12) should be supplemented by the Rankine-Hugoniot conditions written in the linear approximation (4) – (5). analytical solution of this system of equations leads to a very lengthy expressions, and so we don't present it in this work. Quantitative results will be shown below.

We investigate the influence of the presence of viscosity on perturbation field development behind the shock front on specific example. We consider the collision of two semi-infinite plates: impactor(iron) and target (aluminum). Definition of the problem is analogous to that described in section 5. The dependence of imaginary part to real part ratio $\mathrm{Im}\left(k_{2x}\right)/\mathrm{Re}\left(k_{2x}\right)$ for the x - projection of the wave vector of acoustic waves behind the shock front on the first coefficient of viscosity η is shown in Fig. 2. Analogous dependence for entropy-vortex waves is presented in Fig. 3. For both metals we use $\eta\sim 30$ Pa s [13, 14]. From Fig. 2 - 3 it is clear that the effect of viscosity is negligible for perturbation wavelengths more than 20 mkm.

It was shown [20] that the regime of propagation of sound behind the shock front depends on the angle between the axis of anisotropy of the perturbation field and the velocity of the shock wave. There are three types of qualitatively different regimes of this propagation. In this work we consider only the case when sound propagates from the shock wave towards the contact discontinuity. For this reason we should investigate interaction of this wave with the contact discontinuity.

4. Interaction of perturbations with the contact discontinuity

Conditions of the equality of pressures and normal components of the velocity take place on both sides of the contact discontinuity. From this equations it follows:

$$k_{2x}v_2 - \omega_2 = k_{2x}^{ref}v_2 - \omega_2^{ref} = k_{3x}^{tr}v_2 - \omega_3^{tr}, \tag{14}$$

where k_{2x}^{ref}, ω_2^{ref} are wave vector and frequency of the acoustic/entropy-vortex wave, reflected from the contact boundary; k_{3x}^{tr}, ω_3^{tr} are wave vector and frequency of the acoustic/entropy-vortex wave, transmitted through the contact boundary.

The contact boundary values for the entropy-vortex wave is joining with the homogeneous solution in the target. This homogeneous solution decays exponentially with moving away from the contact discontinuity. Thus, the entropy-vortex perturbations exist only in very narrow region near the contact in the target.

For the wave vectors of reflected and transmitted acoustic waves we obtain:

$$\begin{aligned}
k_{2x}^{ref} &= -k_{2x}, \\
k_{3x}^{tr} &= \frac{1}{4\eta\rho_2 c_2^2 (k_2)^2 - \rho_3 c_3^2 \left[4\eta k_2^2 - 3i\rho_2(\omega_2 - k_{2x}v_2)\right]} \times \\
&\times \left[\rho_3 c_3^2 k_{0y}^2 \left(4\eta (k_{2x})^2 - 3i\rho_2(\omega_2 - k_{2x}v_2)\right) - \rho_2 c_2^2 k_2^2 \left(4\eta k_{0y}^2 - 3i\rho_3(\omega_2 - k_{2x}v_2)\right)\right].
\end{aligned}$$

For the pressure amplitudes of reflected and transmitted acoustic waves we receive:

$$\begin{aligned}
\delta P_2^{ref} &= -\frac{\dfrac{k_{2x}}{3\rho_2(\omega_2 - k_{2x}v_2) + 4i\eta(k_{2x}^2 + k_y^2)} - \dfrac{k_{3x}^{tr}}{3\rho_3(\omega_2 - k_{2x}v_2) + 4i\eta((k_{3x}^{tr})^2 + k_y^2)}}{-\dfrac{k_{2x}}{3\rho_2(\omega_2 - k_{2x}v_2) + 4i\eta(k_{2x}^2 + k_y^2)} - \dfrac{k_{3x}^{tr}}{3\rho_3(\omega_2 - k_{2x}v_2) + 4i\eta((k_{3x}^{tr})^2 + k_y^2)}}\delta P_2, \\
\delta P_3^{tr} &= \delta P_2 + \delta P_2^{ref}.
\end{aligned}$$

5. Results of the numerical calculations

The problem of two semi-infinite plates impact is analyzed numerically also. For this task we use hydrocode TIS-2D [19], which is based on the finite-volume method. We consider collision of the iron impactor with initially perturbed density field with homogeneous aluminum target. The initial perturbations amplitude is $\delta\rho_1/\rho_1 \sim 5\%$.

Two different wavelengths of the initial perturbation field was taken into consideration $\lambda_x = 100$ mkm and $\lambda_x = 50$ mkm. The results of numerical calculations for the case $\lambda_x = 100$ mkm at instant 150 ns are presented in Fig. 4 and 5. It is evident that attenuation is negligible and that the numerical data confirms the analytical results.

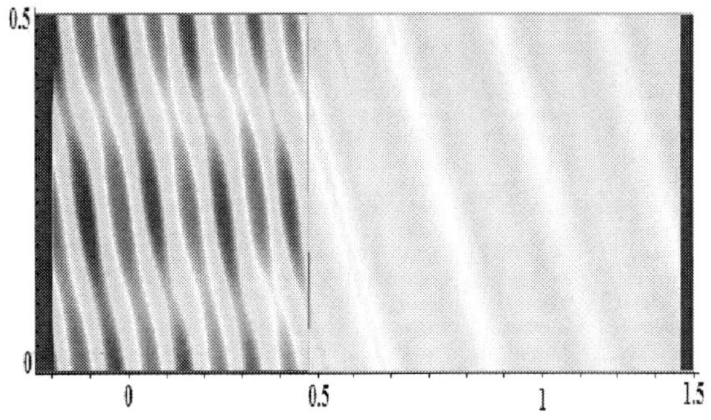

Figure 4: Contour of the pressure for the case $\lambda_x = 100$ mkm at the instant 150 ns after impact.

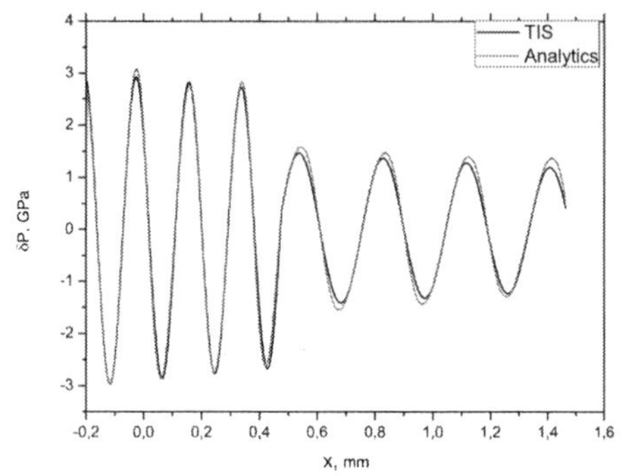

Figure 5: Spatial profile of the pressure for the case $\lambda_x = 100$ mkm at the instant 150 ns after impact.

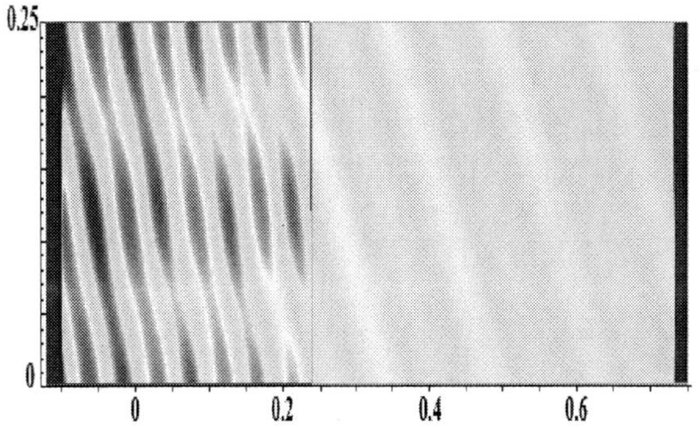

Figure 6: Contour of the pressure for the case $\lambda_x = 50$ mkm at the instant 75 ns after impact.

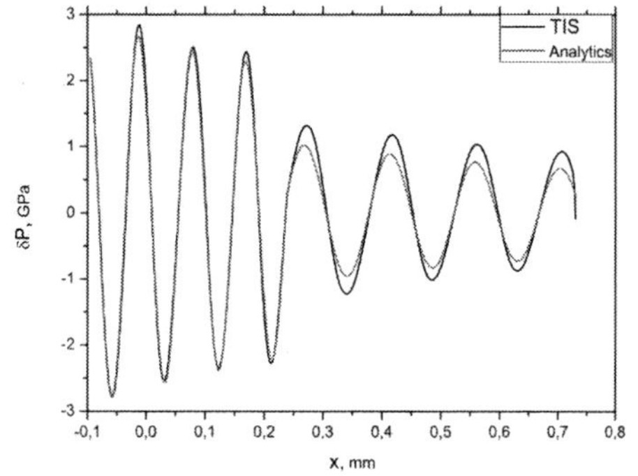

Figure 7: Spatial profile of the pressure for the case $\lambda_x = 50$ mkm at the instant 75 ns after impact.

The results of numerical calculations for the case $\lambda_x = 50$ mkm at instant 75 ns are presented in Fig. 6 and 7. In this case attenuation is significant. The results of theoretical and numerical analysis are in good agreement. For lesser wavelengths attenuation increases and the linear approximation for the analytical consideration is not sufficient.

6. Conclusions

The evolution of the initial density perturbation field in the system of two colliding plates are considered with account for the presence of the viscosity. The parameters of the perturbation field in the compressed region was found analytically. Numerical calculations confirms theoretical results. It is shown, that the presence of viscosity has significant effect at the initial perturbation wavelengths λ_x in order and less than 100 mkm.

Acknowledgments

This work was partially supported(I.D.,I.N.) by the Russian Science Foundation (project no. 14-19-01599).

References

[1] Diakov S P 1954 *Zhur. Exp. i Teor. Fiz.* **27** 288 [in Russian]
[2] Kontorovich V M 1959 Sov. Phys. Acoustics **5** 320
[3] Kuznetsov N M 1989 *Sov. Phys. Usp.* **159** 993
[4] Keller J and Merzkirch W 1990 *Experiments in Fluids* **8** 241
[5] Yang Y and Zhang Q and Sharp D H 1994 *Phys. Fluids* **6** 1856
[6] Wouchuk J G and Huete Ruiz de Lira C and Velikovich A L 2009 *Phys.Rev.E* **79** 066315
[7] Wouchuk J G 2001 *Phys.Rev.E.* **63** 056303
[8] Wouchuk J G and Huete Ruiz de Lira C and Velikovich A L 2009 *Phys.Rev E* **79** 066315
[9] Huete Ruiz de Lira C and Velikovich A L and Wouchuk J G 2011 *Phys.Rev.E* **83** 056320
[10] Nishihara K and Wouchuk J G and Matsuoka C et al 2010 *Phil. Trans. R. Soc. A* **368** 1769
[11] Huete Ruiz de Lira C and Wouchuk J G and Velikovich A L 2012 *Phys.Rev E* **85** 026312
[12] Huete Ruiz de Lira C and Wouchuk J G and Canaud B and Velikovich A L 2012 *Journal of Fluid Mechanics* **700** 214
[13] Kusnezov D and Shubin O N 2009 *Advances in materials science* (Sandia)
[14] Barton N R and Bernier J V and Becker R and Arsenlis A and Cavallo R and Marian J and Rhee M and Park H-S and Remington B A and Olson R T 2011 *Journal of Applied Physics* **109** 073501
[15] Mitra A and Roychoudhury R and Khan M 2016 *Physics of Plasmas* **23** 024503
[16] Gorodnichev K E and Kuratov S E 2013 *VANT ser. Mat. Mod. Fiz. Prots.* **2** 37 [in Russian]
[17] Landau L D and Lifshitz E M 1987 *Fluid Mechanics* (New York: Pergamon)
[18] Zababakhin E I 1997 *Nekotorye voprosy gazodinamiki vzryva* (Snezhinsk) [in Russian]
[19] Menshov I S and Mischenko A V Serejkin A A 2013 *Matem. Mod.* **258** 89
[20] Gorodnichev K E, Kuratov S E, Menshov I S, Serezhkin A A, Zakharov P P 2016 *Math. mod.* **29(3)** 95 [in Russian]

Quantum shell effects in the hydrodynamic equations of a quantum fluid.

S E Kuratov[1], D S Shidlovski[1,2], I S Menshov[1,3] and S I Blinnikov[1,2]

[1] Dukhov Research Institute of Automatics (VNIIA), Sushchevskaya street 22, Moscow, 127055, Russia

[2] NRC "Kurchatov institute" – ITEP, Bol'shaya Cheremushkinskaya street 25, Moscow, 117218, Russia

[3] Keldysh Institute for Applied Mathematics Russian Academy Science

E-mail: ser.evg.kuratov@gmail.com, dmitry.shidlovski@yandex.ru

Abstract. Numerical analysis of the hydrostatic equilibrium of a compressed gas bubble was carried out by the DFT method. Based on this analysis, we modified the hydrodynamic equations of a quantum fluid taking into account quantum shell effects. An external force acting on the electronic subsystem is added to the modified hydrodynamic equations. A numerical algorithm for solving of the system of modified hydrodynamic equations is discussed.

Keywords: hydrodynamic equations, spatial distribution, hydrogen gas bubble, DFT method.

1. Introduction

The effect of the spatial heterogeneity of degenerate electron distribution in a spherical well was analyzed in [1]. The consequence of the effect is the appearance of an electric field acting on the ion system and causing its movement. The characteristic size of the inhomogeneity is of the order of the system size and, accordingly, by several orders of magnitude greater than the distance between the particles. So the relaxation of the ions can be described using the hydrodynamic approximation.

We analyze a spherical mesoscopic system – a compressed gas bubble of submicron size. The characteristic values of the thermodynamic quantities of the compressed gas are the following: $\rho_{gas} \sim$ (10-30) g/cm^3, $N_e \sim 10^{30-31}$ m^{-3}, $E_F \sim$ (20-100) eV, $T_i \sim T_e \sim$ (0.1-1) eV. All electrons are ionized and degenerated, and the ions are the classical gas.

The numerical analysis of the hydrostatic equilibrium of a gas bubble is presented in Section 2. Numerical simulation of the electron distribution was carried out by the DFT method. The analysis allowed us to formulate the simplest version of the hydrodynamic equations of a quantum liquid. The quantum shell effects are taken into account by introducing an external force (the oscillation potential) into the equation of motion of the degenerate electrons. Section 3 presents a numerical algorithm for solving the system of modified hydrodynamic equations. We applied this algorithm for the numerical analysis of ion relaxation in a compressed gas bubble.

2. Hydrostatic equilibrium of compressed gas bubble. DFT calculations.

To calculate distributions of electrons and ions in a spherically symmetrical bubble, we use DFT with spherical jellium background model and hydrostatic equation for ions [1, 2]. In the case of a spherical symmetry the electron density and the density of ion liquid satisfy the system of equations (there we use atomic units, where $\hbar = c = e = 1$):

$$\left(-\frac{1}{2}\frac{\mathrm{d}^2}{\mathrm{d}r^2} + V_{\mathrm{KS}}(r) + \frac{l(l+1)}{2r^2}\right) P_{nl}(r) = \varepsilon_{nl} P_{nl}(r), \quad V_{\mathrm{KS}} = \varphi + V_{xc}, \tag{1}$$

$$n_e(r) = 2\sum_{n,l}(2l+1)\Theta_{nl}\frac{P_{nl}^2(r)}{4\pi r^2}, \tag{2}$$

$$-\frac{1}{n_{\mathrm{ion}}}\nabla P - Z_{\mathrm{ion}}\nabla\varphi = 0, \quad \text{with condition} \int_0^{\infty} 4\pi r^2 n_{\mathrm{ion}}\mathrm{d}r = N \tag{3}$$

$$\varphi(r) = \frac{1}{r}\int_0^r 4\pi s^2(n_{\mathrm{ion}}(s) - n_e(s))\mathrm{d}s + \int_r^{+\infty} 4\pi s(n_{\mathrm{ion}}(s) - n_e(s))\mathrm{d}s. \tag{4}$$

The first equation is the one-dimensional Kohn-Sham equation for radial wavefunctions P_{nl}, where V_{xc} is the exchange-correlation potential in a local density approximation. Equation (3) is the hydrostatic equation for ion density n_{ion}, where $P = n_{\mathrm{ion}}kT$. The last is the expression for the electric potential φ, which can be obtained from the Poisson equation. This system is solved self-consistently using simple iteration method. Equation (3) can be simply integrated

$$\ln(n_{\mathrm{ion}}) = -\frac{Z_{\mathrm{ion}}}{kT}\varphi - \ln T + C. \tag{5}$$

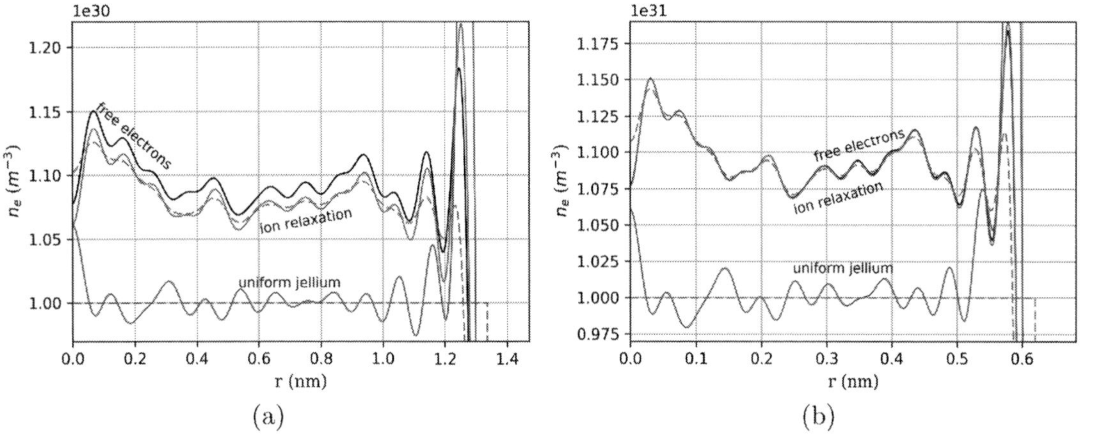

Figure 1. The density distributions of electrons (solid lines) and ions (dashed lines) for 10000 electrons at average density $10^{30}\mathrm{m}^{-3}$ (a) and $10^{31}\mathrm{m}^{-3}$ (b).

We have made several calculations for gas bubbles. Figure 1 shows the electron density distribution and the ion density in equilibrium for 10000 electrons at average density $10^{30}\mathrm{m}^{-3}$ and $10^{31}\mathrm{m}^{-3}$ with infinite potential wall at the border, the temperature of ions is $T_i = 10$ eV. There are also distributions of free electrons and electrons in the case of uniform ion jellium. It can be clearly seen that the density of ions is adjusted to the electron and the difference between density distributions is smaller for higher average density. At the same time the electron density get very close to free electron density and for average density $10^{31}\mathrm{m}^{-3}$ they are almost coincide.

Quantum mechanical calculations can be applied for a relatively small number of particles ($N < 10^5$), so we need to use simpler method such as quantum electron fluid model [3] for larger number of particles. We will consider equation for quantum electron fluid without

IOP Conf. Series: Journal of Physics: Conf. Series **1238** (2019) 012037 doi:10.1088/1742-6596/1238/1/012037

Bohm potential, but with osillation potential U_{osc}, which should reproduce density distribution obtained in DFT calculations

$$-\frac{1}{\rho_e}\nabla p_e + \frac{e}{m_e}\nabla\varphi + \frac{e}{m_e}\nabla U_{\text{osc}} = 0. \tag{6}$$

As we have seen earlier, the electron density in DFT calculations with ions in equilibrium and with an infinite potential border almost equal to the free electron density. Also $\nabla\varphi$ is rather small, because both densities are very close to each other. So the oscillation potential

$$e\nabla U_{\text{osc}} \approx \frac{1}{n_e^{\text{free}}}\nabla p_e(n_e^{\text{free}}). \tag{7}$$

3. Numerical algorithm of hydrodynamic equations

We consider the sphere of the initial radius R_0 consisting of gas of degenerate electrons $n_e(r,t)$ and classical ions $n_i(r,t)$. It is assumed that the electron density satisfies the equilibrium condition at each time moment:

$$-e\frac{\partial}{\partial r}(\varphi + U_{\text{osc}} - U_{\text{bar}}) + \frac{1}{n_e}\frac{\partial p_e}{\partial r} = 0, \tag{8}$$

where U_{osc} and U_{bar} is the oscillation and barrier potentials:

$$U_{\text{bar}} = -\frac{eV_0}{1+\exp[(R_0 - r)/\delta]}, \quad U_{\text{osc}} = C\left(\frac{\bar{n}}{10^{30}}^{1/3}\right)\frac{R_0(0)}{R_0(t)}f(r, R_0(t)) \tag{9}$$

where $\bar{n} = \bar{\rho}/m_i$ is the average electron density, and V_0 and C are the characteristic amplitudes of the barrier and oscillation potentials, δ is the size of the barrier potential, and $f(r, R_0(t))$ the spatial profile of the oscillation potential.

The electron pressure depends on the local electron density and is determined by the following expression:

$$p_e = \frac{(3\pi^2)^{2/3}}{5}\frac{\hbar^2}{m_e}n_e^{5/3}. \tag{10}$$

The motion of the ion gas is a result of the action on the ions of the electrostatic field created by the instantaneous distribution of ions and electrons. The ion gas is assumed to be ideal, i.e., the effects of viscosity and thermal conductivity are negligible, and is calorically perfect with the equation of state $p = (\gamma - 1)\rho\zeta$, where $\gamma = 5/3$ is the adiabatic exponent, p is the pressure, ζ is the specific internal energy.

It can be shown that the influence of nonequilibrium between the distribution of ions and electrons takes place in a very small neighborhood of the wall. Therefore, in the gas dynamics calculations described below, it is assumed that there is a local equilibrium at all points inside the sphere $n_i = n_e$. Then the electron pressure will determine the ion distribution.

The medium inside the sphere is assumed to be electro neutral, therefore at each time moment the distributions of electrons and ions satisfy the condition of zero total charge:

$$\int_0^{R_0(t)} n_e(r,t)r^2 dr = \int_0^{R_0(t)} n_i(r,t)r^2 dr, \quad n_i = \rho/m_i, \tag{11}$$

where m_i is the mass of ion.

The ion motion is described by the standard system of Euler equations for gas dynamics under the assumption of spherical symmetry, which is supplemented by the right-hand side corresponding to the change in momentum and energy due to the electrostatic field produced by charged particles-gas ions and free electrons:

$$\begin{cases} \dfrac{\partial \rho}{\partial t} + \dfrac{\partial \rho U}{\partial r} = \dfrac{2}{r}\rho U, \\ \dfrac{\partial \rho U}{\partial t} + \dfrac{\partial \rho U^2 + p}{\partial r} = \dfrac{2}{r}\rho U^2 - \dfrac{e}{m_i}\rho\dfrac{\partial \varphi}{\partial r}, \\ \dfrac{\partial \rho E}{\partial t} + \dfrac{\partial \rho U H}{\partial r} = \dfrac{2}{r}\rho U H - \frac{e}{m_i}\rho U\dfrac{\partial \varphi}{\partial r}, \end{cases} \tag{12}$$

where e is the electron charge, $E = U^2/2 + \zeta$ is the specific total ion energy, $H = \zeta + p/\rho$ is the specific total enthalpy, $\zeta = p/\rho(\gamma - 1)$ is the specific internal energy of the ions, $\gamma = 5/3$, $0 \leq r \leq R_0(t)$. The potential φ is determined by the equation (8), where we use n_i instead of n_e.

We write the system of equations (8), (12) in dimensionless form. To do this, we choose the scales of the characteristic physical quantities in the following form:

$\rho_* = 10^3 kg/m^3$ – density,
$L_* = 10^{-6} m$ – length,
$U_* = 10^3 m/s$,
$t_* = L_*/U_* = 10^{-9} s$,
$p_* = \rho_* U_*^2 = 10^9 N/m^2$ – preasure,
$\Theta_* = \dfrac{m_i U_*}{e t_*} \approx 1.875 * 10^4 V/m$ – electric field strength,
$\varphi_* = \dfrac{m_i U_*^2}{e} \approx 1.875 * 10^{-2} V$ – electric field potential.

The values of the physical parameters of the model:
$e = 1.6 * 10^{-19} C$ – electron charge,
$m_i = 3 * 10^{-27} kg$ – ion mass,
$\varepsilon_0 = 8.85 * 10^{-12} C/(V \cdot m)$ – dielectric constant of vacuum,
$\varepsilon = 10^4$ – relative dielectric constant of ions.

The system of equations (12) can be written in a conservative form. In dimensionless variables, it has the following form:

$$\begin{cases} \dfrac{\partial r^2 \rho}{\partial t} + \dfrac{\partial r^2 \rho U}{\partial r} = 0, \\ \dfrac{\partial r^2 \rho U}{\partial t} + \dfrac{\partial r^2 (\rho U^2 + p)}{\partial r} = 2rp - \rho r^2 \dfrac{\partial \varphi}{\partial r}, \\ \dfrac{\partial r^2 (\rho E)}{\partial t} + \dfrac{\partial r^2 (\rho U H)}{\partial r} = \rho U r^2 \dfrac{\partial \varphi}{\partial r}, \end{cases} \tag{13}$$

where the gradient of the potential

$$\frac{\partial \varphi}{\partial r} = K_1 \rho^{-1/3} \frac{\partial \rho}{\partial r} + K_3 \left[\frac{R_0(0)}{R_0(t)}\right]^2 \frac{\partial U_{\text{bar}}}{\partial r} - K_4 \frac{R_0(0)}{R_0(t)} \frac{\partial f(r, R_0(t))}{\partial r}. \tag{14}$$

Here K_1, K_3, K_4 are dimensionless constants:

$$K_1 = \frac{(3\pi^2)^{2/3}}{3} \frac{\hbar^2}{e m_e \varphi_*} n_*^{2/3}, \quad n_* = \rho_*/m_i,$$

$$K_3 = \frac{\hbar^2}{2 e m_e \varphi_*} \left(3\pi^2 \frac{\rho_0}{m_i}\right)^{2/3} \left[\frac{R_0(0)}{R_0(t)}\right]^2,$$

$$K_4 = C \frac{10^{-10} m}{\varphi_*} \left(\frac{\rho_0}{m_i}\right)^{1/3}.$$

If we introduce the total pressure $\pi = p + \frac{3}{5}K_1\rho^{5/3}$, the system of equations (13) can be rewritten as follows:

$$\begin{cases} \dfrac{\partial r^2\rho}{\partial t} + \dfrac{\partial r^2\rho U}{\partial r} = 0, \\ \dfrac{\partial r^2\rho U}{\partial t} + \dfrac{\partial r^2(\rho U^2+\pi)}{\partial r} - 2rp = -\rho r^2\left\{K_3\left[\dfrac{R_0(0)}{R_0(t)}\right]^2\dfrac{\partial U_{\text{bar}}}{\partial r} - K_4\dfrac{R_0(0)}{R_0(t)}\dfrac{\partial f(r,R_0(t))}{\partial r}\right\}, \\ \dfrac{\partial r^2(\rho E)}{\partial t} + \dfrac{\partial r^2\rho U(E+\pi/\rho)}{\partial r} = \dfrac{3}{5}K_1\rho^{5/3}\dfrac{\partial r^2 u}{\partial r} - \\ \qquad -\rho r^2\left\{K_3\left[\dfrac{R_0(0)}{R_0(t)}\right]^2\dfrac{\partial U_{\text{bar}}}{\partial r} - K_4\dfrac{R_0(0)}{R_0(t)}\dfrac{\partial f(r,R_0(t))}{\partial r}\right\}, \end{cases} \tag{15}$$

Except for the right-hand side, the system of equations (15) coincides exactly with the classical system of equations of gas dynamics. The EOS of the medium has the form

$$\pi = (\gamma-1)\rho e + \frac{3}{5}K_1\rho^{5/3} \tag{16}$$

Therefore, the numerical integration (15) can be performed by the Godunov method [4]:

$$\mathbf{q}_i^{n+1} = \mathbf{q}_i^n - \frac{3\Delta t}{(r_{i+1/2}^{n+1})^3 - (r_{i-1/2}^{n+1})^3}\left[(r_{i+1/2}^{n+1/2})^2\mathbf{F}_{i+1/2} - (r_{i-1/2}^{n+1/2})^2\mathbf{F}_{i-1/2} + \mathbf{S}_{\pi,i} + \mathbf{S}_{U,i}\right], \tag{17}$$

where

$$\mathbf{S}_{\pi,i} = \begin{bmatrix} 0 \\ \left((r_{i+1/2}^{n+1})^2 - (r_{i-1/2}^{n+1})^2\right)\pi_i \\ 0 \end{bmatrix}, \tag{18}$$

$$\mathbf{S}_{U,i} = \begin{bmatrix} 0 \\ \left((r_{i+1/2}^{n+1})^3 - (r_{i-1/2}^{n+1})^3\right)\rho_i S_{2,i} \\ \left((r_{i+1/2}^{n+1})^2 U_{i+1/2} - (r_{i-1/2}^{n+1})^2 U_{i-1/2}\right)3K_1\rho^{5/3} + \left((r_{i+1/2}^{n+1})^3 - (r_{i-1/2}^{n+1})^3\right)\rho_i U_i S_{3,i} \end{bmatrix},$$

Here, $S_{2,i}$ and $S_{3,i}$ the right-hand sides of the equations connected with the barrier and oscillation potential, and the values at the edges of the cells (half-integral indices) are determined from the solution of the Riemann problem. To solve the Riemann problem on the edges of countable cells, we use the local approximation of the EOS (16)

$$\pi = (\gamma-1)\rho e + c_0^2(\rho-\rho_0) \tag{19}$$

where the parameters of the EOS c_0 and ρ_0 are approximated from the value of the local density by the following relations:

$$\rho_0 = \frac{2}{5}\rho, \quad c_0^2 = K_1\rho^{2/3}, \tag{20}$$

on the basis of splitting equations. At each time step, a hyperbolic system with zero right-hand side is first solved.

The Courant number is determined by the speed of sound of the system (13)

$$c^2 = \frac{\gamma(\pi+\pi_0)\rho e + \frac{3}{5}K_1\rho^{5/3}}{\rho}, \quad \pi_0 = \frac{2}{5\gamma}K_1\rho^{5/3}. \tag{21}$$

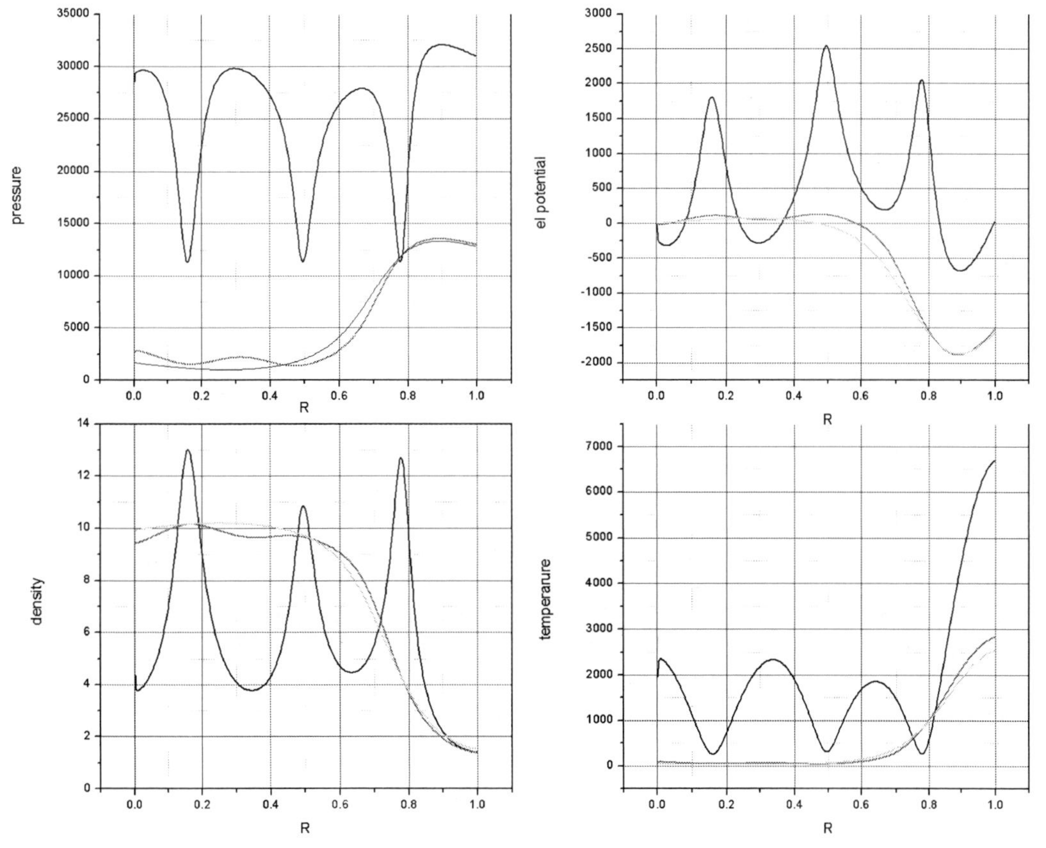

Figure 2. Stationary distributions of hydrodynamic quantities for different values of the oscillation potential amplitude.

The numerical scheme (17), (18) is used to solve the stationary problem of the distribution of the ion-electron system in a spherical region of radius 1 μm. The calculated grid consists of 300 equidistant cells. The number of Courant in the calculations was 0.5. Initial distributions in dimensionless quantities:

$$t = 0: \quad \rho = 5, \quad U = 0, \quad p = 21$$

Calculations were carried out with $\delta = 0.1R_0$ and three variants of the amplitude of the oscillation potential C =10V, 1V, 0.1V. The results of the distributions of hydrodynamic quantities after the system's output to the stationary mode are shown in Fig. 2. Black color shows the results for C = 10V, red – C = 1V and green – C = 0.1V. We can see that the influence of the oscillation potential is significant for C = 10V and is insignificant for C = 1V and C = 1V.

4. Conclusion

Based on the analysis of the hydrostatic equilibrium of a compressed gas bubble, we derived modified hydrodynamic equations taking into account the quantum shell effects. The method is based on the addition of an oscillation potential to the equation of motion of the electrons. A numerical algorithm for solving a system of hydrodynamic equations based on the Godunov method is developed.

The results of numerical analysis of the relaxation of a compressed gas bubble are presented. They demonstrate the significant influence of the oscillation potential at hydrodynamic quantities of compressible gas bubble.

References

[1] Kuratov S E, Shidlovski D S and Blinnikov S I 2018 *Journal of Physics: Conference Series* **1009** 012013
[2] Shidlovski D S and Mukhanov A E 2018 *Journal of Physics: Conference Series* **1009** 012012
[3] Shukla P K and Eliasson B 2010 (in Russian) *Uspekhi Fizicheskikh Nauk* **180**(1) 55–82
[4] Godunov S K, Zabrodin A V and Prokopov G P 1962 (in Russian) *USSR Computational Mathematics and Mathematical Physics* **1**(4) 1187

IOP Conf. Series: Journal of Physics: Conf. Series **1238** (2019) 012038 doi:10.1088/1742-6596/1238/1/012038

Modification of D_2-line absorption contours of alkali atoms in antirelaxation coated cells

E A Tsygankov[1], D E Fomichev[2], D S Chuchelov[1], S A Zibrov[1], V L Velichansky[1,2] and V P Yakovlev[2]

[1]Lebedev Physical Institute, Russian Academy of Sciences, Leninsky Prospect 53, Moscow, 119991 Russia

[2]National Research Nuclear University MEPhI, Kashirskoye Highway 31, Moscow, 115409 Russia

E-mail: selentinthebright@gmail.com

Abstract. Absorption features in cells with antirelaxation coating in D_2-lines of alkali metal atoms due to the competition of cyclic and open transitions lying within the Doppler contour are studied. In cells of this type ensemble of atoms is optically pumped over all velocities and the entire volume of the cell during the characteristic time T, which noticeably exceeds local pumping times related to resonant atomic velocities. If the frequency scanning time is comparable to T, the retardation effect results in significant distortion of the laser wave absorption contour and in emergence of the hysteresis phenomenon. The developed theory for a realistic scheme of atomic levels with one cyclic and two open transitions describes experimentally observed absorption contours, their dependence on the light power and the frequency scanning direction.

Cells with antirelaxation coating are widely used in quantum metrology, frequency standards and magnetometry to obtain narrow metrological resonances [1–4] due to enlarged probability of preserving the atomic inner state after collisions with the cell walls. As it was shown [5–7] the frequency corresponding to the maximum of absorption and the maximum value of the absorption itself in D_2-lines of alkali atoms depend on a sign of the laser frequency scanning direction, if the scanning time is comparable to characteristic relaxation times due to collisions with the cell walls. Absorption contour is found to be significantly distorted — absorption is smaller compared to the cell without coating and this effect is more expressed in the range of open atomic transitions than in the range of cyclic transition.

Cylindrical cell (50 mm in length, 35 mm in diameter) with ^{133}Cs vapor was used in the experiment. Line width of the laser with external resonator was of the order of 1 MHz. Light was linearly polarized and beam size in cross section (5 mm $\times$ 2 mm) was smaller by almost an order of magnitude than the diameter of the cell. Absorption contours obtained in the experiment for scanning rate of about 50 MHz/ms with increasing (1+, 2+) and decreasing (1−, 2−) of frequency for two values of light power are shown in Fig. 1.

The simplest 4-level scheme [7] based on a single cyclic and a single open transition qualitatively explains the characteristic features of the absorption contour — its asymmetry and a kind of hysteresis effect, that is, the dependence on the sign of frequency scanning. However, such a model of levels does not allow to describe dependence of the crossing point for different frequency scanning directions on light power (see the points ω^* and $\omega^\#$ in Fig. 1.)

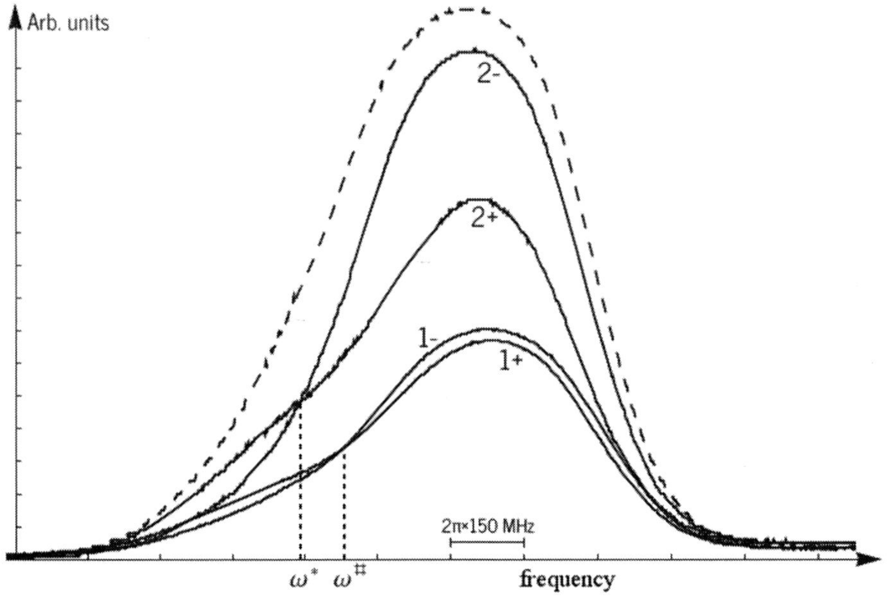

Figure 1. Absorption contours of the ^{133}Cs D$_2$-line long-wavelength components for the light power of 20 μW (curves 1+ and 1−) and 50 μW (curves 2+ and 2−). Absorption contour in the cell without antirelaxation coating is shown by dashed line for comparison. Scanning rate is 50 MHz/ms. Signs $\pm$ correspond to scanning process with increasing and decreasing of the frequency.

In the given paper we consider interaction of the monochromatic laser wave $\mathcal{E}_L \cos\omega_L t$ with an atomic system that has one cyclic ($|g\rangle \leftrightarrows |5\rangle$) and two open ($|g\rangle \leftrightarrows |3\rangle$, $|4\rangle$) transitions (Fig. 2). Such a levels scheme corresponds to the long-wavelength components of an alkali atom D$_2$-line.

Atoms interact with laser radiation during the time τ_0 of flight through the cross section of the beam. Its value is of the order of $\tau_0 \sim 10^{-5}$ s for thermal velocities v_T, so that the parameter $\gamma\tau_0 \gg 1$, where γ is the natural width of the excited state levels ($\gamma/2\pi \approx 5.2$ MHz for ^{133}Cs). In order to register the absorption contour, the laser wave frequency is varied in time according to the law $\omega_L = \omega_0 \pm st$ with some finite rate $s = \omega_D/\tau_{sc}$, where $\omega_D = kv_T$ is the characteristic width of the Doppler contour and τ_{sc} is the scanning time. We consider the latter to be significantly larger than the flight time τ_v, which is, in its turn is noticeably exceeds τ_0. In this case the variation of frequency ω_L during these times of flight can be neglected and the description of the time evolution is considerably simplified.

Detunings of the laser frequency from working atomic transitions with taking into account the Doppler shift due to longitudinal velocity have the form: $\Delta_3(t,v) = \pm st + \Delta/2 - kv$, $\Delta_4(t,v) = \pm st - \Delta/2 - kv$, $\Delta_5(t,v) = \pm st - \Delta_0 - kv$. As far as the Doppler width is much larger than the homogeneous broadening, $\omega_D \gg \gamma$, the resonant interaction for the transitions under consideration is realized for significantly different longitudinal velocities.

When an atom passes through the beam, a change of the local (with respect to velocities) populations $\rho_{gg}(t,v)$ and $\rho_{\tilde{g}\tilde{g}}(t,v)$ of the ground state sublevels

$$\frac{d\rho_{gg}}{dt} = -\frac{d\rho_{\tilde{g}\tilde{g}}}{dt} = -\Gamma(t,v)\rho_{gg} \tag{1}$$

in the low saturation regime is determined by optical pumping rate of the non-absorbing state $|\tilde{g}\rangle$:

IOP Conf. Series: Journal of Physics: Conf. Series **1238** (2019) 012038 doi:10.1088/1742-6596/1238/1/012038

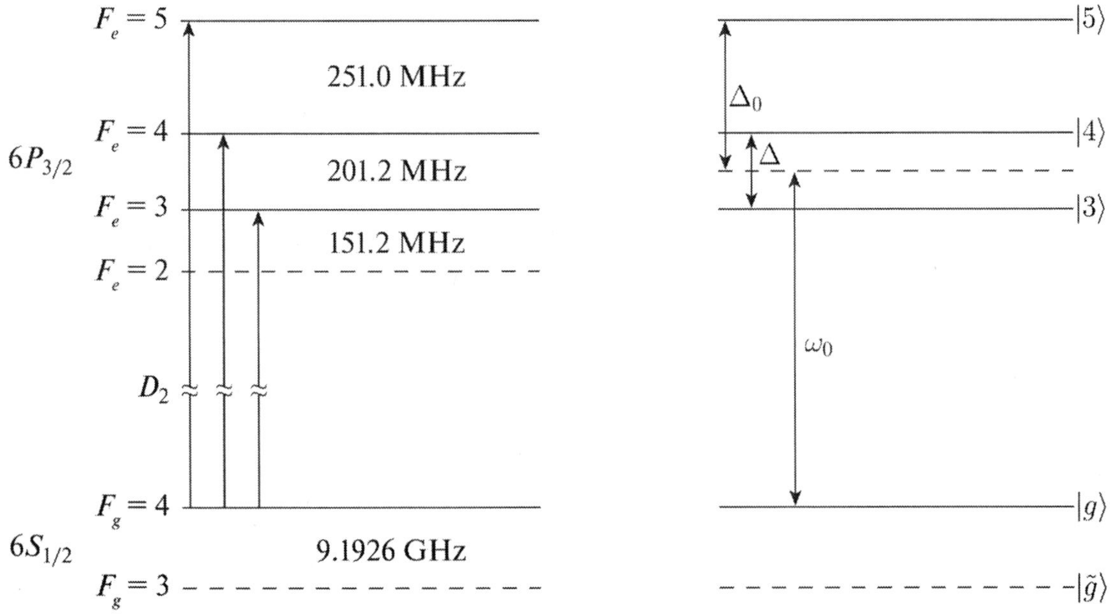

Figure 2. *Left:* Hyperfine structure of the states $6S_{1/2}$ and $6P_{3/2}$ and long-wavelength components of ^{133}Cs D_2-line. Transition $F_g = 4 \rightarrow F_e = 5$ is a cyclic one, transitions $F_g = 4 \rightarrow F_e = 3, 4$ are open ones. Excited state for forbidden transition $F_g = 4 \rightarrow F_e = 2$ is represented by a dashed line. Frequency interval of 9.1926 GHz between levels $F_g = 4$ and $F_g = 3$ is much larger than Doppler width, which is of the order of 380 MHz. Therefore optical pumping of level $F_g = 3$, represented by a dashed line, occurs on the open transitions. *Right:* Five-level scheme of transitions considered in the theory.

$$\Gamma(t,v) = \Gamma_3(t,v) + \Gamma_4(t,v) \equiv \frac{\tilde{\gamma}_3\Omega_3^2}{\gamma^2/4 + \Delta_3^2(t,v)} + \frac{\tilde{\gamma}_4\Omega_4^2}{\gamma^2/4 + \Delta_4^2(t,v)}, \tag{2}$$

where $\Omega_{3,4} = d_{3,4}\mathcal{E}_L/2\hbar$ are the Rabi frequencies for the transitions $|g\rangle \leftrightarrows |3\rangle,\ |4\rangle$, respectively, $\tilde{\gamma}_3$ and $\tilde{\gamma}_4$ are partial spontaneous decay rates of the states $|3\rangle$ and $|4\rangle$ to the non-absorbing state $|\tilde{g}\rangle$, $\gamma = \gamma_5 = \gamma_4 + \tilde{\gamma}_4 = \gamma_3 + \tilde{\gamma}_3$. At each moment of time, i.e. for each value of the laser frequency, the two terms in Eq. (2) look like small narrow peaks in the vicinities of the two significantly different resonant velocities.

The probability for an atom with a given longitudinal velocity to transit into the non-absorbing state $|\tilde{g}\rangle$ during one act of flight is $\tau_0\Gamma(t,v)$. Therefore the equilibrium velocity distribution function $f(v) = \frac{1}{v_T\sqrt{\pi}}e^{-(v/v_T)^2}$ for the states $|g\rangle$, $|\tilde{g}\rangle$ is perturbed in the small vicinities $\delta v \sim \gamma/k \ll v_T$ of the resonant velocities, which are determined by the condition $\Delta_{3,4}(t,v) = 0$. Renewal of the equilibrium velocity distribution for the states $|g\rangle$ and $|\tilde{g}\rangle$ occurs due to collisions with the cell walls in a period of the order of the time τ_v of flight through the cell, which is noticeably larger than the flight time τ_0 through the beam. In other words collisions with the walls extend local perturbation of the velocity distribution functions for the states $|g\rangle$ and $|\tilde{g}\rangle$ to the entire Doppler contour in a stretch of τ_v. Laser field violates the distribution function $f(v)$ only in a small range near resonant velocities. After integration over all velocities we find that the total populations of the ground state sublevels $n_g(t)$ and $n_{\tilde{g}}(t) = 1 - n_g(t)$ change much slower, namely, during the characteristic time T, which has the form

$$\frac{\tau_v}{T(t)} = \langle 1 - exp\left\{-\tau_0 \Gamma(t,v)\right\}\rangle , \tag{3}$$

and generalizes the result obtained in [5] onto the case of two open transitions. Here the angular brackets denote averaging over velocities with the Maxwellian distribution. It should be noted that according to Eqs. (2), (3) the function $T(t)$ depends on time via the frequency scanning law $\omega_L(t)$. This simple result holds true provided the scanning process is slow enough and resonance velocities corresponding to working atomic transitions remain well separated during the relaxation time τ_v, i.e. $k\delta v \sim \tau_v s \sim \tau_v \omega_D/\tau_{sc} \ll \Delta, \Delta_0 < \omega_D$.

The relaxation of the populations of the ground state sublevels to their equilibrium values due to collisions of the atom with the cell walls occurs in a characteristic time τ_g. In antirelaxation coated cells it significantly exceeds not only the flight times, but also the optical pumping time of the whole ensemble of atoms over all velocities: $\tau_g \gg T \sim \tau_v \omega_D/\gamma \gg \tau_v$. Now we focus on the case when $\tau_{sc} \sim T$ and the retardation effect plays an important role. As soon as τ_{sc} is much smaller than τ_g, the process of collisional relaxation of total populations can be neglected, and population of the state $|g\rangle$ is described by expression $n_g(t) = \frac{1}{2} exp\{-\int_{-\infty}^{t} \frac{dt'}{T(t')}\}$, which contains explicitly the retardation effect.

We integrate using the approximate relation $1/\left(\psi^2 + \gamma^2/4\right) = (2\pi/\gamma)\delta(\psi)$ and arive at:

$$n_g^{\pm}(\omega_L) = \frac{1}{2} exp\left\{-\frac{\pi}{4}\frac{\gamma\,\tau_{sc}}{\omega_D\,\tau_v}\left[(1-e^{-N_3})\left(1 \pm \Phi\left(\frac{\Delta_3}{\omega_D}\right)\right) + (1-e^{-N_4})\left(1 \pm \Phi\left(\frac{\Delta_4}{\omega_D}\right)\right)\right]\right\}, \tag{4}$$

where $N_{3,4} = 4\tau_0 \tilde{\gamma}_{3,4} \Omega_{3,4}^2/\gamma^2$, $\Delta_{3,4} = \omega_L - \omega_0 \pm \Delta/2$, Φ is error function: $\Phi(x) = \frac{2}{\sqrt{\pi}} \int\limits_0^x exp\{-u^2\} du$, signs $\pm$ correspond to the two different frequency scanning directions and responsible for the hysteresis phenomenon.

By summarizing the absorption coefficients for the cyclic $k_5 = \left\langle \gamma\Omega_5^2/(\Delta_5^2(\omega_L, v) + \gamma^2/4)\right\rangle n_g^{\pm}(\omega_L)$ and open transitions $k_{3,4} = (\gamma/\tau_0\tilde{\gamma}_{3,4})\langle 1 - exp\{-\tau_0\Gamma_{3,4}(\omega_L, v)\}\rangle n_g^{\pm}(\omega_L)$, we obtain the final expression for the total absorption coefficient:

$$k^{\pm}(\omega_L) = \frac{\sqrt{\pi}\gamma^2}{2\tau_0\omega_D}\left[\frac{(1-e^{-N_3})}{\tilde{\gamma}_3} e^{-\left(\frac{\Delta_3}{\omega_D}\right)^2} + \frac{(1-e^{-N_4})}{\tilde{\gamma}_4} e^{-\left(\frac{\Delta_4}{\omega_D}\right)^2} + \frac{4\tau_0\Omega_5^2}{\gamma^2} e^{-\left(\frac{\Delta_5}{\omega_D}\right)^2}\right] n_g^{\pm}(\omega_L). \tag{5}$$

Here $\Omega_5 = d_5\mathcal{E}_L/2\hbar$ is the Rabi frequency for the transition $|g\rangle \leftrightarrows |5\rangle$, $\Delta_5 = \omega_L - \omega_0 - \Delta_0$. Influence of the antirelaxation coating, including the hysteresis effect, is described by the factor $n_g^{\pm}(\omega_L)$.

If optical pumping is small, i.e. $N_{3,4} \ll 1$, the factors $1 - e^{-N_{3,4}}$ standing in the exponent in Eq. (4) are equal to $N_{3,4}$ in the lowest order approximation and thereby are proportional to the field intensity. The latter appears as a common multiplier for the both frequency scanning directions and does not change the position of the crossing point of the absorption contours $\kappa^{\pm}(\omega_L)$. Dependence of the crossing point position on laser radiation power appears in the higher orders in terms of the field intensity.

Crossing point frequency is determined by the condition:

$$e^{-N_3}\Phi\left(\frac{\Delta/2 + \omega_L - \omega_0}{\omega_D}\right) = e^{-N_4}\Phi\left(\frac{\Delta/2 - \omega_L + \omega_0}{\omega_D}\right). \tag{6}$$

It is obvious that in the case $N_3 = N_4$ the condition (6) is satisfied by $\omega_L = \omega_0$, that is, the crossing occurs when the laser frequency is equal to the half-sum of frequency intervals between levels $|g\rangle$, $|3\rangle$ and $|g\rangle$, $|4\rangle$. For ^{133}Cs atoms we have $d_4 > d_3$. Therefore Eq. (6) is satisfied if $\omega_L < \omega_0$, so that the crossing point shifts toward the level $|3\rangle$ when the intensity increases, as it has been observed in the experiment, see Fig. 1.

Thereby, the theory, generalized to the realistic model with one cyclic and two open transitions, as it takes place in D_2-lines of alkali metal atoms, adequately describes the hysteresis effect and dependence of the crossing point frequency for two opposite scanning directions on laser radiation power, that are observed in experiment with ^{133}Cs atoms.

References

[1] Riehle F *Frequency Standards: Basics and Applications* (Wiley-VCH, Weinheim, 2006)
[2] *Optical Magnetometry*, ed. by Budker D, Kimball D F J (Cambridge Univ. Press, Cambridge, UK, 2013)
[3] Alexandrov E B, Vershovsky A K 2009 *Phys. Usp.* **52** p 573
[4] Shah V, Kitching J 2010 *Adv. At. Mol. Opt. Phys.* **59** p 21
[5] Sevostyanov D I, Yakovlev V P, Kozlov A N, Vassiliev V V, Zibrov S A, Velichansky V L 2013 *Quant. El.* **43** p 638
[6] Sevostyanov D I, Yakovlev V P, Kozlov A N, Vassiliev V V, Zibrov S A, Velichansky V L 2014 *Bull. Lebedev Phys. Inst.* **41** p 89
[7] Zibrov S A, Tsygankov E A, Chucelov D S, Sevostyanov D I, Vassiliev V V, Velichansky V L, Yakovlev V P 2018 *Opt. and Spectr.* **124** p 143

IOP Conf. Series: Journal of Physics: Conf. Series **1238** (2019) 012039 doi:10.1088/1742-6596/1238/1/012039

Investigation of the elemental composition of samples after exposure to θ-pinches in the vicinity of periodic discharges in a flow of liquid

A V Nesterovich

National Nuclear Research University MEPhI, Moscow, Russia

Abstract. The article presents the results of laser mass-spectrometric analysis of surface of the samples, exposed to long-term treatment with high current discharges in the fluid flow (water with admixtures of different composition). It has been found, that the elemental composition of the surface of the metal, in particular, oxygen-free copper, can vary considerably (up to 35 at.%) compared to the original composition (99.8 at.%) to a depth of up to 10 μm. Change of the isotopic composition (isotope concentration of 65 amu may amount to 38.1% while the normal rate is equal to 30.6%) has been indicated. Composition of water (distilled water, tap water, saline water) does not affect the results. The concentration of elements, such as Cl, may exceed the original concentration by 3,000 times. At a depth of 10 μm, the composition is similar to the original one (before the exposure).

1. Introduction

The electro-spark exposure is one of the promising methods of changing the physical properties of materials; it can be implemented by generating periodic high-voltage discharge in the fluid flow [1]. Fluid stream is used as a current-carrying electrode and a current switch, that allows generating a sequence of pulses of electrical discharge from a few nanoseconds to a microsecond with the pulse packet repetition frequency of up to several dozens of Hz.

During the primary discharge which is mostly sparking (partial discharge on the water stream surface) in the inner area of the stream occurs the pinch discharges with all signs of z-pinch, transforming into theta-pinch around the grounded electrode [1,2].

Found surface material modification localized in the tracks of pinch of 30 to 100 μm size inside the main discharge area about 4mm diameter. In the pinch tracks area a surface material became amorphous, indicating extreme pressure and temperature.

The method of laser mass spectrometry was used in order to examine changes in the elemental composition of the surface of various samples.

Let's cite the results of determination of the elemental composition of the surface of the electrode, made of different metals, and then discuss the results.

2. Sample - copper plate, medium change

The conducted research regards the effect of the media, in which the discharge occurs, on the change of qualitative and quantitative composition of admixtures on the target surface. The following media have been selected for experiments: tap water, distilled water and water, specially saturated with sodium salt. The total dose of sparkling remained at the same level in the experiments, but it was impossible to keep the same interaction area. The assessments of the area for the compared experiments are: tap water - 0.5 square centimeters, distilled water - 4 square centimeters, water, saturated with salt - 0.7 square centimeters. The comparison results are presented in table 1.

Measurement was made on the EMAL - mass spectrometer with a laser-plasma ion source. The EMAL characteristic is described below:

-quantum optic generator – ILTI-203;
-crater depth, μm – 0.1 ÷ 1.0;
-crater diameter, μm – 10 ÷ 150;
-absolute lower detection limit, g – $10^{-13} \div 10^{-14}$;
-relative detection limit, % – 3×10^{-7};
-measurable concentration range,% – $10^{-7} \div 10$;
-maximum detectable ion mass, amu – 500.

Table 1. Results for copper plate element analysis for different waters

Element	Content,% wt.		
	Distilled water	Tap water	Saturated salt water
C	**0.4**	**0.2**	**0.2**
N	**0.02**	**0.03**	**0.00**
O	**0.2**	**0.7**	**0.24**
Na$_t$	**0.2**	**0.2**	**0.14**
Mg			
Al	**0.03**	**0.0008**	
Si	**0.07**	**0.003**	
P			
S		**0.009**	
CI	**0.3**	**0.6**	**0.3**
K	**0.06**	**0.07**	**0.08**
Ca	**0.04**	**0.03**	**0.03**
Cu	rest	rest	rest

According to the comparison of the results, normalized to one area, it may be concluded that: the presence of different media does not significantly affect the quantitative composition of the elements on the surface of samples.

3. Sample - copper plate, isotope analysis

The effect of a discharge on the change of isotopic composition of the target base (copper here) has been studied. The results have been obtained on the basis of calculations of the spectra, measured with laser spectroscopy. The measurement was made by ELAN 6000 ICP-MS mass spectrometer accompanied with the sampling Laser Sampler 330. ELAN 6000 characteristic is described below:

-mass range, amu – 1 ÷ 270;
-resolution, amu – 0.3 ÷ 3;
-dynamic range – 10^8;
-relative detection lower limit, % – 10^{-6}.

As a result of the studies, it was found:

1. Isotope ratio in the copper plate without the exposure corresponds to the natural distribution of isotopes namely- portion of 65 amu is equal to 30.6%, while the tabular value is 30.9%; portion of 63 amu is the rest.
2. High isotope concentration of 65 amu - 38.1; 35.2;36.6;37.5 is observed in the first (surface) 2-μm layer.
3. The reduced isotope concentrations of 65 amu are observed in subsequent 2-μm layers: 2-nd layer - 28.1; 28.5; 28.8; 3-rd layer - 25.8; 26.0.
4. Three days later, high isotope concentration of 65 amu: 39.3; 38.3; 35.8; 35.3, also has been observed on a new (not analyzed previously, but exposed to the discharge) area.

4. Sample - copper plate, layer-by-layer analysis

Elemental and quantitative composition was studied both in the area of external exposure and on the unexposed surface of pure copper (99.8 at.%) sample. Areas from both front and back surfaces of exposure have been studied. In addition, a serial layer-by-layer analysis of the composition at intervals of approx. 2 μm (table 2) has been conducted.

Table 2. Results for a copper plate element analysis

Element	Content,% wt.					
	Unexposed surface	Front side surface layer	Layer №2 4 μm	Back side surface layer	Layer №2 4 μm	Layer №3 6 μm
C	0.2	1.0	0.28	2.3	0.6	0.2
N	0.005	0.1	0.03	0.2	0.07	0.04
0	**0.06**	**2.7**	**0.9**	**20**	**2.5**	**1.2**
Na	**0.004**	**0.5**	**0.3**	**0.2**	**1.1**	**0.2**
Mg						
A1	0.003	0.002	Indefined	0.02	0.008	0.004
Si	0.002	0.005	Indefined	0.02	0.008	0.002
P						
S	0.0007	0.02	0.01	0.02	0.01	0,016
CI	**0.006**	**1.0**	0,6	**11.8**	**5.7**	**5.4**
K	**0.005**	**0.1**	**0.1**	**0.3**	**0.2**	**0.1**
Ca	**0.0003**	**0.04**	**0.03**	**0.04**	**0.02**	**0.03**
Cu	rest	rest	rest	rest	rest	rest

As a result of the analyses of the composition, the following features of the studied samples can be pointed out:

1. No formation of new elements, absent in the original sample, is observed.
2. On the surface of samples, in places subjected to external exposure, an increased concentration of admixtures, namely C, N, O, Na, Cl, K, Ca is fixed.
3. The more concentrated exposure is, the greater the concentration of the specified admixtures is. The concentration of exposure determine as estimated energy density to the unit of the irradiated plate surface. Averaging over surface area is determined by density of the craters corresponding size (30 to 100 μm diameter) which is function of discharge intensity, repetition frequency and exposure time.
4. The concentration of certain admixtures, such as Cl, exceeded the content compared to the initial one by 3000 times, and the concentration of Na exceeded the initial concentration by 300 times.
5. The concentration of specified admixtures decreases with deepening in the sample surface.
6. In terms of the majority of the admixtures, concentration of elemental composition of the sample returns to the original one approx 10 microns below the surface layer.
7. Concentration of admixtures in the atomic percentage on the surface, subjected to external exposure, exceeds one third (up to 35 at. %) as compared to the originally pure copper (99.8 at.%).

5. Sample - copper plate in a 3 weeks' interval

Elemental and quantitative composition was studied both in the area of external exposure and on the unexposed surface of the same copper sample. Areas from both front surface and back surfaces of exposure, where the traces of exposure had been observed, have been studied. 3 weeks later the surface analysis was redone. This time new areas, located near the previously analyzed ones, were analyzed (table 3).

Table 3. Results for copper plate in a 3 weeks' interval

Element	Content,% wt.				
	Unexposed surface	Front side surface layer	Front side. Surface layer. After 3 weeks	Back side surface layer	Back side surface layer. After 3 weeks
C	0.2	1.0	0.2	2.3	0.8
N	0.005	0.1	0.04	0.2	0.03
O	**0.06**	**2.7**	**1.5**	**20**	**3.2**
Na	**0.004**	**0.5**	**0.5**	**0.2**	**0.8**
F			**0.001**		**0.008**
Al	0.003	0,002	0,004	0.02	0.004
Si	0.002	0.005	0.005	0.02	0.012
P			**0.01**		0.002
S	0.0007	0.02	0.01	0.02	0.006
Cl	**0.006**	**1.0**	**0.6**	**11.8**	**0.4**
K	**0.005**	**0.1**	**0,09**	**0.3**	**0.2**
Ca	**0.0003**	**0.04**	**0.02**	**0.04**	**0.2**
Cu	rest	Rest	rest	rest	rest

As a result of the composition analyses, the following features of the studied samples can be pointed out:

1. 3 weeks after the external exposure, new elements, not contained in the original sample, such as fluoride and phosphorus, emerged on the sample, subjected to external exposure
2. On the samples. surfaces, in places subjected to the external exposure, the increased concentration of admixtures, namely C, N, O, Cl, has been generally decreased. The number of elements, such as Na, K, Ca, either does not change or increases.
3. As compared to the surface, free from the exposure, 3 weeks after the exposure, there is still a high content of previously specified elements on the surface.

6. Sample - beryllium plate

Elemental and quantitative composition was studied both in the area of external exposure and on the unexposed surface. In addition, a serial layer-by-layer analysis of the composition at intervals of approx. 10 μm has been conducted. Studies of the composition have been resumed 24 hours later at the same place, deeper in the surface (table 4). Layer №4 correspond to original material composition of the plate.

Table 4. Result for a beryllium plate

Element	Content,% wt.				
	Unexposed surface	Layer №1 1 μm	Layer №2 4 μm	Layer №3 10 μm	Layer №4 20 μm
H	0.02	0.02			
Be	rest	rest	rest	rest	rest
B	**5.0**	**0.5**	**0.9**	**0.4**	**0.4**
C	1.0	0.5	1.0	0.4	0.3
N	0.05	0.1			
O	**4.0**	**18**	**10**	**5**	**3**
Na	-	0.006	0.05		
Al	0.1	0.02	0.05		
Si	0.3	0.03	0.07		
S	-	0.02	0.05		
K	0.1	0.1			
Ca		0,03	0,02		
Mn	0.05	0.007			
Fe	0.2	0.1			

As a result of the analyses of the composition, the following features of the studied beryllium samples can be pointed out:

1. No formation of new elements, not contained in the original sample, is observed.

2. The increased concentration of oxygen only was found on the sample surface, subjected to the external exposure, which was reduced to the initial level (free of exposure, i.e. the clean surface) deeper in the surface.
3. The concentration of boron on the surface, subjected to exposure, decreases (by about 10 times).
4. Comparing the balance of content of oxygen and boron (decrease of one element and increase of another element in atomic percents), it can be said, within the limits of the error, that their total number before and after the exposure remains unchanged.
5. It should also be noted, that there are no effects of increased content of admixtures, which were observed after the exposure to the copper plate.

7. Sample - tungsten plate

Elemental and quantitative composition was studied both in the area of external exposure and on the unexposed surface.

Table 5. Result for a tungsten plate.

Element	Content,% wt.				
	Unexposed surface	After exposure, the first layer	After exposure, the second layer	After exposure, the third layer	After exposure, the fourth layer
H	0.3	0.5	-	-	-
C	0.6	0.2	0.1	0.08	0.08
N	0.008	0.02	0.008	0.004	0.001
O	2.6	1.5	0.2	0.08	0.07
Mg	0.08	0.35	0.12	-	
Na	3.0	0.5	0.3	0.04	0.009
F	0.04	0.02	-	-	-
Al	0.1	0,16	0.08	0.012	0.009
Si	0.3	0.20	0.08	0.015	0.009
P	0.01	0.020	-	-	-
S	0.1	0.1	0.02	-	-
Cl	0.2	0.6	0.3	0.05	-
K	2.0	0,25	0,14	0.025	-
Ca	0.4	0.25	0,13	0.020	-
Ti	0.030	0.015	-	-	-
Cr	0.004	0.004	-	-	-
Mn	0.007	0.006	-	-	-
Fe	0.06	0.15	0.09	-	-
Ni	0.015	0.003	-	-	-
Cu	0.01	0.3	0.28	-	-
Zn	**0.01**	**0.2**	**0.1**	-	-
Mo	0.12	-	-	-	-
W	rest	rest	rest	rest	rest

As a result of the analyses of the composition, the following features of the studied samples can be pointed out:

1. On the surface of samples, in places subjected to external exposure, a decrease of concentration of admixtures, namely C, N, O, Na and K was observed.
2. The concentration of Cu and Zn, increases significantly (30 times and 20 times correspondingly).

8. Discussion

The analysis of the obtained results and the results of the materials research and theoretical works, devoted to the study of nature of pinch plasmoids (observed at used in this work periodic high-voltage discharge in water), allows to make a preliminary assumption about possible monopole magnetic charges formation [3,6,7] during such discharges. This monopole charges have been searched earlier in the assumption of their separation by strong magnetic fields. At a pinch base of 30 μm in diameter and currents up to several kA recorded in experiments, the formation of a vertical magnetic field with the induction of several dozen T, separating and accelerating the monopoles, is not excluded [4,5]. The changes of element composition in the surface layer from the back (!) side of the sample (few mm thick) can be assumed to be an indirect confirmation of this hypothesis. The possible explanation of this fact may be the distribution of similar charged monopoles ("northern" monopoles only or "southern" monopoles only), subjected to mutual repulsion. The monopole overrunning the sample would require additional measures (extremely strong magnetic field) [8]. Within the area of monopole localization, the elemental composition should normally be changed presumably due to extra-high density of electrons nearby (which can exceed the electron density in the environment by several orders of magnitude). At such extra high density of electrons the neighboring nuclei fuse together, forming the heavier ones, since the Coulomb barrier for nucleus is decreased under influence of negative charge of these electrons (example - formation of phosphorus and fluorine out of air atoms and ions of hydrogen and oxygen[5,9]). At this conditions neutronization, i.e. fusion of electron and proton in a nucleus, should occur even more easier. Atom can relax over time to its original state with electrons leaving the surface in the process of beta decay [9]. Glowing of the surface in separate spots corresponding to beta decay was reported earlier, or remains in the form of a new element (neutralization of the nearest isotope nucleus), which, presumably, can be facilitated by the probing laser radiation.

Acknowledgements
A.N. Puchkov and S.M. Silnov substantially assisted with performance of works, for which the author expresses gratitude to them.

The work is done under the Agreement No. 17-72-20228 between the Russian Science Foundation, and MEPhI.

References
[1] Alferov P V, Bogdanovich B Yu, Volkov N V, Len N A, Nesterovich A V, Starostin A N 2000 *Structural state of materials after the action of a periodic discharge in a liquid flow.* (*Engineering Physics* vol 4) pp 58 – 61 [in Russian]

[2] Bogdanovich, B.Y., Kalin, B.A. & Nesterovich, A.V. Inorg. Mater. Appl. Res. (2016) 7: 673. https://doi.org/10.1134/S2075113316050038

[3] Bogdanovich B.Yu., Nesterovich A. V., Sukhanova L. A. KHlestkov Yu. A. "The pinch plasmoid description in terms of general relativity". *Izvestiia VUZov. Fizika* [Proceedings of the universities. Physics], 59(10), 109-119, 2016 [in Russian]

[4] Bogdanovich, B. Yu, A. V. Nesterovich, and P. V. Alferov. "Forming of Plants of a Polymer Type in a Zone of Localization of Surfaces Subjected Affecting of an Electromagnetic Collapse of Pinch Discharges." Gravitation and Cosmology 8.S (2002): 213.

[5] Bogdanovich, B. Yu, and A. V. Nesterovich. "Periodic Discharge in a Stream of Liquid and Anomalous Phenomenas in the Post-Discharge Period." 11th International Workshop on Magneto-Plasma Aerodynamics.-April 10-12. 2012.

[6] Meierovich B. E. *Kanal silnogo toka* [Channel of intense current] . – Moscow, FIMA Publ., 1999.

[7] Meierovich B. E. "Equilibrium of intense current channel in general theory of relativity". *ZhETF*, [Journal of experimental and theoretical physics] Vol. 112, No. 2, p. 385, 1997 [in Russian]

[8] Rubakov V. A., Superheavy magnetic monopoles and a proton decay. *Pisma v ZhETF* [Letters to the journal of experimental and theoretical physics] Vol 33 issue 12 p. 658, 1981 [in Russian]

[9] Olkhovatov A. Yu., Rodionov B. U. *Tungusskoe siianie,*[Tunguska Glow]. –Moscow, Base knowledge laboratory Publ., 1999

Impact of plasma nanostructuring on the electrical properties of $Cu(In,Ga)Se_2$ films

S P Zimin[1], D A Mokrov[1], I I Amirov[2], V V Naumov[2], E S Gorlachev[2], V F Gremenok[3,4], V V Khoroshko[4]

[1] Microelectronics and General Physics Department, Yaroslavl State University, Sovetskaya Street 14, Yaroslavl 150003, Russia
[2] Yaroslavl Branch of the Institute of Physics and Technology, Russian Academy of Sciences, Universitetskaya Street 21, Yaroslavl 150007, Russia
[3] Scientific-Practical Materials Research Centre of the National Academy of Sciences of Belarus, P. Brovki Street 19, Minsk 220072, Belarus
[4] Belarusian State University of Informatics and Radioelectronics, P. Brovki Street 6, Minsk 220013, Belarus

E-mail: zimin@uniyar.ac.ru

Abstract. In this work, the impact of the plasma treatment, during the formation of nanostructure arrays on the surface of the $Cu(In,Ga)Se_2$ films on glass substrates, on the conductivity of the films both in the lateral direction and in the direction normal to the substrate surface was studied. The initial $Cu(In,Ga)Se_2$ films with the Ga/(In + Ga) ratio in the range of 0.03–0.12 were obtained by thermal selenization process of stacked metallic precursors and by co-evaporation of all elements from various sources. The plasma treatment was carried out in a high-density low-pressure RF inductively coupled plasma reactor in argon plasma. The average ion energy was 200 eV, the processing time was 60 s. It is shown that the processes of the plasma nanostructuring of the $Cu(In,Ga)Se_2$ film surface lead to the formation of a thin modified near-surface layer with a resistivity of 2–3 orders of magnitude less than for the bulk of the film.

1. Introduction

Semiconductor polycrystalline copper indium gallium diselenide films $Cu(In,Ga)Se_2$ ($CuIn_{1-x}Ga_xSe_2$) are widely used in thin-film solar cells, which are reported to achieve the record solar cell conversion laboratory efficiency of 23.3% [1]. A promising direction for further increasing the efficiency of solar cells based on copper indium gallium diselenide films is the nanostructuring of their surface [2]. We have proposed methods for the effective nanostructuring of the $Cu(In,Ga)Se_2$ surface during the processing of the films in the high-density low-pressure RF inductively coupled argon plasma [3, 4]. As a result, arrays of vertical uniform nanostructures with lateral dimensions of 10–30 nm and a height of 10–90 nm, with a surface density of $(0.8–1.8) \times 10^{11}$ cm^{-2} were formed on the film surface (figure 1), while these geometrical parameters could be controlled by varying the modes of the plasma treatment. It is well known that the processes of thermal, laser, and radiation modification of $Cu(In,Ga)Se_2$ are capable of greatly changing the electrical properties of the films [5, 6]. Therefore, as a further step, it becomes necessary to investigate possible changes in the electrical properties of copper indium gallium diselenide films after the plasma nanostructuring. The aim of this work was to study the electrical conductivity of $CuIn_{1-x}Ga_xSe_2$ (x = 0.03–0.12) thin films in the lateral direction and in the direction normal to the substrate surface in the initial state and after the nanostructure array formation in plasma.

2. Experimental

The initial $Cu(In,Ga)Se_2$ thin films with the Ga/(In + Ga) ratio in the range of 0.03–0.12 were grown on glass substrates by thermal selenization process of stacked metallic precursors and by co-evaporation of all elements from various sources. The Cu/(In + Ga) ratio varied over the wide range of 0.67–2.37. The thickness of the films was 2.0–2.4 μm. The processes of the film growth are described in more detail in [3, 7]. The data on the chemical composition of the obtained films is given in table 1.

Table 1. Impact of the nanostructuring in plasma on the resistivity of the $Cu(In,Ga)Se_2$ thin films

Sample no.	Growth method	Ga/ (In+Ga)	Cu/ (In+Ga)	ρ_{init} Ohm cm	$\rho_{experim}$ Ohm cm	Petritz's model analysis	
						ρ_{surf} Ohm cm	ρ_{init}/ρ_{surf}
Measurements in the lateral direction							
1	Selenization	0.03	0.67	2.2×10^3	1.4×10^1	2.0	1100
2	Selenization	0.07	0.66	9.1×10^2	4.3	4.7×10^{-1}	1936
3	Selenization	0.12	2.37	1.9	7.0×10^{-2}	1.0×10^{-2}	190
Measurements in the direction normal to the surface							
4	Co-evaporation	0.10	0.93	8.0×10^2	2.6×10^2	–	–
5	Selenization	0.11	1.36	1.9×10^4	1.2×10^4	–	–

Experiments on the plasma treatment of the samples were carried out in a high-density low-pressure RF inductively coupled plasma reactor in argon plasma. The plasma discharge was ignited when the RF power of 800 W was applied from the inductor (frequency 13.56 MHz). An independent RF power of 300 W from an individual high-frequency generator was applied to the substrate holder, which provided the average energy of Ar^+ ions of about 200 eV. The processing duration was 60 seconds. Typical images of the pristine surface and of the array of the formed nanostructures, using the sample no. 3 (table 1) as an example, are shown in figure 1.

Measurements of the resistivity in the lateral direction were carried out using a four-probe method for the samples which had the $Cu(In,Ga)Se_2$ film deposited directly on a dielectric glass substrate. Measurements of the resistivity in the direction normal to the surface were carried out by obtaining the current-voltage characteristics of a $Mo/Cu(In,Ga)Se_2/Mo$ structure. In this case, the $CuIn_{1-x}Ga_xSe_2$ film was deposited on a glass substrate with a sublayer of molybdenum, and an ohmic molybdenum contact with an area of about 2 mm^2 was deposited with minimal heating (50 °C) on top of the grown film. The current-voltage characteristics of the $Mo/Cu(In,Ga)Se_2/Mo$ structures were strictly linear over a wide range of voltage bias, which allowed calculating the resistivity value using simple equations.

3. Results and discussion

Performed measurements of the resistivity of the films in the initial state ρ_{init} showed values in the range of 10^0–10^5 Ohm cm (table 1), corresponding to the wide range of resistivity values described in literature [6, 8], which, in the most general case, is determined by the grain sizes, the Cu/(In + Ga) ratio, the presence of pores, the formation of $Cu_{2-x}Se$ inclusions, etc. During the plasma treatment, a surface sputtering was observed, as a result of which the thickness of the films decreased by 100–200 nm and a partial alignment of the microrelief and a formation of nanostructures occurred (figure 1). The values of the resistivity of $Cu(In,Ga)Se_2$ after the plasma treatment $\rho_{experim}$, calculated under the assumption of uniformity of electrical properties along the film thickness, are listed in table 1.

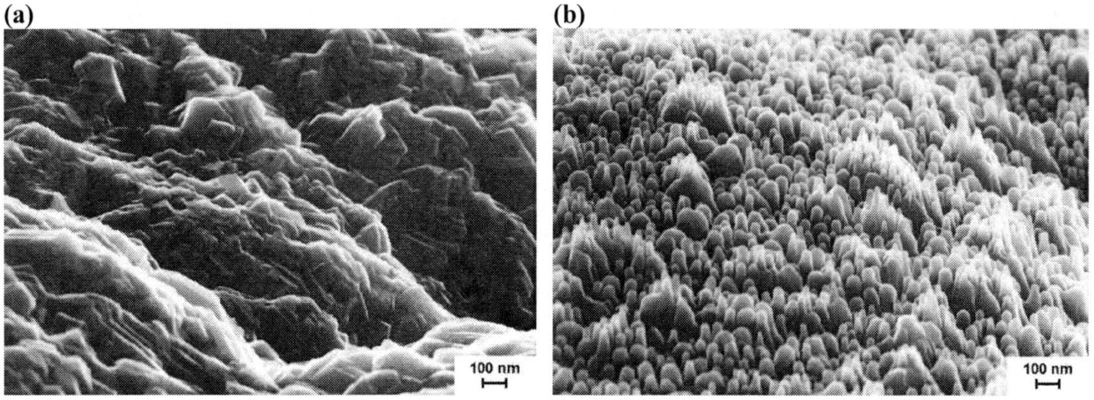

Figure 1. Electron microscopy images of the (a) initial and (b) nanostructured $Cu(In,Ga)Se_2$ film surfaces (shown for the sample no. 3).

A direct analysis of the data in table 1 shows that when the charge carriers flow in the lateral direction, the resistivity decreases sharply by about 2 orders of magnitude. This behaviour is typical for the modification of $Cu(In,Ga)Se_2$ films by rapid thermal annealing and laser processing [5, 6]. However, it should be noted that the electrical conductivity of the $Cu(In,Ga)Se_2$ films in the direction normal to the substrate varies insignificantly – by 1.6–3.1 times.

To explain the obtained experimental results, a model can be applied, according to which the plasma nanostructuring of the $Cu(In,Ga)Se_2$ thin films only modifies the electrical properties of the near-surface part of the films. From a physical point of view, there are no real reasons which, with fairly short-term and low-energy effect of Ar^+ ions, could change the structural and electrical properties of the films over the entire thickness of 2 μm. We presume that as a result of the plasma nanostructuring, a double-layered structure is formed (figure 2), when a layer with high electrical conductivity forms on the surface, and the rest of the bulk of the film does not change its electrical properties.

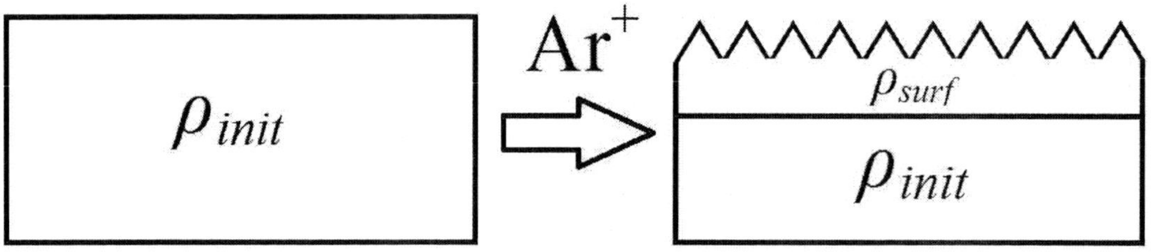

Figure 2. Schematic model of the formation of a double-layered structure of the $Cu(In,Ga)Se_2$ thin films during the plasma treatment.

The results of the calculations of the resistivity of a high-conductive surface layer ρ_{surf} using the mathematical equations of the Petritz's two-layer model [9] under the assumption of the thickness of the modified layer of 200 nm are presented in table 1. Analysis of these data shows that the resistivity of the near-surface layer decreases by 2–3 orders of magnitude, and the minimal changes are typical for the films which initially had high electrical conductivity. The explanation of the decrease in the resistivity of the near-surface layer during the plasma treatment is complex in nature, and includes an increase in the concentration of charge carriers due to radiation defects, the appearance of conductive inclusions on the surface, the filling of voids between the grains due to redeposition effects, etc.

When analyzing the experimental results for the current flowing along the normal to the substrate surface, another contradiction arises: the appearance of a thin near-surface layer can not quantitatively explain the observed decrease in the resistivity by 1.6–3.1 times. In this situation, additional phenomena should be taken into account:

- a possible increase of the upper electrode area due to additional carrier spreading in the low-resistance near-surface layer;
- an increase in the effective surface area of the upper Mo/Cu(In,Ga)Se_2 electrode as a result of the surface nanostructuring. Estimates show that for the actual geometrical parameters of the nanostructures obtained after a 60 s plasma treatment, the effective surface area of the electrode can increase by 1.5–4.0 times. Therefore, to explain the changes in the conductivity in the direction normal to the film, it is necessary to take into account not only the formation of a near-surface layer during the nanostructuring in plasma, but also the increase in the actual electrode surface area.

4. Conclusions

The results of these investigations show that the processes of the plasma nanostructuring of the surface of Cu(In,Ga)Se_2 thin films lead to the formation of a thin modified near-surface layer with a resistivity of 2–3 orders of magnitude less than for the bulk of the film. The presence of this layer should change the parameters of the current-voltage characteristics of the ZnO/CdS/Cu(In,Ga)Se_2/Mo solar cell structure. Our next research will be devoted to the measurements of the current-voltage characteristics of the solar cells with nanostructured Cu(In,Ga)Se_2 layers.

References

[1] Green M A, Hishikawa Y, Dunlop E D, Levi D H, Ebinger J H and Ho-Baillie A 2018 *Prog. Photovolt.: Res. Appl.* **26** 3
[2] Liu C-H, Chen C-H, Chen S-Y, Yen Y-T, Kuo W-C, Liao Y-K, Juang J-Y, Kuo H-C, Lai C-H, Chen L-J and Chueh Y-L 2011 *Nano Lett.* **11** 4443
[3] Zimin S P, Gorlachev E S, Mokrov D A, Amirov I I, Naumov V V, Gremenok V F, Juskenas R, Skapas M, Kim W Y and Bente K 2017 *Semicond. Sci. Technol.* **32** 075014
[4] Zimin S P, Gorlachev E S, Gremenok V F, Bente K, Mokrov D A, Amirov I I, Naumov V V and Kim W Y 2017 *Phys. Status Solidi C* **14** 1600135
[5] Wang X, Li S S, Kim W K, Yoon S, Craciun V, Howard J M, Easwaran S, Manasreh O, Crisalle O D and Anderson T J 2006 *Solar Energy Mater. Solar Cells* **90** 2855
[6] Yoo M H, Ko P J, Kim N-H and Lee H-Y 2017 *J. Korean Phys. Soc.* **71** 1038
[7] Gremenok V F, Zaretskaya E P, Bashkirov S A, Kim W Y, Chai S H, Moon C-B and Jhun C G 2015 *J. Adv. Microsc. Res.* **10** 28
[8] Liu J, Zhuang D-M, Cao M-J, Wang C-Y, Xie M and Li X-L 2012 *Int. J. Photoenergy* **2012** 149210
[9] Petritz R L 1958 *Phys. Rev.* **110** 1254

IOP Conf. Series: Journal of Physics: Conf. Series **1238** (2019) 012041 doi:10.1088/1742-6596/1238/1/012041

Diffusion of light through a disordered ensemble of resonant Mie-particles embedded in a medium with circular dichroism

E E Gorodnichev and D B Rogozkin

National Research Nuclear University MEPhI, Kashirskoe Shosse 31, 115409 Moscow, Russia

E-mail: gorodn@theor.mephi.ru

Abstract. We study transmission of unpolarized light (incoherent superposition of right-hand circular and left-hand circular polarized waves) through an optically active medium doped by scattering Mie-particles. The medium is assumed to possess circular dichroism. Within the spatial diffusion approximation, the degree of circular polarization of the transmitted radiation is calculated. It is shown, that in the presence of scatterers a significant increase in the circular polarization compared to the medium with no scatterers can be observed.

1. Introduction

In an absorbing medium with optical activity, circular dichroism can be observed: the absorption coefficients for the right- and left-handed polarizations differ from each other (see, e.g., [1]). The circular dichroism is very sensitive to the configuration of complex molecules, as well as to conformational transitions in them. Therefore, measuring the circular dichroism is one of the most important methods of stereochemical analysis [2, 3]. The main experimental difficulty in quantifying the circular dichroism characteristics is due to a weakness of the effect [2–4].

In the last decade, in the context of nanophotonics (see, e.g., [5]), the optical properties of particles with a high refractive index in the visible and near infrared ranges have been studied in detail [6–9]. Near the first two Mie-resonances, the so-called first Kerker condition [10] has been shown to be fulfilled (see, e.g., [6, 11]). If the particle parameters satisfy the Kerker condition, scattering in the backward direction is suppressed and, in addition, circularly polarized light does not change its polarization upon scattering [7]. The depolarization of the waves becomes noticeable only after a great number of scattering events [8,9]. In the case of the spatial diffusion of light in such a medium, the attenuation of circular polarization occurs very slowly. The results [8, 9] enable us to expect that the addition of Mie-particles with the optical parameters and sizes satisfying the Kerker condition to a homogeneous medium, on the one hand, should lead to a significant increase in the photon path lengths in the medium, and, on the other hand, should not disrupt the initial circular polarization of the waves propagating in the medium. When the homogeneous medium possesses circular dichroism (the absorption coefficients of right-hand and left-hand polarized waves differ from each other), the addition of Mie-particles should result in a substantial increase in the difference between the intensities of the differently polarized waves, I_R and I_L. The degree of circular polarization of the transmitted radiation in the medium with no scattering $|I_R - I_L| / (I_R + I_L) = \Delta\varkappa L/2$ ($\Delta\varkappa$ is the difference between the corresponding

IOP Conf. Series: Journal of Physics: Conf. Series **1238** (2019) 012041 doi:10.1088/1742-6596/1238/1/012041

absorption coefficients, L is the sample thickness), is replaced by $|I_R - I_L| / (I_R + I_L) = \Delta\varkappa S/2$, where S is the average path length of photons in the sample. In the diffusion regime, the inequality $S \gg L$ is satisfied, and a significant increase in the ratio $|I_R - I_L| / (I_R + I_L)$ should occur. Thus, similarly to such systems as a "random laser" (see, e.g., [12]) and a scattering host medium for measuring weak absorption in liquids and gases [13, 14], a disordered ensemble of nearly resonant Mie-particles embedded in a medium with circular dichroism acts as a spatially distributed resonator which increases the photon path lengths in the sample without violating their circular polarization.

Below we study multiple scattering of initially unpolarized light (incoherent superposition of right-hand and left-hand polarized waves) in a system of Mie-particles embedded in an optically active homogeneous medium. The optical parameters of the Mie-particles are assumed to satisfy the first Kerker condition. Within the diffusion approximation, the degree of circular polarization arising in transmission of light through the medium is calculated. The degree of circular polarization is shown to drastically increase in comparison with the case of a homogeneous medium. The dependence of the degree of polarization and the enhancement factor on the sample thickness are calculated.

2. Transport equations for circularly polarized components of light

As shown in Refs. [8,9], near the first Kerker point, the single-scattering matrix takes the form:

$$\hat{d} = \begin{pmatrix} a_1 & 0 & 0 & 0 \\ 0 & a_1 & 0 & 0 \\ 0 & 0 & a_2 & 0 \\ 0 & 0 & 0 & a_2 \end{pmatrix} \tag{1}$$

where a_1 and a_2 are expressed in terms of the scattering amplitudes $A_\parallel$ and $A_\perp$ of the cross-polarized waves, $a_1 = \left(|A_\parallel|^2 + |A_\perp|^2\right)/2$ and $a_2 = \operatorname{Re} A_\parallel A_\perp^*$ [15]. Under the assumption that the optical activity effect is small, the amplitudes $A_\parallel$ and $A_\perp$ can be calculated without taking into account this effect.

For a medium with circular dichroism, right-hand and left-hand circularly polarized waves are the eigenmodes of the medium [16] and, therefore, multiple scattering of light in such a medium is conveniently described in terms of the Stokes vector in the circular representation [17–19]

$$\hat{I} = \frac{1}{\sqrt{2}} \begin{pmatrix} Q - \mathrm{i}U \\ I - V \\ I + V \\ Q + \mathrm{i}U \end{pmatrix} \tag{2}$$

where I, Q, U V are the Stokes parameters in the standard linear representation [15]. The values $I \pm V$ are proportional to the intensities of right-hand and left-hand polarized waves, $I_{R,L} = (I \mp V)/2$.

The vector radiative transfer equation with the scattering matrix (1) decouples into two independent systems of equations. One system involves the Stokes parameters $Q \mp \mathrm{i}U$ (Q and U characterize the linear polarization of light), the other system describes the circularly polarized components $I_{R,L}$ [8, 9]. So, if the first Kerker condition is satisfied, the linear and circular polarizations evolve independently of each other.

The intensities I_R and I_L of the circularly polarized components obey the system of equations (see, e.g., [18–20])

$$\left[\mu\frac{\partial}{\partial z}+n_0\sigma+\varkappa\right]I_R\left(z,\mu\right)-\frac{\Delta\varkappa}{2}I_R\left(z,\mu\right)= n_0\int d\mathbf{n}'\ a_+\left(\mathbf{n}\mathbf{n}'\right)\cdot I_R\left(z,\mu'\right)+ n_0\int d\mathbf{n}'\ a_-\left(\mathbf{n}\mathbf{n}'\right)\cdot I_L\left(z,\mu'\right) \tag{3}$$

$$\left[\mu\frac{\partial}{\partial z}+n_0\sigma+\varkappa\right]I_L\left(z,\mu\right)+\frac{\Delta\varkappa}{2}I_L\left(z,\mu\right)= n_0\int d\mathbf{n}'\ a_+\left(\mathbf{n}\mathbf{n}'\right)\cdot I_L\left(z,\mu'\right)+ n_0\int d\mathbf{n}'\ a_-\left(\mathbf{n}\mathbf{n}'\right)\cdot I_R\left(z,\mu'\right) \tag{4}$$

where $\mu=\mathbf{n}\mathbf{n}_{int}$, $\mathbf{n}$ is the direction of wave propagation, $\mathbf{n}_{int}$ is the inner normal to the sample surface, $a_\pm\left(\mathbf{n}\mathbf{n}'\right)=\left(a_1\left(\mathbf{n}\mathbf{n}'\right)\pm a_2\left(\mathbf{n}\mathbf{n}'\right)\right)/2$, σ is the cross-section of elastic scattering of light by Mie-particles, n_0 is their number per unit volume, $\varkappa$ is the average absorption coefficient of the medium, $\Delta\varkappa$ is the difference in the absorption coefficients of right-hand and left-hand circularly polarized waves. The value of $\Delta\varkappa$ is the magnitude of circular dichroism in a homogeneous medium [2, 16].

3. Spatial diffusion of light

A solution to the Eqs.(3) and (4) can be written in the form of expansion in a series in the Legendre polynomials (see, e.g., [18, 19]):

$$I_{R,L}\left(z,\mu\right)=\sum_{l=0}^{\infty}\frac{2l+1}{4\pi}\ I_{R,L}\left(z,l\right)P_l\left(\mu\right) \tag{5}$$

The coefficients $I_{R,L}\left(z,l\right)$ appearing in Eq.(5) are subject to the equations

$$\frac{l}{2l+1}\frac{\partial I_R\left(z,l-1\right)}{\partial z}+\frac{l+1}{2l+1}\frac{\partial I_R\left(z,l+1\right)}{\partial z}+\left(\sigma_{tot}^{(R)}-n_0a_+(l)\right)I_R\left(z,l\right)=n_0a_-(l)I_L\left(z,l\right) \tag{6}$$

$$\frac{l}{2l+1}\frac{\partial I_-\left(z,l-1\right)}{\partial z}+\frac{l+1}{2l+1}\frac{\partial I_-\left(z,l+1\right)}{\partial z}+\left(\sigma_{tot}^{(L)}-n_0a_+(l)\right)I_-\left(z,l\right)=n_0a_-(l)I_+\left(z,l\right) \tag{7}$$

where $\sigma_{tot}^{(R,L)}=n_0\sigma+\varkappa\mp\Delta\varkappa/2$ and

$$a_\pm\left(l\right)=\pi\int_{-1}^{1}d\mu\ \left(a_1\left(\mu\right)\pm a_2\left(\mu\right)\right)P_l(\mu) \tag{8}$$

The coefficient $a_-(l=0)$ is proportional to the depolarization cross section σ_{dep} introduced in [21], $a_-(l=0)=\sigma_{dep}/2$, where $\sigma_{dep}=\int d\mathbf{n}'\left(a_1\left(\mathbf{n}\mathbf{n}'\right)-a_2\left(\mathbf{n}\mathbf{n}'\right)\right)$. If the depolarization cross section of σ_{dep} is much smaller than the transport cross section of elastic scattering $\sigma_{dep}\ll\sigma_{tr}=\int d\mathbf{n}'\left(1-\mathbf{n}\mathbf{n}'\right)a_1\left(\mathbf{n}\mathbf{n}'\right)$, the effect of slow decay of circular polarization is observable [21]. In this case, the depolarization of circularly polarized light occurs after the onset of the regime of spatial diffusion of radiation. This effect is most pronounced, if the first Kerker condition is fulfilled and the ratio σ_{dep}/σ_{tr} reaches its minimum value [8, 9].

Under the conditions of spatial diffusion in a weakly absorbing medium (we assume that $\varkappa\ll n_0\sigma_{tr}$), the angular distribution of radiation turns out to be close to isotropic one, and only the first two terms with $l = 0$ and $l = 1$ should be left in the expansion (5) [18, 19, 21]. Then, Eqs.(6) and (7) are reduced to the following system of diffusion equations for $I_{R,L}(z) = I_{R,L}(z, l = 0)$:

$$\left[\frac{\partial^2 I_R(z)}{\partial z^2} - 3n_0\sigma_{tr}\Sigma_R I_R(z)\right] + \frac{3}{2}n_0\sigma_{tr}\sigma_{dep}I_L(z) = 0,$$
$$\left[\frac{\partial^2 I_L(z)}{\partial z^2} - 3n_0\sigma_{tr}\Sigma_L I_L(z)\right] + \frac{3}{2}n_0\sigma_{tr}\sigma_{dep}I_R(z) = 0 \qquad (9)$$

where the effective absorption coefficients appearing in the system (9) are equal to $\Sigma_{R,L} = \varkappa \mp \Delta\varkappa/2 + n_0\sigma_{dep}/2$.

In the limit $L \gg l_{tr}$ the boundary conditions to the diffusion equations (9) can be written as $I_{R,L}(z = 0) = I_{R,L}(z = L) = 0$. For unit flux of unpolarized light (incoherent superposition of right-hand and left-hand polarized light) incident on the boundary $z = 0$, the intensities $I_{R,L}$ of radiation transmitted in the forward direction has the form

$$I_R(L, \mu = 1) = \frac{1}{8\pi(1+\eta^2)}\left[(1+\eta)\frac{\varepsilon_- l_{tr}}{\sinh \varepsilon_- L} - \eta(1-\eta)\frac{\varepsilon_+ l_{tr}}{\sinh \varepsilon_+ L}\right] \qquad (10)$$

$$I_L(L, \mu = 1) = \frac{1}{8\pi(1+\eta^2)}\left[\eta(1+\eta)\frac{\varepsilon_- l_{tr}}{\sinh \varepsilon_- L} + (1-\eta)\frac{\varepsilon_+ l_{tr}}{\sinh \varepsilon_+ L}\right] \qquad (11)$$

where the parameter η entering into Eqs.(10) and (11) is equal to $\eta = \left[\sqrt{(n_0\sigma_{dep})^2 + (\Delta\varkappa)^2} - \Delta\varkappa\right] / (n_0\sigma_{dep})$. The attenuation coefficients ε_- and ε_+ are determined by

$$\varepsilon_\mp = \sqrt{3n_0\sigma_{tr}\left[\varkappa + \frac{1}{2}\left(n_0\sigma_{dep} \mp \sqrt{(n_0\sigma_{dep})^2 + (\Delta\varkappa)^2}\right)\right]} \qquad (12)$$

According to Eq.(12), the quantities $\varkappa + \frac{1}{2}\left(n_0\sigma_{dep} \mp \sqrt{(n_0\sigma_{dep})^2 + (\Delta\varkappa)^2}\right)$ act as the effective absorption coefficients for waves with different polarizations in the medium with circular dichroism. If the depolarization process can be neglected ($n_0\sigma_{dep} \ll \Delta\varkappa$), we return to the initial definition of the absorption coefficients for right-hand and left-hand polarized waves in a homogeneous medium with the circular dichroism, $\varkappa \mp \Delta\varkappa/2$. In the opposite case of noticeable depolarization ($n_0\sigma_{dep} \gg \Delta\varkappa$), the difference between the effective absorption coefficients turns out to be small (the difference is proportional to $(\Delta\varkappa/n_0\sigma_{dep})^2$).

The total intensity $I = I_R + I_L$ and the difference between the intensities of right-hand and left-hand polarized waves $I_L - I_R$ (i.e., the fourth Stokes parameter V) are determined by the relations

$$I(L, \mu = 1) = \frac{1}{4\pi(1+\eta^2)}\left[(1+\eta)^2\frac{\varepsilon_- l_{tr}}{\sinh \varepsilon_- L} + (1-\eta)^2\frac{\varepsilon_+ l_{tr}}{\sinh \varepsilon_+ L}\right] \qquad (13)$$

$$V(L, \mu = 1) = \frac{1-\eta^2}{4\pi(1+\eta^2)}\left[\frac{\varepsilon_- l_{tr}}{\sinh \varepsilon_- L} - \frac{\varepsilon_+ l_{tr}}{\sinh \varepsilon_+ L}\right] \qquad (14)$$

In the absence of circular dichroism ($\Delta\varkappa = 0$) Eq.(13) transforms to the well-known formula for the intensity I of the scalar theory (see, e.g., [21]). In this case the circular polarization does not arise, and the fourth Stokes parameter is zero, $V = 0$.

IOP Conf. Series: Journal of Physics: Conf. Series **1238** (2019) 012041 doi:10.1088/1742-6596/1238/1/012041

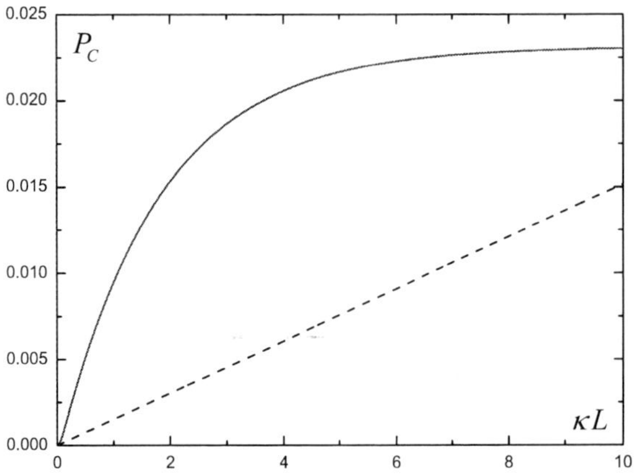

Figure 1. Degree of circular polarization for initially unpolarized light vs sample thickness L. The numerical calculations were carried out for $\Delta\varkappa/\varkappa = 3 \cdot 10^{-3}$, $\sigma_{dep}/\sigma_{tr} = 10^{-3}$, $n_0\sigma_{tr}/\varkappa = 100$. The solid blue and dashed black lines correspond to the scattering medium and the medium with no scattering centers ($n_0 = 0$), respectively.

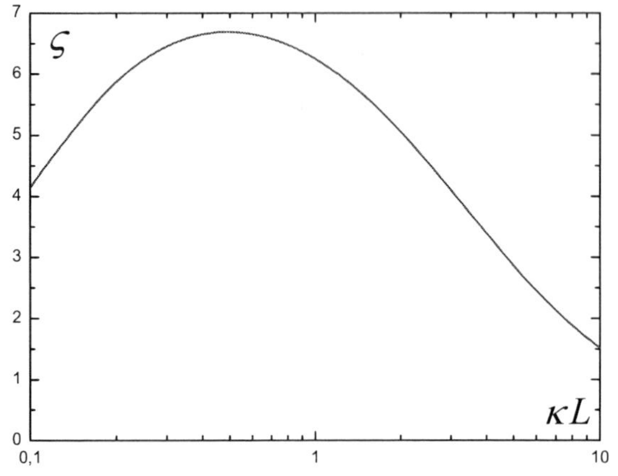

Figure 2. Enhancement factor $\zeta ==$ $P_C/P_C(n_0 = 0)$ vs sample thickness L. The parameters of the scattering medium are the same as in Fig. 1.

4. Discussion

For the initially unpolarized beam, the degree of circular polarization of radiation transmitted through the medium can be calculated with the ratio $P_C = |V|/I$ where V and I are determined by Eqs.(13) and (14). For a relatively thin sample, $l_{tr} \ll L \ll l_D$ ($l_{tr} = (n_0\sigma_{tr})^{-1}$ is the transport mean free path, $l_D = (3n_0\sigma_{tr}\varkappa)^{-1/2}$ is the diffusion length (see, e.g., [22]) the degree of polarization increases with L as

$$P_C = \frac{\Delta\varkappa}{2}\left(\frac{L^2}{l_{tr}}\right) \tag{15}$$

In the limiting case of very thick samples, $L \gg l_{circ}$, ($l_{circ} = (\varepsilon_+ - \varepsilon_-)^{-1}$ is the length of circular polarization attenuation [18,19]), the growth of the degree of polarization saturates and P_C tends to the limiting value

$$P_C = \frac{\Delta\varkappa}{\sqrt{(n_0\sigma_{dep})^2 + (\Delta\varkappa)^2} + n_0\sigma_{dep}} \tag{16}$$

An example of numerical calculations for the L-dependence of P_C is shown in Fig. 1. We use the typical value of the ratio $\Delta\varkappa/\varkappa$ [2] and the value of σ_{dep}/σ_{tr} corresponding to the first Kerker point for Si particles [8,9]. From the figure it follows that the addition of scattering Mie-particles to a homogeneous medium with circular dichroism leads to noticeable increase of the degree of polarization at relatively small thickness L. For characterizing the increase in the degree of polarization compared to the case of a homogeneous medium, we introduce the enhancement factor $\zeta = P_C/P_C(n_0 = 0)$. The L-dependence of this factor is illustrated in Fig. 2 for the same parameters of the medium as in Fig. 1. The enhancement factor peaks at $L \sim 1/\varkappa$. The maximum value of ζ can be estimated as $\zeta_{max} \sim \sqrt{n_0\sigma_{tr}/\varkappa}$.

IOP Conf. Series: Journal of Physics: Conf. Series **1238** (2019) 012041 doi:10.1088/1742-6596/1238/1/012041

5. Conclusions

In conclusion we have studied the propagation of initially unpolarized light (incoherent superposition of right-hand and left-hand polarized waves) in a medium with circular dichroism containing randomly spaced Mie-particles. The degree of circular polarization in the transmitted beam has been shown to increase due to the presence of scatterers. This effect is most pronounced if the first Kerker condition for Mie-particles is fulfilled and the ratio of the depolarization cross section to the transport cross section reaches its minimum value. In this case, the disordered ensemble of Mie-particles acts as an optical resonator which leads to an increase in photon path lengths in the medium without changing the state of their circular polarization.

The results obtained above can be of interest for applications in studies of optically active liquids which can be easily doped by scattering centers.

Acknowledgment. The authors thank A.I. Kuzovlev for their interest in this work and useful discussions. This study was supported by the MEPhI Academic Excellence Project (Contract No.02.a03.21.0005).

References

[1] Barron L D 2004 *Molecular light scattering and optical activity* (Cambridge University Press)
[2] Cantor C R and Schimmel P R 1980 *Biophysical chemistry: Part II "Techniques for the study of biological structure and function"* (WH Freeman and Co, Oxford)
[3] Berova N, Nakanishi K and Woody R W 2000 *Circular Dichroism: Principles and Applications* (Wiley).
[4] Preda F, Perri A, Réhault J, Dutta B, Helbing J, Cerullo G and Polli D 2018 *Opt. Letters* **43** 1882
[5] Kuznetsov A I, Miroshnichenko A E , Brongersma M L , Kivshar Yu S and Lukýanchuk B 2016 *Science* **354** 846
[6] Nieto-Vesperinas M, Gomez-Medina R and Saenz J J 2011 *J. Opt. Soc. Am.* **A28** 54
[7] Schmidt M K, Aizpurua J, Zambrana-Puyalto X, Vidal X, Molina-Terriza G and Saenz J J 2015 *Phys.Rev.Lett.* **114** 113902
[8] Gorodnichev E E, Kuzovlev A I and Rogozkin D B 2016 *JETP Lett.* **104** 157
[9] Gorodnichev E E, Kuzovlev A I and Rogozkin D B 2016 *Quantum Electronics* **46** 947
[10] Kerker M, Wang D-S and Giles C L 1983 *J. Opt. Soc. Am.* **73** 765
[11] Valuckas V, Paniagua-Dom?nguez R, Fu Y H, Lukýanchuk B and Kuznetsov A I 2017 *Appl. Phys. Lett.* **110** 091108
[12] Gottardo S, Sapienza R, Garcia P D, Blanco A, Wiersma D S and López C 2008 *Nat. Photonics* **2** 429
[13] Koman V B, Santschi C and Martin O J F 2015 *Anal. Chem.* **87** 1536
[14] Mupparapu R, Vynck K, Svensson T, Burresi M and Wiersma D S 2015 *Optics Express* **23** 1472
[15] Newton R. G 1982 *Scattering Theory of Waves and Particles* (2nd ed., New York, Springer)
[16] Bohren C F and Huffman D R 2008 *Absorption and Scattering of Light by Small Particles* (Wiley)
[17] Kuscer I and Ribaric M 1959 *Opt. Acta* **6** 42
[18] Gorodnichev E E, Kuzovlev A I and Rogozkin D B 2007 *JETP* **104** 319
[19] Gorodnichev E E, Kuzovlev A I and Rogozkin D B 2014 *Phys.Rev.* E **90** 043205
[20] Kokhanovsky A A 1999 *Phys.Rev.* **E4** 4899
[21] Gorodnichev E E, Kuzovlev A I and Rogozkin D B 1998 *JETP Lett.* **68** 22
[22] Ishimaru A 1997 *Wave Propagation and Scattering in Random Media* (Institute of Electrical and Electronics Engineers)

IOP Conf. Series: Journal of Physics: Conf. Series **1238** (2019) 012042 doi:10.1088/1742-6596/1238/1/012042

On solutions of the Dirac equation in a strong disappearing at infinity electrostatic field

K S Krylov and V D Mur

National Research Nuclear University MEPhI, Kashirskoe Shosse 31, 115409 Moscow, Russia

E-mail: krylov@theor.mephi.ru

Abstract. It is shown that the one-particle radial Dirac equation with a spherically symmetric narrow rectangular well remains valid also in the supercritical region. Positron resonances in scattering arising at low supercriticality are in no way related to the process of spontaneous production of electron-positron pairs from vacuum.

1. Introduction

The question of the validity of the one-particle Dirac equation in a strong electrostatic field disappearing at infinity was apparently first posed in the work of Schiff, Snyder, and Weinberg [1]. Using the example of a spherically symmetric rectangular potential well of depth V_0 they showed, that as the V_0 increases, the electron levels [1] move from the bottom of the upper continuum of the Dirac equation solutions to the upper boundary of the lower continuum. However, the authors of [1] believed that for sufficiently large V_0 the vacuum is not a state with the lowest energy of the system, but differs from it by finite values of the charge and energy.

In greater detail, for any values of $\varkappa$, this problem is considered in the monograph [2], moreover the cases of both a narrow and a wide well (comparing to the Compton wavelength of the electron $l_C = \hbar/mc$) are discussed. Just as in [1], it is noted that the difficulty of interpretation arises when V_0 exceeds the so-called "critical" value $V_{\rm cr}$, at which the given level reaches the boundary of the lower continuum. It is claimed, by analogy with the relativistic Coulomb problem with the charge of a nucleus $Z > Z_{\rm cr}$ [3, 4], that this difficulty is due to spontaneous production by the field at $V_0 > V_{\rm cr}$ of electron-positron pairs, so that the problem "cannot be solved within the framework of the quantum mechanics of a single particle".

At the same time, papers [5–8] show arguments based on the general principles of quantum theory that the one-particle Dirac equation remains valid also in the region of supercritical fields. In particular, at a small supercriticality, resonance scattering of positrons by the nucleus occurs. Nevertheless, the authors of the work [9] disagree with this conclusion and consider that resonances in scattering of positrons by the supercritical nucleus indicate spontaneous production of electron-positron pairs at $Z > Z_{\rm cr}$.

In the present article we show that in the model of a spherical rectangular well used in [1] the one-particle Dirac equation remains valid also in the supercritical region.

[1] The paper [1] discussed levels with a Dirac quantum number $\varkappa = \mp 1$.

IOP Conf. Series: Journal of Physics: Conf. Series **1238** (2019) 012042 doi:10.1088/1742-6596/1238/1/012042

2. The scattering phase and the poles of the scattering matrix of s-states

The radial Dirac equation of states with a conserved Dirac quantum number $\varkappa = \pm 1, \pm 2, \dots$ in units of $\hbar = c = m = 1$ takes the form

$$H_D \Psi_{\varepsilon,\varkappa}(\rho) = \varepsilon \Psi_{\varepsilon,\varkappa}(\rho), \quad \Psi_{\varepsilon,\varkappa} = \begin{pmatrix} F(\rho) \\ G(\rho) \end{pmatrix}, \quad \varepsilon = E/mc^2, \quad \rho = r/l_C,$$
$$H_D = \begin{pmatrix} V(\rho)+1 & \dfrac{d}{d\rho} - \dfrac{\varkappa}{\rho} \\ -\dfrac{d}{d\rho} - \dfrac{\varkappa}{\rho} & V(\rho)-1 \end{pmatrix}, \quad \varkappa = \begin{cases} -(l+1), \ -\widetilde{l}, & j = |\varkappa| - 1/2, \\ l \ , \ \widetilde{l}+1, & j = \varkappa - 1/2. \end{cases} \tag{1}$$

Here, j is the total angular momentum, and l and $\widetilde{l}$ are the orbital angular momenta of the upper and lower components of the Dirac spinor $\Psi_{\varepsilon,\varkappa}$, respectively.

For the potential

$$V(\rho) = -\theta(R-\rho)V, \quad V = V_0/mc^2, \tag{2}$$

where $\theta(x)$ is the Heaviside step function, equation (1) is solved in Bessel functions. So, for example, for the partial elastic scattering phase $\delta_\varkappa(k)$ in states with $\varkappa = -1$ we have

$$\delta_{-1}(k) = \delta^{(s)}(k) - kR, \quad \cot \delta^{(s)}(k) = \frac{1}{kR}\left\{1 - \frac{(\varepsilon+1)}{(V+\varepsilon+1)}\left[1 - \kappa R \cot(\kappa R)\right]\right\}, \tag{3}$$

where $\kappa = \sqrt{(V+\varepsilon)^2 - 1}$ and $k = \sqrt{\varepsilon^2 - 1}$ are the wavevectors of the particle inside and outside the well. The poles of the partial elastic scattering matrix $S_\varkappa(k) = \exp[2i\delta_\varkappa(k)]$ for $k = i\lambda$, i. e. the equation $\cot \delta_\varkappa(i\lambda) = i$, determine the discrete spectrum of the problem, which for $\varkappa = -1$ gives the equation for the spectrum of $ns_{1/2}$-states,

$$\kappa R \cot(\kappa R) = -\lambda R - \frac{V}{(\varepsilon+1)}(1+\lambda R), \quad \lambda = \sqrt{1-\varepsilon^2}, \quad -1 \leqslant \varepsilon \leqslant 1. \tag{4}$$

In the case of a narrow well, $R \ll 1$, this equation can be represented the form:

$$V = \frac{n\pi}{R} - (2\varepsilon+1) + \left[(1+\varepsilon)\sqrt{1-\varepsilon^2} + \frac{1-2\varepsilon(\varepsilon+1)}{2n\pi}\right]R + O(R^2), \tag{5}$$

$n = 1, 2, \dots$ is the radial quantum number. By taking $\varepsilon = -1$ and $n = 1$ in (5), we obtain the critical value of the depth of the well for the ground state,

$$V_{\mathrm{cr}}^{(s)}/mc^2 \equiv V_c^{(s)} = \frac{\pi}{R} + 1 + \frac{1}{2\pi}R + O(R^2), \tag{6}$$

which is consistent with the result [5]. When $V > V_c^{(s)}$, the level dives into the lower continuum, turning into a quasistationary state with complex energy. To determine it, one can use the equation (4), which is convenient to represent near the lower continuum boundary in the form [5]

$$V_c^{(s)} - V = a_2^{(s)}\lambda^2 + a_3^{(s)}\lambda^3, \tag{7}$$

where, $a_2^{(s)} = 1 - R/2\pi$, $a_3^{(s)} = -R/2$ for $n = 1$ and $R \ll 1$.

An analytic continuation to the domain $V > V_c^{(s)}$, $\lambda = -ik$ gives

$$k_\pm = k'_\pm - ik'', \quad k'_\pm = \pm\big(1 + R/4\pi\big)\big(V - V_c^{(s)}\big)^{1/2}, \quad k'' = \tfrac{1}{4}R\big(V - V_c^{(s)}\big). \tag{8}$$

IOP Conf. Series: Journal of Physics: Conf. Series **1238** (2019) 012042 doi:10.1088/1742-6596/1238/1/012042

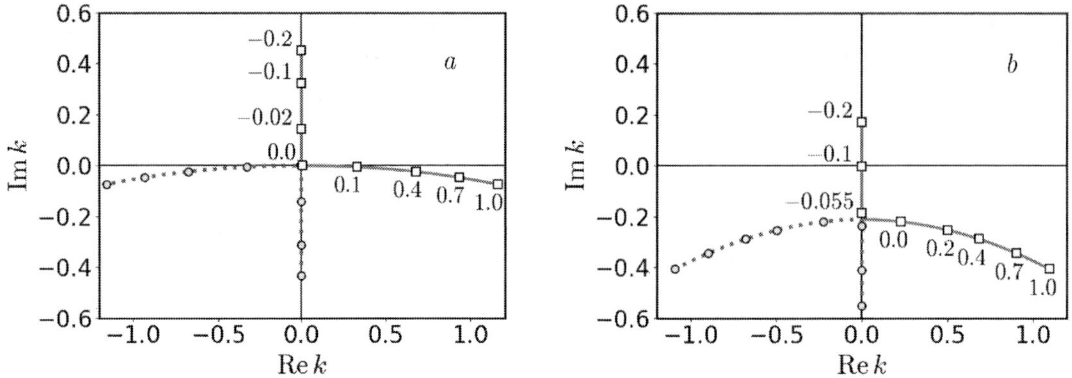

Figure 1. The S-matrix poles motion in the complex k-plane near the boundaries of the lower (a) and upper (b) continua for $\varkappa = -1$ and $R = 1/5$. The marks denotes the values of $(V - V_c^{(s)})$ and $(V_b^{(s)} - V)$, respectively.

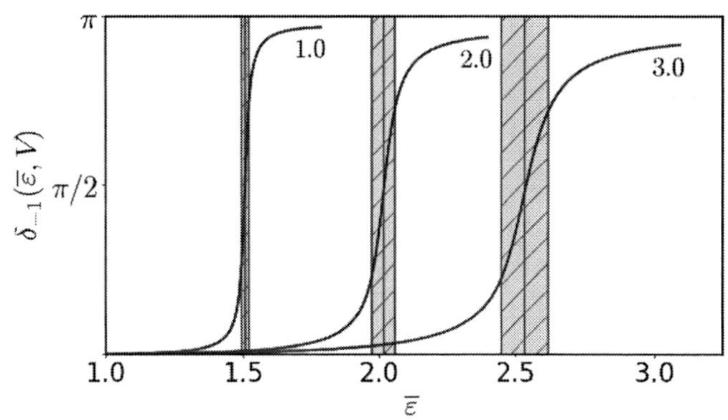

Figure 2. Scattering phases near the boundary of the lower continuum for $\varkappa = -1$ and $R = 1/20$. The shaded area shows the position and the width of the Breit–Wigner resonances. The numbers denotes the values of $(V - V_c^{(s)})$.

The motion of the poles of the matrix $S_{-1}(k)$ in the complex k-plane as a function of $(V - V_c^{(s)})$ is shown in Fig. 1a. The pole closest to the physical region will be called a Breit–Wigner pole, $k_+ = k_{\rm BW}$. The energy of such an $s_{1/2}$-level in the lower continuum, $\varepsilon_{\rm BW} = -\sqrt{k_{\rm BW}^2 + 1}$, is

$$\widetilde{\varepsilon}_{\rm BW}^{(s)} = -\widetilde{\varepsilon}^{(s)} + \tfrac{i}{2}\widetilde{\gamma}^{(s)}, \quad \widetilde{\varepsilon}^{(s)} = 1 + \tfrac{1}{2}\big(1 + R/2\pi\big)\big(V - V_c^{(s)}\big), \quad \widetilde{\gamma}^{(s)} = \tfrac{1}{2}R\big(V - V_c^{(s)}\big)^{3/2}. \tag{9}$$

An unusual sign of the imaginary part, $\widetilde{\gamma}^{(s)} > 0$, is coming from the use of a nonsecond-quantized approach. The Breit–Wigner level $\widetilde{\varepsilon}_{\rm BW}^{(s)}$ in the Dirac sea corresponds to the quasistationary state of the positron with energy

$$\widetilde{\varepsilon}_{\rm qs}^{(s)} = -\widetilde{\varepsilon}_{\rm BW}^{(s)} = \widetilde{\varepsilon}^{(s)} - \tfrac{i}{2}\widetilde{\gamma}^{(s)}, \quad \widetilde{\varepsilon}^{(s)} > 0, \quad \widetilde{\gamma}^{(s)} > 0, \tag{10}$$

which has the correct sign of the imaginary part.

At a small supercriticality, this quasidiscrete level manifests itself as a Breit–Wigner resonance in the elastic scattering of a positron with a width $\widetilde{\gamma}^{(s)}$. The threshold behavior of the width is determined by the permeability of the centrifugal barrier,

$$D \sim k^{2L+1}, \quad k \to 0. \tag{11}$$

IOP Conf. Series: Journal of Physics: Conf. Series **1238** (2019) 012042 doi:10.1088/1742-6596/1238/1/012042

Near the boundary of the lower continuum one must take [5] the orbital angular momentum $\widetilde{l}$ of the lower component of (1) as the orbital angular momentum L. In the case under consideration, $L = \widetilde{l} = 1$, which agrees with the result (9) for the width $\widetilde{\gamma}^{(s)}$. The phase of elastic scattering of the positron as a function of its energy $\overline{\varepsilon} = -\varepsilon$ is shown in Fig. 2. It is seen that at an energy $\overline{\varepsilon}$ close to the position of the resonance $\widetilde{\varepsilon}^{(s)}$, the phase changes abruptly on the width $\widetilde{\gamma}^{(s)}$.

Similarly, one can consider the motion of the poles of $S_{-1}(k)$ near the boundary of the upper continuum. However, as opposed to (7), here the expansion begins with a linear term,

$$V - V_b^{(s)} = c_1^{(s)}\lambda + c_2^{(s)}\lambda^2, \quad c_1^{(s)} = 2R, \quad c_2^{(s)} = 1 + \frac{3}{2\pi}R, \quad V_b^{(s)} = \frac{\pi}{R} - 3 - \frac{3}{2\pi}R, \tag{12}$$

and the first bound level appears at $V = V_b^{(s)}$. Putting $\lambda = -ik$ in (12), we obtain

$$k_\pm^{(s)} = \pm iR\sqrt{\left(1 + \tfrac{3}{2\pi}R\right)\left(V - V_b^{(s)}\right)/R^2 + 1} - iR. \tag{13}$$

As the depth of the well decreases, the discrete level $k_d^{(s)} = k_+^{(s)}$ moves downward along the positive imaginary axis of the k-plane. At $V = V_b^{(s)}$, it reaches the boundary of the upper continuum, and with a further decrease of V it is "pushed" into it, turning into a virtual, rather than quasistationary, level $k_v^{(s)}$, since $l = 0$ and there is no centrifugal barrier. At the same time, a second virtual level $k_{v'}^{(s)}$ moves towards it, for which $k_{v'}^{(s)} = -2iR$ at $V = V_b^{(s)}$. When $V = V_b^{(s)} - R^2$, they collide, then they go to the complex plane, see Fig. 1*b*.

3. The scattering phase and the poles of the scattering matrix of *p*-states

For $np_{1/2}$-states, i.e. at $\varkappa = 1$, the discrete spectrum is determined by the equation

$$\kappa R\cot(\kappa R) = -\lambda R + \frac{V}{(1-\varepsilon)}(1 + \lambda R), \quad \lambda = \sqrt{1-\varepsilon^2}, \quad -1 \leqslant \varepsilon \leqslant 1. \tag{14}$$

which for $R \ll 1$ is equivalent to the equality

$$V = \frac{n\pi}{R} - (2\varepsilon - 1) - \left[(1-\varepsilon)\sqrt{1-\varepsilon^2} - \frac{1 - 2\varepsilon(\varepsilon - 1)}{2n\pi}\right]R + O(R^2). \tag{15}$$

The first bound $1p_{1/2}$-level appears at $V = V_b^{(p)}$,

$$V_b^{(p)} = \frac{\pi}{R} - 1 + \frac{1}{2\pi}R + O(R^2), \tag{16}$$

and equation (14) for $V - V_b^{(p)} \ll 1$ can be represented in the form

$$V - V_b^{(p)} = a_2^{(p)}\lambda^2 + a_3^{(p)}\lambda^3, \quad a_2^{(p)} = 1 - R/2\pi, \quad a_3^{(p)} = -R. \tag{17}$$

Comparison with the expansion (7) shows that the motion of the poles of the S-matrix at $\varkappa = 1$ near the boundary of the upper continuum with decreasing V is similar to the motion of the poles of the S-matrix at $\varkappa = -1$ in the vicinity of $\varepsilon = -1$ with increasing V, cf. Fig. 1*a*.

For the energy of the quasistationary Breit–Wigner state of the electron, we obtain

$$\varepsilon_{\rm qs}^{(p)} = \varepsilon_{\rm BW}^{(p)} = \varepsilon^{(p)} - \tfrac{i}{2}\gamma^{(p)}, \quad \varepsilon^{(p)} = 1 + \tfrac{1}{2}\left(1 - \tfrac{1}{2\pi}R\right)\left(V_b - V\right), \quad \gamma^{(p)} = R\left(V_b - V\right)^{3/2} > 0, \tag{18}$$

i. e. the usual sign in front of the width $\gamma^{(p)}$, as in the nonrelativistic scattering theory. The indicator of the degree of underboundness $V_b - V$ is determined here by the orbital angular momentum of the upper component of the spinor, $l = 1,\ l + 1/2 = 3/2$.

Conversely, the motion of the poles with $\varkappa = 1$ near the boundary of the lower continuum is analogous to the motion of the poles with $\varkappa = -1$ for $\varepsilon \sim 1$. In this case, Eq. (15) gives

$$V_c^{(p)} - V = c_1^{(p)}\lambda + c_2^{(p)}\lambda^2, \quad c_1^{(p)} = 2R, \quad c_2^{(p)} = 1, \quad V_c^{(p)} = \frac{\pi}{R} + 3 - \frac{3}{2\pi}R, \tag{19}$$

and the signs of the coefficients $c_1^{(p)}$ and $c_2^{(p)}$ are just such that for $V > V_c^{(p)}$ the level goes to the second, unphysical sheet [6]. Indeed, setting $\lambda = -ik$, we obtain

$$k_{\pm}^{(p)} = \pm i\sqrt{\left(V_c^{(p)} - V\right) + R^2} - iR. \tag{20}$$

For $V < V_c^{(p)} + R^2$, the real level, with increasing V, moves down the imaginary axis of the k-plane, becoming a virtual level at $V_c^{(p)} < V < V_c^{(p)} + R^2$. The second virtual level moves towards it. When $V = V_c^{(p)} + R^2$, they collide, going to the complex plane. And the level which moves to the right, being closer to the physical region, has the energy

$$\widetilde{\varepsilon}_{-}^{(p)} = -\widetilde{\varepsilon}^{(p)} + \tfrac{i}{2}\widetilde{\gamma}^{(p)}, \quad \widetilde{\varepsilon}^{(p)} = -1 - \tfrac{1}{2}\left(V - V_c^{(p)}\right), \quad \widetilde{\gamma}^{(p)} = R\left(V - V_c^{(p)}\right)^{1/2}. \tag{21}$$

For the energy of the quasidiscrete level of the positron in the state $p_{1/2}$, we get:

$$\overline{\varepsilon}_{v}^{(p)} = -\widetilde{\varepsilon}_{-}^{(p)} = \widetilde{\varepsilon}^{(p)} - \tfrac{i}{2}\widetilde{\gamma}^{(p)}, \quad \widetilde{\varepsilon}^{(p)} > 0, \quad \widetilde{\gamma}^{(p)} > 0. \tag{22}$$

The dependence of the width $\widetilde{\gamma}^{(p)}$ on supercriticality is determined in this case by the orbital angular momentum $\widetilde{l} = 0$, which means the absence of a centrifugal barrier.

4. The case of arbitrary values of the Dirac quantum number $|\varkappa| \geqslant 2$

The results obtained above for the cases $\varkappa = \mp 1$ are generalized to arbitrary values $|\varkappa| \geqslant 2$. For energies of the Breit–Wigner resonances near $\varepsilon = 1$, i. e. for quasistationary electron states, we have

$$\varepsilon_{\text{qs}}^{(\varkappa)} = \varepsilon_{\text{BW}}^{(\varkappa)} = \varepsilon^{(\varkappa)} - \tfrac{i}{2}\gamma^{(\varkappa)}, \quad \varepsilon^{(\varkappa)} > 1, \quad \gamma^{(\varkappa)} > 0, \tag{23}$$

with $\gamma^{(\varkappa)} \sim \left(V_b^{(\varkappa)} - V\right)^{l+1/2}$, $l = j + \frac{1}{2}\,\text{sgn}\,\varkappa$.

At the same time, near the lower continuum boundary for the Breit–Wigner energy follows:

$$\widetilde{\varepsilon}_{\text{BW}}^{(\varkappa)} = -\widetilde{\varepsilon}^{(\varkappa)} + \tfrac{i}{2}\widetilde{\gamma}^{(\varkappa)}, \quad \widetilde{\varepsilon}^{(\varkappa)} > 1, \quad \widetilde{\gamma}^{(\varkappa)} > 0, \tag{24}$$

with the threshold dependency $\widetilde{\gamma}^{(\varkappa)} \sim \left(V - V_c^{(\varkappa)}\right)^{\widetilde{l}+1/2}$, $\widetilde{l} = j - \frac{1}{2}\,\text{sgn}\,\varkappa$. We note that the inequality $\widetilde{\gamma}^{(\varkappa)} > 0$ holds for a well of arbitrary shape, as shown in [6] within the framework of the effective radius approximation for the Dirac equation developed there.

An invalid sign in front of $\widetilde{\gamma}^{(\varkappa)}$ in (24) means the need for a second quantization. The Dirac radial Hamiltonian (1) with potential (2) is a self-adjoint operator whose solutions form a complete system of functions. According to Furry, it can be used to quantize the system [10]. In the Furry picture, the solutions of the Dirac equation in the lower continuum correspond to positron states with energy $\overline{\varepsilon} = -\varepsilon > 1$. Therefore, the Breit–Wigner poles (24) match the quasistationary positron states with energy having a negative imaginary part,

$$\overline{\varepsilon}_{\text{qs}}^{(\varkappa)} = -\widetilde{\varepsilon}_{\text{BW}}^{(\varkappa)} = \widetilde{\varepsilon}^{(\varkappa)} - \tfrac{i}{2}\widetilde{\gamma}^{(\varkappa)}, \quad \widetilde{\varepsilon}^{(\varkappa)} > 1, \quad \widetilde{\gamma}^{(\varkappa)} > 0. \tag{25}$$

With a small supercriticality, this leads to resonance scattering of positrons.

Due to the fact that the partial scattering phases $\delta_\varkappa(k)$ are real, the elastic scattering matrix of positrons $S_\varkappa = \exp[2i\delta_\varkappa(k)]$ is unitary. Thus, inelastic processes in the channel with given $\varkappa$, including spontaneous e^+e^--pair production, are absent.

Thereby, the resonance scattering of positrons at small supercriticality, analogous to the resonance scattering of electrons at small underboundness, can not serve as evidence in favor of the spontaneous production of electron-positron pairs. This is the essential difference between strong electrostatic fields that disappear at infinity and the situation that arises in the Klein paradox [11], or in a constant uniform electric field.

5. Conclusions

As a result, the one-particle approximation for the Dirac equation in the field of a spherically symmetric narrow rectangular well remains valid also in the region of the supercritical depths of such a well. In this model, the discrete levels are well separated from each other near the boundary of both the lower and the upper continua. Therefore, it is not difficult to consider the motion of the poles of the scattering matrix in the vicinity of these boundaries.

At present, a large number of both theoretical and experimental works are devoted to the Coulomb problem in graphene, the dynamics of charge carriers in which is described by the effective two-dimensional Dirac equation in the presence of an impurity with a charge Z, including in the supercritical case, $Z > Z_{\rm cr}$. Moreover, it was shown in Refs. [12, 13] that the positions of the resonances calculated in the gapless equation agree with the experimental data on the spectra of current-voltage characteristics obtained by scanning tunneling microscopy. This means the validity of the one-particle effective gapless two-dimensional Dirac equation for $Z > Z_{\rm cr}$, but the discussion of this question is beyond the scope of this article.

Acknowledgments

We are grateful to V M Kuleshov and A M Fedotov for the useful discussions of the issues raised in the article. The work is partially supported by the Russian Foundation for Basic Research (Grant No. 16-02-00963a).

References

[1] Schiff L I, Snyder H and Weinberg J 1940 *Phys. Rev.* **57** 315–8
[2] Akhiezer A I and Berestetskii V B 1965 *Quantum Electrodynamics* (New York: Wiley)
[3] Pomeranchuk I and Smorodinsky Ya 1945 *J. Phys. USSR* **9** 97–100
[4] Zel'dovich Ya B and Popov V S 1972 *Sov. Phys. Usp.* **14** 673–94
[5] Popov V S and Mur V D 1974 *Sov. J. Nucl. Phys.* **18** 350
[6] Mur V D and Popov V S 1976 *Theor. Math. Phys.* **27** 429–38
[7] Kuleshov V M, Mur V D, Narozhny N B, Fedotov A M, Lozovik Y E and Popov V S 2015 *Physics-Uspekhi* **58** 785–91
[8] Kuleshov V M, Mur V D, Fedotov A M and Lozovik Y E 2017 *JETP* **125** 1144–62
[9] Godunov S I, Machet B and Vysotsky M I 2017 *Eur. Phys. J.* C **77** 782
[10] Furry W H 1951 *Phys. Rev.* **81** 115–24
[11] Klein O 1929 *Z. Phys.* **53** 157–65
[12] Wang Y *et al.* 2013 *Science* **340** 734–7
[13] Mao J, Jiang Y, Moldovan D, Li G, Watanabe K, Taniguchi T, Masir M R, Peeters F M and Andrei E Y 2016 *Nature Physics* **12** 545–9

IOP Conf. Series: Journal of Physics: Conf. Series **1238** (2019) 012043 doi:10.1088/1742-6596/1238/1/012043

Total reflectance of a random medium with the Reynolds-McCormick phase function at grazing incidence of light

V V Marinyuk[1] **and S V Sheberstov**[2]

[1] National Research Nuclear University MEPhI, Kashirskoe Shosse 31, Moscow 115409, Russia
[2] P.P. Shirshov Institute of Oceanology, Nahimovski prospect 36, Moscow 117997, Russia

E-mail: vmarinyuk@mail.ru

Abstract. We study the total reflectance of an absorbing, multiply scattering medium with large (as compared to the light wavelength) inhomogeneities at grazing incidence of light. To model highly forward scattering in the medium, we take advantage of the two-parameter Reynolds-McCormick scattering phase function. Using the scaling analysis for the small-angle radiative transfer equation, we derive simple analytic formulae for the dependence of the reflectance on the medium transport coefficients and the angle of incidence. The results obtained are verified by comparison with results of a direct numerical integration of the radiative transfer equation.

1. Introduction

Reflection of light from a multiply scattering medium with large inhomogeneities is of significant theoretical and practical interest for many problems of atmospheric and oceanic optics [1–5] and biomedical optics [6, 7]. Important information about the bulk optical properties of turbid media can be extracted from the measured reflectances [7–10].

In reflection of light from a highly forward scattering medium, two multiple scattering regimes can occur. The diffusive regime develops over spatial scales exceeding the transport mean free path $l_{\rm tr}$. If the direction of incidence does not differ much from the normal one and absorption is weak (the absorption coefficient κ is much less than the transport scattering coefficient $\sigma_{\rm tr} = l_{\rm tr}^{-1}$), the reflected flux is governed by diffusely scattered light [1]. At grazing angles of incidence ($\pi/2 - \theta_0 \ll 1$, θ_0 is the angle between the direction of light incidence and the inward normal to the medium boundary), the light can be effectively reflected due to small-angle multiple scattering. In this case, even for a medium with rather strong absorption ($\kappa \sim \sigma_{\rm tr}$), the total reflectance can be close to unity [11–13].

Small-angle reflection of light from an absorbing medium at grazing angles of incidence was studied in Refs. [11, 12] (see also Ref. [14]) on the basis of the Fokker-Planck approximation (small-angle diffusion approximation). The latter, however, imposes a restriction on the scattering phase function, the results [11, 12] are valid only for the phase functions decreasing with scattering angle ϑ more rapidly than $1/\vartheta^4$.

In the present work, within the small-angle approximation, we calculate the total reflectance of an absorbing medium with the Reynolds-McCormick phase function. This function decreases with the scattering angle by the power law $1/\vartheta^\alpha$ and unifies a number of realistic scattering models. For grazing angles of incidence, we derive analytic expressions for the total reflectance in the range $2 < \alpha < 4$. To verify the validity of our results, we perform a direct numerical integration of the radiative transfer equation for the Henyey-Greenstein phase function ($\alpha = 3$) by the discrete ordinate method. The obtained analytic expressions are in very good agreement with the numerical data.

2. General relations

Consider a broad collimated beam of light incident on a turbid medium with large-scale inhomogeneities. The medium is assumed to occupy the half-space $z > 0$, the $z-$axis coincides with the inward normal to the surface. The intensity of light $I(z, \mathbf{\Omega})$ inside the medium obeys the radiative transfer equation (see, e.g., Ref. [4])

$$\left(\mu \frac{\partial}{\partial z} + \kappa\right) I(z, \mathbf{\Omega}) = \sigma \int d\mathbf{\Omega}' p(\mathbf{\Omega}\mathbf{\Omega}')[I(z, \mathbf{\Omega}') - I(z, \mathbf{\Omega})] \tag{1}$$

where σ and κ are the scattering and the absorption coefficients, respectively; $\mathbf{\Omega} = (\sin\theta\cos\varphi, \sin\theta\sin\varphi, \cos\theta)$ is the unit vector in the direction of light propagation, $\mu = \Omega_z = \cos\theta$, φ is the azimuthal angle; $p(\cos\vartheta)$ is the single-scattering phase function normalized by the condition $\int_0^\pi 2\pi \sin\vartheta d\vartheta p(\cos\vartheta) = 1$.

Boundary condition for Eq. (1),

$$I(z = 0, \mu, \varphi) = \delta(\mu - \mu_0)\delta(\varphi), \quad \mu > 0, \tag{2}$$

determines the intensity within the half-range of angles $0 < \theta < \pi/2$ which correspond to light incident upon the medium; $\mu_0 = \cos\theta_0$, θ_0 is the angle between the direction of incidence $\mathbf{\Omega}_0$ and the z-axis. The azimuth angle of $\mathbf{\Omega}_0$ is chosen to be zero. We normalize the intensity in such a way that the incident flux is equal to μ_0.

The total reflectance (total reflection coefficient or plane albedo) r is a ratio of the reflected flux to the incident one:

$$r = \mu_0^{-1} \int_{-\pi}^{\pi} d\varphi \int_{-1}^{0} d\mu \, |\mu| I\,(z = 0, \mu, \varphi)\,. \tag{3}$$

In nonabsorbing media ($\kappa = 0$), the total flux is conserved, and $r = 1$.

To model scattering in the medium we take advantage of the two-parameter phase function proposed in Ref. [15],

$$p(\cos\vartheta) = \frac{p_\alpha}{2\pi(\vartheta_0^2 + 2(1 - \cos\vartheta))^{\alpha/2}}, \tag{4}$$

where p_α is the normalization factor. Parameter ϑ_0 is the characteristic angle of single scattering. For large scatterers, the forward scattering dominates and $\vartheta_0 \ll 1$. The phase function (4) unifies a number of scattering models [4], including scattering of light by a fractal [16]. For $\alpha = 3$, Eq. (4) coincides with the widely used Henyey-Greenstein phase function.

The small-angle form of the phase function (4) at $\alpha > 2$ is given by

$$p(\vartheta) = \frac{\alpha - 2}{2\pi} \frac{\vartheta_0^{\alpha-2}}{(\vartheta_0^2 + \vartheta^2)^{\alpha/2}}. \tag{5}$$

For $\alpha < 2$, multiple scattering of light can not be described within the small-angle approximation (rigorously speaking, the regime of small-angle multiple scattering of light exists for $2 - \alpha \ll 1$ [17]). In what follows, we consider only the case $\alpha > 2$.

At grazing angles of incidence, $\pi/2 - \theta_0 \ll 1$, the radiative transfer equation (1) can be simplified using the following relations (see, e.g., [11, 14, 18])

$$\mu = \cos\theta = \sin\zeta \approx \zeta, \qquad 2\,(1 - \mathbf{\Omega\Omega}') \approx (\zeta - \zeta')^2 + (\varphi - \varphi')^2 \tag{6}$$

where $\zeta = \pi/2 - \theta$ is the angle between the direction $\mathbf{\Omega}$ and the medium boundary. With allowance for the explicit form of the phase function (5), the radiative transfer equation (1) takes the form

$$\left(\zeta\frac{\partial}{\partial z} + \kappa\right) I(z,\zeta,\varphi) = \frac{\sigma\vartheta_0^{\alpha-2}}{2\pi}\int_{-\infty}^{\infty} d\zeta' \int_{-\infty}^{\infty} d\varphi' \frac{I(z,\zeta',\varphi') - I(z,\zeta,\varphi)}{(\vartheta_0^2 + (\zeta-\zeta')^2 + (\varphi-\varphi')^2)^{\alpha/2}}, \tag{7}$$

$$I(z=0,\zeta,\varphi) = \delta(\zeta - \zeta_0)\delta(\varphi), \quad \zeta > 0. \tag{8}$$

As is usual for the small-angle approximation (see, e.g., Ref. [4]), the angles ζ and φ are considered to vary within infinite limits provided that the intensity decreases rapidly as the values of $|\zeta|$ and $|\varphi|$ increase.

3. Scaling laws for the total reflectance at grazing angles of incidence

Within the small-angle approximation, the total reflectance (3) takes the form

$$r = \zeta_0^{-1}\int_{-\infty}^{\infty} d\varphi \int_{-\infty}^{0} |\zeta| I\,(z=0,\zeta,\varphi)\, d\zeta. \tag{9}$$

The solution to Eqs.(7) and (8) is expressed as a function of dimensionless variables,

$$I(z,\zeta,\varphi) = \frac{1}{\vartheta_0^2} F_\alpha\left(\frac{\sigma z}{\vartheta_0}, \frac{\zeta}{\vartheta_0}, \frac{\varphi}{\vartheta_0}; \frac{\zeta_0}{\vartheta_0}, \frac{\kappa}{\sigma}\right). \tag{10}$$

From Eqs.(9) and (10) we find the following scaling law for the total reflectance:

$$r = r_\alpha\left(\frac{\zeta_0}{\vartheta_0}, \frac{\kappa}{\sigma}\right). \tag{11}$$

If the angle ζ_0 is greater than the characteristic single-scattering angle, $\zeta_0 > \vartheta_0$, the reflected flux is governed by light scattered through angles exceeding the value ϑ_0, and ϑ_0 can be neglected in the denominator of the RHS of Eq.(7). For $\alpha < 4$, no divergency appears in Eq.(7) because the singularity in the denominator at $\zeta = \zeta'$ and $\varphi = \varphi'$ is canceled by the difference of the intensities in the numerator [18]. Within this approximation, a solution to the radiative transfer equation can be presented in the following self-similar form

$$I(z,\zeta,\varphi) = \frac{1}{\zeta_a^2}\tilde{F}_\alpha\left(\frac{\sigma_{tr} z}{\zeta_a^{\alpha-1}}, \frac{\zeta}{\zeta_a}, \frac{\varphi}{\zeta_a}; \frac{\zeta_0}{\zeta_a}\right), \tag{12}$$

where $\zeta_a = (\sigma_{tr}/\kappa)^{1/(\alpha-2)}$; $\sigma_{tr} = \sigma(1 - \langle\cos\vartheta\rangle) = \frac{2(\alpha-2)}{4-\alpha}(\vartheta_0/2)^{\alpha-2}$ is the transport scattering coefficient. In this case, for a given value α, the total reflectance proves to be a universal function of a single parameter,

$$r = \tilde{r}_\alpha\left(\frac{\kappa\zeta_0^{\alpha-2}}{\sigma_{tr}}\right), \quad 2 < \alpha < 4. \tag{13}$$

The explicit form of the functions $r_\alpha(x,y)$ and $\tilde{r}_\alpha(x)$ appearing in Eqs. (11) and (13) is governed by the specific angular profile of the scattering phase function (i.e. by the exponent α).

For $\alpha = 4$, relation (13) has a logarithmic accuracy. For $\alpha > 4$, the characteristic angle ϑ_0 appearing in Eq.(7) can not be neglected in the denominator. Instead, the intensity $I(z, \zeta', \varphi')$ can be expanded in angles of single-scattering deflection $\zeta' - \zeta$ and $\varphi' - \varphi$, resulting in the Fokker-Planck approximation (the small-angle diffusion approximation) in Eq.(7) [18]. Then, in going to dimensionless variables, we obtain r in the form of Eq.(13), with the only difference that exponent α should be put $\alpha = 4$. So, for rapidly decreasing phase functions, the total reflectance takes the universal form $r = r(\kappa\zeta_0^2/\sigma_{\rm tr})$.

Using the method proposed in Refs. [13,19,20], we derive the following asymptotic expression ($1 - r \ll 1$) for the total reflectance in the range $2 < \alpha \leq 4$:

$$1 - r = f_\alpha\left(\zeta_0/\vartheta_0\right)\left(\kappa/\sigma\right)^{1/\alpha} \tag{14}$$

for

$$\kappa < \min\{\sigma, \sigma(\vartheta_0/\zeta_0)^{\alpha-2}\}. \tag{15}$$

Note that the condition (15) is weaker than the analogous condition at normal incidence of light. In the latter case, the condition $1 - r \ll 1$ implies that absorption is weak, $\kappa < \sigma_{tr}$ [1](i.e. $\kappa < \sigma\vartheta_0^{\alpha-2}$).

For $\zeta_0 > \vartheta_0$, Eq.(14) takes the form (see, Ref. [13]):

$$1 - r = c_\alpha\left(\frac{\kappa\zeta_0^{\alpha-2}}{\sigma_{\rm tr}}\right)^{\frac{1}{\alpha}} \quad \text{for } \kappa < \sigma_{\rm tr}/\zeta_0^{\alpha-2}. \tag{16}$$

where the numeric constant c_α depends only on the parameter α. Comparing Eqs.(14) and (16), we find the asymptotic behavior of the f_α-function: $f_\alpha(x) \sim x^{1-2/\alpha}$ for $x > 1$.

If the phase function decreases more rapidly than $1/\vartheta^4$ ($\alpha > 4$), the radiative transfer equation can be solved analytically within the small-angle diffusion approximation [11, 12]. In this case, $1 - r = 1.42...(\kappa\zeta_0^2/\sigma_{\rm tr})^{1/4}$ for $\kappa < \sigma_{\rm tr}/\zeta_0^2$.

4. Results of a direct numerical integration of the radiative transfer equation

To illustrate the range of validity of the results obtained above we perform a direct numerical integration of the radiative transfer equation for a medium with the Henyey-Greenstein phase function ($\alpha = 3$, $\vartheta_0 = 1 - g$, g is the mean cosine of the single-scattering angle). Without resorting to the small-angle approximation, we calculate the total reflectance r from Eqs.(1)-(3) with a numerical code based on the discrete-ordinate method [21, 22]. Results of our numerical calculations are presented in Fig.1a. The figure shows the dependence of the difference $1 - r$ on the ratio of the absorption coefficient to the scattering one. Figure 1b shows the numerical data of Ref. [1].

The scaling law (11) is well illustrated by the presented numerical data, the graphs for $g = 0.9$, $\mu_0 = 0.1$ and $g = 0.95$, $\mu_0 = 0.05$ are practically indistinguishable in a wide range of values of the absorption coefficient(see Fig.1a), and the total reflectance becomes independent of the anisotropy factor g in the limit $\zeta_0 \to +0$ (see Fig.1b). In all the presented graphs it is easy to distinguish the range of values of the absorption coefficient, where the difference $1 - r$ follows the law

$$1 - r = f(\zeta_0/\vartheta_0)(\kappa/\sigma)^{1/3} \tag{17}$$

in accordance with Eq.(14). Some values of the function $f(\zeta_0/\vartheta_0)$ can be extracted from our numerical data (see Fig.2a). With increasing the ratio ζ_0/ϑ_0, the f-function tends to the dependence $f = 1.25...(\zeta_0/\vartheta_0)^{1/3}$, and Eq.(17) takes the form (see Eq.(16) with $\alpha = 3$)

$$1 - r = 1.25...(\kappa\zeta_0/\sigma_{\rm tr})^{1/3}. \tag{18}$$

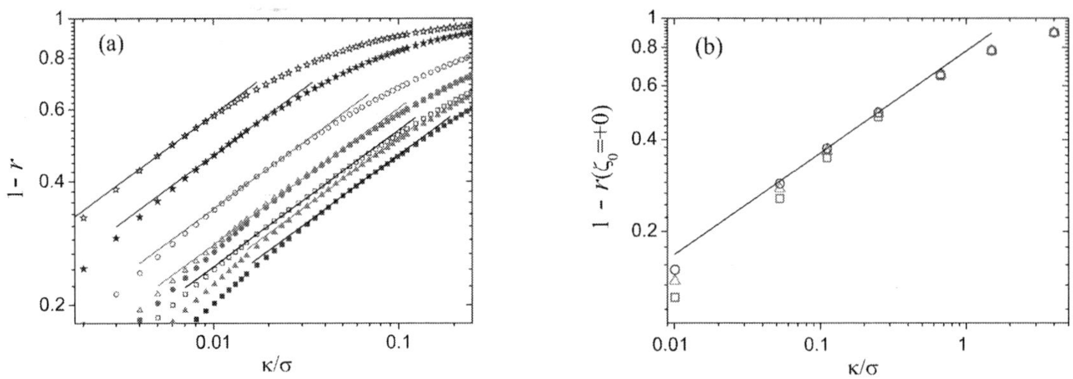

Figure 1. Total reflectance as a function of a ratio of the absorption coefficient to the scattering one. (**a**) The results of our numerical calculations with Eqs.(1) - (3) by the discrete-ordinate method. Empty and filled symbols correspond to $g = 0.95$ and $g = 0.9$, respectively. From lower to upper graphs, μ_0=0.03 (squares), 0.05 (triangles), 0.1 (circles) and 0.5 (stars). (**b**) Total reflectance in the limit $\zeta_0 = +0$. Symbols are the numerical data of Ref. [1] for the Henyey-Greenstein phase function with $g = 0.875$ (circles), $g = 0.75$ (triangles) and $g = 0.5$ (squares). Solid lines correspond to the dependence $1 - r \sim (\kappa/\sigma)^{1/3}$.

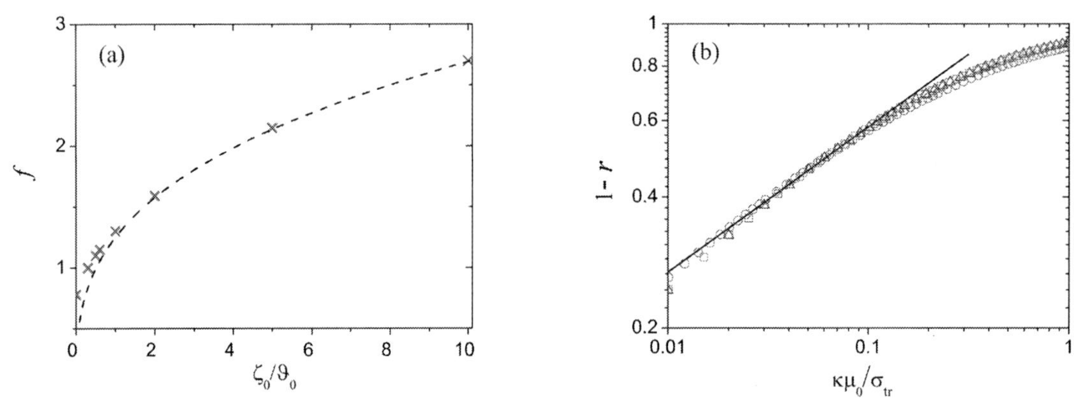

Figure 2. (**a**) The f-function appearing into the relation (17). Crosses are the values of f extracted from the data shown in Fig.1a,b. The dashed curve shows the dependence $f = 1.25(\zeta_0/\vartheta_0)^{1/3}$. (**b**) Total reflectance as a function of dimensionless variable $\kappa\mu_0/\sigma_{\rm tr}$. Symbols are our numerical results for $g = 0.95$, $\mu_0 = 0.1$ (triangles), $g = 0.95$, $\mu_0 = 0.5$ (squares) and $g = 0.9$, $\mu_0 = 0.5$ (circles). The solid line is Eq.(18)

For $\mu_0 > 1 - g$, in accordance with the scaling law (13), the total reflectance proves to be a universal function of a single variable $\kappa\mu_0/\sigma_{\rm tr}$ (see Fig.2b). Remarkably, Eq.(18) turns out to be a good approximation even for rather moderate values of the total reflectance, $1 - r \leq 0.6$ (ratio $\kappa\zeta_0/\sigma_{\rm tr} \leq 0.1$).

The scaling laws (11) and (13) fail in the case of very weak absorption, where the small-angle approximation becomes inapplicable, and the contribution from light scattered through large (of the order of unity) angles should be taken into account.

5. Conclusions

In conclusion, we have presented a theoretical study of light reflection from a turbid medium with highly forward single scattering at grazing angles of incidence. Within the small-angle approximation, we have calculated the total reflectance of an absorbing medium with the Reynolds-McCormick phase function. A range of values of the medium transport coefficients is found where the total reflectance proves to be a universal function of a single parameter which is expressed in terms of the incidence angle, the absorption coefficient and the transport scattering one. The explicit form of this function is governed by the specific angular profile of the scattering phase function. Our results are validated by comparison with results (both our and other authors') of direct numerical integration of the radiative transfer equation.

References

[1] van de Hulst H C 1980 *Multiple Light Scattering: Tables, Formulas, and Applications* (Academic Press)
[2] Mobley C D 1994 *Light and Water - Radiative Transfer in Natural Waters* (Academic Press)
[3] Walker R E, 1994 *Marine Light Field Statistics* (Wiley)
[4] Ishimaru A 1999 *Wave Propagation and Scattering in Random Media* (Wiley-IEEE Press)
[5] Zege E P, Ivanov A P and Katsev I L 2011 *Image Transfer through a Scattering Medium* (Springer)
[6] Tuchin V, Wang L and Zimnyakov D 2006 *Optical Polarization in Biomedical Applications* (Springer)
[7] Yaroslavsky A N, Yaroslavsky I V, Goldbach T and Schwarzmaier H-J 1999, *J. Biomed. Opt.* **4** 47-53
[8] Prahl S A, van Gemert M J C and Welch A J 1993 *Appl. Opt.* **32** 559-568
[9] Leyre S, Meuret Y, Durinck G, Hofkens J, Deconinck G and Hanselaer P 2014 *Appl. Opt.* **53** 2117-2125
[10] Naglic P, Pernus F, Likar B and Burmen M 2016 *J. Biomed. Opt.* **21** 095003
[11] Remizovich V S, Ryazanov M I and Tilinin I S 1981 *Izv. Acad. Sci. USSR Atmos. Oceanic Phys.* **17** 654-656
[12] Remizovich V S 1992 *Las. Phys.* **2** 40-57
[13] Marinyuk V V and Sheberstov S V 2017 *Appl. Opt.* **56** 9105-9113
[14] Remizovich V S, Ryazanov M I and Tilinin I S 1980 *Sov. Phys. JETP* **52** 225-230
[15] Reynolds L O and McCormick N J 1980 *J. Opt. Soc. Am.* **70** 1206-1212
[16] Sheppard C J R 2007 *Opt. Lett.* **32** 142-144
[17] Gorodnichev E E and Rogozkin D B 1995 *JETP* **80** 112-126
[18] Remizovich V S 1984 *Sov. Phys. JETP* **60** 290-299
[19] Marinyuk V V and Sheberstov S V 2016 *Opt. Lett.* **41** 922-925
[20] Marinyuk V V and Sheberstov S V 2017 *J. Phys.: Conf. Series* **788** 012045
[21] Stamnes K, Tsay S C, Wiscombe W and Jayaweera K 1988 *Appl. Opt.* **27** 2502-2509
[22] Marinyuk V V, Rogozkin D B and Sheberstov S V 2014 *Optics and Spectroscopy* **117** 102-110

IOP Conf. Series: Journal of Physics: Conf. Series **1238** (2019) 012044 doi:10.1088/1742-6596/1238/1/012044

About the influences of gas composition and generator frequency on the formation conditions and parameters of vortex tube in the RF inductively coupled plasma

Yu M Grishin, L. Miao*

Bauman Moscow State Technical University, 2- ya Baumanskaya str., bld. 5, 1, Moscow, 105005, Russia

*E-mail: miaolongbmstu@gmail.com

Abstract. The results of numerical simulation of plasma flow in the channel of technological radio frequency inductively coupled plasma (RF–ICP) with three coils are presented. The generator frequency is varied from 1.76 to 13.56 MHz. The pure argon or Ar/H_2 mixture are considered as the working gas. The volume fraction of hydrogen in Ar/H2 mixture is varied from 0 to 10%. The distributions of gas-dynamic parameters of plasma flow are calculated. It is shown that when the amplitude of discharge current exceeds a critical value (depends on volume fraction of hydrogen), the states of plasma flow transform from potential to vortical patterns, in which a toroidal vortex is found in front of the inductor zone. The dependencies of critical current on the generator frequency and volume fraction of hydrogen are established. The influences of generator frequency, volume fraction and discharge current on the intensity and coordinate of vortex tube are determined.

1. Introduction

At present in connection with the development of new technologies for the production of ultrapure substances [1-5], nanopowders [6-9], the interests in the inductively coupled plasma (ICP) are steadily increasing. The basic gas–dynamic parameters of plasma (velocity and temperature) are determined by the flow features in the inductor zone of ICP (exothermic zone). Results [10-12] indicate that there is a specific vortex in front of this zone under high discharge current. This leads to a change of plasma flow state and violates the heating and evaporation processes of the solid particles in high temperature zone. Since those solid particles may be slow down by the reversed flow (vortex) and then bypass the high temperature zone without efficient heating and evaporation.

The implementation of plasma technology requires the use of ICP under different plasma power, generator frequency and within various working gases, which may lead to significantly differences in the parameters of vortex tube. However, until now there is practically no systematic information about the formation conditions and parameters of vortex flow in ICP channel.

The aim of present work is to investigate the features of pure argon or Ar/H_2 plasma flow in the ICP channel, find the conditions of forming vortex and determine the parameters of plasma flow under different volume fraction of hydrogen.

2. Physical and mathematical model.

Figure 1 shows a sketch of the ICP with axial gases supplies and three inductor coils to be modeled. The flow rates of transporting, central and sheath gases are correspondingly set as Q_1=0.5 – 4L/min (R_1=3.7 mm, δ_1=2 mm, Z_1=50 mm), Q_2=3.4L/min and Q_3=35.4L/min. The length of external quartz tube is Z_4 = 400 mm with an internal radius R_3 = 25 mm and wall thickness δ_3 = 3.5 mm. The peripheral annual channel have a length Z_2 = 50 mm and width R_3-R_2-δ_2= 2.2 mm. Three copper coils with the same coil diameter d_{coil}=6 mm and loop radius R_4=33 mm are evenly distributed. The first coil is located at Z_{inf}= 63 mm. The distance between the first and last coils is l_{in}=Z_{inl}- Z_{inf}=60 mm. The amplitude of discharge current J_{coil} and generator frequency ω are varied respectively in the range of 60-300 A and 1.76-13.56 MHz.

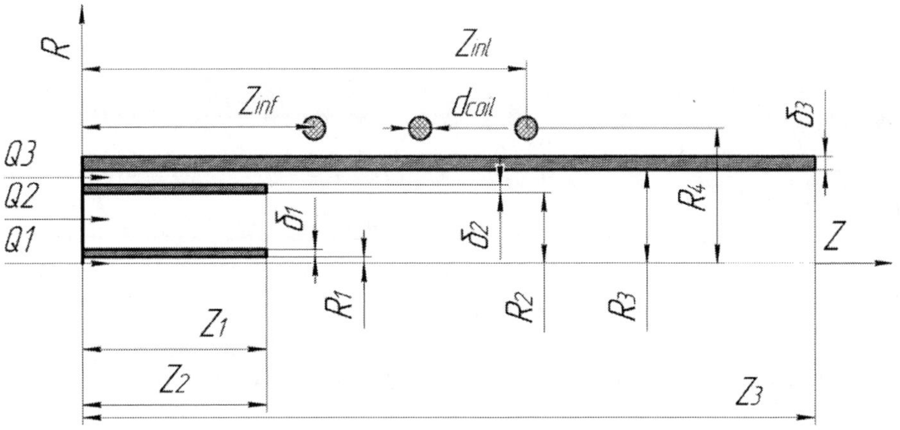

Figure 1. Schematic geometry of the ICP torch: 1 – injection probe channel of transporting gas, 2 – channel of central gas, 3 – channel of sheath gas.

In present calculations the spiral inductor is assumed as three cylindrically symmetric parallel coils. The flows of transporting, central and sheath gases at the inlets are azimuthally symmetrical and stationary. Under the fulfillments of these assumptions the electromagnetic and gas dynamic equations can be written under 2*D* – model in the cylindrical coordinate (*r, z*) system. Besides that, it is calculated that plasma flow in the ICP channels is laminar and subsonic with average *Re* number $Re<$ 1000.

The gas dynamic parameters of plasma flow are calculated on the basis of plasma dynamic system equations. The effect of electromagnetic forces, Joule heating and radiation losses in the gas dynamics equations have been taken into account. The electromagnetic filed were calculated using Maxwell equations [12, 13].

In numerical calculations the working gas is either pure argon or Ar/H_2 mixture with a variation of hydrogen volume fraction α from 0 to 10%. The thermophysical parameters and electrical conductivity of plasma are obtained in the approximation of local thermodynamic equilibrium (LTE) from calculated data [11, 12, 14].

At the inlets (z = 0) the temperature and the axial velocities of the transporting, central and sheath gases (in corresponding with Q_1, Q_2 and Q_3) are set constant. The velocities of gases at the walls are zero. The gas temperature on the inner surface of external quartz wall is equal to 400 K. At the outlet (Z_4=400mm) the pressure is set constant $p = 10^5$ Pa. The boundary conditions for the Maxwell equations are given as in accordance with [12].

The numerical solutions of plasma dynamic and Maxwell equations are based on the finite volume method in ANSYS CFX. The meshes in the calculating regions are built by ANSYS ICEM CFD package using HEXA_8 hexagonal structural blocks.

IOP Conf. Series: Journal of Physics: Conf. Series **1238** (2019) 012044 doi:10.1088/1742-6596/1238/1/012044

3. Results and Discussion

The main reason for the arose of vortex tube in front of inductor zone is due to the formation of high pressure zone [15, 16] near the axis of ICP torch, which is connected with the radial component of electromagnetic force of induced current in the plasma.

It is found that when the relative pressure is above a certain value, the working gas flows into the high pressure zone and forms a toroidal vortex around the cross section of first coil (Figure 2). The value and distribution of plasma pressure are determined by the discharge current J_{coil}.

It is found that there is a critical current J_{cr}, which separates the states of plasma flow into potential (subcritical, $J_{coil} \leq J_{cr}$) and vortical (supercritical, $J_{coil} \geq J_{cr}$) regimes.

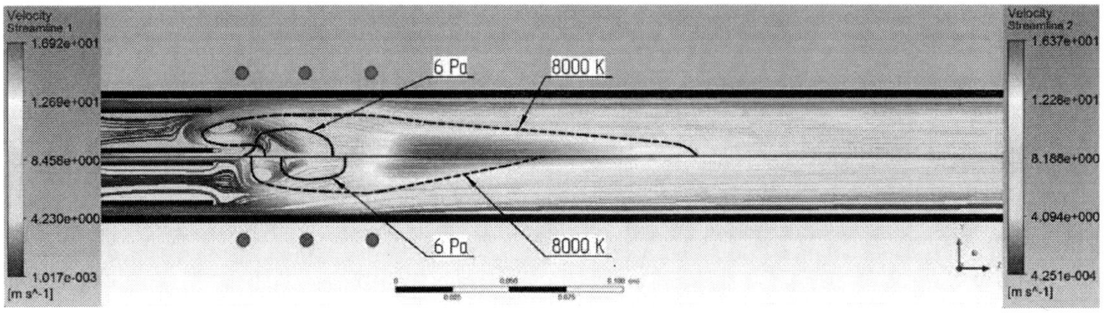

Figure 2. The streamlines (colored lines), temperature contour T = 8000 K (dashed lines) and relative pressure contour Δp = 6 Pa (solid lines) under J_{coil} = 170 A: α = 0% – upper half-plane; α = 5% – lower half-plane.

As calculations shown, the critical current for ICP with three coils depends on the properties of working gas and generator frequency under the given dimension parameters and mass flow rate of working gas.

The position and shape (axial cross section) of vortex tube in supercritical regime are demonstrated by the distribution of spatial velocity, which are shown in Figure 2 (color lines). When pure argon is used as working gas, the critical current J_{cr} (under ω= 3 MHz) is found at 90–100 A. The axial cross section of vortex tube have an ellipsoidal shape. The vortex center locates at the left side of the first coil.

However, the use of Ar/H_2 mixtures as working gas leads to a change in the heat-gas dynamic parameters of plasma flow. Since the Ar/H_2 mixtures are characterized by lower electrical conductivity, higher heat capacity and thermal conductivity [14] compared to the pure argon plasma.

It is found that at the same discharge current J_{coil} the Joule heating zone and its volumetric energy q_v are reduced in the Ar/H_2 plasma in comparison to that in the pure argon plasma. As can be seen in Figure 2, this leads to a decrease of plasma zone (the dashed lines of temperature contours T = 8000 K are reduced) and maximum plasma temperature. Note that, with the increase of volume fraction α the distribution of temperature in the plasma zone becomes more homogeneous because of higher thermal conductivity. What's more, the transitions of plasma flow states from potential to vortical regimes are observed at higher discharge current. As a result, the dependence of critical discharge current J_{cr} on the volume fraction of hydrogen α (under ω =3 MHz) could be approximated by the linear function $J_{cr} \approx 100 + 10\alpha(\%)$, A.

The dependencies of axial coordinate Z_v of vortex center on the discharge current (under ω =3 MHz and various α) are shown in Figure 3 by dashes lines. In pure argon plasma (α = 0) Z_v is practically located at the cross section of first coil (Z_{inf}=63 mm) when the discharge current $J_{coil} = J_{cr} \approx 100$ A. With the increase of discharge current J_{coil} the axial coordinate of vortex center monotonous decreases to the minimum limit value $Z_{v,\,min} \approx$ 50 mm (i.e. the vortex center moves left to the upstream of central gas).

IOP Conf. Series: Journal of Physics: Conf. Series **1238** (2019) 012044 doi:10.1088/1742-6596/1238/1/012044

The dependencies of vortex intensity ε_v on the discharge current J_{coil} are shown in Figure 3 by solid lines. The vortex intensity ε_v is defined as in [10]: $\varepsilon_v = G_v/(G_1 + G_2)$, where G_v is the backflow rate of working gas in the vortex zone, G_1 and G_2 are the mass flow rates of transporting and central gases. In all cases ε_v is a monotonically increasing function of the discharge current. Under significant high discharge current $J_{coil} \geq 200 - 250$ A the intensity of vortex tube ε_v becomes more than unit ($\varepsilon_v > 1$). This can be explained by that the sheath gas begins to play a role in the formation of vortex, part of its stream flows into the vortex tube. It is investigated that in the vortical regime the intensity of vortex tube in pure Ar plasma ($\alpha = 0$) is either greater or equal to that in the Ar/H_2 plasma ($\alpha \neq 0$) at the same discharge current. That is, the values of ε_v and Z_v under $\alpha = 0$ are the upper limits of the $\varepsilon_v(J_{coil})$ and $Z_v(J_{coil})$ functions under all other different compositions of Ar/H_2 plasma ($\alpha \neq 0$).

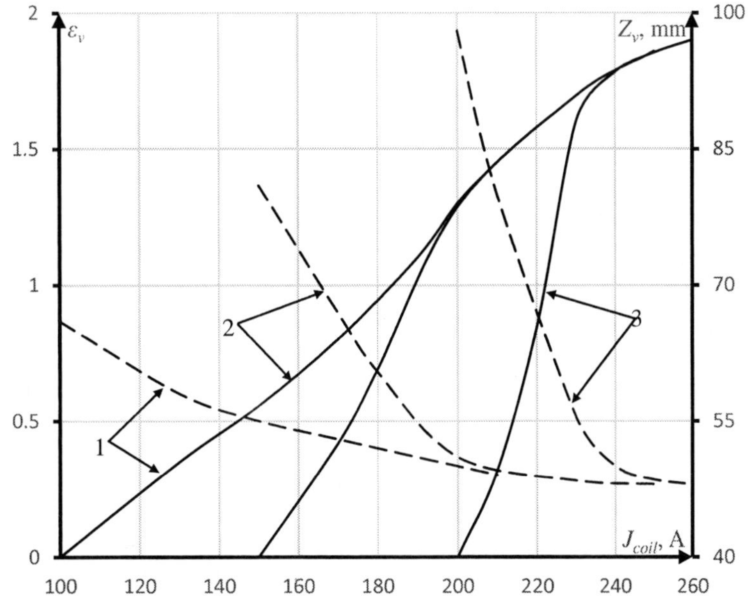

Figure 3. The dependencies of vortex intensity ε_v (solid lines) and axial coordinate of vortex center Z_v (dashed lines) on the discharge current J_{coil} under different volume fractions of hydrogen α: 1 – α = 0%; 2 – α = 5%; 3 – α = 10%.

In order to investigate the influences of generator frequency on the features and parameters of vortex tube, the generator frequency is chosen in the range of 1.76 – 13.56 MHz. The pure argon is considered as working gas. It is found that the critical current J_{cr} is a decreasing function of generator frequency and can be approximated by formula: $J_{cr} = 145\omega^{-0.35}$.

In the vortical regime the intensity of vortex is an increasing function of discharge current. It is found that under a given discharge current, the increase of generator frequency from 1 to 5 MHz leads to a significantly rise of vortex intensity. While the increase of generator frequency from 5 to 8 MHz has little influences on the vortex intensity (Fig.4, solid lines). This may be due to the changes in the distribution of magneto-hydrodynamic (MHD) forces, which are located far more away from axis under high generator frequency as a result of the "skin effect" in the discharge zone.

Besides that, the axial coordinate of vortex centre Z_v also depended on the amplitude and generator frequency of discharge current, it is a decreasing function of discharge current. As can be seen that under a given amplitude of discharge current, the axial coordinate Z_v increases with the decrease of discharge frequency. When the discharge current $J_{cr} \geq 200$A the vortex center almost doesn't changes, it is located around $Z_v \approx 53$ mm (Fig.4, dashed lines).

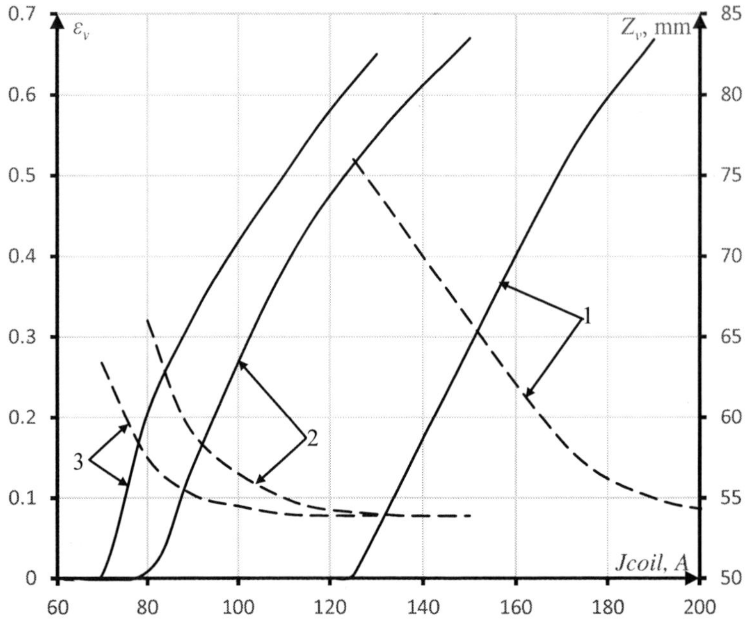

Figure 4. The dependencies of vortex intensity ε_v (solid lines) and axial coordinate of vortex center Z_v (dashed lines) on the discharge current under different generator frequencies: 1 – ω = 1.76 MHz; 2 – ω = 5.28 MHz; 3 – ω = 8 MHz.

4. Conclusions

The gas dynamic parameters of structures of plasma flow in the ICP torch are obtained. The features of the distribution of velocity and temperature fields are found in the ICP torch under the presence of toroidal vortical flow before the high temperature zone. It is established that the necessary conditions for the form of vortex tube are that the discharge current exceeds an critical value J_{cr}, which depends on the volume fraction of hydrogen α (under ω =3 MHz): $J_{cr} \approx 100 + 10\alpha(\%)$, A. The dependence of critical current (for pure Ar plasma) on generator frequency is shown by the following formula: $J_{cr} = 145\omega^{-0.35}$, A. The intensity and position of vortex tube in supercritical flow regimes are determined as a function of the discharge current J_{coil}, volume fraction α and current frequency ω.

5. References

[1] Grishin Yu M, Kozlov N P and Skryabin A S 2012 *High Temperature* **50** 459
[2] Grishin Yu M, Kozlov N P and Skryabin A S 2016 *High Temperature* **54** 619.
[3] Kuzenov V V and Ryzhkov S V 2016 *Bulletin of the Russian Academy of Sciences. Physics* **80** 598
[4] Ameya B, Christopher R P, Steven A C and Carter C B 2003 *J. Appl. Phys.* **94** 1969.
[5] Colombo V, Ghedini E and Sanibondi P 2010 *Plasma Sources Science and Technology* **19** 065024.
[6] Kuzenov V V and Ryzhkov S V 2013 *Problems of Atomic Science and Technology* **4** (86) 103
[7] Punjabi S B, Sahasrabudhe S N, Joshi N K, Mangalvedekar H A, Das A K and Kothari D C 2014 *Physics of Plasmas* **21** 013506.
[8] Grishin Yu M and Miao L 2017 *Journal of Physics: Conf. Series.* **891**, 012302.
[9] Qian L J, Lin J Z and Yu M Z 2013 *Journal of Thermal Spray Technology* **22** 1024.
[10] Grishin Yu M and Miao L 2016 *Prikladnaya Fizika*, No. 4, 33.
[11] Boulos M I, Fauchais P and Pfender E 1994 *Thermal plasmas: fundamentals and applications, Volume 1* (New York).

[12] Holik E F 2008 *Simulation results of an inductively coupled rf plasma torch in two and three dimensions for producing a metal matrix composite for nuclear fuel cladding* (m.s. thesis: Texas A&M University).
[13] Ryzhkov S V 2011 *Plasma Physics Reports* **37** 1075
[14] Murphy A B 2000 *Plasma Chemistry and Plasma Processing* **20** 279.
[15] Chase J D 1971 *J. Appl. Phys.* **42**, 4870.
[16] Grishin Yu M and Miao L 2017 *Prikladnaya Fizika*, No. 3, 31.

The acceleration of laser plasma in a strong non-stationary magnetic field

A. Shikanov, E. Vovchenko, K. Kozlovskii, A. Isaev, A. Plekhanov and M. Lisovskii
National Nuclear Research University "MEPhI", 115409, Russia, Moscow, Kashirskoe highway, 31

edvovchenko@mail.ru

Abstract. Based on experimental and computer simulation the acceleration of deuterons from laser plasma in a strong non-stationary magnetic field was studied. The possibility of reaching an energy of ~100 keV, corresponding to the effective course of the nuclear reactions D $(d, n)\,^3$He and T $(d, n)\,^4$He, was demonstrated. YAG: Nd^{3+} laser ($W \leq 0.85$ J, $\tau \approx 10$ ns) was used in the experiment with focusing laser radiation on a deuterated polyethylene target. The high voltage pulse generator with a conical spiral coil was used to generate a high-speed magnetic field ($2\cdot10^7$ T/s). A mathematical model of the process is proposed. According to this model, the acceleration of a laser plasma is analyzed by means of a computer. The algorithm is based on a numerical solution of the system of Newton-Lorentz equations.

1. General

It is possible to use various methods of accelerating deuterons to create portable pulse neutron generators (PNG) based on the nuclear reactions D $(d, n)^3$He and T $(d, n)^4$He. The most popular methods so far are methods of direct acceleration of deuterons, in which the additional increase energy is realized in a single pass in an electric field. This field is formed with the help of high voltage pulse generators. In this case, the main efforts are aimed at reducing the influence of the space charge and the electron emission current, respectively, which limit the deuteron current density and the efficiency of their acceleration.

An alternative method of the acceleration of deuterons without these shortcomings, based on the expansion of a quasineutral laser plasma in a nonstationary strong magnetic field, is considered in the article. Experimental results and theoretical estimates are given. The prospect of such studies is related to the possibility of their application in development of PNG, plasma engine and ion injection systems for accelerators. The main goal of the research is to accelerate the deuterons of the laser plasma to velocities of $\sim 3\cdot10^6$ m/s and more. This corresponds to deuteron energies $\geq$100 keV and allows efficient initiation of the nuclear reaction D $(d, n)^3$He and T $(d, n)^4$He.

The idea of such acceleration is not new and was known at the end of the last century [1]. Later, it was developed in works [2, 3], where the principle of the acceleration of a laser plasma in a high-speed (~100 ns) inhomogeneous magnetic field is used. It field is formed when a storage capacitance is discharged onto a conducting ring with a radius of ~0.01 m. At that time, a qualitative model of an acceleration of laser plasma is proposed. Based on it the requirements for the experimental design is formulated. A new interest in such studies is caused by the appearance and availability of a compact high-power lasers which provide wider experimental possibilities.

2. Physical principle and experiment

The induction mechanism of acceleration is based on the interaction of an external variable magnetic field with a laser plasma, as a conducting medium. Laser plasma is created by the action of intense laser radiation on a solid-state dielectric target. The laser target is surrounds by an induction coil with a current passes through it. This current pulse is synchronized with the laser pulse.

A rapidly increasing magnetic field $\vec{B}$ induces an eddy current $\vec{j}_B$ in a laser plasma. The azimuthal component of this current j_φ interacts with the radial component of the magnetic field B_r and creates the electromagnetic Ampere-Lorentz force $F_z \sim [\vec{j},\vec{B}]_z = j_\varphi \cdot B_r$. This force accelerates the plasma in the axial direction. We note that in this physical model the laser plasma is consider as a continuous medium in which a quasineutrality violation does not occur at a macroscopic level. This removes the limitations associated with the space charge. Ions are accelerated in the presence of electrons. However, at the same directed velocity, the main energy of the flow due to the mass difference is determined by the ions. The energy of accelerated deuterons in such a system can reach ~100 keV, which is necessary for efficient neutron generation.

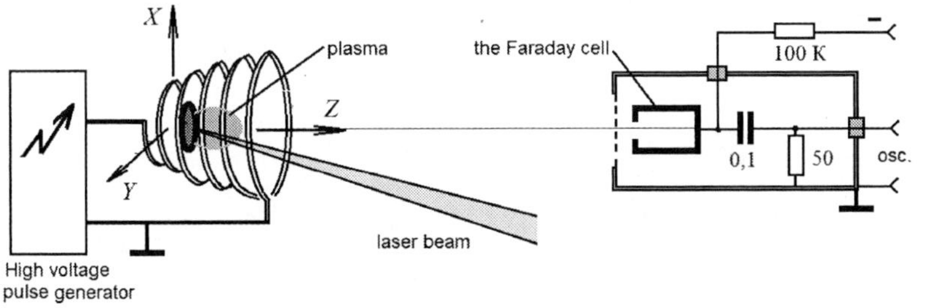

Fig. 1.
The scheme of the deuterons accelerating experiment

The scheme of the deuterons accelerating experiment with using laser plasma in high-speed magnetic field and the time-of-flight collector measurements of the velocity of these deuterons is shown in Fig. 1. Nd: YAG laser (λ = 1.06 μm) with energy per pulse $W \leq 0{,}85$ J and pulse duration $\tau \approx 10$ ns was used to generate laser plasma. A power density of the order of $5 \cdot 10^{15}$ W·m^{-2} was achieved by focusing its radiation on a dielectric target of deuterated polyethylene $(CD_2)_n$. The moment of appearance of the laser pulse on the target was registered with a coaxial photocell FEK–22SPU.

A high-speed magnetic field in the laser target was created using a pulsed current through an induction coil. The coil was made in the form of a conical helix with the vertex angle of the cone $2\alpha \approx 20°$. Coil parameters: $r_{min} = 7.5 \cdot 10^{-3}$ m, $r_{max} = 12.5 \cdot 10^{-3}$ m, length $3 \cdot 10^{-2}$ m, number of turns $N = 6$, inductance $L = 0.65$ μH. Two type of pulse high voltage generators were used to excite current in the induction coil: "fast" generator with high voltage ($U \sim 400$ kV) and current in load less then 1.5 kA; "slow" generator with more lower voltage (U ~ 100 kV) but a current up to 10 kA. The ion velocity was determined by the time-of-flight method with registration of the ion current on the collector as Faraday cylinder. The collector was installed at a distance of 0.5 m from the laser target.

In the first series of experiments, the Marx generator [4] was used. This device allows receiving a high voltage pulse with an amplitude of up to 450 kV with a stored energy of up to 50 J and, as noted above, a current in the load up to 1.5 kA. The results of these preliminary experiments was described in [5] and are given here for comparison and analysis. The pulse duration of the current flowing through the induction coil is 30 ns. The velocity of increase of the magnetic field reached up to $2 \cdot 10^7$ T/s. For these conditions, the velocity of the fastest component of the ion flux was ~10^6 m/s, but the bulk of the deuterons had a lower velocity of ~$5 \cdot 10^5$ m/s. In this case, up to 10^{13} ions arrive at the collector, and the total number of accelerated ions is 10^{15}. It should be noted a high level of electromagnetic interference, caused by a cascade breakdown of air spark gaps in the scheme Marx generator. Since this markedly limited the accuracy of the measurements, all the measured values should be considered only an estimate. As an illustration, Fig. 2a shows the oscillogram of the current from the collector (upper beam) and the current in the induction coil (lower beam).

IOP Conf. Series: Journal of Physics: Conf. Series **1238** (2019) 012045 doi:10.1088/1742-6596/1238/1/012045

New results were obtained using a more powerful generator, performed based on pulse charging of the storage capacitance from a high-voltage pulse transformer. When the voltage is reached the value about 100 kV, a breakdown of the spark gap and the discharge of the capacitance onto the induction coil generating a magnetic field occurs.

The maximum current can be estimated by neglecting the active resistance using the ratio $I_{max} = U \cdot (C_n / L)^{1/2} \approx 10$ kA, where $C_n = 4{,}4$ nF is the total capacitance of high voltage storage and $L = 0{,}65$ μH is the coil inductance. The capacitance C_n has 12 sections, and each section consists of six 2.2 nF capacitors connected in series. The sections are arranged symmetrically with respect to the axis of the cylinder, on which the spark gap is installed. This design has significantly reduced the level of electromagnetic interference.

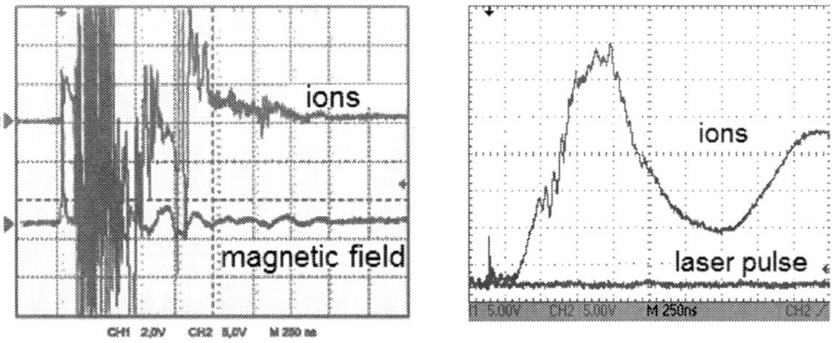

Fig. 2.
The collector measurements of ions: Marx generator (a); generator with the pulse transformer (b)

The oscillogram of the current from the collector (upper beam) and the laser pulse (lower beam) are shown in Fig. 2b. It can be seen that a sufficiently large group of fast deuterons (~ 30% of the total number) reaches the collector trough 500 ns after the laser pulse is has come to the $(CD_2)_n$ target, and the slowest group of deuterons – is no later than 1000 ns. Their velocities are respectively is equals to $\sim 10^6$ m/s and $5 \cdot 10^5$ m/s. The second maximum on the collector current pulse is due to a group of slower carbon ions.

3. The model of acceleration of ring current

A mathematical model of the behavior of an axially symmetric laser plasma in a nonstationary magnetic field for the purpose of analyzing and selecting the optimal conditions for a physical experiment was developed. Axial symmetry makes it possible to represent a laser plasma as a superposition of toroidal plasmoids.

The basis of the computational model was the problem of accelerating an individual plasma toroid surrounded by a single conductive ring with a current that creates a magnetic field [6]. To estimate the energy, we use the notion of the acceleration field $f(r, z, t)$. Variable parameters: r is the transverse size of the plasma; z is the distance between the conductive ring and the laser target with displacement toward the incident laser radiation; t is the time delay of plasma formation in relation to magnetic field build-up.

The time-increasing magnetic field produced by the current $I(t)$ when the capacitor is discharged to the single conductive ring can be determined through the components of the magnetic induction vector [7]

$$B_r(r,z,t) = I(t) \cdot B_{0r}(r,z) \text{ и } B_z(r,z,t) = I(t) \cdot B_{0z}(r,z), \tag{1}$$

where

$$B_{0r}(r,z) = \frac{\mu_0}{2\pi r}\left[E(k)\frac{a^2 + r^2 + z^2}{(a-r)^2 + z^2} - K(k)\right]\frac{z}{\sqrt{(a+r)^2 + z^2}}, \tag{2}$$

$$B_{0z}(r,z) = \frac{\mu_0}{2\pi}\left[E(k)\frac{a^2 - r^2 - z^2}{(a-r)^2 + z^2} + K(k)\right]\frac{1}{\sqrt{(a+r)^2 + z^2}}, \tag{3}$$

μ_0 is the magnetic constant, a is the radius of the conductive ring with current, $K(k)$ and $E(k)$ are complete elliptic integrals with argument

$$k(r,z) = \sqrt{\frac{4ar}{(a+r)^2+z^2}}\,, \tag{4}$$

and the current in the discharge circuit with capacitance C, resistance R and inductance L is equal to

$$I(t) = U\sqrt{C/_{2L}}\exp\left(-\frac{R}{2L}t\right)sin\left(\frac{t}{\sqrt{LC}}\right). \tag{5}$$

The azimuthal vortex electric field arises because of a temporal changing of the magnetic field. This vortex field is primarily determined by the rate of current rise in the circuit

$$E_\varphi(r,z,t) \approx -\left(\frac{r}{2}\right)\frac{d}{dt}[B_z(r,z,t)]. \tag{6}$$

In this case, an azimuth current with a density $j\,(r,\,z,\,t,\,\theta) = \sigma\,(\theta)\cdot E_\varphi\,(r,\,z,\,t)$ is excited in the plasma, which interacts with the radial component of the field $B_r\,(r,\,z,\,t)$ and creates ponderomotive force, that acts on an element of unit volume of the plasma

$$F_z(r,z,t) = \sigma(\theta)E_\varphi(r,z,t)B_r(r,z,t), \tag{7}$$

where $\sigma(\theta)$ is the conductivity of the plasma, which depends on the temperature θ. The relation determines the acceleration field corresponding to this force

$$f(r,z,t) = \frac{\sigma(\theta)E_\varphi(r,z,t)B_r(r,z,t)}{Mn}, \tag{8}$$

where M is the deuteron mass, and n is the electron density.

The algorithm for calculating the acceleration field $f\,(r,\,z,\,t)$ was developed on the basis of a mathematical model, a numerical experiment was performed to determine the optimal conditions for the acceleration of a laser plasma in a rapidly growing magnetic field, and recommendations were obtained for setting up a real physical experiment. The main conclusions are given below:

– optimal conditions for interaction of the plasma with the magnetic field are achieved when the plasma-forming target is located at a distance $z_0 \approx (0.15 - 0.20)a$ from the plane of the conductive ring with a current in the direction of the plasma expansion;

– the acceleration of deuterons increases monotonically with increasing radius of the focusing ring and reaches a maximum at $r_0 \approx 0.9\cdot a$;

– the decrease in the transverse dimension of the plasma, due to the focusing properties of the magnetic field, leads to reducing the influence of $B_r(r,\,z,\,t)$, and means to reduction of acceleration force $F_z(r,\,z,\,t)$ in the axial direction and the decreasing of characteristic length of acceleration;

– the acceleration of deuterons increases monotonically with increasing parameter $(U^2\cdot C^{1/2})\,/L^{3/2}$.

4. The theoretical model of acceleration in geometry of conical spiral

The next step in the development of the model (1) - (8) was the transition from the single conductive ring with a current to the model in which the magnetic field is formed by an induction coil in the form of a conical spiral of N turns. This coil expands along the Z axis in accordance with the following geometric with ratios:

$$b_i = b_0 + \frac{i}{N-1}(b_{N-1} - b_0)\,, \qquad h_i = \frac{i}{N-1}h_{N-1}\,, \tag{9}$$

b_i is the radius of the i-th turn, h_i is the distance from the point of focusing to the plane of the i-th coil. In this case, to determine the components of the vector of induction instead of formula (1) is necessary to consider the total contribution of each coil

$$B_r(r,z,t) = I(t)\sum_{i=0}^{N-1} B_{rN}\,(b_i,r,z-h_i) \quad \text{и} \quad B_z(r,z,t) = I(t)\sum_{i=0}^{N-1} B_{zN}\,(b_i,r,z-h_i) \tag{10}$$

IOP Conf. Series: Journal of Physics: Conf. Series **1238** (2019) 012045 doi:10.1088/1742-6596/1238/1/012045

$B_{rN}(b_i, r, z-h_i)$ and $B_{zN}(b_i, r, z-h_i)$ are defined by formulas (2) - (4), replacing the radius a by b_i and replacing the current coordinate z by $z-h_i$. $I(t)$ is the current in the induction coil. In contrast to (5), we will be considered current $I(t)$ is harmonic in order to simplify the calculations

$$I(t) = I_0 sin\left(\pi\frac{t+\tau}{\sqrt{LC}}\right);$$

where τ is the laser pulse delay relative to the beginning of the current in the induction coil. Vortex electric field $E_\varphi(r, z, t)$ (6) occurs as the result of high-speed increasing of the magnetic field. Due to this processes the azimuthal current is excited in the laser plasma. The density of current is

$$j(r,z,t) \approx \sigma\left(\frac{r}{2}\right)\sum_{i=0}^{N-1} B_{zN}(b_i,r,z-h_i)\cdot\left(\frac{\pi I_0}{\sqrt{LC}}\right) I_0 cos\left(\pi\frac{t+\tau}{\sqrt{LC}}\right). \qquad (11)$$

The corresponding magnetic moment is equal to

$$L(r_c,z_c,t) = \frac{1}{2}\int_V [r,j]\,dV \approx \frac{1}{2}\iiint j\cdot r^2 dr\cdot dz\cdot d\varphi \approx \pi \int_0^{r_0+z_c} dz \int_0^{r_c} j\cdot r^2\, dr\,, \qquad (12)$$

z_c is the longitudinal coordinate of the center of gravity of the plasma, r_c is the mean radial deviation of the front of the laser plasma from the axis, and r_0 is the initial radius of expansion of the laser plasma. The force acts on the deuterons of laser plasma in the axial direction

$$F(r_c,z_c,t) \approx -\nabla\left(\vec{L},\vec{B}\right)_z \approx -L(r_c,z_c,t)\frac{dB_z(r_c,z_c,t)}{dz}. \qquad (13)$$

According to experiment, analysis of the kinematics of the expansion of a laser plasma has shown that this process can be described as an expanding plasma sphere whose center moves in the direction perpendicular to the surface of the laser target with the propagation velocity of the plasma front V_f, that is acceptable for estimate calculations. This is true for a plasma, which is formed by the interaction of a short laser pulse with a solid target. The velocity field in this case will be determined by the following expression

$$\mathbf{V}(\mathbf{r},t) = \mathbf{e}_r\frac{rV_f}{V_f t + r_0} + \mathbf{e}_z V_f\left(\frac{z}{V_f t + r_0} + 1\right) \qquad (14)$$

The velocity field becomes «narrower» when a longitudinal magnetic field arises with the induction value $B(t)$. The velocity field is defined by the following expression

$$\mathbf{V}(\mathbf{r},t) = \mathbf{e}_r\frac{rV_\perp(t)}{V_\perp(t)t + r_0} + \mathbf{e}_z V_f\left(\frac{z}{V_f t + r_0} + 1\right)$$
$$V_\perp(t) \approx 0.7\cdot V_f[1+1.9B(t)^{1.2}]^{-1/2}, \qquad (15)$$

where $B(t)$ is the value of the longitudinal component of the magnetic field induction vector at the center of gravity of the plasma [8].

The system of equations of motion of the center of gravity of a deuteron avalanche with total mass Δm has the following form

$$\frac{dV_c}{dt} = \frac{F(t,z_c,r_c)}{\Delta m},$$
$$\frac{dz_c}{dt} = V_c, \qquad (16)$$
$$\frac{dr_c}{dt} = V_\perp(t),$$

with the initial conditions $V_c(0) = V_0$; $z_c(0) = r_0$; $r_c(0) = r_0$. The average energy of an accelerated deuteron in a bunch is determined by the following formula

$$T(t) = \frac{10^{-3}M}{2e}V_c^{\,2}\ [\text{кэВ}]. \qquad (17)$$

The computer program is designed to solve the system of equations (16), (17) numerically, taking into account conditions (11) - (15). The calculated dependence of the average kinetic energy of an accelerated deuteron on time is shown in Fig. 3 as an illustration. These results corresponds by time delay τ = 60 ns between the laser pulse and the beginning the current in the induction coil. The calculation is made for $N = 6$; $h_5 = 0.03$ m; $b_0 = 0.005$ m; $V_0 = 2\cdot10^5$ m/s; $I_0 = 10^4$ A.

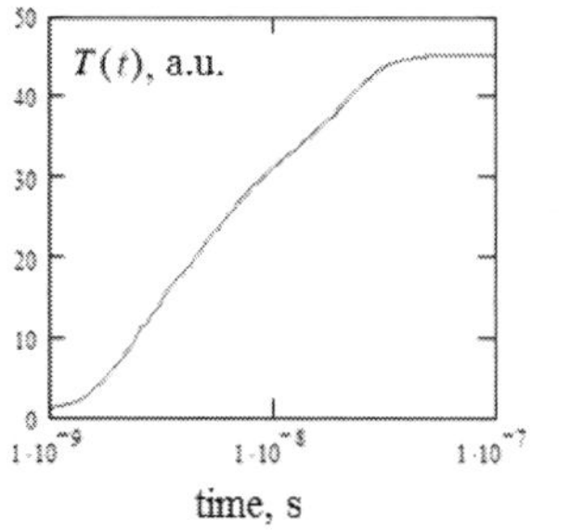

Fig. 3.
Modeling of kinetic energy of deuterons

5. Summary

Based on experimental and computer simulation the acceleration of deuterons from laser plasma in a strong non-stationary magnetic field was studied. The results of the modeling calculations showed that the energy of the directed motion of deuterons increases approximately by 40–50 times during a time of 60 ns. This corresponds to an increase in speed to 10^6 m/s. In this case, the deuterons have an energy of ~100 keV, and this agrees with the experiment. It is acceptable for the generation of neutrons in the synthesis reactions of heavy hydrogen nuclides.

Acknowledgment
This work was carried out with the financial support of the Ministry of Science and Education of the Russian Federation under the agreement № 14.575.21.0169 (RFMEFI57517X0169).

References

[1] Kozlovskii K. I., Tsybin A. S., Kozyrev Yu. P. et al. Formation of intense ion fluxes from laser plasma. *Proceedings of the 17th International Conference on phenomena in ionized gases, Budapest,* vol. 2., (1985), pp. 945-947.

[2] Tsybin A. S., Kozlovskii K. I., Kuznetsov A. Yu., Shikanov A. E. The laser plasma accelerator for radiation-physical investigations. *Proceedings of the 12th International Connference on Radiation Physics and Chemistry of Inorganic Materials, Russia, Tomsk*, (2003), P. 511 – 514.

[3] Kozlovskii K. I., Shikanov A.E. et al. The model of laser plasma acceleration by nonstationary magnetic fields. *Proceedings of the Scientific session of MEPhI*, vol. 4, (2003), pp. 65-66.

[4] Vovchenko E. D., Kozlovskii K. I., Shikanov A. E. et al. An Accelerating Voltage Generator for Compact Pulsed Neutron Sources. *Instruments and Experimental Techniques*, vol. 60, № 3, (2017), pp. 362–366.

[5] Isaev A. A., Kozlovskii K. I., Shikanov A. E. et al. Collective acceleration of laser plasma in a nonstationary and nonuniform magnetic field. *Journal of Physics: Conf. Series* 941 (2017) 012016, doi: 10.1088/1742-6596/941/1/012016.

[6] Isaev A. A., Kozlovskii K. I., Shikanov A.E. et al. On the acceleration of the particle flux of laser deuterium plasma in a fast-growing magnetic field. *Poster session. 12th International Scientific Workshop in Memory of Professor V.P. Sarantsev «Problems of colliders and Charged Particle Accelerators»*, 3-8 September 2017, Alushta, Crimea.

[7] Landau L. D., Lifshitz E. M. Electrodynamics of Continuous Media (*Volume 8 of A Course of Theoretical Physics) Pergamon Press* 1960, P – 429.

[8] Didenko A. N., Shikanov A. E., Kozlovskii K.I. et al. Small-size magnetically insulated plasma diodes for neutron generation. *Plasma Physics Report*, 2014. vol. 40, № 11, P. 910 – 918.

IOP Conf. Series: Journal of Physics: Conf. Series **1238** (2019) 012046 doi:10.1088/1742-6596/1238/1/012046

Production of carbon nanofibers by plasma jet of helium and hydrocarbons

M B Shavelkina, R Kh Amirov, V I Kiselev and V A Katarzhis

Joint Institute for High Temperatures of Russian Academy of Sciences, Izhorskaya Street 13/2, 125412 Moscow, Russia c

E-mail: mshavelkina@gmail.com

Abstract. The parameters of a non-catalytic process for synthesizing carbon nanostructures during the conversion of methane / helium and propane-butane / helium jets have been studied. The spectral features of the radiation of mixed systems have been investigated. Analysis of the spectra showed that the main part of carbon dimers is withdrawn from the hot region of the plasma in the outlet plasma stream. Synthesized products were been characterized by scanning electron microscopy, Raman spectroscopy, thermal analysis and X-ray photoelectron spectroscopy. It is shown that carbon nanostructures have a bamboo-like morphology and similar properties when using both methane and propane-butane mixture as a carbon source. It has been established that the morphology of synthesized carbon nanofibers is not affected by the pressure of the medium.

1. Introduction

Thermal plasma processing of carbon sources using a plasma jet is one of the universal methods for the synthesis of advanced materials [1-3]. The advanced materials, at present, include various carbon nanostructures (fullerenes, carbon nanotubes (CNTs), carbon fibers, graphene, onions, etc.). Due to their unique electronic and extraordinary mechanical properties, carbon nanostructures have been receiving much attention in a wide variety of applications: electronics and computing, field emitter devices, sensors, electrodes, high strength composites, and storage of hydrogen, lithium and other metals [4,5].

Among the existing types of carbon nanostructures, filamentary carbon nanotubes and nanofibers are the most preferred as potential modifiers of elastomeric compositions. Unlike in a flat graphene sheet, in carbon nanotubes, the common p-electron cloud is deformed, compressed inside the tube and is stretched over the outer surface. The smaller the nanotube diameter the higher is the deformation degree. This affects the strength of the bonds: the greater the curvature of nanotube surface, the smaller degree of the p-orbitals overlapping, the higher is a nanotube capability to chemical reactivity.

In terms of geometry, CNTs are structures with a high aspect ratio, i.e. ratio L / D (length to diameter), because diameter of CNTs is from one to several tens of nanometers, and the length amounts to tens of microns. Therefore, aspect ratio of CNT can reach very high values.

CNTs and nanofibers are mainly produced by CVD technology (the most acceptable in terms of target product yield, energy costs, etc.). The principle of CVD is the decomposition of various hydrocarbons over transition metal supported catalyst. As a result of the above technology application, composition of the filamentary particles contains catalytic particles (metals of variable valence), elements of a catalyst substrate, and also various carbon impurities. In addition, nanostructures synthesized by CVD technology, at the supramolecular level form very strong agglomerates and have various kinds of defects. It means that CNTs and nanofibers have poor technological properties, are characterized by the complexity of dispersing and uniform distribution in the structuring material matrix.

In this paper, our purpose is to investigate the possibility of direct synthesis of carbon nanofibers in a plasma jet of hydrocarbon / helium without the use of catalysts.

2. Methods

Synthesis of carbon nanostructures was carried out in a flow-type gas reactor. The setup consists of a vacuum chamber, a water ring pump, a plasma generator, a water cooling system, an energy supply system (figure 1). The thermal plasma generator which is a high current divergent anode-channel DC plasma torch was used for the synthesis of graphene materials earlier. When this plasma torch is used for synthesis of filamentary structures its design is adjusted so that the distance between cathode and nozzle is extended to 2 mm in order to increase the flow rate of helium-hydrocarbon mixture (figure 2).

A detailed description of the experimental setup was given in the study [6]. The experiment involved an input of mixture of hydrocarbons with a the plasma forming gas into the plasma torch, wherein heating and decompositions occurred in the region of the arc discharge, followed by condensation of the synthesis product on the metal surface of the collector, see figure 1. The rate of hydrocarbons, plasma gas flow and the power of the plasma torch varied independently of each other. For the experimental conditions the electric power of plasma torch was set up to 50 kW. Helium was used as plasma forming gases. Carbon precursors were propane-butane mixture, methane.

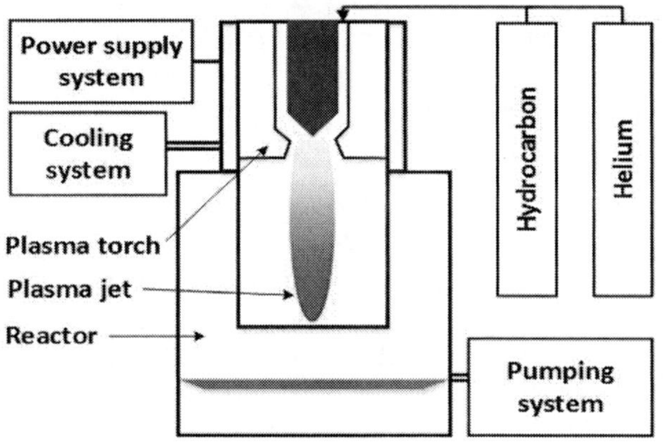

Figure 1. Schematic view of a reactor.

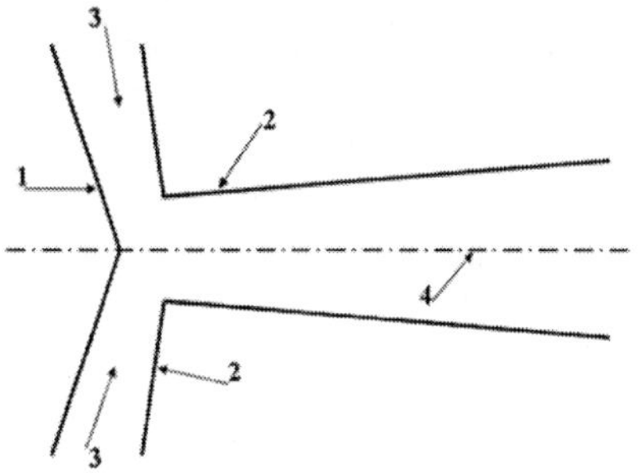

Figure 2. Geometry of the plasma generator: 1 – a cathode, 2 – a water cooled anode with a diverging duct, 3 – plasma forming gas/ hydrocarbon, 4 – symmetry axis.

For each experiment, input of hydrocarbons was carried out after establishment of the temperature field in a collector (determined by chromel-alumel thermocouple). The current value of the plasma torch was constant during the experiment and equaled to 400 A. Arc voltage was changed from 60 to 110 V depending on gas pressure. Helium (class A) flow rate was 0.75 g/s.

Experiments were conducted at various pressures and flow rates of hydrocarbons. Pressure changed from 150 to 740 Torr, and hydrocarbon (propane-butane technical (PBT) GOST R 52087-2003), methane TU 51-841-87) flow rate varied from 0.05 to 0.37 g/s. Duration of the experiments was 5 - 20 min. The experimental conditions are presented in table 1.

Table 1. Technological conditions.

Power (kW)	Current (A)	Voltage (V)	Gas pressure (Torr)	Helium flow rate (g/s)	Hydro-carbons flow rate (g/s)
30-45	300 -400	60-110	150-740	0.5-0.9	0.05-0.37

Direct registration of the dimensional parameters and images is carried out using a scanning electron microscopy (SEM) - microscope Hitachi S5500 with in-lens technology for obtaining both ultra-high resolution and high sensitivity of EDS analysis. The used modes are DF-STEM, BF-STEM and SE.

The synthesis efficiency, phase composition and thermal stability of carbon products were evaluated by thermogravimetry combined with differential scanning calorimetry (STA 449 Netzsch F3 Jupiter platform) with linear heating in argon in steps of 10 K / min in the temperature range of 20-1000 ° C.

Raman spectroscopy (Ntegra Spectra, a laser with 532 nm wavelength) was used to study the structure of nanoscale structures.

A quantitative study of the elemental composition and electronic state of atoms was carried out using X-ray photoelectron spectroscopy (XPS) at room temperature using a Theta Probe spectrometer of ThermoFisher Scientific (UK).

The spectra of hydrocarbons and helium plasma were taken on a three-channel fiber-optic spectrometer AvaSpec 2048 with a spectral resolution of 0.1 ÷ 0.4 nm and spectral range 200 ÷ 1100 nm. The electron temperature was considered for helium plasma, using the method of relative intensities of the same lines of ionization particles (method Boltzmann exhibitors).

3. Results and discussions

All experiments on mixed plasma conversion were carried out in a reactor preheated by nitrogen plasma for 1-2 minutes. After the temperature on the collector surface reached 823 K, pure helium followed by helium – hydrocarbon mixture are tangentially supplied into the plasma torch, as shown in figure 2. Immediately after the input, the temperature of the collector drops to 600 K, but then is restored within 5-6 minutes. The arc voltage with hydrocarbon addition increases by 20 - 30 V. Figures 3-4 show the emission spectra of pure helium and helium plasma with the addition of propane-butane. Figure 3 shows that the spectrum of pure helium recombining plasma jet at atmospheric pressure and an electron temperature of 20,000-30,000 K represents a set of strong atomic lines in the range of 300-800 nm .When propane-butane is added (figure 4), the number of helium lines decreases and lines typical of C_2 are observed in the spectrum, whose number grows with distance from the nozzle. The highest intensity of the atomic C lines corresponds to the range 200-400 nm. At the same time both vibrational and rotational temperatures for C_2 do not vary over the entire length of the jet. The presence of the CN line is due to the fact that the measurements were carried out in an air atmosphere.

The synthesized products are black color powder. Investigation of the powder by scanning electron microscopy shows that these are filamentary nanoscale structures. Figures 5 - 7 show the morphology of the samples obtained by changing the medium pressure and the precursor type. It can be seen from the figures that the structure in all images is the same and has a "bamboo" appearance with approximately same length of "bamboo fragments" (100 nm) and a diameter of 30-50 nm. This may indicate a periodic nature of the growth process of these carbon nanofibers, which according to [7] is usually associated with the effect of a catalyst.

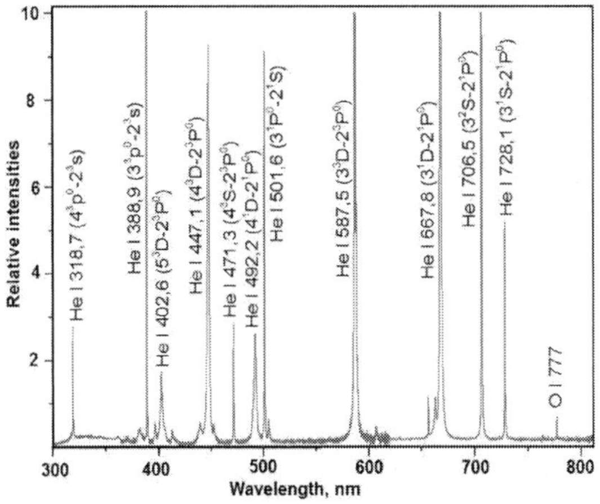

Figure 3. The emission spectrum of helium plasma without the addition of a hydrocarbon.

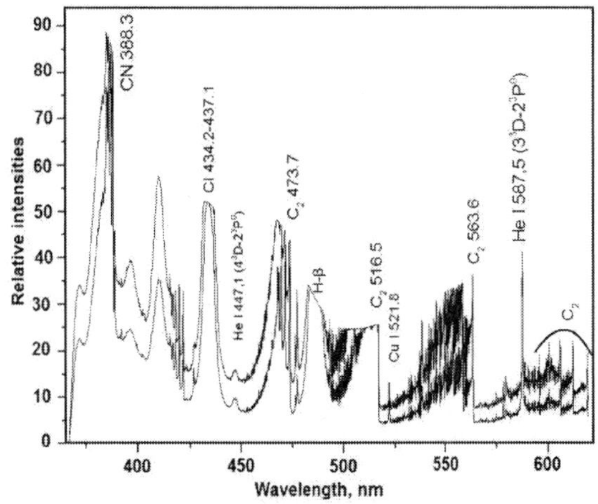

Figure 4. The emission spectrum of helium plasma with the addition of propane-butane.

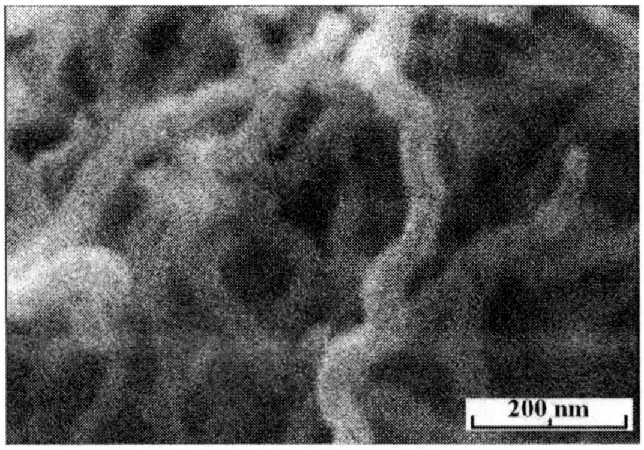

Figure 5. Image of carbon nanofibers obtained by conversion of propane-butane / helium plasma at 710 Torr.

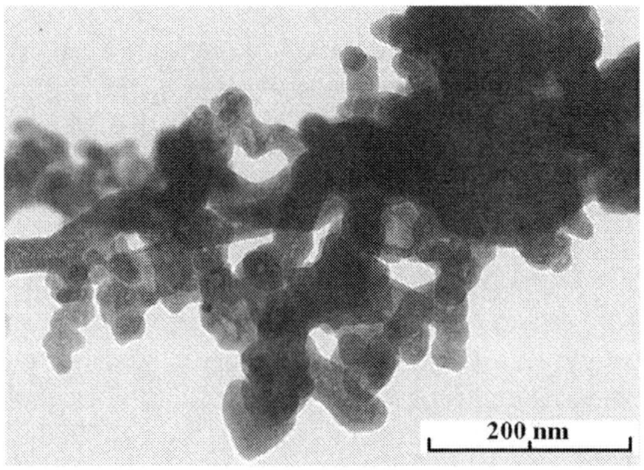

Figure 6. Image of carbon nanofibers obtained by conversion of propane-butane / helium plasma at 350 Torr.

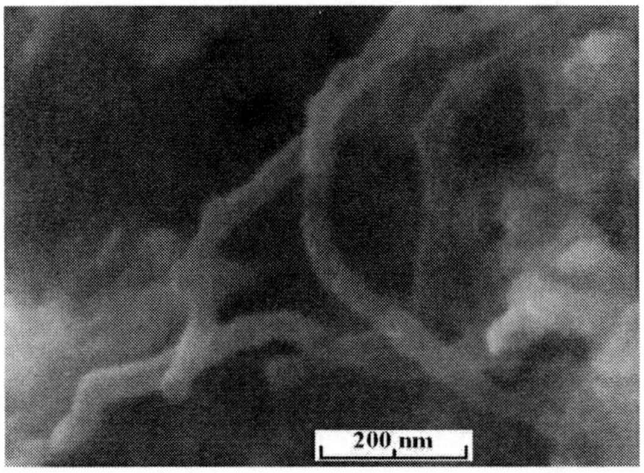

Figure 7. Image of carbon nanofibers obtained during the conversion of methane/helium plasma at 350 Torr.

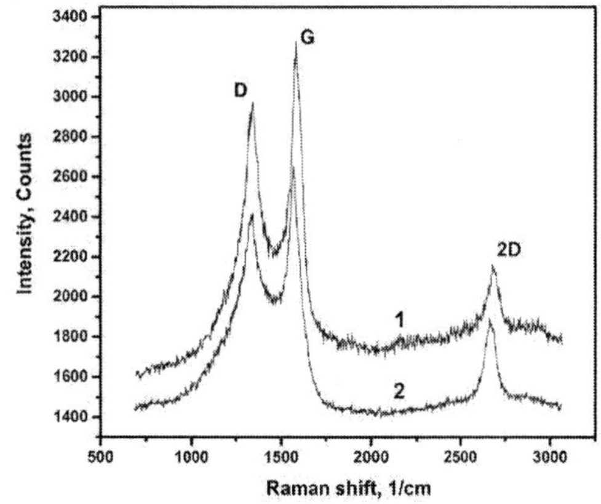

Figure 8. The Raman spectrum of carbon nanofibers, obtained by conversion of methane/ helium plasma at 350 Torr (1), propane-butane / helium plasma at 710 Torr (2).

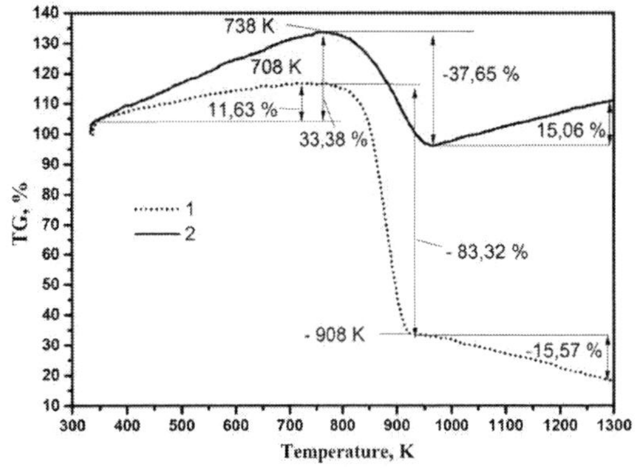

Figure 9. Thermogravimetry of carbon nanofibers obtained by conversion of propane-butane/ helium plasma at 350 Torr (1), methane / helium plasma at 350 Torr (2).

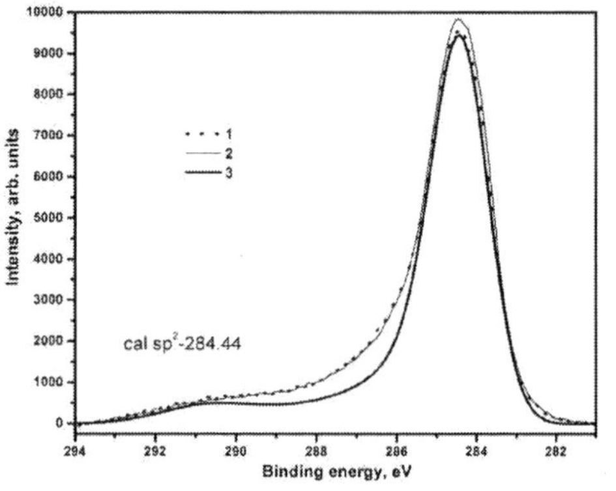

Figure 10. XPS C 1s spectra of carbon nanofibers obtained by conversion of propane-butane/helium plasma at 350 Torr (1), methane / helium plasma at 350 Tor (2), HORG (3).

Catalyst was not used during the experiments in this work. Heterogeneity of the surface, apparently, is associated with formation of internal overlaps. Such structure corresponds to nanofiber [8].

Figure 7 shows a typical Raman spectrum for carbon nanofibers samples synthesized in a helium plasma jet system. The spectrum of the samples reflects a unique feature of the G-band: because of localization and curvature of graphene layers in formation of nanofibres [9].

Studies of the thermal stability of the samples show that the mass loss occurs in the temperature range of 673-873 K which is typical for purified carbon nanostructures. However, the amount of mass lost in the oxidation of samples synthesized in conversion of propane-butane / helium plasma at 350 Torr and in conversion of methane-helium plasma at 350 Torr, differs by 46 mass % by weight in air (figure 9). This is due to the different surface structure formed in the gas phases of different composition. This result agrees with the X-ray photoelectron spectroscopy data.

As can be seen from figure 10, some difference in the peak widths of the samples is caused both by the samples inhomogeneity and by the presence of several chemical states of the carbon atoms. For their interpretation and determination of relative values, account of surface charge has been conducted with isolation of a state with sp^2-hybridization, to which energy of 284.44 eV is attributed. For the samples synthesized from methane-helium plasma, the fraction of sp^2-hybridization is slightly less than for the samples synthesized from propane-butane / helium plasma.

Thus, we have developed a method for synthesizing carbon nanostructured materials by plasma jet without the use of catalysts. The resulting carbon fibers have fewer defects than those generated using chemical methods [11].

4. Conclusion

The possibility of non-catalytic synthesis of carbon nanofibers in the conversion of a mixed plasma: helium / hydrocarbon, where propane-butane or methane is used as a hydrocarbon has been demonstrated experimentally. Analysis of the emission spectra of these systems showed that the amount of carbon dimers increases at the exit of the stream from the zone of arc discharge. The study using physicochemical methods of synthesis products obtained in different helium plasma systems has been showed that the bamboo-shaped morphology of carbon nanofibers and their properties do not change when the pressure of the medium and the type of carbon source change.

5. Acknowledgements

The authors gratefully acknowledge Russian Foundation for Basic Research for the support by grants No 16-08-00145, 16-08-00057, 16-08-00081 and 18-08-00306.

6. References

[1] Ostrikov K and Murphy A B 2007 *J. of Phys. D: Appl. Phys.* **40** 2223

[2] Gonzalez-Aguilar J, Moreno M and Fulcheri L 2007 *J. Phys. D: Appl. Phys.* **40** 2361

[3] Boulos M I 1991 *IEEE Trans. Plasma Sci.* **19** 1078

[4] Mostofizadeh A, Li Y, Song J and Huang Y 2011 *J. of Nanomater.* **2011** Article ID 685081 21 pp

[5] *Carbon Nanostructures* Series Ed.: Araujo Paulo Springer: Berlin, Germany 2018

[6] Shavelkina M B, Amirov R H, Katarzhis V A, Kiselev V I 2017 *J. of Phys.: Conf. Ser.* **748** 012021

[7] Ivanov V, Nagy J B, Lambin P, Lukas A, Zhang X B, Zhang X F, Bernaerts D, Tendeloo G, Amelinckx S and Landuyt J. 1994 *Chem. Phys. Lett.* **223** 329

[8] Volodin A A, Fursikov P V, Kasumov Y A et al. 2006 *Russ Chem Bull* **55** 1425

[9] Ferrari A C 2007 *Solid State Commun.* **143** 47

[10] Naumkin A V, Kraut-Vass A, Gaarenstroom S W and Powell C J 2012 *NIST Standard Reference Database 20*, version 4.1

[11] Che G, Lakshmi B B, Martin C R and Fisher E R 1998 *Chem. Mater.* **10** 260

IOP Conf. Series: Journal of Physics: Conf. Series **1238** (2019) 012047 doi:10.1088/1742-6596/1238/1/012047

Interaction of corona discharge with surface layer of metal

D L Kirko*, A S Savjolov

National Research Nuclear University MEPhI (Moscow Engineering Physics Institute), Kashirskoe shosse 31, 115409 Moscow, Russia

E-mail: *dlkirko@mephi.ru

Abstract. Properties of the corona discharge containing a conic electrode were researched. The main modes and characteristics of this discharge were researched. There are current oscillations of the discharge within the range of 1 kHz - 150 MHz. Possible oscillation and wave processes are considered to explain these oscillations. Emergence of rotation motion when the discharge affects the dielectrics is being studied. The effect of this discharge on metal surface was explored.

1. Introduction

The corona discharges normally have weak currents within the microampere units' range at voltages on the order of tens of kilovolts, and have quite subtle glowing. One of the unusual properties of the corona discharge is appearance of high-voltage oscillations within the kilohertz range first identified in work [1]. Oscillation processes with frequencies up to hundreds of megahertz display during the works on the study of the corona discharges [2,3]. When the current is increased up to 100 μA, the discharge acquires new properties that are not inherent to the typical corona discharge. The corona discharge in the system of conic and flat electrodes was studied in this work. When the discharge is burning, there are areas that are in immediate contact with the metal surface. The possibility of using corona discharges for modification of surfaces of various materials was assessed in the works [4]. The rotation properties of the dielectrics were studied the corona discharge environment.

2. Experimental setup

The experimental unit for research of the corona discharge is reviewed in figure 1. The conic electrode (diameter - 2-6 mm, cone angle - α=15^0-25^0) was made of steel or copper, and the cylindrical electrode was made of aluminum or steel (diameter - 35-45 mm, thickness - 10-15 mm). The electrodes were positioned at the distance of 0.5-45 mm. A high-voltage power source with 1-27 kV voltage and up to 2 mA current was used. The discharge had more stable combustion when positive polarity was applied to the conic electrode. The discharge voltage was measured using a static kilovoltmeter, and the current in the circuit was measured with a microammeter. A Nikon 1 G camera (time resolution is 1 ms) was used for recording of the discharge image. The corona discharge images for the two main discharge modes are provided in figure 2. The glow normally has dark light blue or violet color, and the intensity of the glow is quite feeble. Within the current range of I=1-40 μA, this corona discharge has a practically cylindrical shape (figure 2a) with a diameter of 1-2 mm and a length equal to the inter-electrode space of 0.5-5 mm. At great currents within the area of I=60-150 μA, the discharge

IOP Conf. Series: Journal of Physics: Conf. Series **1238** (2019) 012047 doi:10.1088/1742-6596/1238/1/012047

may have a branched shape (figure 2b). These corona discharge modes have various electrical characteristics.

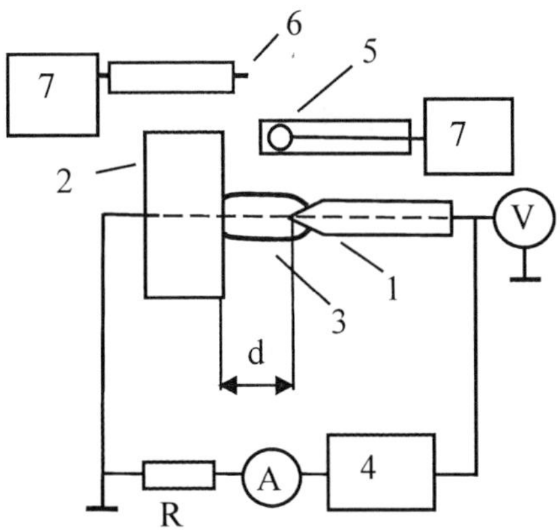

Figure 1. Experimental setup: 1 - anode, 2 - cathode, 3 - corona discharge, 4 - power source, 5 - magnetic probe, 6 - electrostatic probe, 7 - frequency analyzer

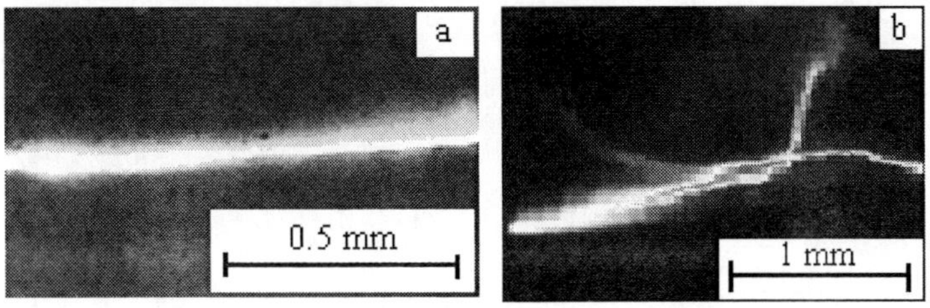

Figure 2. Photo of corona discharge (exposure time 20 ms)

The dependence of voltage U on distance between the electrodes d within the value range of U=1.5-3.5 kV and d=0.5-5 mm is approximately linear. In this case, the current of the discharge falls within the range of I=1-40 μA. The volt-ampere characteristics of the discharge in this mode is described through a quadratic dependence [5]: $I = (8\pi\varepsilon_0 \mu U(U - U_K))/(d^2 \ln(d/r_0))$. The value of U_K is found from the formulas: $U_K \approx E_K r_0$, $E_K \approx 31\delta(1+0{,}308/(\delta r_0))^{1/2}$ kV/cm ($\delta = p/p_n$, p-pressure, p_n-pressure in normal conditions) and equal to U_K=1.9 kV at the anode curvature radius of r_0=0.05 mm.

As voltage increases in the discharge gap within the range of the values U=4-20 kV and if the distance between the electrodes is d=5-40 mm, the amperage of the discharge rises up to the values within the range of I=70-150 μA. The form of the volt-ampere characteristics in this range is approximately linear. The structure of the discharge in this mode acquires a branched form consisting of two or more thin channels. These spark channels come out of the anode tip and branch across the cathode area of about 10-15 mm in diameter. The combustion of the discharge may acquire unstable behavior, and then the current experiences pulsing at a frequency of 0.5-1 Hz.

3. Research of electrical oscillations

In order to study the oscillation processes, electrical and magnetic probes, signals from which were delivered to Tektronix TDS 2024V oscilloscope running in the frequency analyzer mode, were used in the work. Electric probes were metal rods 5-10 cm long and 1-5 mm in diameter. Coils (3-5 mm in diameter, the number of turns is 150-250, wire thickness is 0.1 mm) placed in protective dielectric cases were used as magnetic probes. Electric probes were placed at the distance of 0.3-1.0 m from the discharge. The magnetic probes were positioned both in immediate proximity to the discharge, 0.2-0.5 m away from the discharge. According to these measurements, the current in the discharge circuit is a sequence of pulses following at frequency of ν=1-20 kHz. In order to get the measured pulse frequency distribution, the frequency analyzer mode of the used oscilloscope was utilized. The following frequencies were the main ones: 4.8±0.2 kHz, 39±2 kHz, 87±4 kHz, 228±11 kHz, 3.5±0.2 MHz, 12±0.6 MHz, 26±1 MHz, 32±2 MHz, 63±3 MHz, 85±5 MHz, 91±5 MHz. The spectrum of electric oscillations of the corona discharge is presented in figure 3a. An separate pulse has a peek form with the leading front having the duration of 10-20 μs and a decay on the trailing edge having a duration of 150-350 μs (figure 3b). An oscillation process with a higher frequency of ν=100 kHz - 200 MHz emerges in the background of the pulse.

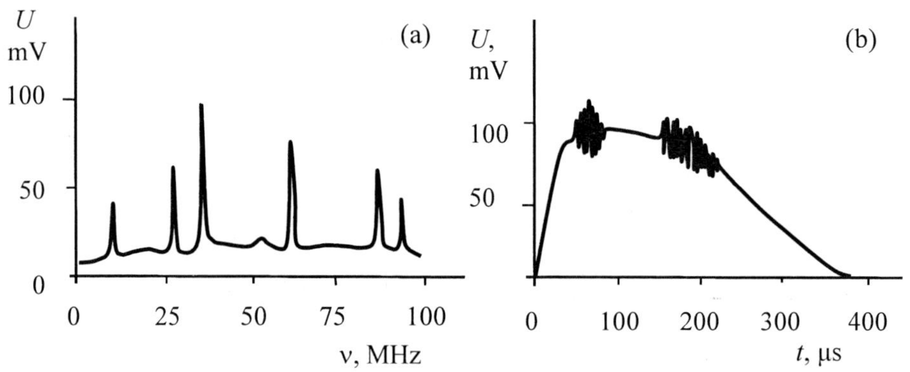

Figure 3. Electric oscillations of the corona discharge: a) the spectrum of current oscillations, b) the characteristic time dependence of the current pulse

For the values of plasma concentration in the corona discharge n_e=10^4-10^{10} cm^{-3}, the range of plasma frequency values $\omega_p=(4\pi n_e e^2/m_e)^{1/2}$ will be in the following range: ω_p=5,5·10^6–5,5·10^9 s^{-1}. According to the conducted measurements, the lower concentrations of values of electronic concentration n_e=10^4-10^6 cm^{-3} exist near the flat surface of the electronic cathode [2]. Higher concentration values n_e=10^7-10^{10} cm^{-3} are present in the area that surrounds the tip of the conic anode where electron avalanche start. In view of this, the originating oscillations of this discharge within the range of ν=1-100 MHz can be associated with excitation of a high-frequency plasma wave in the environment of the corona discharge [6].

Great interest has been currently shown to materials having new reflective properties in the optic and high-frequency ranges of electromagnetic radiation [7]. These characteristics are connected with consideration of effects taking place in thin micron films on the surface of the solid. Due to the dispersion law for electromagnetic waves plasmon waves can exist in the metal at frequencies that a lower than the plasma frequency $\omega_{p0}=(4\pi n_e e^2/m_e)^{1/2}$ (n_e – concentration of electrons in metal). These waves can spread in thin films with a thickness of 1-50 μm. According to these ideas, excitation of the plasmon waves with the frequencies in the area of ν=40 kHz-250 MHz is possible in the thin superficial layer of the conic anode in case of the corona discharge. Lower frequencies within the range ν=1 kHz -150 kHz can correspond to the sound plasmons area.

4. Investigation of the surface of electrodes

The maximum density of current in the construction of this corona discharge is achieved near the tip of the conic anode and on the flat surface of the cathode contacting with the discharge. Gray deposits are formed on the anode surface (material: steel, copper) near the tip during the work of the corona discharge. The surface of the electrodes was researched using Hitachi TM1000 electronic microscope. The image of the steel anode surface at a distance of about 0.2 mm from the tip is presented in figure 4. The effect of the corona discharge is expressed in appearance of hollows and stripe-like structures on the surface. The hollows have a medium size of 0.5-1.0 μm and an irregular elongated shape with the average quantity of 15-20 in the area of 100 μm^2. The stripes are located on the surface in the direction that is predominantly perpendicular to the generator of the cone and have thickness of 0.2-0.4 μm and length of 5-10 μm. Due to the hollows on the surface of the anode, one can assume that current channels with a size of about 1 μm form near the surface of metal in this corona discharge. Density of current in the channels increases due to compressing (pinching) of the current and the metal evaporates in interaction with the surface. As a result, the corona discharge near the anode surface can consist of numerous separate current channels.

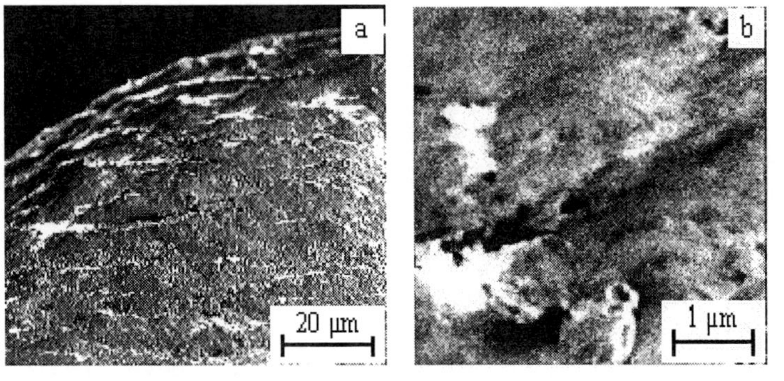

Figure 4. Image of anode surface: a) view of surface near the anode tip, b) micropores on the anode surface

5. Research of the rotation of dielectric materials

Rotation of dielectrics in permanent electric field was first discovered in works [8,9]. Rotation of one of the electrodes or the dielectric cylinder improves the spatial uniformity of the corona discharge [4]. In this work the study was carried out using a dielectric cylindrical rotor. A unit with a rotor made of organic glass (Plexiglas) with a diameter of 4-6 cm, height of 2-4 cm, weight of 50-90 g was used to study the rotation motion. The axis of the rotor was made of steel (diameter 3 mm) and installed on friction bearings (the material is brass), which were fixed in the dielectric case. The electrodes: anode and cathode were made of wire 2-3 mm in diameter (material: copper, steel) and located at a distance of 1-2 mm from the rotor surface at a tangent to the surface. When high voltage is applied to the electrodes, starting from 10-12 kV, the rotor starts to rotate in the direction of the electrode installation.

Due to high voltage, a breakdown (voltage of breakdown is $E \approx 2$ kV/cm) may develop on the surface of the organic glass. The generated discharge in closest to the corona discharge in its parameters. The volt-ampere characteristics of this discharge represent a growing dependency. Current and voltage measurement errors are: $\Delta I = \pm 1$ μA, $\Delta U = \pm 0.1$ kV. The dependence of rotor rotation frequency on voltage is shown in figure 5. The discharge starts at the minimal current of I=1-2 μA. The discharge current is practically unchanged before the rotation frequency of $\nu \approx 20$ Hz. The rotation frequency measurement error is equal to $\Delta\nu = \pm 0.2$ Hz. Then, amperage suddenly changes for voltages U=15.5-17.5 kV. Dependence of the rotation frequency on voltage has constant sections. The direction of the rotor rotation in these experiments was chosen to be counter-clockwise. The power characteristics occurring during rotation were received as the rotor slowed down. The maximum value of the torque is as follows: $M_{max}≈9.2·10^{-5}$ N·m.

Rotation of the dielectric creates a near-surface layer of air making circular motions and drawing the corona discharge. The corona discharge area in this case increases and its spatial uniformity increases.

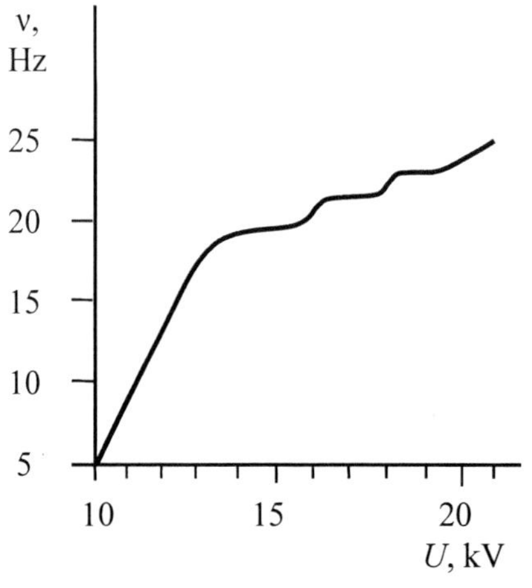

Figure 5. Dependence of the rotation frequency of the rotor on voltage

6. Conclusion

Properties of the corona discharge using a conic electrode were researched in the work. The discharge is characterized by two modes depending on amperage. The first mode at lower current is closer to the classic corona discharge. This discharge, which we have studied, is characterized by occurrence of electric oscillations within the range of 4 kHz-110 MHz. The possibility of plasma waves occurring in the near-electrode plasma with a thickness of 1 mm around the cathode was considered. Along with that, the plasmon waves, specific for the solid, may appear in the 1-50 μm thin superficial layer of the cathode metal. Microstructures form on the surface of the cathode as a result of the corona discharge effect. This can be connected with formation of discharge microchannels. Discrete values of dielectric rotation speed were found when testing the rotation properties of the dielectrics in presence of the corona discharge.

References

[1] Trichel G W 1938 *Phys. Rev.* **54** 1078-1084
[2] Nikitin A I, Leipunsky I O, Nikitina T F 2010 *Prikladnaya Fizika* **2** 15-22 (in Russian)
[3] Lavrinenko M, Biktashev E, Kirko D 2016 *Journal of Physics: Conf. Series* **666** 012031
[4] Zhuravlev A A, Ivchenko V A, Bakulin I A and Strelnikov A Yu 2010 *Bar'ernaja korona postojannogo toka: process formirovanija i primery primenenija* (Samara: Izdatel'stvo Samarskogo Gosudarstvennogo Ajerokosmicheskogo Universiteta) (in Russian)
[5] Raizer Yu P 2009 *Physics of gas discharge* (Dolgoprudny: Intellect) (in Russian)
[6] Kroll N A and Trivelpiece A W 1973 *Principles of plasma physics* (New York: McGrawHill)
[7] Klimov V V 2010 *Nanoplasmonica* (Moscow: Fizmatlit) (in Russian)
[8] Hertz H 1881 Annalen der Physik **13** 266-275
[9] Quincke G 1896 Annalen der Physik **59** 417-486

Application of innovative materials with stochastic nano- and microstructure of the surface: control of turbulent flows in plasma and aerodynamics

Budaev V.P. [1,2,3]

[1] National Research University "MPEI", Moscow, Russia
[2] NRC Kurchatov Institute, Moscow, Russia
[3] Institute of Space Research, Russian Academy of Sciences, Moscow, Russia
E-mail: budaev@mail.ru

Abstract. It is proposed to control plasma turbulence near the LCMS of a tokamak with a system of electrodes made from porous tungsten. We propose to apply the RF modulation at the ion-cyclotron resonance frequency for driving of fluctuations near the LCMS. Porous tungsten plates used as biasing electrodes are the advantage to increase the emissivity of the porous surface and to reduce a surface erosion. To reduce the drag and thermal load on streamlined surfaces at supersonic and hypersonic speeds, it is proposed to cover the aircraft with materials of high porosity. The first experiments were carried by using model with fractal surfaces having non-Gaussian height statistics from ~ 500 nanometers to ~ 200 micrometers obtained by plasma treatment. The advantage of such a surface is the coincidence of the spectral and statistical characteristics of the stochastic topography of the surface with the turbulence characteristics of the flow.

1. Introduction

In recent years, the effects of extreme thermal plasma loads on the material have revealed universal effects of stochastic clustering of a surface with hierarchical granularity and statistical self-similarity - fractality [1-4]. Such a structure of the solid surface is formed in fusion devices after the plasma testing of materials when multiple mechanisms of erosion and redeposition of eroded material, melting and melt movement of surface layers effect simultaneously at scales from tens of nanometers up to hundreds of micrometers. The properties of the unique scale invariance (self-similarity) of the structure of such highly porous materials (see the examples in Fig. 1-3) [1-4] produce hierarchical granularity and the formation of a percolation cluster of defects and dissipative structures over a wide range of scales, starting from submicron scales. Refractory metals with a high-porosity nanostructured surface are in demand for operation under extreme thermal and plasma-beam loads [1-4], in a fusion reactor, to cover the streamlined surfaces of aircraft in order to reduce aerodynamic drag at supersonic and hypersonic speeds, in biomedical applications.

2. Porous tungsten for controlling the plasma sheath layer and the edge plasma turbulence in a tokamak

Edge plasma turbulence near the last closed magnetic surface (LCMS) in a tokamak drives enhanced plasma losses. Plasma flow along magnetic field leads to the narrow width layer (which is of several ion gyro-radiuses) onto the divertor plates. In the ITER, this width is only a few millimetres; extremely high loads are expected in this zone leading to melting and destruction of the divertor plates. To achieve a steady-state operation of a fusion thermonuclear reactor, it is required to develop

IOP Conf. Series: Journal of Physics: Conf. Series **1238** (2019) 012048 doi:10.1088/1742-6596/1238/1/012048

effective control of the plasma sheath layer and the edge plasma turbulence in a tokamak with the aim to extend the radial scale of the high-heat load layer.

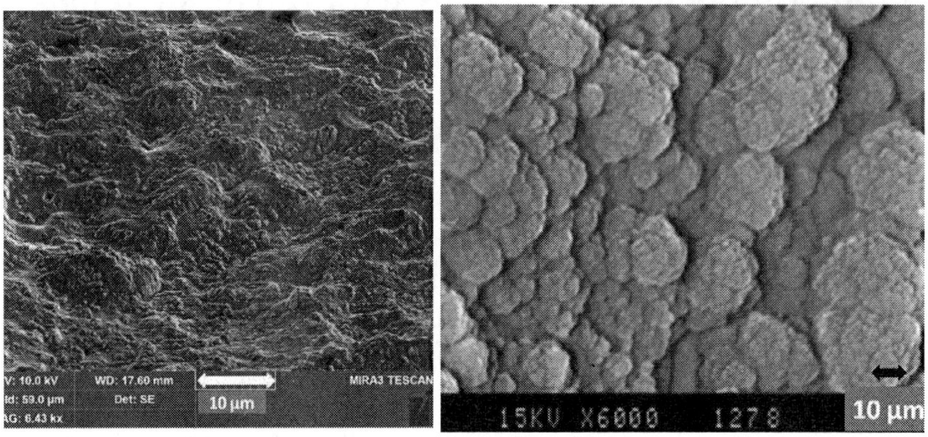

(a) (b)

Figure 1. SEM micrographs of tungsten surface after plasma irradiation: (a) in the T-10 tokamak, (b) in the QSPA-T facility.

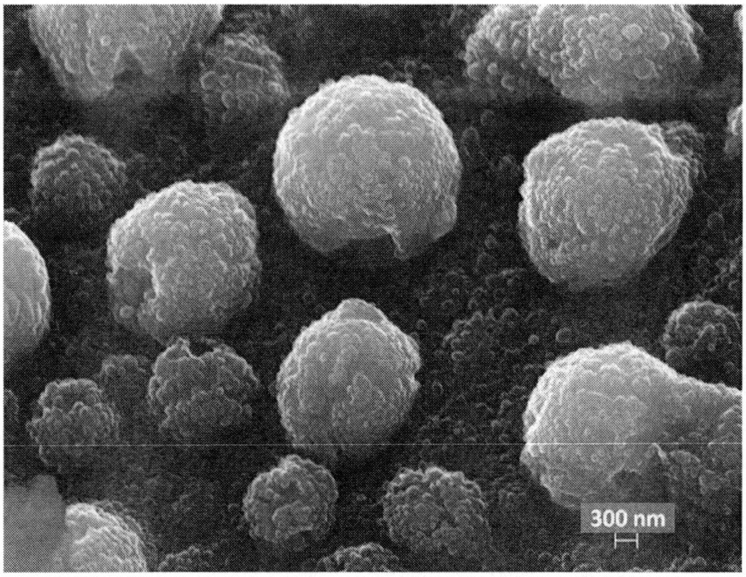

Figure 2. SEM micrograph of beryllium surface after plasma irradiation in the QSPA-Be facility.

Figure 3. SEM micrographs of lithium deposits on the first-wall components in the T-10 tokamak after Li – limiter operation

In this paper, we propose a controlling of plasma turbulence [5,6] near the LCMS with a system of electrodes made from tungsten with stochastic surface and porous structure tungsten, Fig.4. The electrodes biasing forces a modulation of plasma instabilities by a longitudinal electric field. In such a scheme, a driving of ion-acoustic instability is formed under the effect of a longitudinal electric field as considered by the theory of V.P. Silin [8, 9] who predicted the growth of the plasma temperature T with an electric field E as $T \sim E^2$ under developing of ion-acoustic turbulence. The scheme of the experiment proposed for the T-11M tokamak is shown in Fig. 4b. The proposed radial width of the electrodes is of 15 mm; the poloidal width is of 10 cm; the toroidal width is of ~5 cm. The DC biasing voltage Udc between electrodes is proposed to be in the range from -200V to +200V. We propose to apply the RF modulation by U_{RF} voltage amplitude at the ion-cyclotron resonance frequency for driving of fluctuations near the LCMS. In such a scheme, it is expected that the SOL plasma turbulence and related cross-field turbulent plasma transport will be affected locally in the vicinity of the LCMS. This scheme is improved previous biasing schemes used earlier, namely: the biasing voltage is applied between electrodes along the magnetic field with no radial current; in addition, the RF modulation of turbulence and a local change in the cross-field plasma flux near the LCMS are expected. Porous tungsten plates used as biasing electrodes have an additional advantage of increasing the emissivity of the porous surface and reducing surface erosion. A self-sustaining process of maintaining tungsten porosity at the long-term operation in a tokamak fusion reactor is expected [1-4].

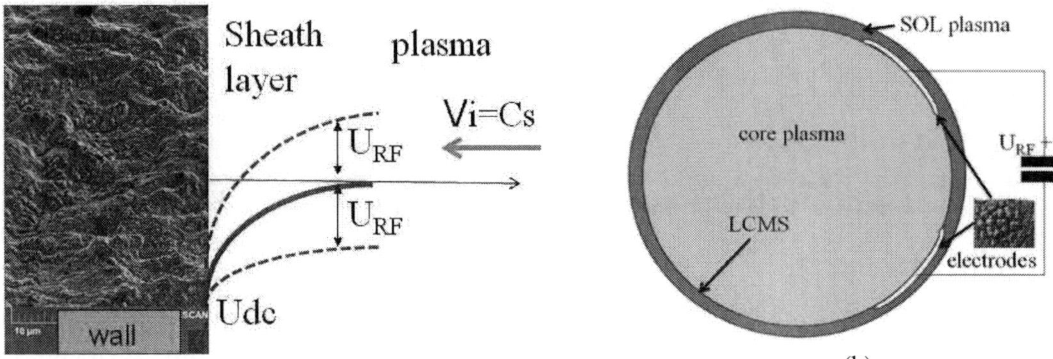

(a) (b)

Figure 4. (a) The RF modulation of the plasma potential in the SOL using the biasing electrodes made from tungsten with stochastic surface structure. (b) Schematic of biasing electrodes for plasma control near the LCMS in a tokamak with a circular cross section.

3. Control of the turbulent boundary layer of aerodynamic flows

To reduce the drag and thermal load on streamlined surfaces at supersonic and hypersonic speeds, it is required to cover the aircraft with materials of high porosity [10,11]. In the wind tunnel [11], the first experiments (Fig. 5b) were carried by using model (Fig.5a) with fractal surfaces having non-Gaussian height statistics (Fig.6a) from ~ 500 nanometers to ~ 200 micrometers obtained by plasma treatment in QSPA-T facility. The advantage of such a surface is the coincidence of the spectral and statistical characteristics of the stochastic topography of the surface with the turbulence characteristics of the flow, Fig.6b. (it relates with turbulent eddy turn-over time scale of ~ 10^{-4} seconds estimated as scale of velocity fluctuations ~1 m/s over roughness scale of 0.1 mm).A significant attenuation of the low-frequency spectral range and a change in the drag force were observed in experiments with a fractal surface plate [10] as a boundary of a turbulent flow. In a wide range of Reynolds numbers Re, it was registered a decrease in aerodynamic drag over the fractal plate compared to the abrasive surface (with Gaussian height statistics) of the same roughness (Fig.7a). For a fractal surface, the scaling index ν of the drag coefficient $c_x \sim Re^{-\nu}$ is close to ν for a smooth plate (made from glass) [11] (Fig.7b) . This demonstrates the potential of a fractal surface to reduce aerodynamic drag at supersonic and hypersonic speeds.

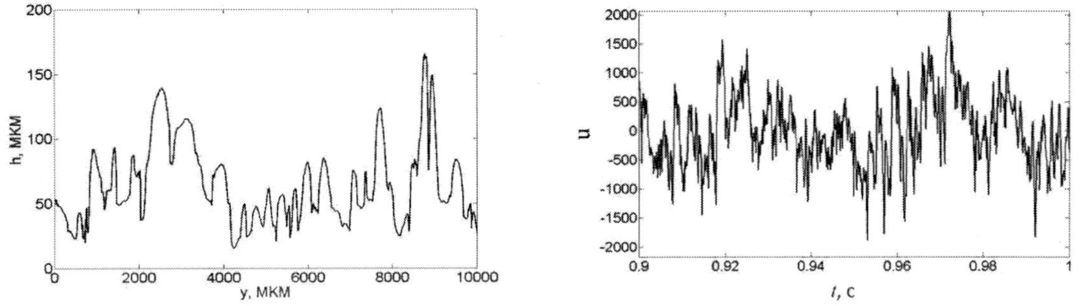

(a) (b)

Figure 5. (a) Fractal surface profile formed in fusion device. (b) Velocity fluctuations in the turbulent boundary layer in the wind tunnel experiments.

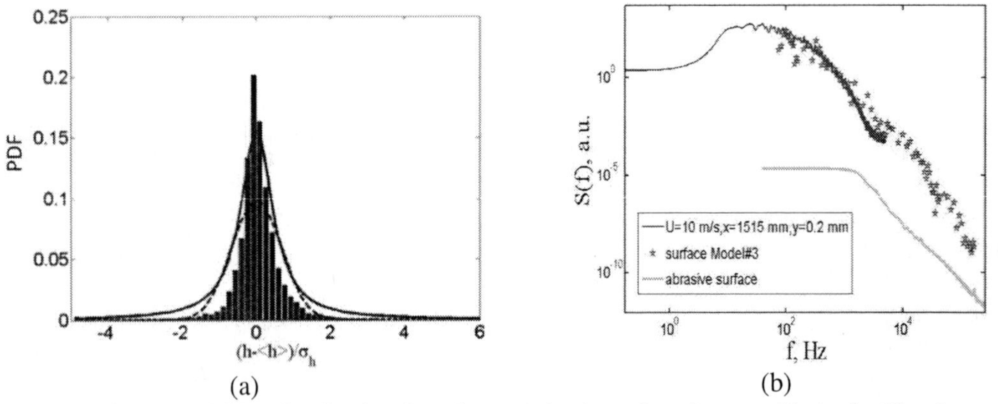

(a) (b)

Figure 6. Probability distribution functions of the fractal surface profile in the Fig. 5a. Gaussian law (dashed line) and Cauchy law (solid line) for the reference. (b) Spectra of velocity fluctuations in the turbulent boundary layer and surface heights fluctuations, experiments in subsonic flows.

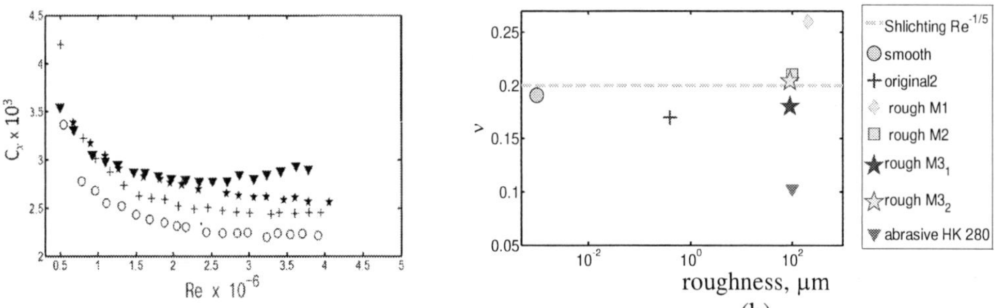

(a) (b)

Figure 7. (a) Drag reduction c_x vs. Reynolds numbers Re; smooth plate (glass) – circles, virgin surface - crosses, fractal surface – black stars , abrasive surface – black triangles. (b) The scaling index ν of the drag coefficient $c_x \sim Re^{-\nu}$ for a smooth and rough plates vs. roughness.

Conclusions

Stochastic clustering of a surface with hierarchical granularity and statistical self-similarity (fractality) is observed in fusion devices after the plasma treatment in fusion device when multiple mechanisms of erosion and redeposition of eroded material, melting and melt movement of surface layers effect simultaneously at scales from tens of nanometers up to hundreds of micrometers. It is proposed to control plasma turbulence near the LCMS of a tokamak with a system of electrodes made from porous tungsten. We propose to apply the RF modulation at the ion-cyclotron resonance frequency for driving of fluctuations near the LCMS. Porous tungsten plates used as biasing electrodes are the advantage to increase the emissivity of the porous surface and to reduce surface erosion. To reduce the drag and thermal load on streamlined surfaces at supersonic and hypersonic speeds, it is proposed to cover the aircraft with materials of high porosity. The first experiments were carried by using model with fractal surfaces having non-Gaussian height statistics from ~ 500 nanometers to ~ 200 micrometers obtained by plasma treatment. The advantage of such a surface is the coincidence of the spectral and statistical characteristics of the stochastic topography of the surface with the turbulence characteristics of the flow. In a wide range of Reynolds numbers, it was registered a decrease in aerodynamic drag over the fractal plate compared to the abrasive surface (with Gaussian height statistics) of the same roughness.

References

[1] Budaev V P Physics of Atomic Nuclei 79(7) 1137-1162 (2016)
[2] Budaev V P JETP Letters 105, 5 (2017)
[3] Budaev V.P., Physics Letters A, 381, 43, 3706 (2017).
[4] Budaev V P, et. al. JETP Letters 95(2) 78-84 (2012)
[5] Budaev V. P., Savin S. P., and Zelenyi L. M. Phys. Usp. **54** 875 (2011)
[6] Budaev V. P., Zelenyi L.M., Savin S.P. J.Plasma Physics 81 6 395810602 10 (2015)
[7] Budaev V P, et. al. Plasma Physics Reports 39(11) 910-924 (2013)
[8] Silin V.P., Plasma Physics Reports, 37, 739 (2011)
[9] Silin V.P. e a, Bulletin of the Lebedev Physics Institute , 43, 4, 132 (2016)
[10] Budaev V P, et. al. 2007 Physica A: Statistical Mechanics and its Applications 382 359-377
[11] Brutyan M. A., Budaev V. P., Volkov A. V., et al , Uch. Zap. TsAGI **44** (4) 15 (2013)

Acknowledgments

The work was supported by the grant of the RSF No. 16-19-10531.

Portable neutron generator with microwave ion source for wide applications

M A Gorbunov, D S Stepanov and E Y Shkolnikov

National Research Nuclear University MEPhI (Moscow Engineering Physics Institute), Kashirskoe shosse 31, 115409 Moscow, Russia

DSStepanov@mephi.ru

Abstract. The paper presents the issues related to the development of a portable neutron generator with a microwave source of heavy hydrogen ions for a wide range of applications. The paper analyzes the effect of electromagnetic microwave field strength and the magnetic induction value of a constant external magnetic field on the ionization rate of molecular deuterium and determines their optimal ratios. It also describes the configuration of the magnetic system of permanent magnets made of NdFeB material, which provide a homogeneous magnetic field with an induction of 850 Gs necessary for the effective operation of the device using the electron cyclotron resonance (ECR). The dynamics of the primary deuteron flux, as well as secondary particles in the ion-optical system are analyzed. Elements of a conceptual design of a portable neutron generator are described.

1. Introduction

At present, there is a strong tendency to expand the use of portable neutron generators and create new promising technologies based on them. At the same time, the requirements to the technical and operational characteristics of such generators are substantially being increased, that essentially means the development of their new generation.

A special place in experimental development is occupied by neutron generators with a microwave ion source using the ECR. The absence of electrodes in the microwave discharge makes it possible to avoid contamination on insulators and electrodes in the ion extraction and acceleration system. The effect of ECR significantly increases the energy deposition into the plasma, which leads to a high degree of ionization and energy efficiency, and also to the formation of a high fraction of monoatomic ions. Experimental samples of portable neutron generators show high values of the operational life of ~ 5000 hours and neutron fluxes from 10^{10} n/s with the D $(d,n)^3$ He reaction and up to 10^{12} n/s with the T $(d,n)^4$He reaction [1].

2. Microwave resonator

The basis of a neutron generator is a microwave ion source, which is a resonator filled with molecular deuterium. The resonator itself has a prismatic shape and is powered by a microwave generator with a frequency of 2.45 GHz in the H_{101} mode. When microwave power enters the resonator, the "background" electrons in it acquire the energy necessary for ionization of molecular deuterium. However, without using additional effects that contribute to the absorption of microwave energy by electrons, tens of kilowatts of power are required to ignite such a discharge, that makes it impossible to build the portable design of such a generator. To increase the efficiency of microwave power absorption, the ECR effect is used [2]. The effect of the magnitude of the magnetic induction on the efficiency of the absorption of the microwave field energy by electrons is given in [3]. To determine the requirements for the magnetic system that forms the distribution of the magnetic field, a detailed analysis of the ionization processes in the microwave plasma is carried out below.

To analyze the concentration dynamics and temperature of plasma components of a microwave discharge in the resonator, a kinetic scheme of nonequilibrium processes in deuterium was used, as well as the energy balance equations that take into account the heating of electrons by the microwave field, as well as the transfer of energy from electrons to heavy particles. This scheme takes into account the following reactions:

$$D_2 + e \rightarrow \begin{cases} e + D(G) + D(G) \\ e + D(G) + D(2s) \\ e + D(G) + D(2p) \\ e + D(G) + D(3l) \end{cases}$$

$$D + e \rightarrow e + e + D^+$$

$$D_2 + e \rightarrow e + e + D_2^+$$

$$D_2 + e \rightarrow e + e + D + D^+$$

$$D_2^+ + e \rightarrow e + D(G) + D^+$$

$$D_2^+ + e \rightarrow e + e + D^+ + D^+$$

$$D_2^+ + e \rightarrow D_2$$

The solution of the corresponding set of equations makes it possible to relate the electric field strength in the resonator and the magnitude of the magnetic field with the plasma parameters. Thus, for example, the analysis of these solutions for the electric microwave field strength equal to $2{\cdot}10^4$ V/m (minimum value at which the discharge can evolve) showed (see figure 1) that the values of the magnetic field induction of about 850 Gs correspond to the most optimal mode of operation. At the same time, electrons energy losses in inelastic collisions limit the growth of their temperature at a level of 30 eV. Exceeding this value will involve excessive heating of the electronic plasma component and a decrease in the ionization rate.

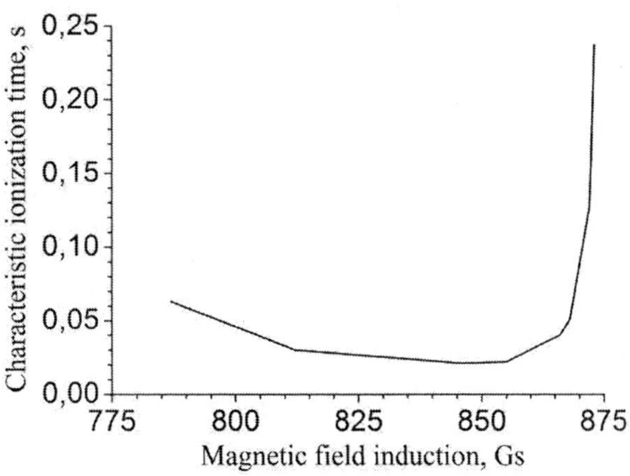

Figure 1. The dependence of the characteristic ionization time on the magnitude of the magnetic induction

To create the required magnetic field distribution, a magnetic system consisting of 8 plates of the permanent NdFeB magnets and two adjusting iron plates was developed (figure 2 (a)). Such an assembly provides a fairly uniform distribution of the magnitude of the magnetic induction B ~ 850 Gs in most of the resonator volume (figure 2 (b)). The distribution of the amplitude of the electric field in the H_{101} mode has a sinusoidal nature, so if the region of strong electric fields and weak cyclotron resonance (the reverse situation is also possible) is chosen as a mode of operation of the resonator, then there is no particular reason to maintain a uniform magnetic field in the entire volume of the resonator, as this will require increasing the size of the magnetic system itself. Under these conditions, it is sufficient to use the region of the maximum of the electric field lying in the range from 20 to 60 mm along the resonator.

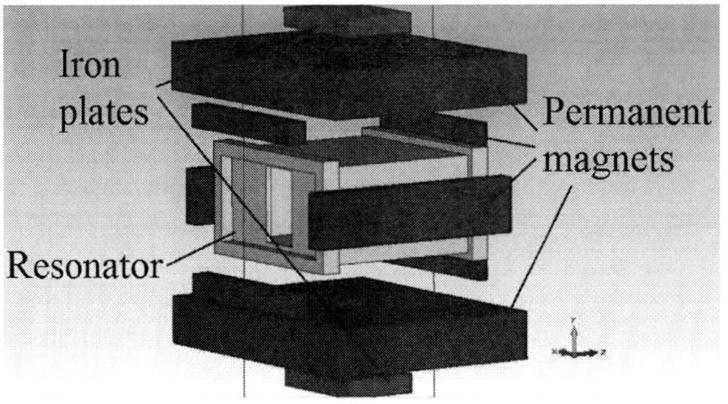

(a)

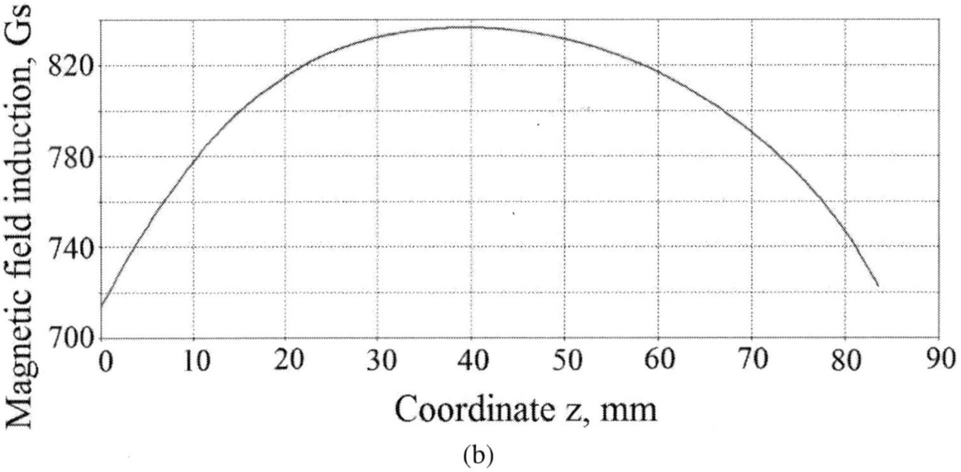

(b)

Figure 2. The magnetic system of the microwave resonator: (a) - the relative location of the magnets, (b) the distribution of the magnetic field induction along the axis of the resonator

The residual induction of permanent magnets varies from 0.8 kGs for correcting magnets and up to 1.4 kGs for main magnets. The magnetic field is adjusted by vertical displacement of the ferromagnetic plates. Such a mechanism is necessary for finding the operating conditions of a discharge.

3. Ion-optical system

The process of accelerating and transporting a neutron-forming beam from the ion source to the target is very important for the quality of the operating characteristics of a neutron generator. The dynamics of secondary beams produced in the volume of the ion-optical system (IOS) and in the target assembly is also of considerable importance here. The lack of proper focusing of the ion beam entails sputtering the electrodes of the IOS, which leads, among other things, to the appearance of a conducting layer on the surface of the high-voltage insulator and to further electrical breakdown. In addition, a strong inhomogeneity in the cross section of the beam leads to local overheating of the target and a decrease in the resource of its stable operation. Also, the design of the IOS should include the locking of secondary electrons that arise when the target is bombarded with the ion beam.

To analyze the dynamics of charged particle flows in the IOS, the KARAT package was used [4]. The voltage at the pulling electrode is -45 kV and at the target assembly -100 kV. The injection of the deuterium ion beam occurs with an equipotential of -500 V and located on the boundary of the resonator. The deuteron flux with a current of 30 mA shown in figure 3 does not deposit on the electrodes, and also creates a fairly uniform spot on the target surface (figure 4).

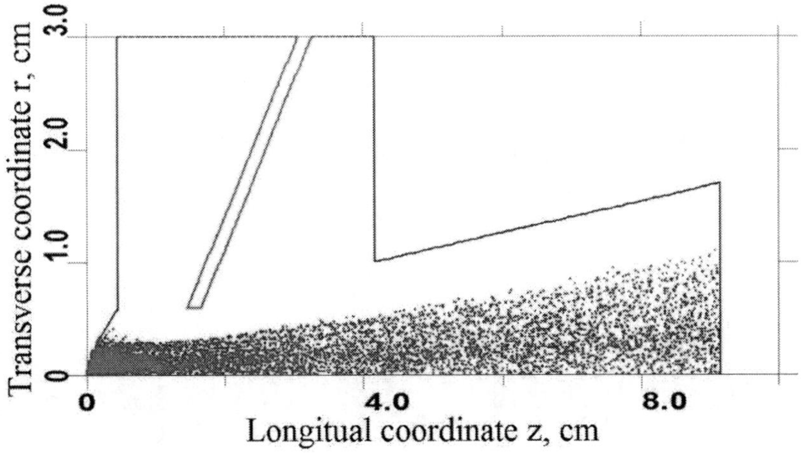

Figure 3. The deuteron flux in the ion-optical system

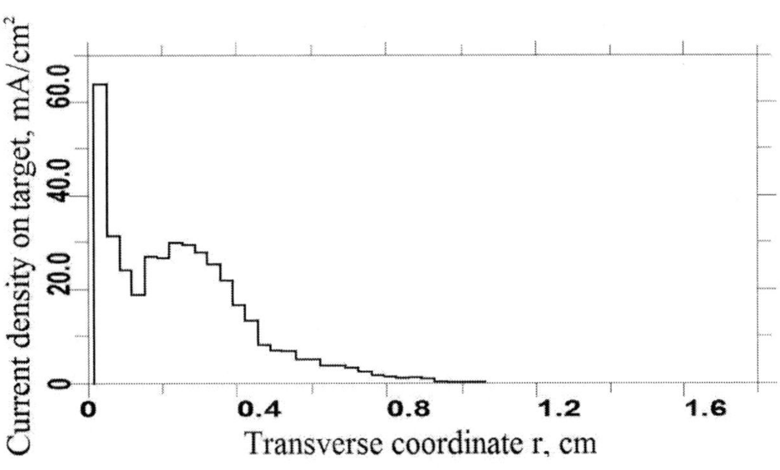

Figure 4. Distribution of the current density on the target

Suppression of secondary electrons knocked out of the target by the ion beam occurs by means of a magnetic field generated by permanent magnets located behind the target assembly. The secondary electrons emerging from the target settle on the inner surface of the target electrode under the action of a magnetic field and under the condition of a weak electric field. The secondary electrons formed as a result of ionization by the deuterium particle beam of the residual gas do not fall under the action of a magnetic field and form a beam with a current equal to 0.4 mA and an average energy of 50 keV at the resonator output port.

4. Design of a neutron generator

The designed model of a neutron generator is shown in figure 5. The prismatic microwave resonator at a frequency of 2.45 GHz is mounted on the supporting flange. Microwave power enters the resonator through a rectangular waveguide with a special vacuum microwave window. In the waveguide section, vacuum fixtures are mounted, which also ensure the flow of molecular deuterium. The ECR is provided by a system of permanent magnets mounted around the resonator together with a system for their adjustment. The system is pumped to a pressure of 10^{-3} Torr through the hole in the supporting flange. The ion-optical system with a target assembly is mounted on the opposite side of the flange. Behind the target assembly, outside the vacuum volume, there is a system of magnets that suppresses secondary electron emission from the surface of the target. The system of water or air cooling is also integrated in the magnetic system. The entire system has an overall dimensions of 300x600 mm.

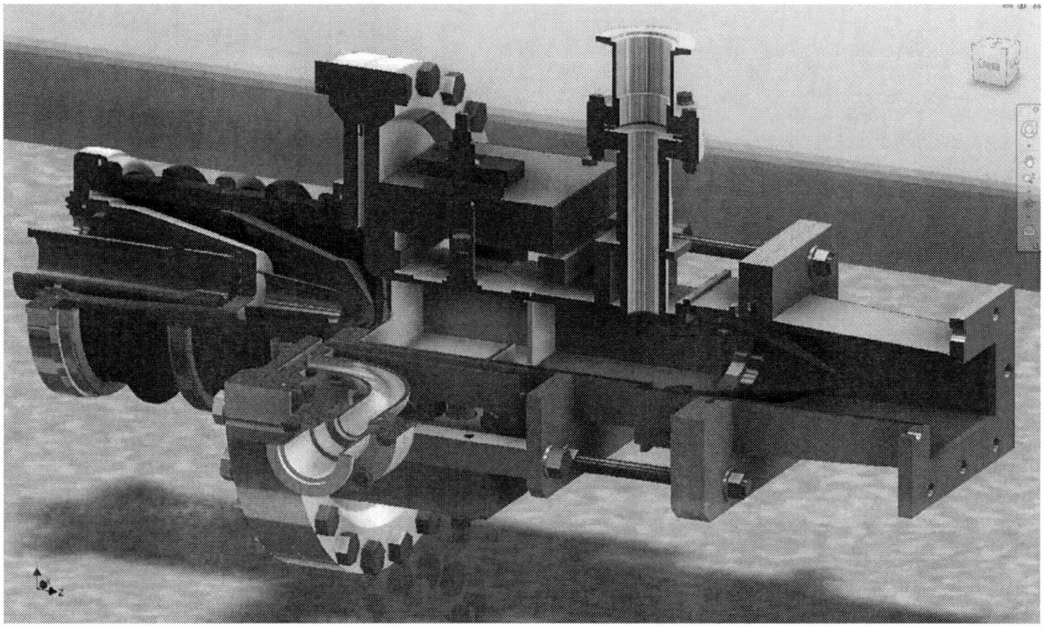

Figure 5. The designed model of a neutron generator

5. Conclusion

The results presented above made it possible to develop a conceptual design documentation for an experimental sample of a portable neutron generator with a microwave ion source in the electron cyclotron resonance mode. The neutron generator has connection ports to all necessary auxiliary devices: microwave path at 2.45 GHz, vacuum system, external source of molecular deuterium as well as air or liquid cooling systems. The design also includes adjusting the resonator to the operating frequency in the conditions of burning a microwave discharge and the variation in the distribution of the magnetic field in order to find the optimal operating mode.

Acknowledgement

The work was supported by the Ministry of Science and Education of the Russian Federation under the agreement № 14.575.21.0169 (RFMEFI57517X0169).

References

[1] Bogdanovich B Yu, Didenko A N, Nesterovich A V et al. Neutron generators based on superhigh-frequency sources of heavy hydrogen nuclides (review). *2015 Insts. and Exper. Tech.* **4** p 5-13
[2] Ginzburg V L, Rukhadze A A 1975 *Waves in magnetoactive plasma* (Moscow: Nauka) p 254
[3] Stepanov D S, Chebotarev A V, Shkolnikov E Ya Electromagnetic Field in the Resonator of Neutron Generator's Hydrogen Ion Source *2016 Proc. Int. Conf. APEDE* **2** 7879076.

[4] Tarakanov V P «User's Manual for Code KARAT», BRA Inc., Va, USA,(1992)

IOP Conf. Series: Journal of Physics: Conf. Series **1238** (2019) 012050 doi:10.1088/1742-6596/1238/1/012050

Effect of low temperature plasma of atmospheric pressure on single-cell model organisms of ciliate *Paramecium caudatum*

A N Inozemtsev[1], O V Karpukhina[1], I V Melikhov[2], V A Riaby[3], V P Savinov[4], V Y Timoshenko[4] and V G Yakunin[4]

[1] Faculty of Biology Lomonosov Moscow State University, Leninskie gori 1/12Moscow, 119234, Russia
[2] Chemical faculty Lomonosov Moscow State University, Leninskie gori 1/3, GSP-1 Moscow, 119991 Russia
[3] Moscow Aviation Institute (National Research University), Volokolamskoe shosse GSP-3, Moscow, 125993, Russia
[4] Faculty of Physics Lomonosov Moscow State University Moscow, Leninskie gori 1/2, GSP-1, Moscow, 119991, Russia

e-mail: karpukhina.msu@yandex.ru

Abstract. We studied low-temperature plasma effects on Paramecium caudatum using the original electric arc plasmatron. The decrease in the number of experimental cells has been shown, which may indicate membrane destructive pathology due to oxidative stress induced by the plasma impact. It is concluded that reactions at the population level of Paramecium caudatum are effective indices of primary information about plasma radiation cytotoxicity.

1. Introduction

The modern state of medicine indicates the need for radical changes in the medical process, because many traditional medicines leave in the body toxic substances, pharmacological preparations have a number of side effects (toxic and allergic reactions, change immunobiological properties of the body, etc.), and then, viruses, pathogenic biological entities exhibit greater adaptability to recently highly effective antibiotics [1, 2]. In the process of chemo- and antibiotic therapy it is necessary to know well antimicrobial activity applied medicines, in addition there is a danger of secondary resistance to the drug used, resulting from spontaneous mutations of viruses and pathogenic microorganisms.

Recently, developing rapidly an extensive interdisciplinary area of plasma medicine, free of the aforementional shortcomings, because of the simultaneous presence of plasma many medical factors leading to a synergistic effect of exposure [3, 4]. Naturally, consideration of this problem it is important to lead on three levels: cell, tissue and organism. Choice of the object of study of this work is due precisely to these considerations.

Choice Paramecium caudatum as on object of this study is due to the following criteria. Thanks to a combination in Paramecium signs of the cell and organism it can be studied as cellular and organismic form of reactions to external influence. Compared to others protozoa groups these infusoria have the most complex structure and are distinguished by the variety and complexity of functions.

It is a great importance of type of plasma source used for creating a technological environment haracterized by many parameters distributions of which in importance to a large extent depends on the specific biomedical impact object.

The source of plasma in this work was electric arc plasmatron of direct current a new type with a high service life and a plasma jet practically free from metal particles of electrodes detailed in the work [5] and references cited therein.

2. Materials and Methods

The basic design components of the plasmatron are water-cooled rod cathode and nozzle anode. To protect the cathode from oxidation into the gap between the cathode and the anode an inert gas is supplied (argon) in the atmosphere of which an arc discharge is excited. Copper anode structure provides distributed nature anodic arc binding by filing in the gas channel of the anode process gas (oxygen, air, freons, etc.). This gas arrives in a circular slot moving to the axis of the plasmatron in radial direction. In the zone of its mixing with the shielding gas of the cathode there is a vortex zone in which is formed fixing arc length distributed anode binding surrounding anode from the outside anode nozzle channel. This feature leads to a sharp decrease in the current density of the anode and reduce the rate of its erosion.

To get information on characteristics technological plasma jet in the functional zone diagnostic methods were used: optical emission spectroscopy, thermal imager and chemical barbater, The following data were obtained [5]: plasma radiation spectrum in the range 200 – 700 nm, presence of chemical radicals NO-, OH-, N^{2-} and their density, gas temperature distribution in the functional zone of the plasmatron.

The study of the low-temperature plasma effect was carried out on the cell culture of *Paramecium caudatum*, which showed effectiveness in determining the role of oxidative stress [6]. In the experiment, test -tubes with ciliates were placed under the plasma jet at a distance of 15 cm. Time of exposure of plasma radiation on the cells was 0.5 min, 1 min, 3 min and 5 min. After the action of the plasma jet, we implemented a microscopy with a video recording of morphological and functional changes in cells from the test samples. 2.5 h and 24 h after irradiation, we determined the number of living cells. Cells that have been outside the plasma impact zone during the equivalent time and non-irradiated intact cells from the vial with stock culture served as a control. Ph measurements were carried out at all stages of the experiment. Selection of culture from one vial and filling of experimental vessel with sample with cells was carried out up to 5 times.

The data wasprocessed statistically with the calculation of mean for the whole sample, mean error, variance using the functions included in the Excel 2013 software package. After checking the distribution for normality, the significance of the differences between the groups was assessed by t-criteria using Statistica 10.0. The differences were considered statistically significant at $p < 0.05$.

3. Results and Discussion

It is established that the action of the plasma jet decreases the number of living cells, the severity of the cytotoxic effect depends on the time of exposure to the plasma flow. The effect of low-temperature plasma for 0.5 and 1 min did not cause a statistically significant decrease in the number of living cells (Fig. 1).

Immediately after irradiation for 3 minutes, the number of living cells decreased by 2 times ($p<0.01$), and after 5 minutes – by 14.5 times compared with intact cells ($p< 0.005$). After 2.5 hours of experience in the tubes under the action of atmospheric plasma 3 min and 5 min, the cell survival was 0 %.

In addition, changes in cell shape have been recorded in all experimental samples. To the greatest extent, the number of atypical cells appeared at 3 and 5 minutes exposure of the plasma jet. Apparently, charged plasma particles being in direct contact with the cell can accumulate on its surface and cause electrostatic potential. This can lead to changes in morphology, changing strength and stretching the cell membrane [7, 8]. In addition to the morphological changes of cells, after exposure to low-temperature plasma, there were disturbance in the mobility of *Paramecium caudatum*. Infusoria were moving slowly, rotational movement differed from the normal. It is known that the motor activity of Paramecia is largely formed on the basis of the ion channels built into the membrane of the motion organs (cilia), and reflects the functional state of the cell. It is known that *Paramecium caudatum* functions to maintain the membrane potential. As a result of reduction of membrane potential, cells move more slowly or rotate in one place around one end.

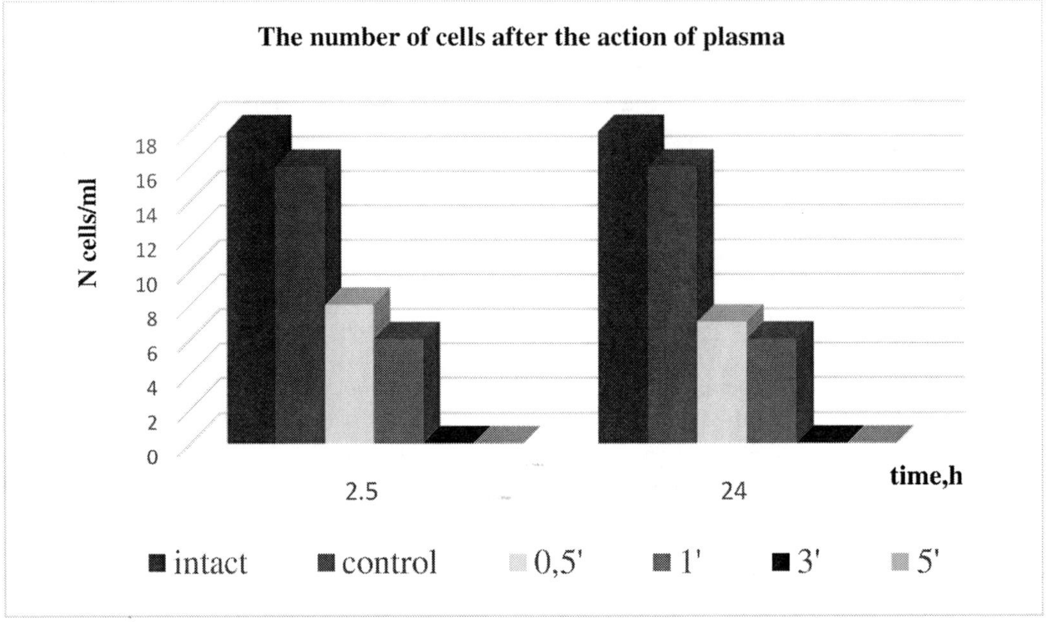

Fig. 1. The effect of plasma on the number of Paramecium caudatum cells through 2.5 h and 24 h (experimental samples exposure time is 0.5 min, 1 min, 3 min, 5 min; the distance of exposure to cells - 15cm from the plasmatron outlet)

The number reduction of *Paramecium caudatum* and morphological changes in cells indicate a destructive membrane pathology, in the occurrence of which the processes of free radical oxidation also play an important role. Free radicals can attack directly polyunsaturated fatty acids in membranes and initiate lipid peroxidation. The main effect of lipid peroxidation is a decrease in membrane fluidity, which significantly changes the functional properties of the cell membrane. It is possible that bioactive radicals NO^-, OH^-, N^{2-} trigger a cascade of chain reactions, which lead to the development of oxidative stress, and as a consequence to the death of cells *Paramecium caudatum*.

To date, the direct mechanisms of action of low-temperature plasma have not been studied in detail, however, the works of Laroussi M. et al., 2003; Mogul R. Et al., 2003; Kalghatgi, et al., 2011 [8-10] indicate changes in the structure of membranes, DNA and enzymatic activity of cells.

4. Conclusions

Identification and analysis of biological effects, which are determined by physical effects at the cellular level and the body as a whole are of great interest. Our studies have shown that the reactions at the population level of *Paramecium caudatum* are informative enough to obtain primary data on the cytotoxicity of plasma radiation, which can be used in the future.

IOP Conf. Series: Journal of Physics: Conf. Series **1238** (2019) 012050 doi:10.1088/1742-6596/1238/1/012050

References

[1] Kong MG, Kroesen G, Morfill G, Nosenko T, Shimizu T, van Dijk J, Zimmermann J.L. 2009 *New J. of Physics* **11**, 115012

[2] Wertheimer MR., Coulombe S. 2012 *La Physique au Canada.* **68:4,** 189

[3] Spellberg B, Bartlett JG, Gilbert DN. 2013 *N Engl J Med* **368:4,** 299

[4] Stoffels E. 2006 *Journal of Physics D: Applied Physics* **39(16),** 3450

[5] Riaby VA, Savinov VP, Yakunin VG et al. 2017 *Med. Biomed. Eng.* **51:3**, 195

[6] Morgunov IG, Karpukhina OV, Kamzolova SV, Samoilenko VA and Inozemtsev AN 2018 *Preparative Biochem &Biotech.* **48:1,** 1

[7] Mendis D Rosenberg M and Azam F 2000 *IEEE Trans. Plasma Sci.* **28**, 1304

[8] Laroussi M. Mendis DA and Rosenberg M 2003 *New J. Phys.* **5**, 41

[9] Mogul R., Bolshakov AA, Chan SL, Stevens RM et al. 2003 *Biotechnol. Prog.* **19**, 776

[10] Kalghatgi S, Kelly CM, Cerchar E, Torabi B, Alekseev O, Fridman A, Friedman G, Azizkhan-Clifford J. 2011 *PLoS One*, **6** (e16270.10.1371/journal.pone.0016270)

Formation of nanoscale structures in thin films deposited by reactive magnetron sputtering of binary metal targets in $Ar+O_2+N_2$ gas mixture

D Bernt[1], V Ponomarenko[1], S Leshchev[2] and A Pisarev[2]

[1] Pilkington Glass, 1 Stekolnaya Str, Zhukovo village Chulkovsky s/o Ramensky District, 140125 Moscow region, Russia

[2] National Research Nuclear University MEPhI (Moscow Engineering Physics Institute), Kashirskoe shosse 31, 115409 Moscow, Russia

E-mail: Dmitriy.Bernt@spglass.ru

Abstract. Bi-metal targets of ZnSn and SiAl were sputtered in $Ar+O_2+N_2$ gas mixtures at various N_2/O_2 ratios, and the deposited films were investigated by AFM and XPS. No nitrogen was detected in the films even at the highest nitrogen partial pressure in the gas, and the films were oxides. The concentration of the doping metal component in the deposited film increases with the increase of nitrogen partial pressure in the working gas. These experimental observations were supported by numerical calculations of the sputter-deposition process using a SRIM/SIMTRA+RSD software. It was shown that added nitrogen promotes sputtering of doping metal components of the targets, and the concentration of oxygen remains high enough to substitute the less active nitrogen in the growing film. The films consisted of nanograins with the average diameter, which decreased from 18 nm to 3 nm with increase of nitrogen content in the working gas. This effect was explained by the mechanism of abnormal grain growth: the film consisted of grains of oxides of two metals with different concentrations. The doping metal forms very fine nanograins of oxides, which suppress the growth of oxides of the main metal. With increase of nitrogen in the working gas mixture, the sputtering rate of doping metals from the cathodes and their respective concentration in the growing films increases, so the average diameter of oxide grains forming the coating decreases.

1. Introduction

In recent years, oxinitride thin films have attracted much attention, because of their versatile properties important for optics, photonics, microelectronics, oleophobic, catalytic, cutting tools, battery cells, and other applications [1-5]. Oxides and nitrides often have opposite behaviors with respect to film properties, therefore oxinitrides sometimes offer a compromise between the extreme behaviors, and controlled introduction of nitrogen to oxides leads to a good control of designed properties [6-13].

Functional properties of oxides, nitrides, and oxinitrides are connected with their structure, which, in turn, depends on many factors. Films can be amorphous like in [10] and crystalline like in [11]. The grain sizes may become less with the increase of the film thickness [12]. Films deposited at room temperature [6, 10, 11] were amorphous but became crystalline at annealing with the lattice parameters dependent on the nitrogen content. Phenomenological study [3-5, 13] of oxinitride films deposited by magnetron sputtering demonstrated dependence of the film properties on O_2/N_2 flow rate in the reaction chamber.

Among the properties of interest for many applications, oleophobicity - which in general states for the surface repelency of viscous liquids - must be mentioned. It was demonstrated in [13, 14] that addition of N_2 to O_2 during magnetron sputtering of binary metal Zn-Sn and Si-Al targets leads to up to 45% increase of the oleophobicity of deposited thin films according to the droplet spread diameter and contact angle measurements tests. Also it was observed that nanoscale grains become smaller due to N_2 addition, and this can result in the increase of the surface oleopobicity [15].

The aim of this work is to study features and mechanisms of formation of such films and elucidate the role that nitrogen plays in deposition of films sputtered from bi-metallic targets in $Ar+O_2+N_2$ gas mixture.

2. Experimental

Deposition of films was performed in an industrial line VonArdenne GC330H for architectural glass coating. A dual rotating cylindrical magnetron system, driven by a TRUMPF Hüttinger TruPlasma power supply, installed in the line chamber was used for sputtering. It operated in pulsed DC mode at the frequency of 34 kHz and the power limit of 70±4 kW. The sputter targets were made of binary ZnSn and SiAl alloys. Substrates for test depositions were made of quartz microscope glass with the thickness of 1mm and dimensions of 75×25 mm. They were cleaned in deionized water and then dried with compressed dry air. No heating was applied towards the substrates during the deposition.

Three series of reactive sputter depositions were performed. The first and the second series were performed with ZnSn target with two different alloying concentrations (10wt% Sn and 50wt% Sn respectively) In the third series SiAl (10wt% Al) sputtering target was used. The minimal purity for the alloys used was 99.7% with the alloy concentration stability of ±2 wt.%. Sputtering was performed in a mixed gas atmosphere $Ar+O_2+N_2$. The gas supply was controlled by MKS mass flow controllers. Oxygen and argon flows were set to operate magnetrons in the transition mode of the cathode poisoning hysteresis, which resulted in O_2 flow of approximately 900 sccm through the whole set of experiments. The total resulting gas pressure in the chamber was maintained at the level of approximately 3.5×10^{-3} mbar. Nitrogen flow varied from 0 to 800 sccm with the step of 100 sccm within every series. Deposition experiments were performed in the so called transition reactive sputtering mode, which is the intermediate operating regime between the metallic and fully poisoned modes on the reactive sputtering hysteresis. The operation regime was controlled through optical emission of exited atoms of the gas and metal in plasma of the magnetron discharge with a feed-back regulation of the gas flow rates.

The thicknesses of the deposited coatings were in the range of 120-135 nm. Their thicknesses were established and then maintained at the given range via indirect control based on the spectrophotometric ex-situ measurement in the range of wavelengths between 350 and 1300 nm. Spectra were obtained in transmission and film-side reflection from the samples.

Elemental analysis of coatings was performed using X-ray Photoelectron Spectrometer Kratos XSAM-800. Exitation was made using the AlKα (1486.6 eV) radiation. Bruker FastScan AFM with a silicon nitride cantilever (silicon tip ~1nm in diameter) operated in peak force taping mode was used to determine the topography of the coatings.

3. Results

AFM analyses demonstrates that deposited films consist of very fine rounded grains (approximately 5-20 nm in diameter) in all experiments (Fig. 1-a). Nitrogen addition in the gas always leads to decrease of the grain size and respective smoothening of the film surface. The same effect was also observed previously [14].

The average grain diameter of the coatings surface as a function of the nitrogen flow demonstrates prominent decrease of the grain diameters at the nitrogen flow rate of about 350-400 sccm, which is approximately a half of the oxygen flow rate used. At higher values of the nitrogen addition, the average grain diameter decreases much slowly remaining at the level of the resolution limit of the AFM tip (Fig. 1-b).

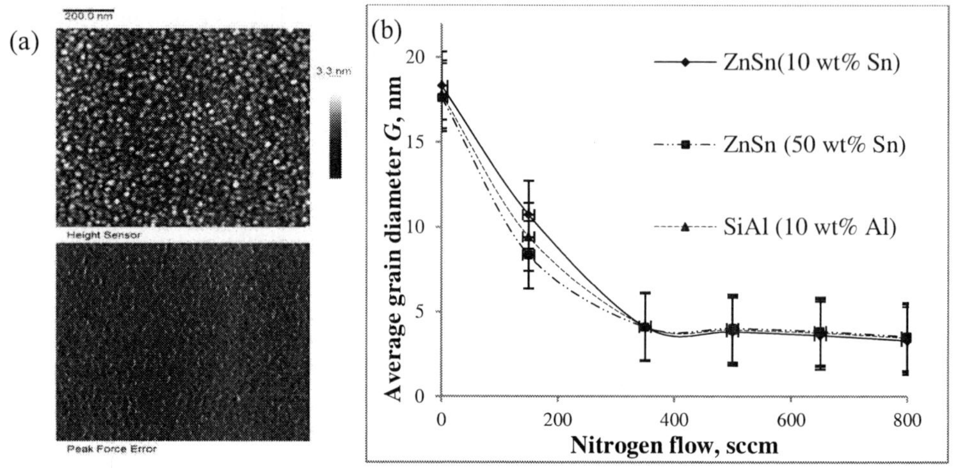

Figure 1. AFM surface morphology analysis: a) height and peak force error images of the film surface deposited by sputtering of the Zn+10wt.%Sn target at the nitrogen gas flow of 350 sccm ; (b) dependence of the average diameter of surface grains *G* on the nitrogen flow rate for three targets used for sputtering.

Fig. 2 shows XPS spectra obtained from the virgin film surface and from the film sputtered by Ar^+ ion beam just before XPS measurements. These measurements demonstrate no nitrogen both on the surface of the film and in its bulk. No nitrogen was found for all films independent on the sputter target composition and nitrogen-to-oxygen flow ratios used. One must note that the peak of Sn in the film is higher if the flux of N_2 is higher. This is well seen for two ultimate cases represented in Fig. 2: no nitrogen and 800 sccm of nitrogen.

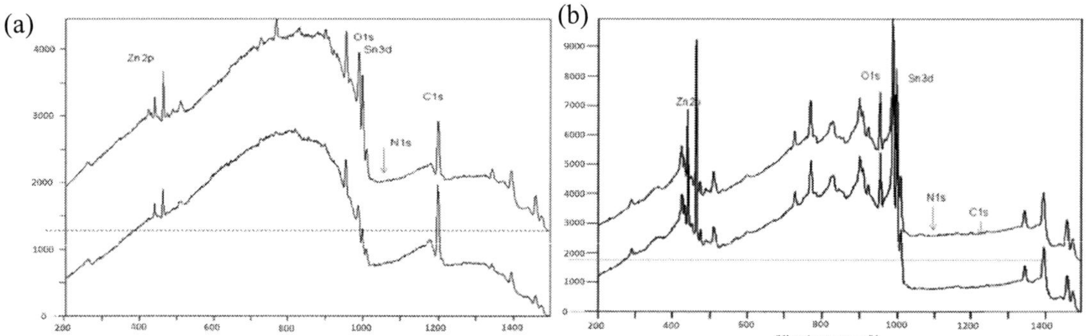

Figure 2. Example of the results of XPS analysis for ZnSn (50wt% Sn) coatings. Frame (a) XPS from as deposited film, frame (b) XPS after etching by Ar+ ions for 6 min. Bottom spectra are for 0 sccm of N_2 flow , top spectra are for 800 sccm of N_2 flow.

4. Discussion

The feature of the films deposited by sputtering of bimetal targets is nanoscale dimensions of their grains. One may suggest that such grains grow according to the so-called abnormal grain growth process, which is also referred to as exaggerated or secondary recrystallization grain growth mechanism. This mechanism suggests that atoms of the more reactive elements in relation to oxygen (like doping component or an impurity) that are deposited on the substrate form fine oxide grains. Those define the formation of the oxide grains of the main element upon them, and as a result restrict the dimensions of the final granular structure [16]. The abnormal grain growth process is encountered in metallic or ceramic systems exhibiting secondary phase inclusions above a certain threshold concentration. It is quite well-known especially for the ceramic materials, as this phenomenon can result in improved fracture toughness through the impedance of crack propagation [17].

In our case, atoms of the doping components on the cathode and on the substrate are more active in regard to oxidation, therefore they form oxides more actively than atoms of the main metal components. Due to high oxidation rate, the concentration of oxides of the active component is very high, while the dimensions of oxide grains of these components is respectively small. Oxidation of the main metal component takes place slower, and the respective oxide grains grow in the matrix of fine grains of the more active component. Therefore the dimensions of the oxide grains of the less active component are restricted by fine grains of the more active component. The higher is the concentration of the active component, the higher are the concentration and smaller sizes of its oxide grains, and respectively the smaller are dimensions of oxide grains of the less active component.

The resulting average grains size of the multielemental layer, which grow due to the abnormal grain growth process, can be described with the Clarke equation (1) [18]:

$$\delta = \frac{1}{3}\left(\frac{x_i - x_i^0}{x_i^L}\right)G \qquad (1)$$

where x_i is the concentration of the more active component introduced in the system, x^0_i is the solubility limit, x^L_i is the equilibrium concentration of the active component in the liquid phase, δ is the thickness of the forming film and G is an average grains diameter. It is assumed that the individual elements are distributed uniformly within the growing coating layer and all grains are all spherical [19].

Fluxes of metal atoms depositing on the substrate due to the reactive sputtering of the bimetal can be quantitatively simulated by SRIM/SIMTRA+RSD software package [20]. Time evolution of the target and film compositions is calculated. Modeling is performed taking into account configuration of the experimental system and parameters of the experiment. Non uniform ion flux on the cathode, re-deposition and re-sputtering, as well as the chemical reactions on the cathode and on the substrate are taken into account in the code.

Computer simulation verified the experimental observation that in conditions of this work no nitrides are formed in films deposited on the substrate, and only oxides are formed. This conclusion was true for all bimetal targets and nitrogen flow rates used in the experiments. This is apparently connected with a higher reactivity of oxygen which tends to substitute nitrogen both on the sputtered target and in the growing film. Formation of oxinitrides and nitrides during the reactive sputtering O_2+N_2 mixture is possible only in the case of very high nitrogen partial content.

A support of this conclusion can be obtained from auxiliary experiments on observation of optical emission of plasma components in the transition reactive sputtering state, which was used in experiments. As it was shown in [21], two extreme regimes are realized during sputtering of Ti in O_2+N_2 mixture depending on the balance between O_2 and N_2. In the case of relatively low concentration of nitrogen, only oxides are formed on the substrate, and this mode is called "oxide" mode. In the case the concentration of nitrogen is high, oxinitrides are formed on the substrate, and this mode is called "nitride" mode. If during operation the regime shifts from the metallic state to the "oxide" mode of transition state, the characteristic emission of oxygen increases, while the characteristic emission of the metallic components decreases. If during operation the regime shifts to the "nitride" mode, the characteristic emission of nitrogen increases, while the characteristic emission of the metallic components changes either moderately or negligibly.

Just this behavior was observed in our experiments demonstrated in Fig.3. One can see that even at the highest nitrogen flow rate used in our experiments (800 sccm in Fig.3a), the sputter deposition is in the "oxide" mode, and only if the nitrogen flow rate is 10-20 times higher than the oxygen flow rate (4500 sccm in Fig.3b) the deposition is in the "nitrogen" mode. This explains why no nitrogen was observed in our experiments even at the highest nitrogen flow rates.

The experimental regimes were simulated numerically, and the result agrees with the experiment. For example, observation of nitride components in the film was predicted to be possible at the value of N_2 to O_2 ratio of 16:1 for Si+10wt% Al target.

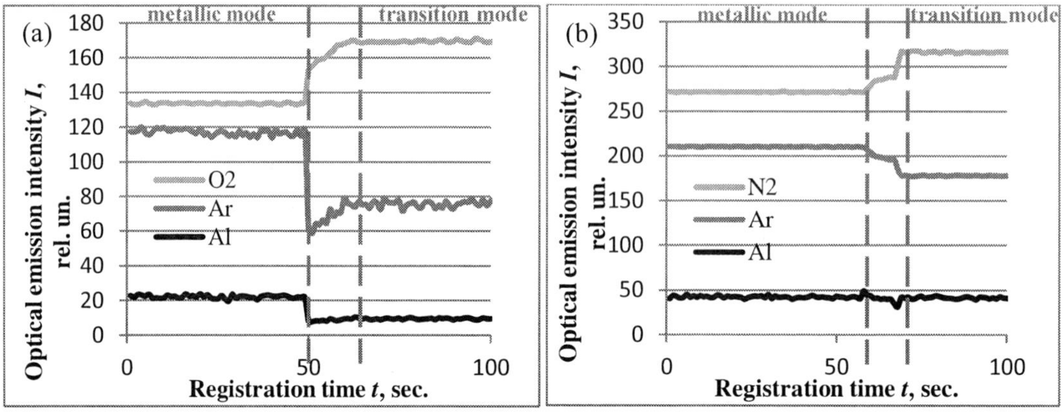

Figure 3. Time trends of plasma discharge optical emissions monitoring for the according elements' characteristic emission (as noted in the trends) during process switching from metallic into the transition reactive sputtering regime at 800 sccm of N_2 flow (a) and 4500 sccm N_2 flow (b).

Experimental observation of the increase of the dopant oxide content on the substrate with the increase of nitrogen flow rate was confirmed by numerical simulations. Fig. 4-a demonstrates the increase of the concentration of the doping component with increase of the nitrogen flow rate for the three targets used in the experiment. It can be assumed based on the obtained results, that the state of the target surface poisoning achieved with the addition of a certain N2 flow energetically favors the sputtering of the doing component in regards to the main metal.

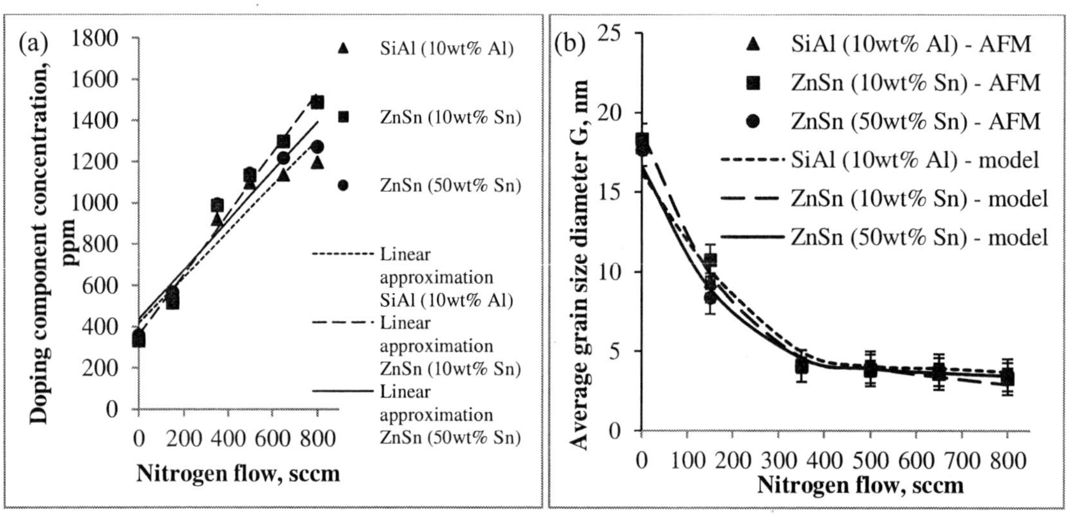

Figure 4. Dependency from the N2 introduction flow for a) – concentration of the doping component in the forming thin film; and b) – average grains diameter for the case of all target materials used.

The calculated concentrations of elements shown in Fig.4a were used to calculate the dimensions of crystallites according to formula (1). The results are shown in Fig.4b. One can see that the calculations perfectly describe the experimental data shown in Fig.1b.

5. Conclusion

Composite Zn+10wt. % Sn, Zn+50wt. % Sn, and Si+10wt. % Al targets were magnetron sputtered in the reactive O_2+N_2 gas mixture. Thin films deposited on the substrate consisted of fine round-shaped nanograins. The increase of nitrogen content in the gas mixture led to decrease in the average diameter of the grains. No nitrogen was observed in XPS for all the coatings deposited in a wide range of the O_2/N_2 ratio. It was suggested that oxygen substitutes nitrogen on the target and on the substrate due to its higher chemical activity in conditions of the experiment. Auxiliary experiments on optical emission of atoms sputtered from the target and then exited in plasma demonstrated that the process in the so-called "oxide" mode, where no nitrides are expected on the substrate. Nitrides can be expected at nitrogen flow rates an order of magnitude higher than those used in the experiments performed.

Sputter-deposition process was modelled numerically by SRIM/SIMTRA+RSD software package, which demonstrated increase of the deposition rate of Sn and Al with increase of the nitrogen content in the working gas.

The calculated deposition rates were used for calculations of the grain dimensions according to the Clarke equation that describe the so-called abnormal grain growth process, and the calculated dependences of the grain dimensions against the nitrogen flow rate perfectly described experimental dependences measured by AFM.

One can conclude that the mechanism of the abnormal grain growth can be applied to description of the grain dimensions in films sputtered in O_2+N_2 gas mixture from bimetal cathodes in magnetron discharge.

Acknowledgments

The authors acknowledge Dr Yu. Lebedinski of Moscow Institute of Physics and Technology for XPS analysis, and Dr. D. Depla and the team of DRAFT group of Ghent University for access to the PVD process simulation software packages. The work was partially supported by Grant № 14.Y26.31.0008 from the Ministry of Education and Science of Russia.

References

[1] Le Dréo H, Banakh O and Keppner H 2006 *Thin Solid Films* **515** 952
[2] Marlec F, Le Paven C and Le Gendre L 2017 *Surf. and Coat. Technol.* 2017 **324** 607
[3] Ali S, Paul B and Magnusson R 2016 *Vacuum* **131** 1
[4] Haye E, Capon F and Barrat S 2016 *Surf. and Coat. Technol.* **298** 39
[5] Rawal S, Chawla A and Jayaganthan R 2014 *Materials Sc. and Eng.* **181** 16
[6] Wang Y, Wang H and Zhang J 2009 *Appl. Phys. Lett.* **95** 032905
[7] Lei P, Guo S and Zhu J 2016 *Surf. Eng.* **32** 585
[8] Murdoch B, Ganesan R and McKenzie D 2015 *Appl. Phys. Lett.* **107** 112903
[9] Xu L, Nishimura T and Shibayama S 2016 *Appl. Phys. Express.* **9** 091501
[10] Haye E, Bruyère S, André E and Boulet P 2017 *Journal of Alloys and Compounds* **724** 74
[11] Zhaozhe Y, Huarui X and Guisheng Z 2016 *Applied Surface Science* **368** 173
[12] Le Paven C, Benzerga R and Ferri A 2017 *Materials Research Bulletin* **96(2)** 126
[13] Bernt D, Ponomarenko V and Pisarev A 2017 *Surf. and Coat. Technol.* **330** 211
[14] Bernt D, Ponomarenko V and Pisarev A 2016 *Journal of Physics: Conference Series* **748**
[15] Steele A, Bayer I and Loth E 2009 *NanoLetters* **9(1)** 501
[16] Kang S-J L 2005 *Sintering: Densification, Grain Growth, and Microstructure* (Amsterdam; Boston; London: Elsevier Butterworth-Heinemann)
[17] Padture N and Lawn B 1994 *J. Am. Ceram. Soc.* **77** 2518
[18] Clarke D R 1987 *J. Am. Ceram. Soc.* **70(1)** 15
[19] Bae I-J and Baik S 1997 *J. Am. Ceram. Soc.* **80(5)** 1149
[20] Strijckmans K and Depla D 2014 *J. Phys.: Appl. Phys.* **47(23)** 235302
[21] Pinaev V and Shapovalov V 2009 *Vakuum. Tekhn. Tekhnol.* **19** 7

Silicon and germanium nanostructures formed by spark discharge plasma

V G Yakunin [1], D E Presnov [1,2], M V Stepikhova [3,4], A N Yablonskiy [3,4], R B Assilbaeva [5], T Yu Grevtseva [6], Z Zh Zhanabaev [6], V P Savinov [1] and V Yu Timoshenko [1,7,8]

[1] Lomonosov Moscow State University, Faculty of Physics, Moscow, 119991, Russia
[2] Skobeltsyn Institute of Nuclear Physics, Lomonosov Moscow State University, Moscow, 119991, Russia
[3] Institute for Physics of Microstructures of RAS, 603087 Nizhniy Novgorod, Russia
[4] Lobachevsky State University, 603950 Nizhny Novgorod, Russia
[5] Satpayev Kazakh National Technical University, 050013, Almaty, Kazakhstan
[6] al-Farabi Kazakh National University, Almaty, Kazakhstan
[7] National Research Nuclear University "MEPhI", Phys-Bio Institute, 115409 Moscow, Russia
[8] Lebedev Physical Institute of RAS, 119333 Moscow, Russia

E-mail: vtimoshe@gmail.com

Abstract. Formation of semiconductor nanostructures on the surface of single crystalline silicon and germanium wafers by spark discharge plasma in air was investigated. The prepared nanostructures were analyzed by means of the scanning and transmission electron microscopy and optical spectroscopy of the photoluminescence and Raman scattering. The formed nanostructures exhibit a fractal-like morphology with interconnected nanocrystals of 2-200 nm sizes that is explained by repeated processes of spark ablation and subsequent condensation. While the size and morphology of the nanostructure depend on power sources of the spark discharge, short interaction times of spark discharge plasma and target determine a relatively low efficiency of the chemical oxidation of germanium and silicon, as well as low ionic temperatures of the plasma.

Keywords: spark discharge, plasma, ablation, nanocrystals, fractals, Raman scattering, photoluminescence

1. Introduction

Different types of spark discharge were used to produce nanoparticles of semiconductors and metals [1-4]. Spark ablation represents a process in which the substance is removed from the electrodes due to the spark plasma excitation and then it is deposited in the form of nanoparticles or nanoclusters. If the electrodes, which are usually a tip and substrate, are made of the same material, it is possible to obtain nanoparticles without external impurities, since the reaction rate and the conditions of the ablation process exclude the possibility of interaction of evaporated particles with the environment. The size and type of the nanostructure obtained by ablation are determined by the concentration of particles in the gas cloud of the evaporated substance, since it is the concentration that regulates the process of coagulation and condensation of particles. Reducing the concentration of ablated material in gas phase or liquids makes it possible to reduce the size of deposited nanoparticles of conventional semiconductors, metals and metal oxides [1, 2. 5].

Silicon (Si) nanocrystals obtained during spark ablation were found to exhibit photoluminescence (PL) in the visible range [3]. Spark-discharge treatment of crystalline germanium (c-Ge) in air resulted in formation of strongly oxidized Ge nanostructures with PL properties, which were attributed to the radiative transitions between defect levels [4]. Our recent work is devoted to study the formation of nanostructures on the surface of c-Si and c-Ge wafers during their processing by spark discharge in air ambient.

2. Experimental

Experiments were carried out with one-side polished low doped c-Si and c-Ge wafers, which were subjected to spark discharge processing by using different power sources. The power supply unit of ILGA-503 type nitrogen laser was used in the first source, and self-made generators based on the high – voltage ignition coil, e.g. model 27.3705, and so-called Tesla spark discharge generator were also used. In all experiments, the total duration of radio pulse did not exceed 1 μs, the repetition rate of high – frequency pulses was about 100 ns, and the repetition rate of pulse series was about 10 Hz. The average current in the first source was in the range of 1-10 mA. In the self-made sources the average current was always less than 1 mA, so the power of spark discharge was much less. The experiments were carried out at atmospheric pressure in air. Exposure time was varied from 1 to 40 min. A scheme of the experimental set-up and the photographic image of the spark processing of a semiconductor plate are shown in Figure1.

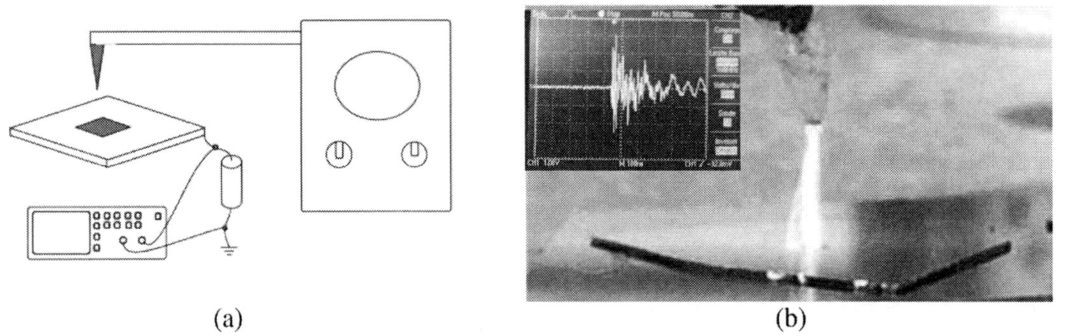

(a) (b)

Figure 1. Scheme of the experimental set-up (a) and a photograph of the spark processing of a semiconductor wafer (b). Inset in panel (b) shows a typical transient (scale bar 100 ns) of the electrical current under spark discharge.

3. Results and Discussion

Figure 2 shows typical scanning electron microscopy (SEM) images of a c-Si wafer after spark-processing by high and medium power supplies for 30 min. The images reveal the presence of nanoparticles with sizes from about 10 nm up to 500 nm. Using the approach developed in Refs.[6,7], we analyzed the SEM c-Si plates after prolonged treatment in spark discharge (Fig. 2), which gave the value of the fractal dimension of 1.8 - 1.9 of the spark ablated nanostructures. The latter value is close to 2D dimensionality that indicates the role of internal surface of the formed nanostructeres.

(a) (b)

Figure 2. SEM images with smaller (a) and larger (b) magnification of the surface of c-Si wafers after prolonged treatment with high power spark discharge.

Experiments have shown that structures of bound nanoparticles of different sizes are formed on the surface of c-Ge plates after long treatment with a spark discharge of low power (see Figure3), reminiscent of the fractal structure of silicon, discussed above.

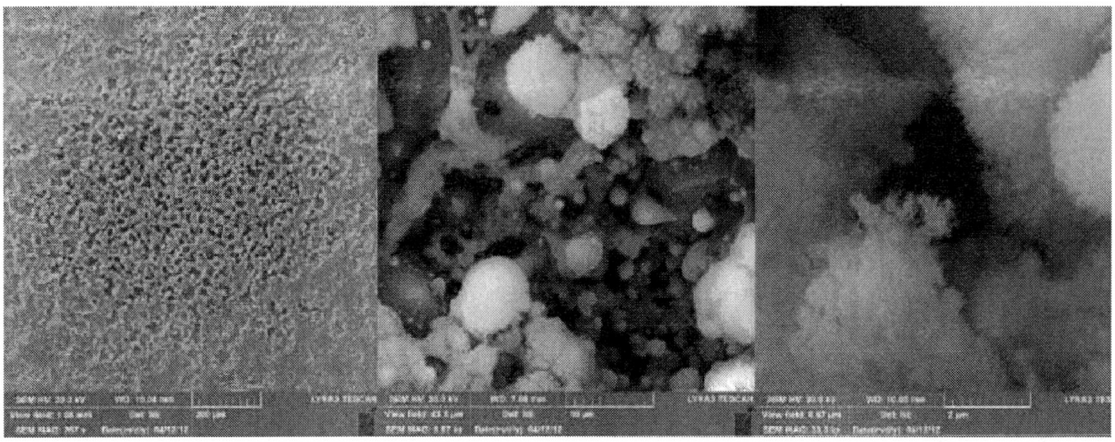

(a) (b) (c)

Figure 3. SEM image of the surface of c-Ge wafer after treatment with a low-power spark discharge for 20 min. The scale bars in panels (a), (b) and (c) are 200, 20 and 2 μm, respectively.

It was found that the nanoparticles formed by spark discharge near the surface of c-Ge are mainly Ge nanocrystals with fractal morphology (Figure 3). Individual Ge nanocrystals with sizes of about several nanometers are observed in TEM pictures (Figure 4). At the same time, the maximum distribution lies in the region of 5-7 nm, but there are also nanoparticles with dimensions of more than 10 nm (see inset in Figure 4a) Study of the nanostructures obtained by means of the high resolution TEM allowed us to identify an individual isolated nanocrystals with sizes from 5 to 20 nm (see Figure 4b).

IOP Conf. Series: Journal of Physics: Conf. Series **1238** (2019) 012052 doi:10.1088/1742-6596/1238/1/012052

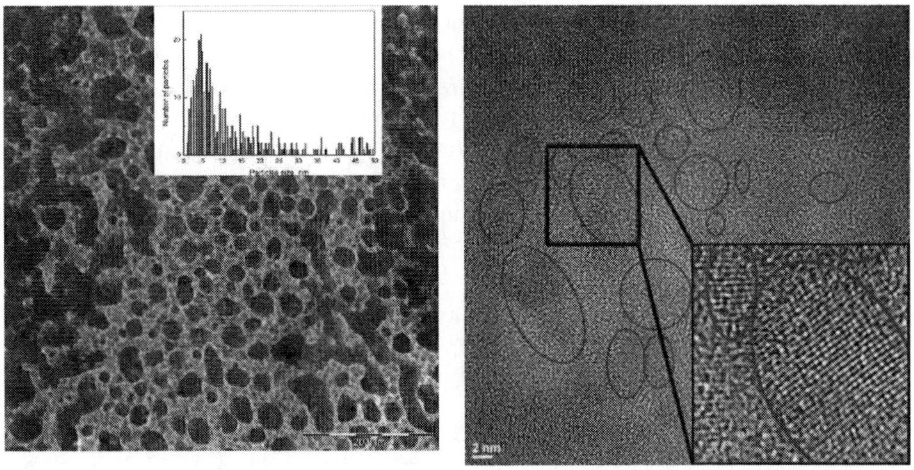

(a) (b)

Figure 4. TEM images of nanoparticles obtained by processing a low-power c-Ge spark discharge for 20 min at a scale of 600x600 nm^2 (a) and high resolution TEM (b). Insert in panel (a) shows the size distribution of nanoparticles. Panel (b) shows the boundaries of nanocrystals with red oval lines, and the black straight lines and box indicate a separate Si nanocrystal.

Figure 5 shows Raman and PL spectra of a c-Si wafer after treatment with a spark discharge of high power. It is seen that for the treated part of the wafer, the spark discharge causes a multiple increase in the Raman scattering intensity, which can be explained by multiple elastic scattering of the excitation radiation in randomly located Si nanostructures, similar to that observed in layers of silicon nanowires [8,9]. The Raman line shape of the resulting nanostructures was close to that of initial c-Si wafer that indicate a negligible effect of the phonon confinement [10] (see Fig. 5a). This fact is also in good agreement with the observed SEM images, which revealed nanoparticles with sizes above 10 nm. At the same time, the PL spectra of silicon nanostructures formed by spark discharge exhibit a weak PL band in the region of 400-600 nm (see Fig. 5b), which can be associated with radiative transitions in the oxide layer on the surface of Si nanocrystals with dimensions large enough to realize the quantum confinement effect [11-13].

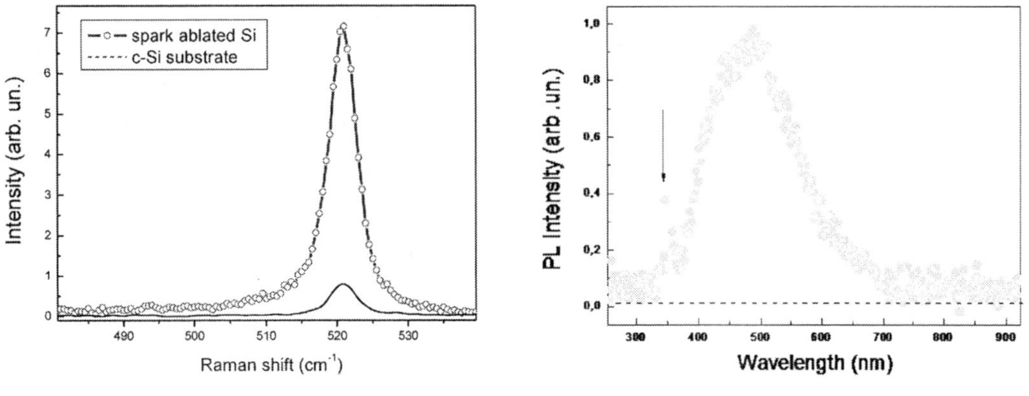

(a) (b)

Figure 5. Raman (a) and PL (b) spectra of c-Si surface after spark discharge treatment. The lower curve in part (b) shows the Raman spectrum for c-Si reference plate. Vertical arrow in part (a) indicates excitation wavelength.

Small nanoparticles with sizes less than 10 nm were observed on the surface of c-Ge wafers after treatment with a low-power spark discharge (Tesla generator) and more powerful sources. As can be seen from Figure 6a, in such nanostructures there is a modification of the Raman spectrum, which indicates an effect of the spatial confinements for phonons in good agreement with the generally accepted model [10]. The experiments showed that the spark-discharge generated Ge nanostructures have a maximum of the PL spectrum near 1 μm and lifetime of about 1 μs (see Fig.6 b). These PL properties are in good agreement with the literature data known for Ge nanocrystals with sizes of about 8-10 nm [13,14].

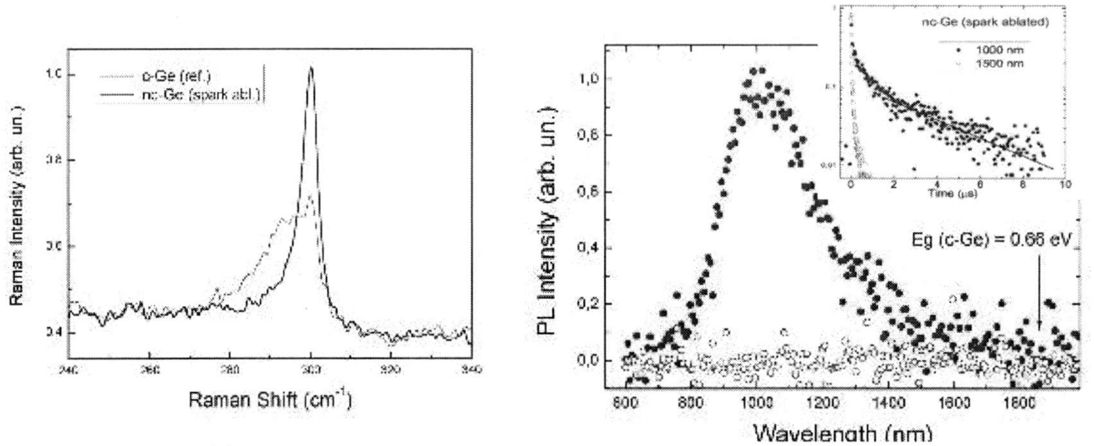

(a) (b)

Figure 6. Raman (a) and PL spectra (b) of c-Ge surface before and after spark discharge treatment. Vertical arrow in section (a) indicates the band gap of c-Ge, and the inset shows decay curves of the PL intensity for different wavelengths after pulsed laser excitation.

4. Conclusions

The experiments demonstrated the possibility of formation of small silicon and germanium nanocrystals in the process of spark ablation of the corresponding c-Si and c-Ge wafers in air ambient. The nanocrystal formation is explained by the short time of interaction of spark discharge plasma with the surface of a solid target. Fractal-like morphology of the formed nanostructures is probably caused by repeated processes of the spark ablation and subsequent condensation of the ablated materials. While the short interaction time of spark discharge plasma and target determines a relatively low probability of the oxidation of germanium and silicon, the formed nanostructures possess nanocrystalline origin. The mean nanocrystal sizes were found to be larger than 10 nm and about 5-10 nm for Si and Ge structures, respectively. The spark-ablated Ge nanocrystals exhibit photoluminescence in the near infrared spectral region, which can be attributed to the radiative recombination of excitons in Ge nanocrystals. This fact indicates a low density of the structure defects in the prepared Ge nanocrystals. The observed formation of rather small Si and Ge nanocrystals by using the spark-discharge in air with low power sources can be a good alternative for the laser-ablative synthesis of ultrapure nanoparticles for biomedical applications in both therapy and diagnostics (theranostics) [15]. The spark-discharge synthesis seems to be cheap and simple method to prepare luminescent nanoparticles of Si and Ge, which can be used in bioimaging of cancer cells under excitation with visible and near-infrared radiation sources. While our results confirm the possibility to prepare a small amount of the nanoparticles, the spark processing can be scalable and optimized to obtain required amount of the nanoparticles for biomedical purposes.

Acknowledgments

The authors thank A. Ermishin, A. Fedorenko and D. Petrov for their assistance in the preparation of samples, measurements of TEM and SEM, respectively. The work was partially supported by the

Russian Foundation for Basic Research (project 16-02-00668) and by the Ministry of Education and Science of the Russian Federation (state project 16.2969.2017/4.6). SEM measurements were partially carried out by using the equipment of the "Educational and Methodical Center of Lithography and Microscopy," M.V. Lomonosov Moscow State University Research Facilities Sharing Centre.

References

[1] Berkowitz A E, Walter J L 1987 Spark erosion: A method for producing rapidly quenched fine powders *J. Mater. Res.* **2** 277–288
[2] Ivanov A V, Nikiforov V N, Shevchenko S V, Timoshenko V Y, Pryadun V V, Bulychev N A, Bychenko A B, Kazaryan M A 2017 Properties of metal oxide nanoparticles prepared by plasma discharge in water with ultrasonic cavitation *Int. J. Nanotech.* **14** (7-8) 618-626
[3] Saunders W A, Sercel P C, Lee R B, Atwater H A 1993 Synthesis of luminescent silicon clusters by spark ablation *Appl. Phys. Lett.* 63 (11) 1549 – 1551
[4] Chang S-S, Choi G J, Hummel R E 2000 Optical properties of spark-processed Ge 2000 *Materials Science and Engineering B* **76** 237-240
[5] Vons V A, Anastasopol A, Legerstee W, Mulder F, Eijt S, Schmidt-Ott A 2011 Low-temperature hydrogen desorption and the structural properties of spark discharge generated Mg nanoparticles *Acta Mater.* **59** 3070
[6] Zhanabaev Z Zh, Grevtseva T Yu 2014 Physical fractal phenomena in nanostructured semiconductors *Rev. Theor. Sci.* **2** (3) 211-259
[7] Zhanabaev Z Zh, Grevtseva T Yu, Ibraimov M K 2016 Morphology and electrical properties of silicon films with vertical nanowires *J. Computat. & Theor. Nanosci.* **13** 615-618
[8] Osminkina L A, Gonchar K A, Marshov V S, Bunkov K V, Petrov D V, Golovan L A, Talkenberg F, Sivakov V A, Timoshenko V Yu 2012 Optical properties of silicon nanowire arrays formed by metal-assisted chemical etching: evidences for light localization effect. *Nanosc. Reseach Lett.* **7** (1) 524
[9] Gonchar K A, Osminkina L A, Sivakov V, Lysenko V, Timoshenko V Y 2014 Optical properties of nanowire structures produced by the metal-assisted chemical etching of lightly doped silicon crystal wafers *Semiconductors* **48** (12) 1613-1618
[10] Richter H, Wang Z P, Ley L 1981 The one phonon Raman spectrum in microcrystalline silicon *Sol. Stat. Comm.* **39** 625 – 629
[11] Kovalev D, Heckler H, Polisski G, Koch F 1999 Optical properties of Si nanocrystals *Phys. Stat. Sol. (b)* **215** 871-932
[12] Saunders W A, Sercel P C, Lee R B, Atwater H A 1993 Synthesis of luminescent silicon clusters by spark ablation. *Appl. Phys. Lett.* **63**. (11) 1549 – 1551
[13] Chang S S, Choi G J, Hummel R E 2000 Optical properties of spark-processed Ge. *Mat. Sci. & Eng. B* **76** 237-240
[14] Niquet Y M, Allan G, Delerue C, Lannoo M 2000 Quantum confinement in germanium nanocrystals *Appl. Phys. Lett.* **77** (8) 1182-1184.
[15] Kabashin A V, Timoshenko V Y 2016 What theranostic applications could ultrapure laser-synthesized Si nanoparticles have in cancer? *Nanomedicine* **11** (17) 2247-2250.

Time- and spatially resolved spectra of X-ray radiation of Z-pinch of tungsten multiwire arrays

Gritsuk A.N.[1], Aleksandrov V.V.[1], Grabovskiy E.V.[1], Mitrofanov K.N.[1], Oleinik G.M.[1], Sasorov P. V.[2], Shevelko A. P.[3]

[1]SRC RF TRINITI, Troitsk, Moscow, Russia
[2]Keldysh Institute of Applied Mathematics of the Russian Academy of Sciences, Moscow, Russia
[3]Lebedev Physical Institute RAS, Moscow, Russia

griar@triniti.ru

Abstract. Studies of the spectra of the radiation source based on the Z-pinch of multiwire arrays on the Angara-5-1 facility are presented. Their intensity and character change drastically during the radiation pulse. The instantaneous spectra of Z-pinch radiation differ substantially from time-integrated spectra close in shape to Planck's, their shape depends on the time of their recording and is characterized by the presence of bands which, according to calculations, correspond to the 4d-4f transitions (in the wavelength range greater than 120 Å) and 4d-5p (less than 120 Å) in tungsten plasma with an average ionization ~14 and a temperature of ~40 eV.

1. Introduction.

It has been shown that a thermal X-ray source with a power of hundreds of TW is created in the pulsed multimega-ampere range during the current implosion of cylindrical wire arrays in the fast Z-pinch discharge mode [1]. Such a source has found wide application in studies on inertial confinement fusion [2] and in high energy density physics [3].

The significant non-stationarity of the processes during the implosion of wire arrays leads to a non-equilibrium of the radiating plasma, deviation of the ionization composition and the level populations from the local thermodynamic equilibrium. In such plasma, the radiation transfer is determined by the absorption in the arrays of overlapping spectral lines. Because of the large number of transitions in multicharged ions and mutual overlapping of the lines, an almost continuous spectrum of radiation is formed. This radiation is nonequilibrium, and its spectrum is not Planckian. To describe such plasma, it is necessary to use kinetic models in which level-by-level kinetics and radiation transfer are taken into account [4, 5].

An experimental study of the spectral characteristics of the X-ray emission of such sources is extremely important, since it allows us to use experimental data on the radiative properties of Z-pinch plasma for comparison with the results of calculations simulating spectral opacity and emissivity of a dense high-temperature plasma from substances with a high atomic number in order to verify initial conditions and calculation codes. The data can also be used to calibrate programs designed to obtain spectral opacity and emissivity of dense high-temperature plasma. Measurements of the soft x-ray spectra with temporal and spatial resolution give unique information about the plasma characteristics of the central part of the pinch and the physical processes taking place within it. To interpret these experimental data, it is necessary to improve the theoretical model of pinch implosion in order to simulate non-stationary spectra and emission power of soft x-ray pinch with spatial resolution.

2. Results

2.1. Experimental Layout

At the Angara-5-1 facility [6], experiments were carried out to study the spectra of Z-pinch radiation from tungsten plasma with temporal and spatial resolution. Wire arrays with a diameter of 10 mm consisting of 6 μm W wires were used as generator loads. The amplitude of the current pulses of the generator was 3.5-4 MA at a current rise time of about 100 ns. In this case, pulses of soft x-ray radiation (SXR) with a power of 6-7 TW and a duration of 9-10 ns were obtained.

To study the spectral composition of the soft X-ray radiation (SXR) in the wavelength range 20-400 Å, the grazing incidence spectrograph GIS-3d was used. It had an entrance slit of 50 μm and a concave grating of 20 × 30 mm (curvature radius R=1 m, grazing angle 4°, 600 grooves/mm, W/Re coating) as a dispersion element of the spectrograph [7]. It was placed 170 cm away from the axis of the facility. An off-Rowland scheme is applied where the spectrum is recorded in the plane normal to the diffracted beams. To provide spatial resolution, a 70-μm slit was placed between the spectrograph and the radiation source. The slit produced a demagnified image of the source on the grating. To record the spectra of radiation with a time resolution in the GIS-3d spectrograph, a system is used based on the three-stripe micro-channel plate (MCP). The spatial resolution of the spectrograph was about ~300 μm, the exposure time of the frames was ~1.5 ns.

2.2. Experimental results and discussion

The experiments carried out at the Angara-5-1 and Z [8, 9, 10] facilities show that the emission spectrum of a wire tungsten array is a superposition of two sources: a compact high-temperature pinch and a "halo" of a substantially larger transverse dimension producing a much softer emission spectrum. The large effective size of the radiation source is explained by the fact that the radiation from a compressed compact pinch re-emits on the trailing plasma located outside the compressed region [8].

The emission spectra of the Z-pinch of a tungsten wire array were obtained at different times relative to the maximum of the radiation. Fig. 1a shows emission spectra of the Z-pinch at times -9 ns, +1 ns and +14 ns with respect to the maximum SXR. They are characterized by the presence of bands which, as follows from the calculations of Refs [4, 5], correspond to the 4d-4f transitions (in the wavelength range greater than 120 Å) and 4d-5p (less than 120 Å) in tungsten ions with average ionization ~14, which corresponds to electron temperature ~40 eV.

A comparison of the instantaneous spectra of the Z-pinch emission with a time-integrated spectrum is shown in Fig. 2.

IOP Conf. Series: Journal of Physics: Conf. Series **1238** (2019) 012053 doi:10.1088/1742-6596/1238/1/012053

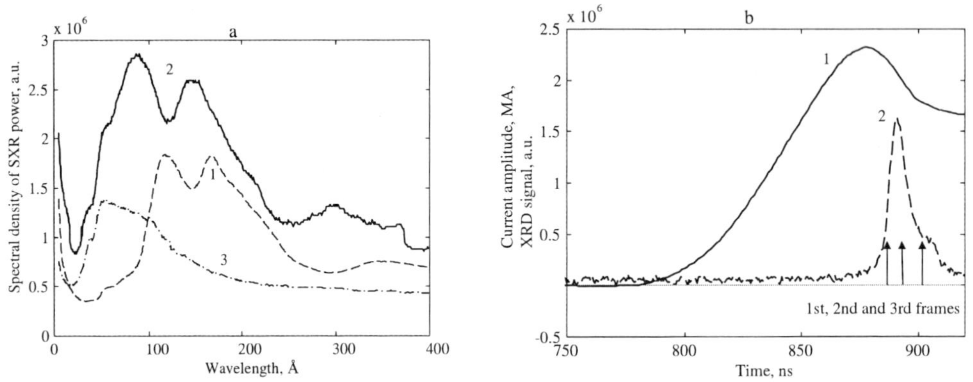

Fig. 1. a) The emission spectra at time -9 ns (curve 1), 0 ns (curve 2), and +14 ns (curve 3). b) curve 1 is the time profile of the current, curve 2 is the time profile of the X-ray signal of the XRD (> 100 eV). The arrows indicate the time of registration of the spectra shown in Fig. 1a.

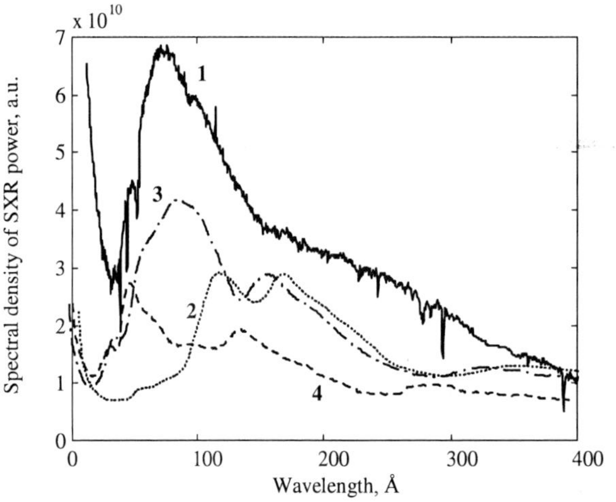

Fig. 2. Comparison of the emission spectra measured at -9 ns (curve 2), 0 ns (curve 3), and +14 ns (curve 4) relative to the peak of the SXR pulse with the time-integrated emission spectrum (curve 1) of the Z pinch of tungsten wire arrays. Exposure time of frames at registration of time resolved spectra 1.5-2 ns.

A comparison of the time-resolved spectra of the Z-pinch emission with a time-integrated spectrum shows that the "dips" between the emission bands in the time-resolved spectra are closed due to a change in their position with time due to a change in the temperature of the radiating plasma.

The time-resolved spectra were compared with the calculations of [4] (Fig. 3).

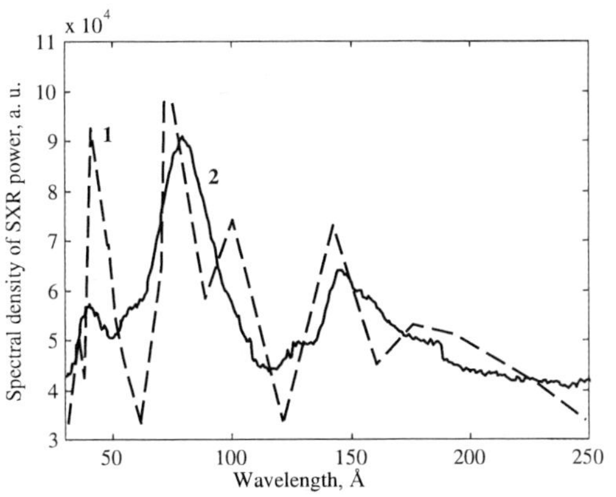

Fig. 3. The Z-pinch spectrum (curve 1) at the maximum of the SXR pulse calculated for plasma with an electron density $N_e = 3\times10^{20}$ cm^{-3} and a temperature $T_e = 40$ eV, and an experimental Z-pinch spectrum (curve 2), also obtained at the moment of maximum SXR.

Thus, the spectral information on the X-ray source, obtained for the first time with a time and spatial resolution, indicates a change in the spectral composition of the Z-pinch plasma during a short radiation pulse. The obtained features on the spectra and their time variation are important for the verification of computational codes that take into account the kinetics and transport of radiation.

This work was supported by the Russian Foundation for Basic Research under Project 16-02-00084, 16-02-00112, 16-02-00491, 17-02-00167 and 18-02-00170.

References

[1] Deeney C., Douglas M.R., Spielman R.B., Nash T.J., Peterson D.L., L'Eplattenier P., Chandler G.A., Seaman J.F., Struve K.W. // Phys. Rev. Lett. 1998. V. 81. 4883.

[2] Hammer J.H., Tabak M., Wilks S.C., Lindl J.D., Bailey D.S., Rambo P.W., Toor A., Zimmerman G.B., Porter J.L., Jr. // Phys. Plasmas. 1999. V. 6. P. 2129.

[3] Remington B.A., Drake R.P., Ryutov D.D. // Rev. Mod. Phys. 2006. V. 78. P. 755.

[4] I.Y. Vichev, V.G. Novikov, A.D. Solomyannaya // Modeling of the emission spectra of tungsten plasma // Mathematical Models and Computer Simulations 2009, v.1, no.4, pp.470-481.

[5] O. G. Olhovskaya, V. A. Gasilov, M. M. Basko, P. V. Sasorov, V. G. Novikov, I. Yu. Vitchev, I. I. Galiguzova Calculation of output power and X-ray spectrum of Z-pinches based on multiwire arrays // Mathematical Models and Computer Simulations 2016, v.8, no.4, pp.422-437.

[6] Z. A. Al'bikov, E. P. Velikhov, A. I. Veretennikov, V. A. Glukhikh, E. V. Grabovski, G. M. Gryaznov, et al., "Experimental facility Angara-5-1," Atmos. Energy, 1990. v. 68, no. 1, pp. 1–26,.

[7] Gritsuk A, N, (2017) Spatio-temporal and spectral characteristics of X-ray radiation emitted by the Z-pinch during current implosion of quasi-spherical wire arrays (PhD Thesis), Moscow, http://ods.mephi.ru/docs/GritsukAN_20170330_dissertation.pdf

[8] A.N. Gritsuk, V.V. Aleksandrov, E.V. Grabovskiy, Y. Laukhin, K.N. Mitrofanov, G.M. Oleinik, G.S. Volkov, I.N. Frolov, A.P. Shevel'ko, IEEE Transactions on Plasma Science 2013. V. 41, P. 3184.

[9] M. E. Cuneo, E. M. Waisman, S. V. Lebedev, J. P. Chittenden, W. A. Stygar, G. A. Chandler, et al., Phys. Rev. E, 2005 V. 71, P. 046406.

[10] V. V. Alexandrov, G. S. Volkov, E. V. Grabovsky, A. N. Gritsuk, I. G. Malyutin, K. N. Mitrofanov and G. M. Oleinik, Journal of Physics: Conference Series 2016 V. 666, P. 012008.

IOP Conf. Series: Journal of Physics: Conf. Series **1238** (2019) 012054 doi:10.1088/1742-6596/1238/1/012054

Structure and tribological behavior of electro-spark deposited TiC based coatings

G De Medts[1,2], M I Petrzhik[1] and P V Kiryukhantsev-Korneev[1]

[1] Department of Powder Metallurgy and Functional Coatings, National University of Science and Technology MISiS, 4 Leninsky prospect, 119049, Moscow, Russia
[2] European Engineering School in Materials Science EEIGM, 6 rue Bastien-Lepage, 54000, Nancy, France

E-mail : g.de-medts@hotmail.fr

Abstract. This work gives an analysis of the coating obtained by electro-spark alloying using TiCNiCr electrode regarding their composition, structure, morphology, roughness, and mechanical properties. Tribological properties were studied by pin-on-disk test against different counterparts without lubricant. Soft materials like Cu, 100Cr6 steel are not damaging the coating and hard materials are either a good couple like with WC-Co, or characterized by a aggressive wear like with SiC, Si_3N_4, Al_2O_3. Detailed microscopic investigations of the tracks were done after tribo-tests.

1. Introduction

Raw materials present basic tribological characteristics. Coating deposition is an efficient and economical way using a low quantity of matters and energy [1] to improve the wear and the corrosion resistance for metallic alloys. As an example, a lifetime of cutting tools can be improved by 1.5 to 1.8 [2].

Three major processes exist: Physical Vapor Deposition (PVD), Chemical Vapor Deposition (CVD) and Electro-Spark Alloying, also known as Electro-Spark Deposition (ESD). PVD and CVD are two technics adopted by companies for high quality coatings. The process required an inert atmosphere or in vacuum to produce a coating, resistant to delamination and fractures [3]. Otherwise, without the special atmosphere, it will produce thick coatings and a fragile interface [4].

ESD is cheaper than PVD and CVD, and industrially attractive because of its easier process. It works under any atmosphere. Using Argon is faster but optional [5]. Its process uses alternative current. The coating material is transferred from the electrode to the surface with an electrical pulse leading to a local melt (elevated temperature and high pressure) [6] [7]. A frequency sometimes bellows $\tau = 100$ µs conducts to the extreme rapid heating and cooling (up to 10^9 K/s [8]) of the coating creating a specific microstructure in the coating. First the solidification is parallel to the surface characterized by a high tensile strength then perpendicular [9].

The TiC is commonly used as coating agent because of its high hardness. Its melting point is 3 257 °C, reached locally with EDS [10].

A wide combination of parameters is possible, letting the porosity and the size of grains definable. For example, the shape of the electrode is related to its electrical resistance and its electric field intensity [11] [12]. Thus, a 6 mm diameter electrode TiC based produces specific form of 5 µm ball surrounded by Fe [13]. The size of grains composing the electrode affects the hardness of the coating. Nano and micro-structured electrode increase the hardness by three in comparison with standard electrode; the Young modulus by 2.4 times for the nano-structured electrode; the density increase to 98 and 90 % for nano and micro-structure electrodes respectively, and the thickness of the coating to 40 and 30 µm respectively [14].

The composition of the electrode also impacts the coating. Introducing Ni, acting like a binder, in the electrode decrease the number of defect in the coating. The deposition of a multilayer of Ni / (TiC/Ni) / Ni also reduces the number of cracks but also the hardness from 1000 to 500 Hv in comparison with (TiC/Ni).

Laser treatment can also reduce the number of cracks and improve the toughness of coating [15]. Tribological tests revealed a lower level of mechanical properties for the interface between coating and substrate, since it is the case for the interface between MAX phase Cr_2AlC based coatings on titanium alloy [5,16].

TiC based electrodes with Chromium are not enough investigated for now, and this work will bring light over the correlation of the morphology of the wear-resistant coating and its tribological behavior against varied materials.

2. Materials and methods

The synthesis of electrodes was carried out by the force Self-propagating High-temperature Synthesis (SHS)-pressing technology starting from the precursor powder (in mass percent: 57.6 % Ti, 16.8 % C, 15.6 % Cr and 10 % Ni). Force action of 15 tonnes upon hot SHS products (immediately on completing the gas-free combustion process) followed by cooling in sand. Polished substrate was made of the Russian steel 40X with a composition in mass percent: 97 % Fe, 0.95 % Cr, 0.65 % Mn, 0.4 % C, 0.3 % Cu and Ni, 0.27 % Si, 0.035 % S and P.

Element (mass %)	Fe	C	Ti	Cr	Ni	Mn	Others
Substrate	97,00	0,40	0,00	0,95	0,30	0,65	0,70
Coating	49,7	22,4	19,5	5,2	2,8	0,4	/

Table 1. EDS composition analysis.

The Alier-303-Metal stands for the ESD. It is used in experiments afforded regulating the following parameters: intensity I = 120 A, duration of pulse τ = 100 µs and frequency f = 640 Hz. Depositions of the coating were carried out in a 0.25-liter box under argon. ESD was made in manual mode with an opposite polarity current with a rectangular rod of 2.5x2.5x30 mm^3.

Sliding tests were made with a Tribometer (CSM Instruments, Switzerland) [17] in a reciprocating linear cycle of 4 mm with maximal speed of 5 cm/s applying a constant force of 5 N. Counterparts, used during 20 000 cycles, are 3 mm balls made of hard metal (94 % WC, 6 % Co), SiC, Al_2O_3, Si_3N_4, during 10 000 cycles for Steel (100Cr6) and during 4 000 cycles for Copper.

X-Ray Diffraction (XRD) was carried out with a D8 Advanced (Bruker, US) with Copper K-α radiations. Images of counterparts and the cross-section were made thanks an Axiovert CA25 optical microscope (Carl Zeiss, Germany). Microstructure and elemental compositions of coatings were characterized by a S-3400N scanning electron microscope (SEM) (Hitachi, Japan) equipped with an UltraDry energy-dispersive detector (EDS) and Silicon Drift X-ray detector (Thermo Fisher Scientific, US). Profilometry was made by a Wyko NT1100 optical profiling system (Veeco, US).

Hardness are calculated thanks imprints made with Nano-hardness Tester (CSM, Switzerland) over the cross-section mirror polished. Each imprint is separated by 8 µm and sequences are separated by 10 µm. Approach speed is 2 000 nm/min with a maximum load of 30 mN with a program of loading/unloading rate of 60 mN/min with a pause of 5 s. Microhardness of coatings was measured via a DuraScan70 microhardness tester (Emco-Test Prüfmaschinen GmbH, Austria) with loads of 25, 50, 100, 300 and 500 g.

3. Results and discussion

3.1. Structure and phase composition of coating

IOP Conf. Series: Journal of Physics: Conf. Series **1238** (2019) 012054 doi:10.1088/1742-6596/1238/1/012054

The constraint apply to the coating during fast heating and fast cooling will affect the coating in two ways: the surface appears molten, with holes, cracks from 80 to 200 μm and high points, comparable to lumps (Figure 1). Roughness is Ra = 3.84 μm, Rq = 4.69 μm with maximum high points of 50 μm. The composition of the coating is measured with EDS (Table 1).

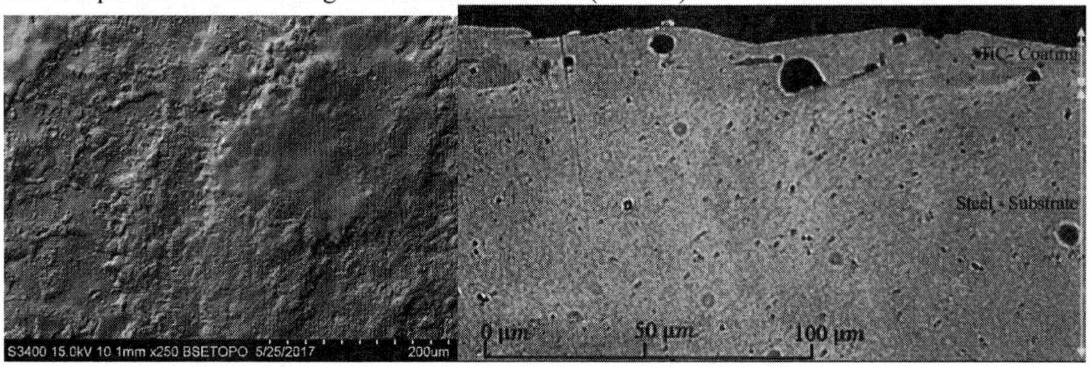

Figure 1. Top view and cross-section of the Fe-Ti-Cr-Ni-C coating.

XRD analysis (Figure 2) shows that the deposition affects the morphology of the substrate. First, it is composed of two phases, α-Fe in majority, the stable phase of iron, and γ-Fe, its metastable form. After deposition, the amount of α-Fe reduces from 79.5 to 1.1 % (volume %) with a cell parameter of 0.288 nm to an undetermined value; and the amount of γ-Fe rises from 20.5 to 66.2 % (volume %) with a cell parameter from 0.3609 to 0.3620 nm. The coating is composed of 32.7 % (volume %) of TiC with a cell parameter of 0.4283 nm, corresponding to a concentration of TiC < 0.4 [18].

Although that the coating is formed of phases of the electrode, the substrate and the combination of both, the XRD analysis with a grazer angle of measurement confirms the absence of other phases. In the case of a WC92–Co8 electrode used, Fe, absent in the electrode, is found in phases of the coating [19].

IOP Conf. Series: Journal of Physics: Conf. Series **1238** (2019) 012054 doi:10.1088/1742-6596/1238/1/012054

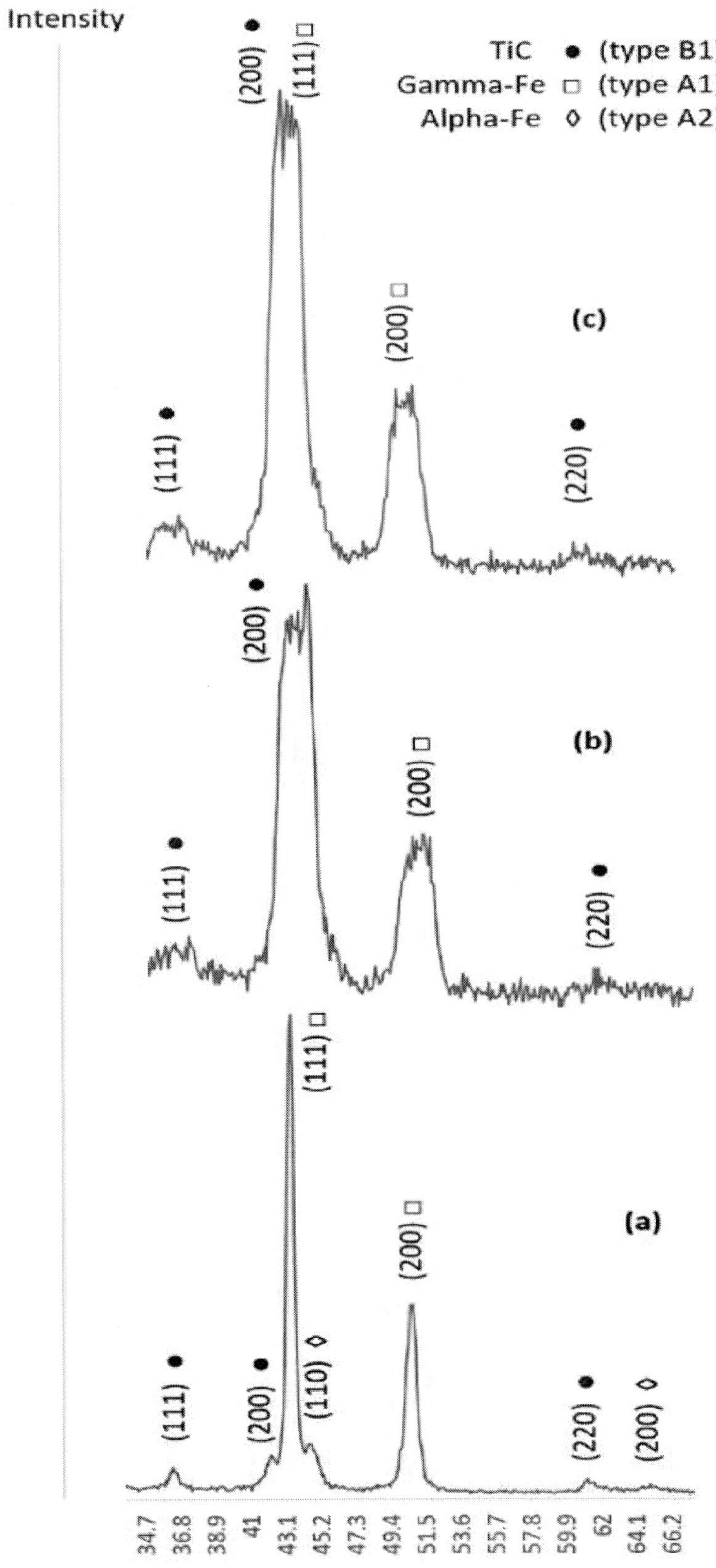

Figure 2. XRD of the sample varying angle of attack from frontal (a) to grazer angles (b)(c).

3.2 Tribological tests and properties investigations

	From literature [20]				From experiences			
Counter-part	*HV* (GPa)	*E* (GPa)	*ν*	Hertz stresses (GPa)	Wear rate ($mm^3/N/m$)		CoF	
					Ball (10^{-6})	Disk (10^{-3})	Start	Max
WC	15.5	610	0.24	3.16	7.06	0.10	0.10	0.22
SiC	24-28	310	0.14	2.57	7.89	0.13	0.12	0.35
Al_2O_3	19	380	0.22	2.71	1.52	0.10	0.11	0.64
Si_3N_4	15.5-16	320	0.27	2.57	7.89	0.13	0.11	0.66
Copper	2.1	121	0.35	/	487	No track	0.28	0.67
Steel	0.832	210	0.30	/	33.1	No track	0.06	0.51

Table 2. Tribological data of wear test.

Nano-indentation gives a Young modulus evolving according to the depth with 245 GPa for the substrate, from 150 to 200 GPa for the interface and a mean value of 205 GPa for the coating (Figure 3). Hardness seems to follow a schematic behavior (Figure 3c) with a minimum at interphase layer. Hardness is 1200 Hv for the substrate, 500 Hv for the interface and 800 Hv for the coating (Figure 3).

Micro-indentaion shows that top-layer has got a hardness of 800 HV, with values decreasing with increasing of the penetration depth (Fig 3d). In the literature, TiCNi coating is harder than this TiCCrNi coating [15,21].

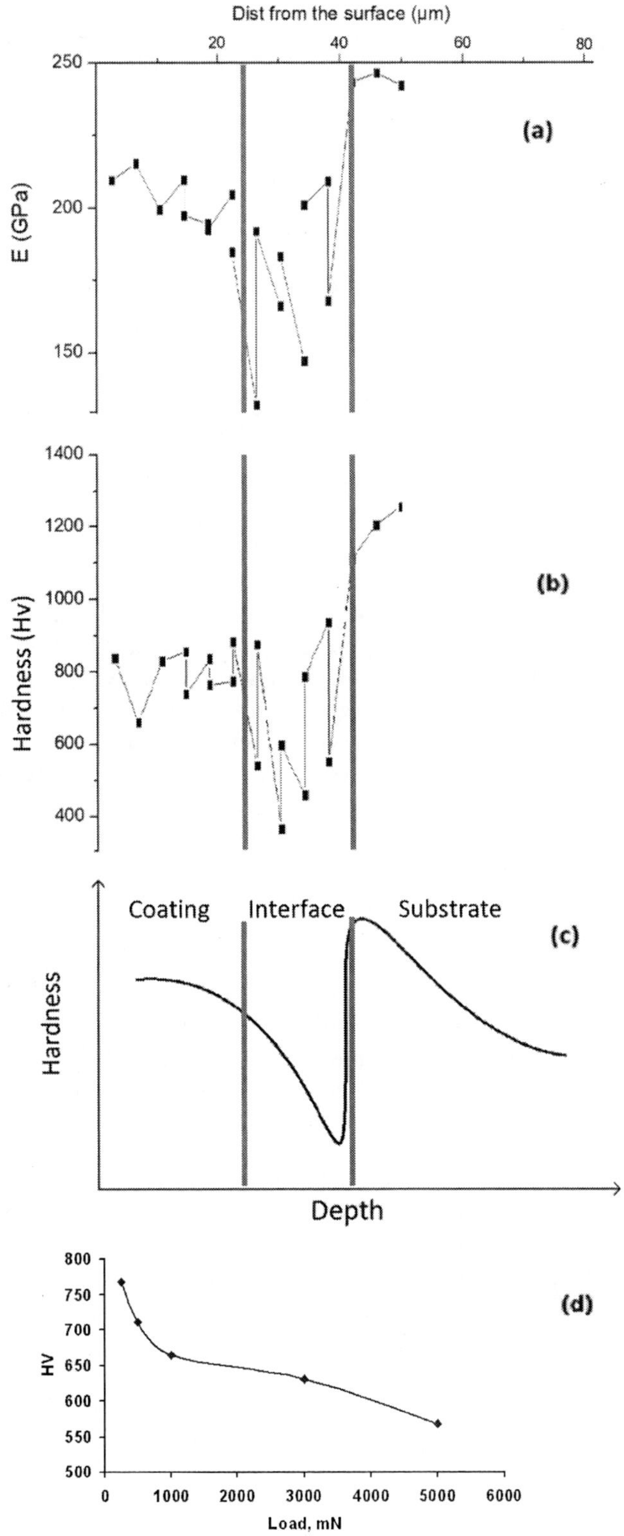

Figure 3. Nanoindentation of the cross section (a),(b). Schematic dependence of hardness from surface to bulk (c). Microindentation data (plan) (d).

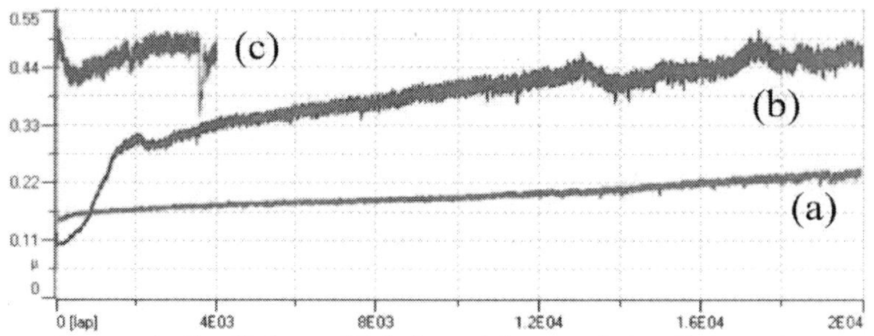

Figure 4. Coefficient of Friction of the Fe-Ti-Cr-Ni-C coating against WC (a), Al_2O_3 (b) and Copper (c).

The response of the coating to the sliding test is characterized in three distinct types of behavior with damages either one-sided or equitable distributed (figure 4).

SiC, Al_2O_3 and Si_3N_4 are defined by a highly scratched coating.
The CoF of Si_3N_4 evolves linearly, which makes its evolution more predictable. Cross-section of the sample shows that Si_3N_4 sands the coating without chemical adherence showed by a round shape of the imprint of the counterpart. Si_3N_4 counterpart did a 600 μm groove with a depth of 10 μm. Si atoms can be found in the coating. The coating is ripped off and oxides of iron are glutted on the counterpart. The present of black oxides and the acceleration of the CoF proves that the friction creates a rapid heating. This leads to a cascade of events responsible for a rapid abrasive behavior.

SiC counterpart pulled away the entire coating over a width of 315 μm and with a maximal depth of 5 μm. As Si_3N_4 test, the shape of the scratch is a perfect round ball trace, which shows no adhesion of the counterpart. No oxide is formed in SiC counterpart showing no increase of temperature. The damage of the test is the lowest of all the three.

Al_2O_3 counterpart, relatively intact after test, results the most destructive one. Two stages of groove are visible: one with a width of 500 μm with a maximal depth of 10 μm and one central one thinner of 74 μm, 2 μm deeper. The coating appears teared off, showing either a strong chemical interaction or an elevated temperature, transforming debris into red oxides of iron.

Their wear test starts with the same coefficient of friction (CoF) as counterpart not affecting the coating like WC and Steel, but it grows up faster to a high value (Figure 4b). First acceleration will appear after 1 000 cycles with SiC counterpart, 2 000 cycles with Al_2O_3, and 10 000 cycles with Si_3N_4. After 10 000 cycles, the CoF of Al_2O_3 and Si_3N_4 are over 0.4, while the CoF of SiC is about 0.24. Then the CoF grows slowly to 0.47 and 0.50 respectively. This behavior shows a scratching as three steps: first a small CoF, due to the surface of contact increasing because of the increase of depth of groove and slowed by the roughness of the coating; then a rapid growth of CoF, corresponding to a minimum roughness and a surface of contact increasing with depth of groove. This stage continues until the complete suppression of coating; and finally, a slow growth.

The second behavior corresponds to a good couple of friction with few damages over the coating and the counterpart. WC/TiC based coating does not show any visible groove according to microscopy despite the profilometry shows one. WC appears also a good friction partner because of the low CoF (Figure 4a). The third body is not built of the counterpart as proved with an absence of W and Co over the coating. It is formed with oxidation of iron high concentrated in Oxygen (58 atoms%) showing absorption of the humidity, coatings or high carbon concentrated particles (92 atoms%). MgO was observed over the scratched ball.

Two different areas are visible; the Ti-coating sanded surrounded by dusts and oxidations accumulated. Because of the heat, the uncoated part becomes oxidised and brittle as has been highlighted by cracks seen by the SEM. Cracks in the coating show that the substrate also absorbs a part of the movement.

And the third behavior shows no damage in the coating and the deep sanded counterpart (Figure 5c). Neither profilometry nor cross-section show any track for Copper test and almost invisible one for Steel test of 126 μm.

Steel wear test shows low CoF from 0.6 to 0.18 in 2 000 cycles and stays constant. Highest value corresponds to the moment until high points become enough flat to have a stable surface of contact against the ball, becoming able to sand the counterpart without variation. In comparison, WC counterpart evolves slowly but constantly growing to 0.18. There is no signal indicating an eventual future high constant value of CoF.

According to profilometry analysis, one high point of 400 μm damaged is the only visible trace of the track of the Steel. This counterpart is the one most affected with a high surface of contact in only 10 000 cycles or two times less than other experiences. This coating can also be a viable alternative to the cutting of steel.

Copper wear test presents a stable instantly high CoF of 0.48 during all the test (Figure 4c). A stable third body uniquely composed of oxides of copper due to a high heating is immediately created. Debris is under the form of little balls or parts ripped, comparable as rubber debris. After 4 000 cycles, or 5 times less cycles than other experiences, the counterpart is already deeply sanded.

Their wear tests define uniquely the behavior of the counterpart because no change is observable in the coating, thus only heating and time of the formation of the third body is observable.

Other publications, because of the use of lubricant present significant lower CoF with lower damages of the coating [19].

IOP Conf. Series: Journal of Physics: Conf. Series **1238** (2019) 012054 doi:10.1088/1742-6596/1238/1/012054

Figure 5. Wear track of the Fe-Ti-Cr-Ni-C coating and contact spot of counterpart for SiC (a), Al_2O_3 (b) and Copper (c).

4. Conclusion

The coating is composed of 66.2 % of γ-Fe and 32.7 % (volume%) of TiC with interstitial Cr, Ni, and Mn. The coating presents cracks, points higher than 5 µm and holes smaller than a few micrometers; is rough with Ra = 3.84 µm and its thickness ranges from 10 to 25 µm.

The hardness of the coating is about 800 Hv and Young modulus about 200 GPa. Tribological behavior of the coating against hard metal (WC-Co) is very good with a CoF under 0.2 during 20 000 cycles with negligible wear rate.

Indeed, the coating cannot support friction against Al_2O_3. 20 000 cycles are enough to remove the coating with advanced teared off damages. Si_3N_4 and SiC counterparts damage corresponds to an intermediate behavior with damages over the counterpart and the coating.

Soft metals made of Steel and Copper are destroyed after less than 4 000 cycles of the same wear test, and do not hurt at all the coating.

Acknowledgements

This work was carried out with financial support from the Ministry of Education and Science of the Russian Federation in the framework of state assignment No.11.7172.2017/8.9.

References

[1] Sheldon G L and Johnson R N 1984 Conference on Electro-spark deposition: A technique for producing wear resistant coatings pp 388-396
[2] Podchernyaeva I A, Yurechko D V, Bochko A V, Sedlyar G A and Kostenko L M 2012 *Powder Metallurgy and Metal Ceramics* vol 51 issue 3-4 pp 198–203
[3] Johnson R N and Sheldon G L, 1986 *Journal of Vacuum Science & Technology* vol 4 issue 6 pp 2740–2746
[4] Padilla K, Velasquez A, Berrios J A and Cabrera E S P 2002 *Surface and Coatings Technology vol* 150 issue 2-3 pp 151–162
[5] Zamulaeva E I, Levashov E A, Sviridova T A, Shvyndina N V and Petrzhik M I, 2013 *Surface and Coatings Technology vol 235* pp 454-460
[6] Tang S K, Nguyenand T C and Zhou Y 2010 *Welding Journal* vol 89 pp 172-180
[7] Radek N and Bartkowiak K 2010 *Physics Procedia* vol 5 part A pp 417-423
[8] Gould J 2011 Welding Journal vol 90, pp 191-197
[9] Xie Y J and Wang M C 2006 *Surface and Coating Technology* vol 201 pp 691–698
[10] Samsonov G V and Vinnitskii I M 1976 *Handbook Refractory Compounds* (Moscow)
[11] Katz Z and Tibbles C J 2005 The International Journal of Advanced Manufacturing Technology vol 25 pp 923–928
[12] Muttamara A, Fukuzawa Y, Mohri N et al. 2003 *Journal of Materials Processing Technology* vol 140 pp 243–247
[13] Panteleenko F I, Sarantsev V V, Stolin A M, Bazhin P M and Azarenko E L *2011 Surface Engineering and Applied Electrochemistry* no 4 pp 106-115
[14] Levashov E A, Vakaev P V, Zamulaeva E I, Kudryashov A E, Kurbatkina V V, Shtansky D V, Voevodin A A and Sanz A 2007 *Surface and Coatings Technology* vol 201 issue 13 pp 6176-6181
[15] Zhen Chen and Zhou Y, 2006 *Surface and Coating Technology* vol 201 issues 3–4 pp1503-1510
[16] Zamulaeva E I. et al. 2016 *Surface and Coatings Technology* vol 298 pp 15-23
[17] Levashov E A et al. 2010 *Russian Metallurgy (Metally),* No. 10, pp. 917–935.
[18] Courleux A 2011 *Thesis* pp 27-41
[19] Padgurskas J, Kreivaitis R, Rukuiža R, Mihailov V, Agafii V, Kriūkienė R and Baltušnikas A 2017 *Surface and Coatings Technology* vol 311 pp 90–97
[20] http://www.matweb.com 2017

Experimental study of environment ionization in the zone of a periodic discharge in a flow of liquid (PDFL)

B Yu Bogdanovich, G O Buyanov, A V Nesterovich

National Research Nuclear University MEPhI (Moscow Engineering Physics Institute), Kashirskoe shosse 31, 115409 Moscow, Russia

Abstract. The study of the ionized environment in the area surrounding PDFL has showed the presence of visible plasmoids moving in the horizontal plane. The study of traces on the X-ray film made it possible to detect a significant number of traces of an identical shape and size that look like the "birds" which were found earlier in other facilities. A preliminary analysis has been carried out for their identification, which led to a conclusion about the nature of these formations, in particular, their identity with the Dirac monopole (or similar object) having a "tail" and "wings" formed by a flow of electrons from the surroundings.

1. Introduction

From the very beginning of work with PDFL in 1995-1996 [1], the appearance of unusual pinch plasmoids in the zone adjacent to the discharge was noted, Fig. 1, 2, 3. In particular, there were the luminous filaments up to a few centimeters long, having an increased luminosity at the ends. In some cases, these filaments "hung" in the air, in other cases one of the "filament" ends extended to the grounded substrate, and the overall pattern resembled a "firework". Luminous long-lived plasmoids were observed in experiments in the following years [2], and their distance from the discharge could be several centimeters or more, which excludes the possibility of explaining their appearance by the presence of liquid droplets sprayed by shock waves. To clarify the nature of these phenomena, the authors have carried out studies to determine the properties of plasma formations in the vicinity of the discharge, in particular, their ionizing ability, namely, the effect on the X-ray film.

Fig.1. A periodic discharge in a flow of liquid. A white circle identifies a plasmoid in the air.

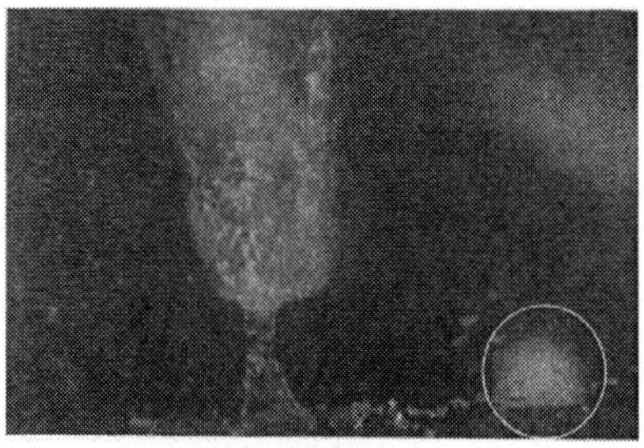

Fig.2. A periodic discharge in a flow of liquid. A white circle identifies a plasmoid located on the surface of a treated sample after the discharge.

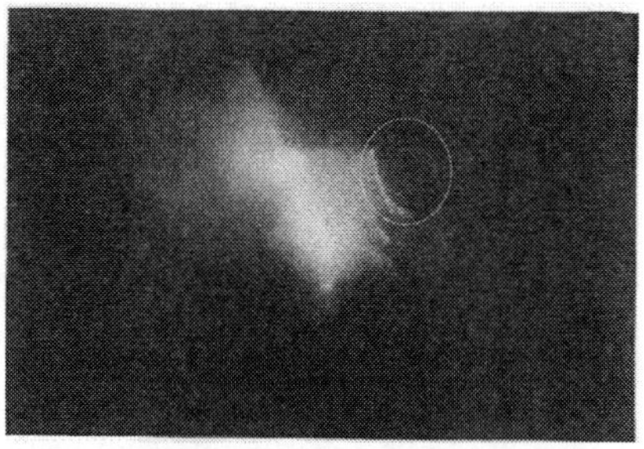

Fig. 3. A periodic discharge in a flow of liquid. A white circle identifies a luminous filament.

2. Experimental scheme

The experimental scheme is shown in Fig. 4.

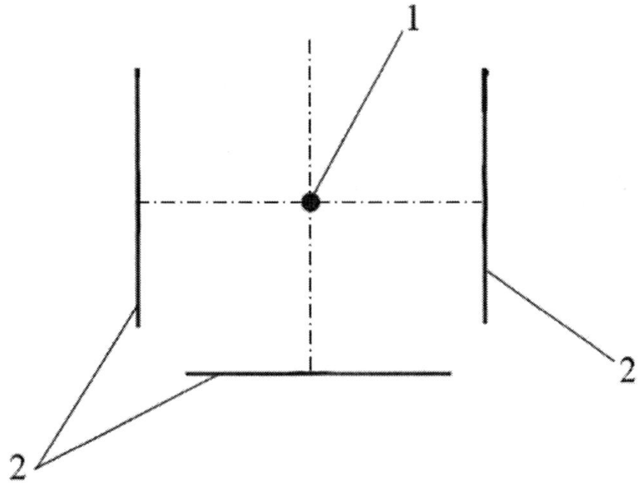

Fig. 4. The experimental scheme on the registration of objects in the vicinity of PDFL on the x-ray film. 1 - discharge electrodes; 2 - X-ray film. View from above.

The film was placed at a distance of 19 cm from the axis of the facility (the film size was 18 cm by 24 cm, the films were placed vertically (suspended on special holders in double polyethylene envelopes, opaque to light radiation). The film and envelopes were manufactured by Retina X-Ray XBE Blue-sensitive film, USA.

Several sessions were held, each lasting about 5 minutes. The discharge parameters: voltage on a high-voltage electrode (maximum) is 7.5 kV; duration of current pulses is (2 ÷ 2.5) μs; pulse repetition frequency is ~ 50 Hz; the maximum current in the pulse is up to 5 kA. A detailed description of the facility operation, the measurement technique, the discharge parameters, and the measurement procedure have been repeatedly given earlier [3 ÷ 10].

The film was developed by specialists in an accredited laboratory with industrial equipment designed for this purpose.

2.1 Experimental results

The analysis of traces showed the following.

IOP Conf. Series: Journal of Physics: Conf. Series **1238** (2019) 012055 doi:10.1088/1742-6596/1238/1/012055

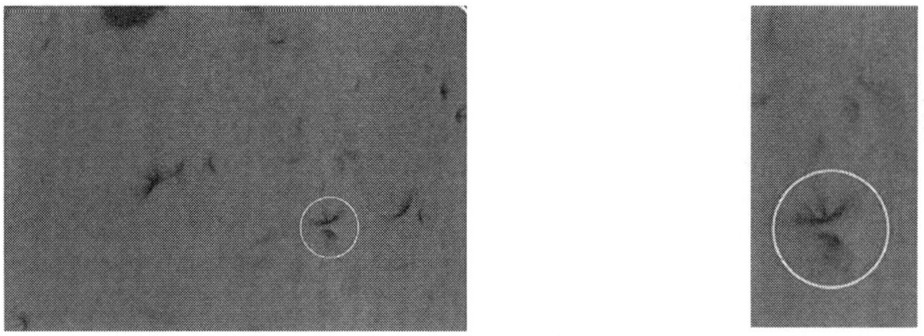

Fig. 5. A film with traces. The most visible traces are marked in white.

All films, without exception, showed a large number of the traces of objects which looked like birds (swallows, swifts) with a "wingspread" up to 1 cm (Fig. 5). As a rule, the length of the straight section is from 3 to 5 mm. The most surprising was that in a significant number of cases, dents on the film and even spots were detected in the "head" part of the traces drawing suspicion to thermal effects (as if from burning), which may indicate a thermal impact, but not a mechanical one. There are corresponding bulges on the back side of the film.

No less surprising is the fact that there are similar dents on the back side of the film.

At the same time, no such traces were found on the envelopes (neither external nor internal), which indicates the selective (resonant) nature of the radiation effect on the substance which the film consists of, in particular, on the chemical agents of the photosensitive coating.

The orientation of the position of the "birds" is mostly chaotic (clarification of the systematic general features depending on the parameters of the environment is being carried out at the present time). The total number of such traces at the time of writing the paper was at least a hundred. To exclude the likelihood of the appearance of traces as a result of breaking the conditions for developing or storing the film, a number of control sessions (without a discharge) were carried out, eliminating possible interpretations and suspicions.

3. Photographic recording of objects in the electronic synchrotron "Pakhra"

By coincidence, during the period of the experiments with PDFL the authors had to conduct similar photographic recording sessions on the same film in double envelopes in another facility, namely, in the electron accelerator "Pakhra" (synchrotron with energy up to 1.2 GeV). The experimental scheme is shown in Fig. 6.

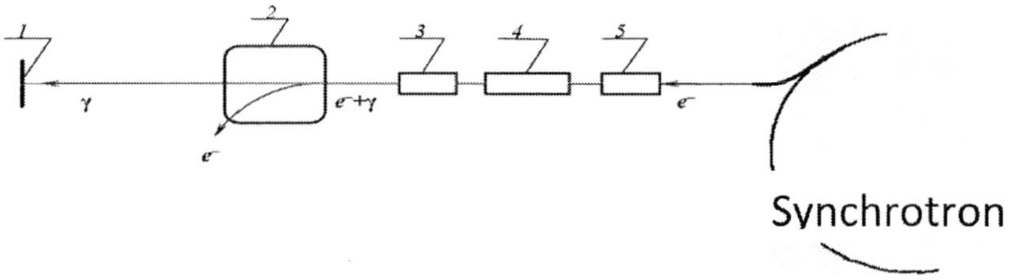

Fig. 6. Experimental scheme in the synchrotron of LPI (Lebedev Physical Institute). 1 – an X-ray film; 2 - a magnet deflecting the electron beam; 3, 4, 5 - sections of the conversion target.

The objective of the experiment was to determine the "trace" of a bremsstrahlung beam from the conversion target after the passage of electrons through it. The electron beam was then deflected in a magnet with a vertical magnetic field with an induction of up to 1.5 T and a length of about 1 m. The film was located about 1 m from the edge of the magnet. The electrons reaching the film are excluded (the control measurements were conducted).

In the zone of film localization, the presence of magnetic scattering fields from a magnet is estimated to be of the order of 10^{-2} T. At the same time, the presence of ionization in the air from an electron beam and a beam of bremsstrahlung (gamma) radiation is recorded.

In the developed photographs, as one would expect, traces of a bremsstrahlung beam were detected. The edges of the magnet can be seen (the gap between the poles of the magnet is ~ 6 cm). To the greatest surprise, the same "birds" were found, (Fig. 7), as well as during the experiments with PDFL. Previously, the same facts were not presumably taken into consideration, even if they had occurred. Comparison of experiments in both cases allows us, first of all, to note the similarity of the presence of the following factors:

1. high degree of ionization in the zone of film localization;
2. impulse nature of the effect of ionization factors, namely, pulse duration ~ (2 ± 3) μs, repetition frequency ~ 50 Hz;
3. magnetic field, directed mainly parallel to the plane of the film.

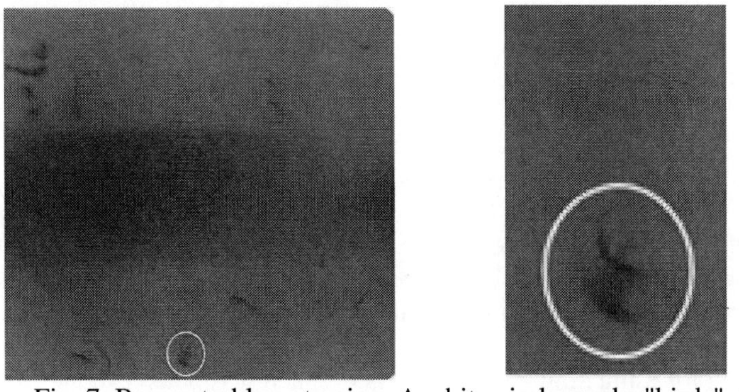

Fig. 7. Bremsstrahlung tracing. A white circle marks "birds". On the right there is an enlarged image.

Ionization of the air in the PDFL zone is caused both by the expansion of the discharge plasma and by soft X-rays (photoionization is not excluded).

The electromagnetic field is generated in the zone of the LPI accelerator from the accelerator and the output beam and has an amplitude of the same order as the EMP from PDFL.

4. Analysis of the results

As a working hypothesis on the formation of traces, we consider a version of the formation of microobjects in the ionized air, having all the features similar to the Dirac monopole [11] with the accompanying cone-shaped electron trail consisting of Cooper pairs with high penetrating power in the environment. This version is confirmed by the analysis of traces representing in their curved part the cross-section of the conical surface identified with the electron trail, whereas a rectilinear trace can be a monopole track (this version requires clarification). The different orientation of the "birds" can be explained by the dynamics of the objects under consideration having both a magnetic and an electric charge. In non-homogeneous fields in space, when moving both longitudinal and transverse components are affected on them, so that the movement of the leading "center" (monopole or quasimonopole) can be overlapped by the rotation of the electronic trail, which in turn affects the movement of the "monopole- trail "in general.

Currently, other versions are being under consideration, and experiments are being planned to refine the nature and properties of the detected physical objects.

The appearance of monopoles in the zone of PDFL localization can be explained by the nonzero probability of their location in the incoming water. When discharged along the surface of the jet, a part of the water evaporates, decomposes into components (hydrogen, oxygen) and ionizes, which presumably contributes to weakening of the monopole bonds (if they are present there). Since the force lines of the magnetic field have the form of circles symmetrical with regard to the axis of the discharge, the trajectories of the monopoles should look like unfolding flat spirals, at least at the beginning of acceleration (if the monopoles are of the two types - "northern" and "southern" - the

IOP Conf. Series: Journal of Physics: Conf. Series **1238** (2019) 012055 doi:10.1088/1742-6596/1238/1/012055

directions of motion are oncoming). As they accelerate, they "build up" electrons from the plasma surrounding the main "body" of the discharge, which contributes to their radial acceleration. If not all, then at least some of these monopoles with a trail of electrons (presumably from Cooper pairs) penetrate through the envelope onto the surface of photographic paper, where they stop, leaving a trace in the form of "birds".

In case of the synchrotron experiment, it can be assumed that the monopoles were originally located on the magnet's surface facing the median plane where the electron beam moves. The "halo" of the electron beam and a significant part of the bremsstrahlung photons fall on these surfaces and can initiate the output of monopoles due to the photoelectric effect and secondary emission, especially since the output is facilitated by the magnetic field of the magnet.

In the pole gap of the magnet, the air is strongly ionized, which facilitates the "accumulation" of monopoles with an electron trail. The dispersion of monopoles with a trail is likely to be in the direction of the motion of photons (by virtue of the law of conservation of momentum). This is also facilitated by charging the envelope with the film up to a positive potential due to the emission of photoelectrons and the difference in the range of electrons and positrons when the substance passes after the formation of electron-positron pairs by a beam of bremsstrahlung photons.

The hypothesis for the presence of magnetic charges in the PDFL area is supported by the formation of pinch discharges, which in their structure are practically similar to the structure of a monopole with a trail of rotating electrons (Fig. 8, 9). A distinctive feature can be the sizes (length up to several millimeters), which, in this case, can be provided by the presence in the base of each pinch of not one but a significant number of monopoles. To find them in a quasistable state in a single cluster, it is sufficient to fulfill the conditions for the charge quasineutrality arising from the presence of not only electrons but also positive ions in the zone of their localization, as well as a certain ratio of the parameters (density, cluster size, etc.). These parameters are determined theoretically on the basis of the gravitational model for estimating the parameters of pinch plasmoids [3].

Fig. 8. Pinch formations in the form of rotating electron "tornadoes" of a vertical type.

Fig. 9. Pinch formations in the form of rotating electron "tornadoes" of a horizontal type.

5. Conclusion

The results obtained in the part related to the detection of traces in the X-ray film in the form of "birds" indicate that we are dealing with a fundamental property of matter, which is confirmed by their detection in the experiments of other authors [12].

Acknowledgements

The work was performed as a part of the Grant Agreement # 17-72-20228 between the Russian Science Foundation, "National Research Nuclear University "MEPhI"" and D.N. Sinelnikov, Head of the project, to conduct fundamental and pilot studies.

References

[1] Alferov P V, Bogdanovich B Yu, Volkov N V, Len N A, Nesterovich A V, Starostin A N, Silnov S M 2000 *Features of the flow of a periodic discharge in a liquid flow and the specificity of its effect on the surface of electrodes* (*Engineering Physics* vol 4) pp 19 – 22 [In Rus]

[2] Alferov P V, Bogdanovich B Yu, Volkov N V, Len N A, Nesterovich A V, Starostin A N 2000 *Structural state of materials after the action of a periodic discharge in a liquid flow.* (*Engineering Physics* vol 4) pp 58 – 61 [In Rus]

[3] Bogdanovich B Yu, Nesterovich A V, Sukhanova L A, Khlestkov Yu A 2014 *The use of the gravitational model for estimating the parameters of pinch plasmoids* (6th International Scientific and Practical Conference "*Theoretical and practical problems of the development of modern science*" (Place and date of the Makhachkala, 30. 11. 2014)) [In Rus]

[4] Baev V K, Bogdanovich B Yu, Nesterovich A V 2015 *Analytical description of the dynamics of charged particles in the field of a magnetic spherical dipole* (*Journal of Applied Physics* vol 85, issue 7) pp 119-124 [In Rus]

[5] Bogdanovich B Yu and Nesterovich A V 2015 *Increase in corrosion resistance of technological and structural metals during treatment with a periodic discharge in a liquid flow (PDLF) (Scientific and technical collection "Defense technology")* [In Rus]

[6] Bogdanovich B Yu, Kalin B A, Nesterovich A V 2016 *Modification of the ferrous and nonferrous metals in the process of pinch discharges in the water vapor* (*Inorganic Materials: Applied Research* vol 7 Issue 5) pp 673–681

[7] Bogdanovich B Yu and Nesterovich A V 2016 *Analysis of main regularities of periodic discharge passing in fluid stream* (*Journal of Physics: Conference Series* vol 747) UNSP 012001

[8] Bogdanovich B Yu, Nesterovich A V, Sukhanova L A, Khlestkov Yu A 2016 *Internal structure of pinch plasmoids in the gravitational-electrodynamic model of general relativity* (*Russian Physics Journal* vol 59., issue 7) pp 152-158 [In Rus]

[9] Bogdanovich B Yu, Nesterovich A V, Sukhanova L A, Khlestkov Yu A 2016 *Description of pinch plasmoids within the framework of the general theory of relativity* (*Russian Physics Journal* vol 59., issue 7) pp 109-119 [In Rus]

[10] Khlestkov Yu A and Sukhanova L A 2016 *Exact Solutions to the Einstein–Maxwell Equations Describing Wormholes and Handles* (*Foundations of Physics* vol 46, Issue 6) pp 668–688

[11] Vishnevsky R P Reference date: 01/29/2018 *Superconductivity of the Dirac monopole* (*electronic resource, link to the material*) http://lightdynamics.narod.ru/Sverhprovodimost_monopolya_Dirka.htm [In Rus]

[12] Zhigalov V A 2015 *Tracks on a film of strange radiation: replication* (*International Journal of unconventional science* vol 9, issue 3) pp 55-62 [In Rus]

IOP Conf. Series: Journal of Physics: Conf. Series **1238** (2019) 012056 doi:10.1088/1742-6596/1238/1/012056

Investigation of periodic discharge in liquid flow

D. Sinelnikov, D. Bulgadaryan and G. Buyanov

National Research Nuclear University MEPhI (Moscow Engineering Physics Institute), Kashirskoe shosse 31, 115409 Moscow, Russia

Abstract. Periodic discharge in water flow was investigated using optic spectrometer and 1000 fps video camera with the discharge pulse up to several milliseconds long regulated by water flow speed. Copper drops were found on the cathode surface in the case of copper-tube anode. In spite of water cooling anode can reach melting point. Optic spectra were compared for several materials of electrodes. For copper electrodes it was found that anode is the main contributor of copper in discharge plasma.

1. Introduction

Impulse discharge in a liquid flow at atmospheric pressure can significantly modify the surface of the electrodes, changing both the morphology of the surface and its mechanical characteristics[1]. Cooling of the sample with a liquid during discharge does not allow it to be overheated up to temperatures close to melting point, while the power deposited from the discharge can be as high as several kW per several cm^2 area of a treated surface.

2. Experimental setup and samples

In figure 1 one can see experimental setup scheme. Water is supplied from the reservoir 1 through the tubular anode 2, so that a discharge gap of 0.5-1 cm long it passes in a drip mode with a droplet drop frequency of several Hz. The voltage up to 7 kV is applied to the anode using BP-138 power supply. When a drop of water during the fall commutates the discharge gap, a breakdown occurs leading to the sprinkling of water. After the end of the breakdown pulse the capacitor 5 with a capacity of 0.5 μF at 20 kV starts recharging until the next triggering by water drop.

Since there is a lot of noise in the discharge circuit preventing the measurement of frequency, current, discharge voltage and other electrical parameters, we focused on spectroscopic measurements of the discharge plasma composition. Four samples and two tubes with different materials were used to investigate how electrode materials influence on discharge plasma. Sample materials were copper, aluminum, titanium and graphite, and tube materials were copper and stainless steel.

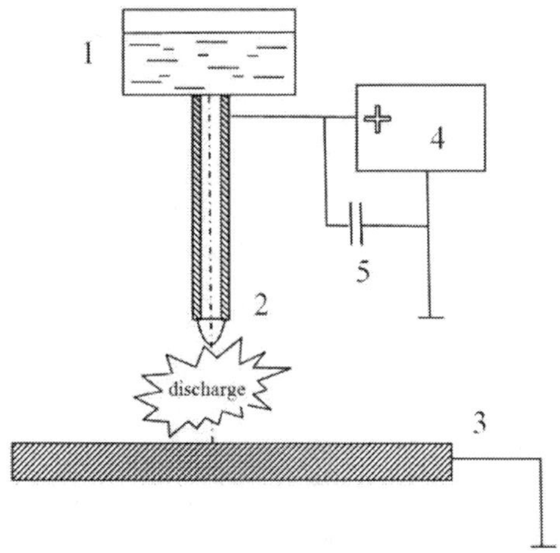

Figure 1. Experimental setup scheme: 1- reservoir with water, 2- tubular electrode (anode), 3- processed sample (cathode), 4-power supply unit BP-138, 5- capacitor

3. Energy-dispersive X-ray spectroscopy analysis of the samples

Scanning electron microscopy with energy-dispersive X-ray spectroscopy (EDS) was used to check material transfer from one electrode to another during the discharge. In the case of copper tubular anode on the surface of all cathodes (with the exception of the copper one, where it cannot be distinguishable) small copper balls were found with ~10 μm diameter shown in figure 2. That means that anode temperature reaches the melting point even for short discharge pulse period.

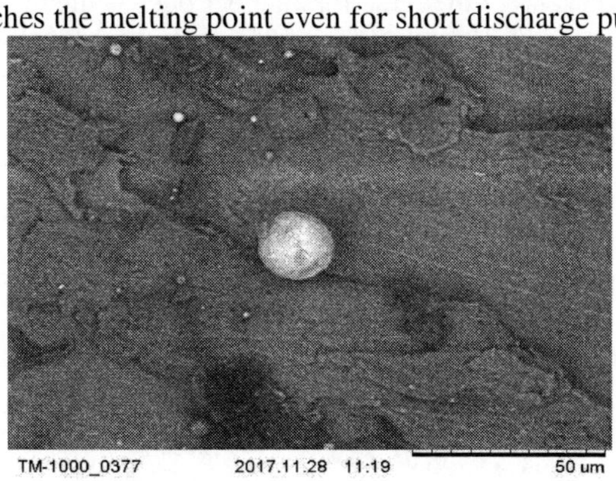

Figure 2. Copper drop on the aluminum sample surface after the discharge

4. Discharge dynamics

For investigation of the discharge dynamics 1000 fps camera was used. Only in some cases one can find by analyzing the video 3-5 frames, where discharge shining cloud dimensions increase or decrease slightly changing its shape. This means that the discharge duration can be up to several ms long. Taking into account 1A current of power supply it can charge 0.5 μF, 20 kV capacitor for 1 ms, but if the time between two pulses is shorter, than maximum discharge voltage is lower than 6 kV. However, it cannot be lower than several hundred volts, minimum voltage sufficient for providing conductive channel in water vapor in 1 cm gap [2]. That could mean that shortest pulses should be several tens of microsecond duration.

5. Optical spectroscopy of the discharge

Optical spectroscopy of the plasma was made using AvaSpec 2048x14 spectrometer, and typical spectrum is shown in fig. 3. Maximum intensity in it was for hydrogen (H_α 656 nm and H_β 486 nm lines) and atomic oxygen due to water molecules decomposition. Since discharge was in the air, the products of nitrogen molecules decomposition are also presented. One can also see copper lines in the left part of the spectrum due to electrode erosion.

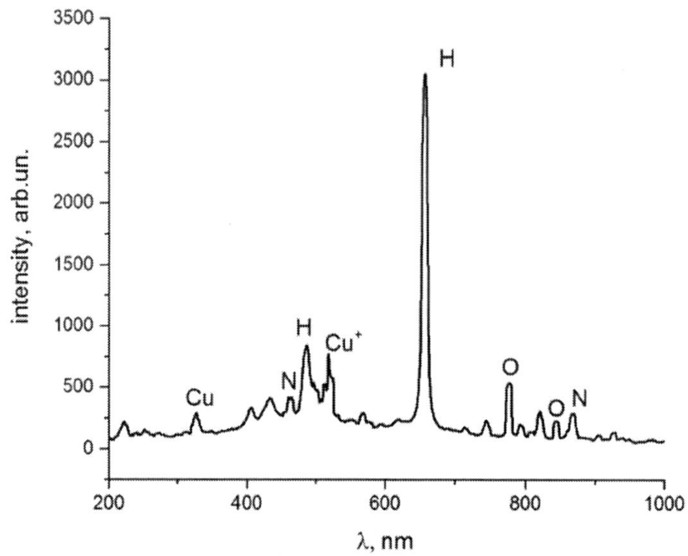

Figure 3. Optical spectrum for copper-copper electrodes

Experiment was made for several pairs of electrode materials and measured spectra are shown in fig. 4. The right parts of the spectra were the same for all electrodes while the left parts were different.

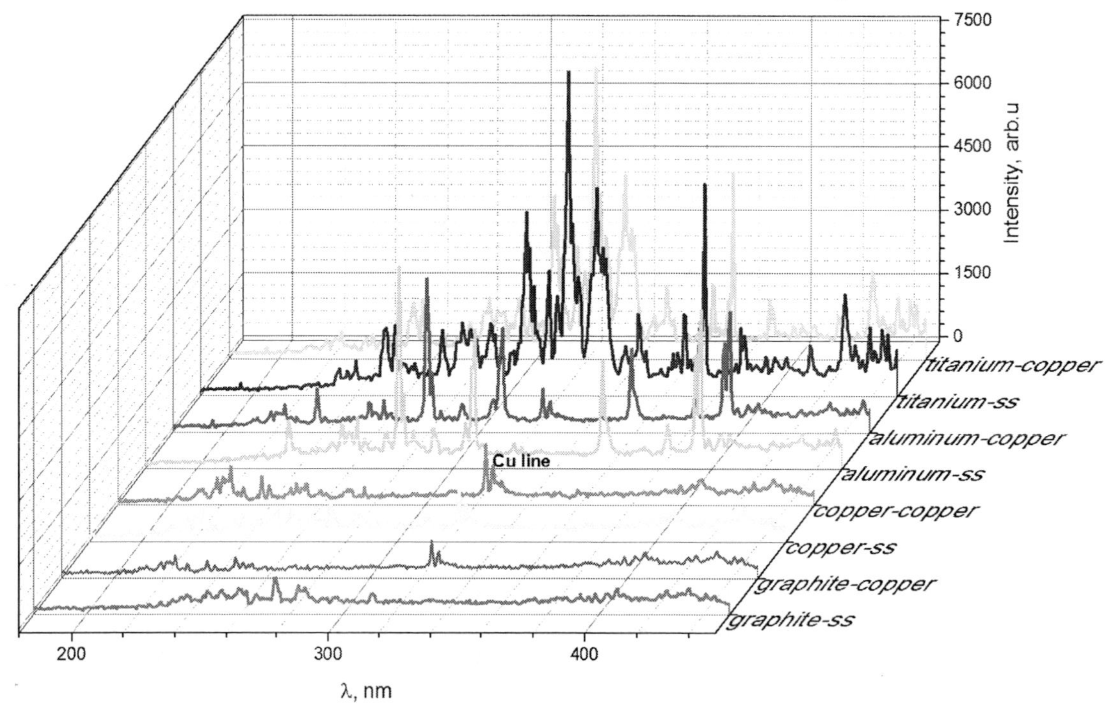

Figure 4. Optical spectra for various pairs of electrodes

Two spectra with maximum intensity were measured for titanium cathode and the most intense lines correspond to atomic and ionized titanium. The spectra for graphite cathode have minimum intensity and comparing spectra with stainless steel and copper anode one can see the deposition of the anode material in the spectrum. Comparing copper-copper and copper-stainless steel spectra one can see that copper intensity is similar in them that could be because of the cooper anode was the main contributor of copper in plasma.

6. Conclusion

Periodic discharge in liquid flow was investigated using optical spectroscopy. Maximum intensity was for hydrogen (H_α 656nm and H_β 486nm lines) and atomic oxygen due to water molecules decomposition. Comparing copper-copper and copper-stainless steel spectra copper anode was the main contributor of copper in plasma. It was shown that discharge pulse can be up to several milliseconds long, but more often shorter and can be varied from several tens of microseconds. EDS analysis of the cathodes demonstrates copper drops in the case of usage of copper anode, meaning that anode temperature reaches melting point even for short discharge pulse period.

References

[1] Bogdanovich, B.Yu., Nesterovich, A.V. Analysis of main regularities of periodic discharge passing in fluid stream (2016) J. of Phys.: Conf. Ser., 747 (1), статья№012001.
[2] N. Škoro, D. Marić, G. Malović, W. G. Graham, and Z. Lj. Petrović Electrical Breakdown in Water Vapor (2011) Phys. Rev. E 84, 055401(R).

Acknowledgments
This work has been funded by the RSF grant № 17-72-20228.

IOP Conf. Series: Journal of Physics: Conf. Series **1238** (2019) 012057 doi:10.1088/1742-6596/1238/1/012057

Multielemental alloying of steel surface layer under the action of compression plasma flows

N N Cherenda[1,2], V V Uglov[1,2], G M Dzagnidze[1], V M Astashynski[3], A M Kusmitski[3]

[1]Belarusian State University, Nezavisimosti ave., 4, Minsk 220030, Belarus
[2]National Research Tomsk Polytechnic University, Lenin ave., 2a, Tomsk 634028, Russia
[3]Heat and Mass Transfer Institute of the National Academy of Sciences of Belarus, P. Brovki str., 15, Minsk 220072, Belarus

Corresponding author's e-mail: cherenda@bsu.by

Abstract. The microstructure, phase and element composition, microhardness of the carbon steel surface layer simultaneously alloyed with molybdenum and chromium atoms under the impact of compression plasma flows were investigated in this work. X-ray diffraction analysis, scanning electron microscopy, energy dispersion microanalysis and microhardness measurements were used for the investigation of the alloyed layer structure and properties. The findings showed that an increase of the plasma pulses number led to a decrease of the molybdenum and chromium atoms concentration in the alloyed layer as well as to the growth of alloying elements distribution homogeneity. The alloyed layer contained solid solutions on the basis of α-Fe and γ-Fe according to the data of the phase composition analysis. Plasma impact resulted in 3.5 times microhardness increase due to structure refinement.

1. Introduction

High energy particles can be effectively used for materials surface layer alloying by elements resulting in the material exploitation characteristics increase. Alloying is carried out by preliminary deposition of an alloying element film or a coating and subsequent treatment by ion, electron, laser and plasma beams [1-9]. Convective mass transfer in the melt is the main mechanism of coating and main material mixing when treatment results in melting of the surface layer. Mixing homogeneity will be increased with the growth of melt life time. Compression plasma flows (CPF) generated by quasi-stationary plasma accelerators are of special interest in this field due to relatively high pulse duration ~ 100 μs [8, 9]. A large number of experimental data have been accumulated on the elemental and phase composition, structure, properties of the metals and alloys surface layer formed by CPF treatment of the coating/material system. But most of the data were received when the coating consisted of one element only. At the same time simultaneous alloying by a few elements is of special interest for the application in industry, too. That is why it is necessary to investigate the main regularities of the elements concentration change in this case.

The investigation of element and phase composition, structure and strength properties of the plain carbon steel surface layer simultaneously alloyed by Mo and Cr atoms under the action of compression plasma flows was the main aim of this work. The choice of alloying elements was made due to the fact that these elements dissolve in each other without limits [10]. Thus, Mo-Cr coating should consist of one phase and possess equal thermal characteristics in each local area, providing greater homogeneity of Mo and Cr distribution inside the alloyed layer. Besides being the main alloying elements in steel these elements are widely used in surface alloying realized by impact of high energy particles on coating/substrate system [6-9].

2. Experimental

The samples of plain carbon steel (concentration of the alloying elements in wt.%: 0,2 C, 0,2 Si, 0,5 Mn) were investigated objects. Mo-Cr coating with the thickness of 2 μm was deposited on samples before CPF treatment. The coatings were deposited using the vacuum arc vapor deposition technique with two cathodes. The operating parameters were asfollows: the arc current was 100 A, the negative bias voltage on the samples was – 120 V, the deposition time was 5 minutes. Chromium concentration in the coating was 15 at. %. The samples with the coating were treated by compression plasma flows generated in the magneto-plasma compressor of compact geometry. Treatment was carried out by 1, 3 and 5 pulses at the interval of ~ 5 s. The energy density of heat flux absorbed by the sample was 13 J/cm^2 per pulse (registered by calorimetric measurements). The experiments were performed in the "residual gas" mode, i.e. the vacuum chamber was filled with the nitrogen gas at the pressure of 400 Pa.

The structure and phase composition of the surface layer were investigated by means of the X-ray diffraction (XRD) method uthing the Ultima IV RIGAKU diffractometer in Bragg-Brentano geometry with parallel beams in Cu Kα radiation. The analysis of cross-section morphology was carried out by means of scanning-electron microscopy (SEM) using the LEO 1455 VP microscope. The element composition of the modified surface layers of the samples was determined by the energy dispersion X-ray microanalysis using the Oxford X-ray detector. The microhardness was measured by the Vickers diamond indentor under the load of 0.98 N using the PMT-3 device.

3. Results and discussion

The analysis of the samples cross-section structure carried out by SEM showed that CPF impact on the (Mo,Cr)/steel system resulted in the formation of the steel surface layer alloyed by Mo and Cr atoms (Figure 1). The thickness of the melted layer after one pulse of treatment was 5.3-7.5 μm (Figure 1c). It increased up to 10.0-13.3 μm with the growth of the pulses number (Fig. 1e). The action of convection processes provided alloying elements distribution in the whole volume of the melt. The thickness of the alloyed layer after one pulse of CPF treatment was equal to 8 μm and after five pulses of treatment increased up to 10 μm according to the data presented in Figures 1d and 1f. These values correspond to the thicknesses of the melted layer in the areas chosen for the element analysis. Thus, an increase of the pulses number at the used value of energy absorbed by the surface layer led to the growth of the melted and, hence, alloyed layer in the investigated system. Besides, the pulses number growth led to an increase of elements distribution homogeneity in the alloyed layer (Figures 1d and 1f). This effect occurred due to the growth of the melt life time and thus due to an increase of convective processes duration that led to elements distribution homogenization.

The pulses number growth also resulted in diminishing the molybdenum and chromium concentration in the alloyed layer (Figure 2) that can be caused by two main reasons: redistribution of alloying elements in the melt with greater thickness, surface erosion during plasma impact. Hydrodynamic flow of the melt under pressure of plasma flow spreading along the surface is the main mechanism of erosion [11]. It should be noted that the ratio of chromium and molybdenum concentrations in the alloyed layer corresponded (within the measurement error) to that in the coating (Figure 2). T

The results of the phase composition analysis before and after CPF treatment are presented in Figure 3. One can see that a solid solution Mo(Cr) was the main phase in the coating deposited on the steel surface. Besides, the diffraction line is observed in the area of 39.5 degree that can be attributed to molybdenum oxide. This oxide can be formed due to interaction with molecules of residual atmosphere in the vacuum chamber during deposition.

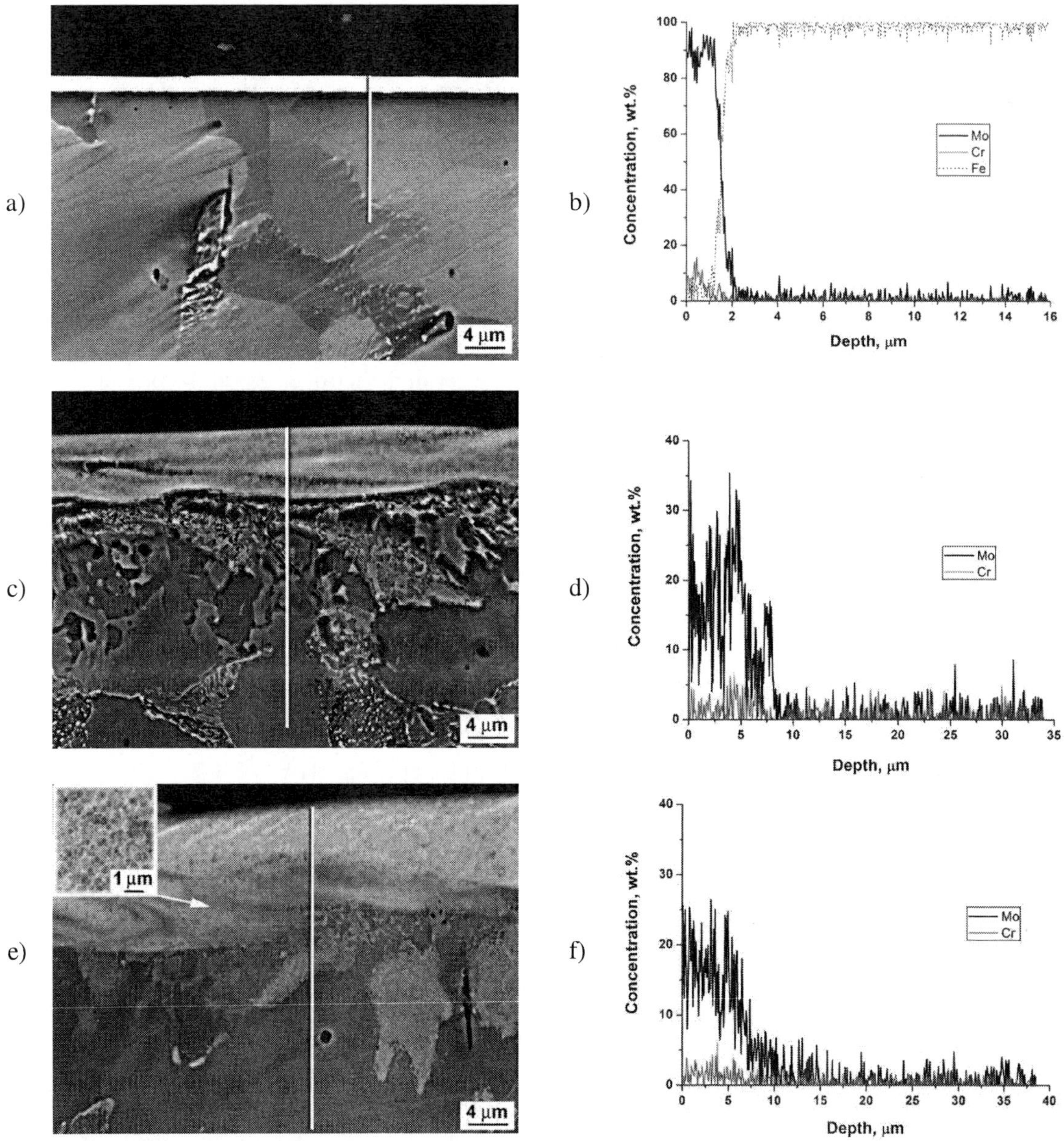

Figure 1. Cross-section morphology (a, c, e) and distribution of elements along the line (b, d, f) in the sample with the coating (a, b) and the samples treated by CPF by one pulse (c, d) and five pulses (e, f).

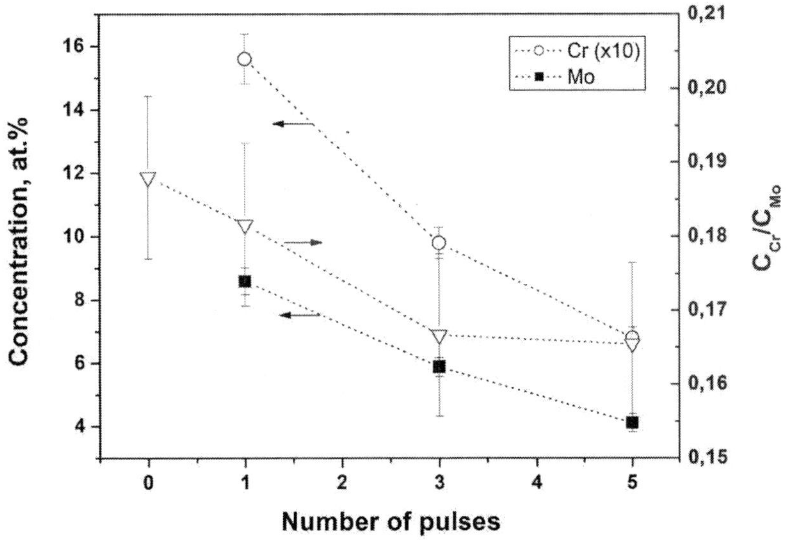

Figure 2. Dependence of the alloying elements concentration and their ratio in the steel surface layer on the number of pulses.

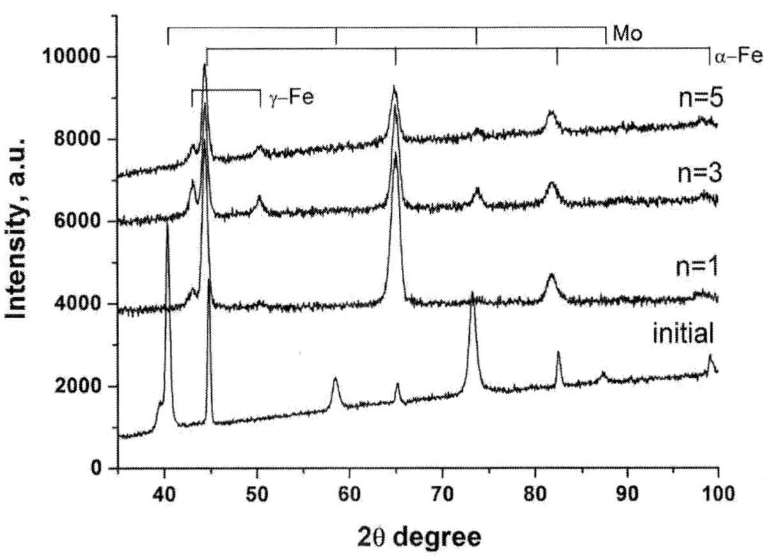

Figure 3. Diffraction patterns of the steel sample with the coating and the samples treated by CPF with different number of pulses.

The diffraction lines of Mo(Cr) at the diffraction patterns of the samples treated by CPF are not observed. α-Fe and γ-Fe solid solutions containing alloying elements were the main phases found in the alloyed layer. α-Fe diffraction lines are shifted to the area of smaller angles due to the lattice parameter increase, thus indicating formation of substitution solid solution α-Fe(Mo,Cr). α-Fe lattice parameter derived using (211) diffraction line was equal to 0.2865±0.0003 nm before CPF treatment. After one pulse of treatment it increased up to 0.2886±0.0003 nm and after five pulses - 0.2883±0.0003 nm. According to the equilibrium phase diagrams of Fe-Cr and Fe-Mo and the elements composition analysis data (Figure 2) one can expect that all chromium atoms will take part in the formation of the solid solution on the base of α-Fe while molybdenum atoms can take part both in the formation of a solid solution and Fe_2Mo intermetallide though Fe_2Mo diffraction lines are not found in the diffraction patterns. One can see that α-Fe lattice parameter does not depend on the number of pulses. It means that molybdenum solid solubility limit in α-Fe can be reached after five pulses of treatment. Thus dispersed Fe_2Mo precipitates irresolvable by XRD analysis can exist in the alloyed layer. Earlier it was found [12] that diffusion of nitrogen atoms from residual atmosphere of vacuum chamber into the surface layer was the main reason of γ-Fe existence.

The surface layer structure change led to the the modification of its strength characteristics (Figure 4). Microhardness of the initial steel sample (without the coating) was 210±50 HV. The increase of the pulses number led to the microhardness increase. After five pulses of CPF treatment the microhardness of the surface layer was 3.5 times higher than that of steel before treatment.

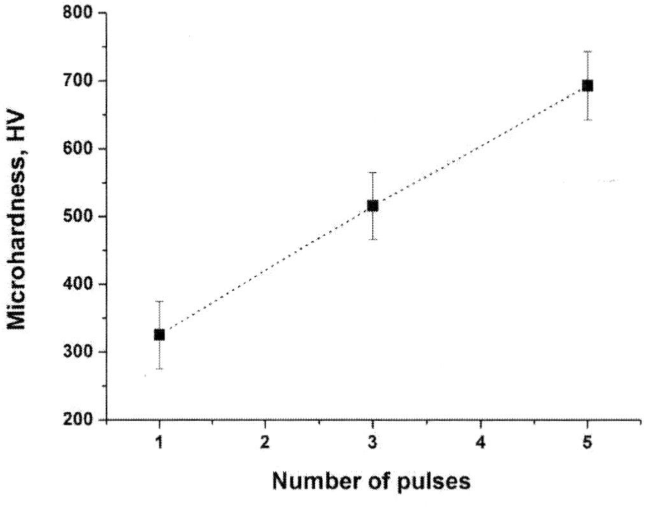

Figure 4. Dependence of samples microhardness on the number of pulses.

The increase of the pulses number resulted in the alloying elements concentration decrease which in turn, can lead to diminishing solid solution hardening and disperse hardening (due to possible formation of dispersed intermetallide precipitates) contributions to the microhardness increase. Therefore, structure refinement that occurred due to repeated plasma treatment can be the main hardening mechanism. This supposition is confirmed by SEM data. The mean size of α-Fe grains in the alloyed layer after one pulse of treatment was 0.75 μm, while after five pulses of treatment it became 0.25 μm (insert in Figure 1e).

4. Conclusions

The findings showed that compression plasma flows impact on the (Mo, Cr)/steel system led to the formation of the steel surface layer alloyed by Mo and Cr atoms with the thickness up to 13 μm. The growth of pulses number resulted in an increase of the alloyed layer thickness, an increase of elements distribution homogeneity and a decrease of Mo and Cr concentration inside the alloyed layer. The ratio of Cr and Mo concentration in the alloyed layer remained constant within the measurements error. α-Fe and γ-Fe solid solutions containing alloying elements were the main phases found in the alloyed layer. The microhardness of the alloyed layer after five CPF pulses treatment became 3,5 times higher in comparison with the initial steel because of structure refinement.

References

[1] Kalin B A, Yakushin V L, Vasiliev V I, Tserevitinov S S 1997 *Surf. Coat. Tech.* **96** 110
[2] Zhang C, Lv P, Cai J, Peng C-T, Jin Y, Guan Q 2017 *Applied Surface Science* **422** 582
[3] Zhang C, Lv P, Cai J, Zhang Y, Xia H, Guan Q 2017 *J. Alloy Compd.* **723** 258
[4] Rotshtein V P, Ivanov Yu F, Markov A B, Proskurovsky V I, Karlik K V, Oskomov K V, Uglov V V, Kuleshov A K, Novitskaya M V, Dub S N, Pauleaue Y, Shulepov I A 2006 *Surf. Coat. Tech.* **200** 6378
[5] Leivi A Ya, Yalovets A P, Krasnikov V S 2013 *High Temp. Mater.* **17(1)** 15
[6] Makhlay V A, Garkusha I E, Bandura A N, Byrka O V, Chebotarev V V, Fedorchenko V D, Medvedev A V, Tereshin V I 2009 *Eur. Phys. J. D* **54** 185
[7] Pogrebnjak A D, Bakharev O G, Sushko V V, Bratushka S, Mikhaliov A D, Ivanov Yu F, Markov A B, Proskurovskiy D I, Rotstein V P, Valyaev A N 1998 *Surf. Coat. Tech.* **99** 98
[8] Uglov V V, Anishchik V M, Cherenda N N, Stalmashonak A K, Astashinski V M, Kuzmickii A M, Kostyukevich E A, Kovyazo A V 2005 *Vacuum* **78** 489
[9] Uglov V V, Cherenda N N, Anishchik V M, Stalmashonak A K, Astashinski V M, Mishchuk A A 2007 *Vacuum* **81** 1341
[10] Lyakishev N P 1996 *Equilibrium diagrams of binary metal systems* (Moscow: Mashinostroenie) vol. 2 p 1024
[11] Tereshin V I, Garkusha I E, Bandura A N, 2003 *J. Nucl. Mater.* **313–316** 685
[12] Cherenda N N, Shimanskii V I, Uglov V V, Astashinskii V M, Ukhov V A 2012 *J. Surf. Invest.-X-ray+* **6(2)** 319

IOP Conf. Series: Journal of Physics: Conf. Series **1238** (2019) 012058 doi:10.1088/1742-6596/1238/1/012058

Optical radiation from plasma of abnormal glow discharge in various gas mixtures.

T V Stepanova*[1], Yu V Borisyuk[1], D V Mozgrin[1], P S Nenashev[1], V S Norakidze[1], N M Oreshnikova[1], T. Müller[2], L. Misozhnikov[3], and A A Pisarev[1]

[1] National Research Nuclear University MEPhI (Moscow Engineering Physics Institute), Kashirskoe shosse 31, 115409 Moscow, Russia

[2] Rübig GMBH, Rübig GmbH&CoKG, Durisolstrasse 12, 4600, Wells, Austria

[3] SKTB KASKAD, 109147, Marxist street, 34/4, Moscow, Russia

E-mail: shtanya@mail.ru

Abstract. Optical radiation from plasma of abnormal glow discharge in Ar+N_2, He+ N_2 and N_2+H_2 was investigated. Lines of N_2 N_2^+, N, N^+, Ar^+, Ar, He, Fe, H, H_2 and OH were detected by optical spectrometry. Discharge current and intensities of spectral lines of N_2, N_2^+, N, and N^+ versus the concentration of Ar, He, and H_2 in the gas mixtures were measured at various compositions of the working gas.

1. Introduction

Abnormal glow discharge is often used for plasma nitriding [1,2] to improve mechanical and chemical sustainability of parts of various machines. Well-known drawback of the glow discharge at elevated pressures is its sudden transition in the arc mode. Nevertheless, if to operate the discharge at impulse-periodical regime at a high frequency, the arc has no time to develop [3], which makes this regime attractive for technological applications.

Nitriding is usually performed in mixtures of gases, which are added to nitrogen to accelerate the process of nitriding and control the structure of the nitrided layer. For example, addition of hydrogen removes oxides due to chemical interaction, while addition of argon cleans the surface due to physical sputtering, and both activate nitrogen penetration into the material. Some authors suggest using helium instead of argon in the process of nitriding [4]. The addition of helium leads to a smooth heating of the sample surface without local overheating and surface destruction. However, the composition of plasma, particularly the concentration of ions and exited neutrals, which participate in nitriding, depend on the composition of the working gas, and this is also a factor of significant influence on the characteristics of the nitrided surface. Therefore, various investigations of the discharge are performed to choose optimum processing parameters and plasma composition.

This work describes results of investigation of discharge current and content of nitrogen ions (N_2^+, N^+) as well as excited atoms and molecules (N_2, N) in plasma of abnormal glow discharge in various N_2+Ar, N_2+He, and N_2+H_2 gas mixtures.

2. Experimental setup

The scheme of the installation operating with the abnormal glow discharge is shown in Figure 1a. It consists of a vacuum chamber, a pumping system, a gas inlet system, and a power supplier [1,2]. The cathode is a rod 100 mm in lengh and 12 mm in diameter coaxially installed inside a cylindrical anode 180 mm in diameter and 420 mm in length. The cathode was made of stainless steel AISI 321.

The discharge is powered by a generator of rectangular pulses with the amplitude of 400-650V, maximum power of 12 kW, pulse repetition frequency 1 - 100 kHz, and the pulse duty factor 10 - 80%. The discharge voltage and current are measured by a four-channel digital oscilloscope Tektronix TPC 2024B using a voltage divider and a shunt. Typical dependencies of the discharge voltage and current are shown in Figure 1b.

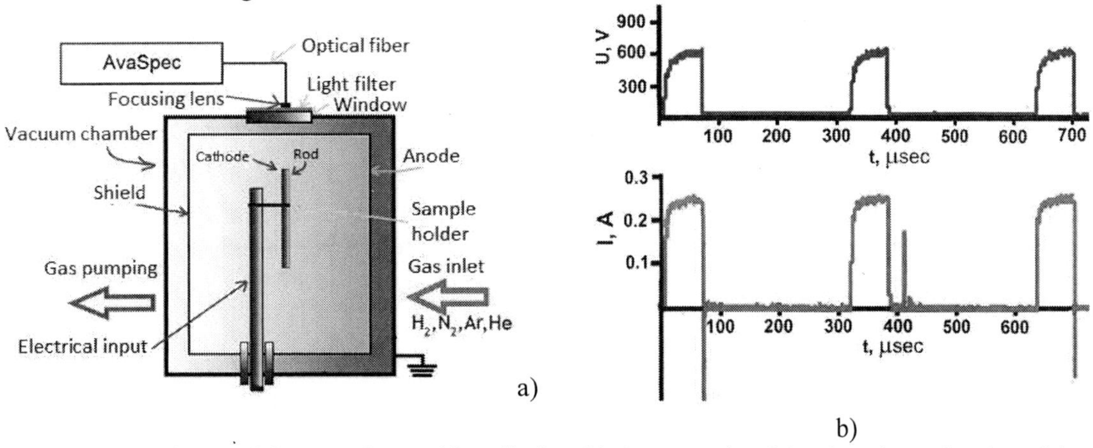

Figure 1. a) Scheme of the experimental installation, b) An example of the time dependencies of the discharge voltage and current for the discharge ignited in N_2 at 1 Torr, 600V, 3 kHz.

Discharges in Ar, N_2, H_2, He, Ar+N_2, N_2+He, and N_2+H_2 at a pressure of 1.0, 1.5, and 2.0 Torr were investigated. The electron temperature of the discharge and the plasma density were previously measured to be 2-5 eV and $1\text{-}3\times10^{10}$ cm^{-3}, respectively [5].

Optical radiation from plasma was measured through a quartz glass window and neutral grey light filter by a 3-channel spectrometer AvaSpec-ULS2048L-USB2-RM. The wavelength ranges and optical resolutions of the three channels are: 200-365 nm and 0.12 nm for the first channel, 364-603 nm and 0.18 nm for the second channel, and 600-810 nm and 0.15 nm for the third channel. The integration times used for data accumulation were 400 ms, 50-200 ms, and 40-100 ms for the three channels, respectively. The signal was analyzed by a standard software package Avantes and identified using data bases [6-8]. The start of spectroscopic measurements was synchronized with application of the discharge voltage.

3. Experiment results

Figure 2 shows examples of panoramic optical radiation spectra from the discharges in Ar+N_2 (1:1), N_2+H_2 (1:1), and N_2+He (1:1), ignited at a pressure of 2.0 Torr, voltage of 600V, pulse frequency of 3 kHz, and pulse duty factor of 80%. Panoramic optical radiation spectra ignited at a pressure of 1.0 and 1.5 Torr are similar but differ in the intensity of spectral lines and band of lines. Intensities of spectral lines increase with the gas pressure. By comparison of the experimental spectra with reference spectra, the following lines and bands of lines were identified in the Ar+N_2 discharge: Ar^+, Ar, N_2, N_2^+, N, N^+, Fe, H, OH. Lines and bands of lines in mixture of N_2+H_2 were identified as N_2, N_2^+, N, N^+, Fe, H, H_2, and OH. Similar results were observed in [3]. Lines and bands of lines in mixture of N_2+He were identified as N_2, N_2^+, N, N^+, He, H, Fe, and OH. The intensity of radiation depends on the pressure and the applied voltage. The energy of electrons mainly depends on the applied potential and on the gas pressure. Optical emission increases with the gas pressure due to increase of the probability of inelastic collisions [9]. Lines OH and H in nitrogen plasma appear due to water vapor, while lines of Fe appear due to sputtering of the cathode.

Figure 3 shows dependencies of intensities of several optical lines (in arbitrary units) on gas composition, and the discharge current as a function of the composition is given for comparison. These dependencies were obtained at a pressure of 1.0 Torr, voltage of 600 V, pulse frequency of 3 kHz, and a pulse duty factor of 80%. The intensities of the following lines and bands of lines are shown: H_α (656,1 nm), $N_2(C^3\Pi_u \rightarrow B^3\Pi_g$, 380.5 nm) , N (2p3s→2p3s, 648.4 nm), $N_2^+(B^2\Sigma_u^+ \rightarrow X^2\Sigma_g^+$, 391.4 nm), N^+(2p3s→2p3p, 566.6 nm), Ar $(3s^23p^5(^2P^o_{3/2})4s \rightarrow 3s^23p^5(^2P^o_{1/2})4p$, 696.4 nm), $Ar^+(3s^23p^4(^3P)4s \rightarrow 3s^23p^4(^3P)4p$, 488.0 nm), and He (1s2p→1s3d, 587.5 nm). Intensities of molecular lines $N_2(C^3\Pi_u \rightarrow B^3\Pi_g$, 337.1 nm) and N_2^+ $(B^2\Sigma_u^+ \rightarrow X^2\Sigma_g^+$, 427.8 nm) were also measured, and their features were the same as those of the intensities of respective neutral and ion lines given in the figure. Experiments at the gas pressures of 1.5, and 2.0 Torr demonstrated similar tendencies.

Figure 3a is given for the discharge in the N_2+Ar mixture. One can see that the discharge current and intensity of Ar radiation (both ions and neutrals) increase with increase of the Ar content at the fixed total pressure of the N_2+Ar mixture. The discharge current increases with the pressure of Ar due to the increase of the secondary electron emission from the cathode under impact of heavier Ar ions and due to increase of ionization degree in plasma with increase of the electron emission. The concentration of Ar^+ ions increases due to increase of the number of electrons that ionize neutrals. The concentration of N_2^+ and N^+ increases with increase of nitrogen content in mixture, though the current decreases, and this may be connected with smaller ionization potential of N_2 (14.53V) with respect to argon (15.76V). N_2 and N radiations have maxima: a faint maximum for atoms and a bright maximum for molecules (right branch is due to decrease of the N_2 concentration, while the left branch is due to variation of the current with Ar pressure). The intensities of excited N_2 and N have maxima (the right branch decreases due to a decrease of N_2 content in mixture, and left branch increases due to Penning exitation [10] where excited Ar transfers its energy to N_2 molecule in inelastic collision and excites it. So, general trends in the current and line intensities in Ar+N_2 mixture are qualitatively described by features of electron emission from the cathode and ionization and exitation of gas components by electron impact. Penning excitation of N_2 by exited Ar atoms is important at Ar content below 50%.

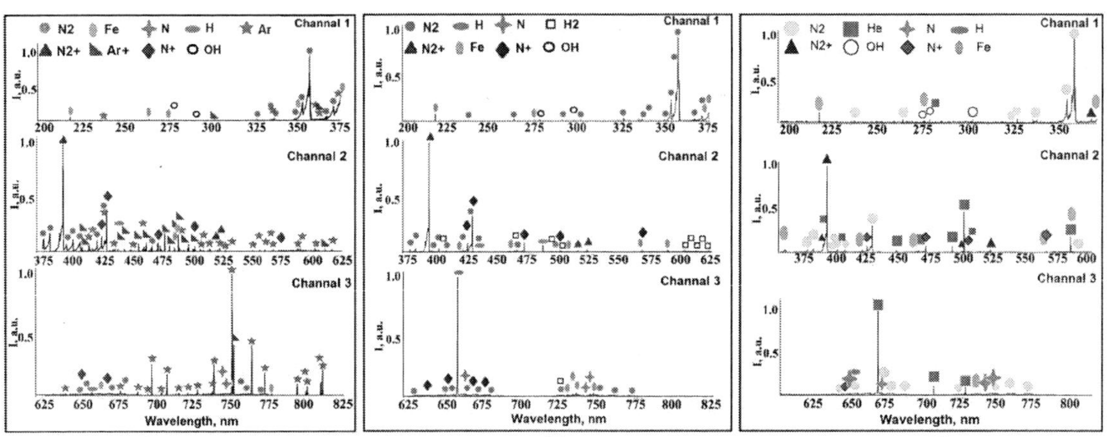

(a) (b) (c)

Figure 2. Emission spectra of an abnormal glow discharge in: a) Ar +N_2 (1:1), b) N_2+H_2 (1:1), c) He+N_2(1:1)

Figure 3b is given for the discharge in the N_2+H_2 mixture. One can see that the current decreases at high and low concentrations of H_2 content in mixture. The increase of left branch of the current is due to smaller ionization potential of H_2 (13.6 eV for H instead of 14.53 eV for N), while the decreasing right branch is due to decrease of the secondary electron emission from the cathode. H lines follow the current qualitatively as the dissociation, excitation, and ionization are proportional to electron density in plasma. Besides, the increasing left branch is associated with increase of H_2 content in the gas, and therefore increase of the H line is faster than increase of the current. All nitrogen lines (N, N^+) and bands (N_2, N_2^+) behave similarly to each other. A small increase of intensity is observed at 10-20% H_2 content. Then intensity of all lines and bands decreases with the increase of the H_2 content in mixture. Increasing left branch of all nitrogen lines is due to smaller ionization potential of hydrogen. Fast general decrease of nitrogen lines at the H_2 concentration above 20% is due to two factors: decrease of N_2 content in gas and decrease of the current, which is connected with the decrease of the secondary emission from the cathode. The nitrogen lines decrease faster than the current because of the contribution of the first factor.

Figure 3c is given for the discharge in the N_2+He mixture. One can see that the current decreases rather rapidly with increase of He content in the mixture due to very high ionization potential of He (24.58 eV) and decrease of the secondary electron emission from the cathode. All nitrogen lines and bands generally follow the decrease of the current with increase of He content. Increase of He content leads to increase of the number of He atoms in metastable state in plasma. According to [11], the faint increase in the intensity of N_2^+, which is observed at low concentration of He, can be due to the Penning ionization process where He in metastable state transfers its energy to N_2 molecule in inelastic collision and ionizes it.

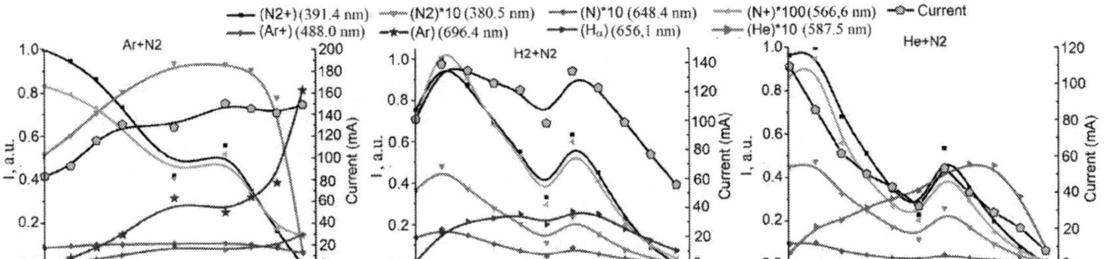

Figure 3 . Discharge current and intensities of selected optical lines at various compositions of the working gas: a) Ar +N_2, b) N_2+H_2, c) He+N_2 at U=600V, ν=3 kHz, D=80%, p=1 Torr

Figure 4 shows the compilation of the dependences of the discharge current and line intensities (in arbitrary units) of ions N^+ and excited neutrals N on the pressures of N_2 and supplementary gases in various gas mixtures at U=600V, ν=3 kHz, D=80%, p=1 Torr.

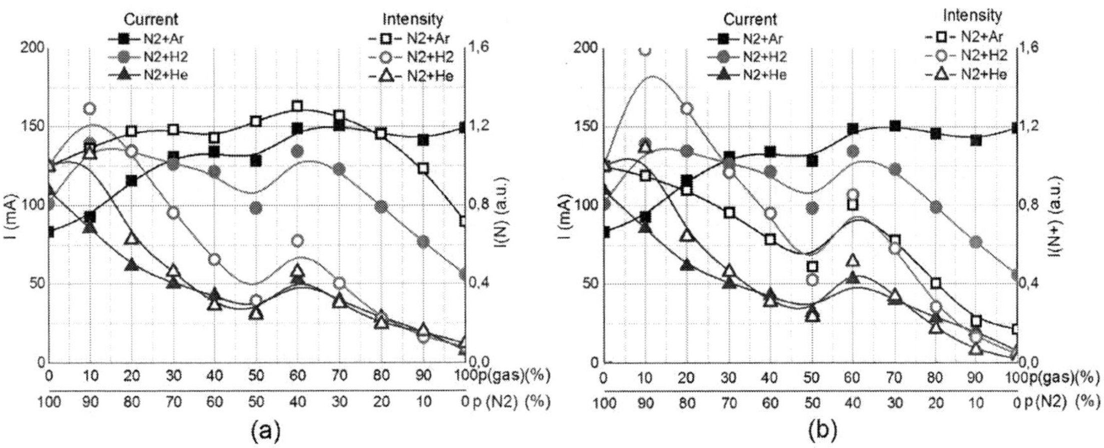

Figure 4. Discharge current and radiation of a) neutrals N (648.4 nm) and b) ions N^+ (566.6 nm) in various gas mixtures at U=600V, ν=3 kHz, D=80%, p=1 Torr

One can see that with increase of the N_2 content, the discharge current decreases in mixture with Ar, increases in general in mixture with He, and has an intermediate behavior in mixture with H_2. In the technologically useful region of N_2 content below 40%, the current is approximately constant in mixture with Ar, and increases with increase of N_2 content in mixtures with H_2 and He. With increase of N_2 content in gas mixtures, intensities of neutral lines generally increase in mixtures with light gases following the current trends, but in mixture with heavy Ar they have flat maxima. Concentrations of N_2^+ and N^+ generally increase with increase of N_2 content in all gas mixtures. In the technologically useful region of N_2 content below 40%, all neutral and all ion lines increase with increase of the N_2 content in all working gases. At high concentrations of N_2 above 90%, concentrations of N_2^+ and N^+ decrease in mixture with light gases (H_2 and He) and increase in mixture with heavy Ar. Maximum current at 90% N_2 content is in mixture with H_2.

4. Conclusion

Variation of the current and optical plasma radiation of neutrals and ions in the abnormal glow discharge powered with the frequency of 3 kHz in N_2+Ar, N_2+He, and N_2+H_2 were investigated at pressures of 1.0 - 2.0 Torr. All main lines of N_2, N_2^+, N, and N^+ particles, which are active in nitriding of steels and alloys, have been found in optical spectra. At the N_2 content in the working gases below 40%, which is important for technological applications, the discharge current is invariable in mixture N_2+Ar, while it decreases with decrease of the N_2 pressure in mixtures N_2+He, and N_2+H_2. Concentrations of all exited nitrogen neutrals and nitrogen ions decrease with decrease of the N_2 pressure in this pressure range in all gas mixtures. In the region of high N_2 content, behavior of line intensities is rather complicated, but in general, increase of the N_2 pressure leads to the following features: the neutral lines (N) in mixture N_2+Ar slightly decrease, while neutral lines in N_2+He, and N_2+H_2, increase and have a not well pronounced maximum at about 90%N_2. All ion lines at high N_2 concentration increase with the N_2 pressure in general, and lines in mixtures N_2+He, and N_2+H_2 have a not well pronounced maximum at about 90% N_2.

Acknowledgements
This work was partially supported by Ministry of Education and Science of the Russian Federation (Contract № 14.Y26.31.0008 and Contract № 3.9750.2017/8).

References

[1] Borisyuk Yu V, Oreshnikova N M, Berdnikova M A, Tumarkin A V, Khodachenko G V and Pisarev A A 2015 *Phys. Procedia* **71** 105-109
[2] Borisyuk Yu V, Oreshnikova and Pisarev A A 2017 *J. of Surface Investigation: X-ray, Synchrotron and Neutron Techniques* **11(1)** 259-264
[3] Granovsky V D 1974 *Electric current in gases. The steady current* (Moscow: Nauka)
[4] Arzamasov B N, Gromov V I 1998 *J. of Friction and Wear* **19(2)** 224-226
[5] Borisyuk Yu V, Oreshnikova N M, Mozgrin D V, Stepanova T V and Pisarev A A 2016 *Journal of Physics: Conference Series* **747** 012019
[6] Zaidel A N, Prokofiev V K and Raisk S M 1962 *Tables of Spectral Lines* (Moscow).
[7] Pearse R W B and Gaydon A G 1975 *The Indentification of Molrecular spectra* (London, New York: Chapman and Hall)
[8] Ochkin V N 2009 *Spectroscopy of low temperature plasma* (Wiley-VCH)
[9] Saeed A, Khan A W, Jan F, et al. 2014 *Mat. Res.* **17** 857
[10] Qayyum A, Shaista Zeb, Naveed M A, Rehman N U, Ghauri S A and Zakaullah M 2007 *J. Quant. Spectrosc. Radiat. Transfer* **107** 361–371
[11] Shakhatov V A, Lebedev Yu A 2012 *High Temperature* **50** 658-681

Hydrodynamic model of plasma evolution under heating by high-energy ion flux

S Yu Gus'kov[1,2], Ph A Korneev[1,2], M S Solyanikova[2]

[1] P. N. Lebedev Physical Institute, RAS, Moscow
[2] National Research Nuclear University MEPhI (Moscow Engineering Physics Institute), Moscow, Russia

Solyanikova_marina@mail.ru

Abstract. The analytical solution is found for thermodynamic state of plasma created when the half-space is heating by high-energy ion flow. The dependence of free path length of heating ions on plasma temperature is taken into account. Besides, an analysis of characteristic parameters of shock wave and efficiency of energy transfer to shock wave from heating ion flow is given.

1. Introduction

The heating of matter by high-energy charged particles flow is effective method to create ultrahigh pressure and to generate powerful shock wave in the laboratory experiment. Applied interest is associated with development of the promising method of ignition of inertial confined fusion target - ignition of pre-compressed target fuel under the action of powerful converging shock wave (shock ignition) [1, 2], as well as with a possibility to study of equation of state of matter in laboratory conditions at pressure of gigabar level, which is more than an order of magnitude higher in comparison with recent research using shock wave [3].

Special interest of matter heating by high-energy ions is connected with development of heavy ion inertial fusion.. The peculiarity of plasma heating by high-energy ions, in comparison with fast electrons, is that ion free path length depends on plasma temperature.

Non-stationary hydrodynamic model of plasma evolution during the heating a half-space by flow monoenergetic fast electrons with constant intensity was constructed in [4]. The analytical solution was found as a superposition of solutions for isothermal rarefaction wave and self-similar solution for isothermal expansion of given mass of matter [5] under conditions when the particle free path length of heating fast electrons is independent on temperature of heated region and remained unchanged during the all time of process of heating.

This paper is devoted to solution of similar problem for the heating a half-space by flow monoenergetic ions with the only difference that heating ion free path length λ_j is not independent value on temperature (and, therefore, from time) and increases with temperature of heated region as $\lambda_j \propto T^{3/2}$.

2. Isothermal expansion of plasma heated by high-energy ion flow

To obtain an analytical solution we use simplified dependence of mass range of ion deceleration (which below is named as ion mass range) $\mu_j \equiv \int \lambda_j d\rho$ on its initial energy in the form of matching path dependences at large $\left(\mu_j \propto E_j^2\right)$ and low ion energies $\left(\mu_j \propto E_j^{1/2} \cdot T^{3/2}\right)$ [6]. The dependencies are matched at the so-called energy thermal threshold, the value of which is close to the characteristic value the ion energy when its velocity is equal to the velocity the thermal electron.

$$\mu_j \approx \mu_s, \qquad \mathrm{E}_j \geq E_s$$

$$\mu_j \approx \mu_s \left(\frac{E_s}{E_j}\right)^{3/2}, \quad \mathrm{E}_j \leq E_s ,$$

where λ_j is the length the total deceleration of the ion with the initial energy E_j; T and ρ - average values density and temperature of substance;

$$\mu_s \approx \frac{0.002 E_j^2}{Z_j^2 A_j (1-0.1\Lambda)}, \quad g/cm^2,$$

E_s is energy thermal threshold is equal to the ion energy at which its velocity is 1.5 v_{T_e} (v_{T_e} - velocity of thermal electrons)

$$E_s = \frac{m_j}{m_e} T \approx 4.1 \cdot 10^3 A_j T ,$$

where Z_j and A_j is charge and atomic weight ion; E_j and E_s is measured in MeV, $\Lambda = ln\left(\rho\mu_j^2 / \mathrm{E}_j^2\right)$. The approximation used is in good agreement with the data numerical calculations for the range of fast ions [7-9].

Solution [4] for fast electrons is superposition of two solutions: solution for initial heating stage - at $0 \leq t \leq t_h$, when rarefaction wave reaches the internal boundary of heated region - where matter is considered to be at rest, and the temperature increases linearly with time at the fixed intensity of the heating flow, and self-similar solution for $t \geq t_h$ for plane isothermal expansion of given mass of matter, according to which the density decreases with time, according to the law $\rho \propto t^{-3/2}$, and the temperature still grows linearly with time. The maximum pressure in the heated layer of matter is reached at the time t_h, after which it begins to decrease with time (according to the law $P \propto t^{-1/2}$).

The first heating stage will always correspond to the case when heating ion range is independent from plasma temperature and is equal to the value μ_*. In this there is characteristic time t_s, during which the substance is heated to such an extent that the thermal threshold reaches the initial energy the ion $E_s = E_j$, after which further heating occurs on the ion mass range , which increases with increasing temperature, according to the law $\mu_* \propto T^{3/2}$. The solution type and value the maximum pressure that can be reached under high-energy ion heating depends on the ratio times t_s and t_h, i.e. it will be different situation in the cases when increasing of ion mass range with temperature begins before or after the beginning of hydrodynamic matter expansion.

If expansion of the heated layer occurs before increasing of ion mass range, then there are the following stages:

1) heating of immobile of given mass layer corresponding to initial and fixed ion mass range (the step continues until the rarefaction wave spreads to the entire heated mass);

2) heating of given mass layer corresponding to initial fast ions range during the isothermal expansion of the layer (the step continues until the value of thermal threshold reaches the value of the initial ion energy);

3) heating of the layer with increasing mass, corresponding to increasing mass range of ions during the isothermal expansion of layer.

The solution in this case is:

$$T = T_h \begin{cases} \dfrac{t}{t_h}, 0 \le t \le t_s \\ \left(\dfrac{t_s}{t_h}\right)^{3/5} \left(\dfrac{t}{t_s}\right)^{2/5}, t_s \le t \end{cases}, \mathrm{P} = \mathrm{P}_h \begin{cases} \dfrac{t}{t_h}, t \le t_h \\ \left(\dfrac{t_h}{t}\right)^{1/2}, t_h \le t \le t_s \\ \left(\dfrac{t_h}{t_s}\right)^{3/10} \left(\dfrac{t_h}{t}\right)^{1/5}, t_s \le t \end{cases}, \rho = \rho_0 \begin{cases} 1, t \le t_h \\ \left(\dfrac{t_h}{t}\right)^{3/2}, t_h \le t \le t_s \\ \left(\dfrac{t_h}{t_s}\right)^{9/10} \left(\dfrac{t_h}{t}\right)^{3/5}, t_s \le t \end{cases},$$

t_h, t_s - are the time moments corresponding to endings of the first and second stages; T_h, P_h are defined by formulas:

$$T_h = \frac{1}{C_V}\left[\frac{9a_{th}^2}{4(\gamma-1)}\right]^{1/3}\left(\frac{I_j}{\rho_0}\right)^{2/3} \quad \text{and} \quad P_h = \left[\frac{3(\gamma-1)a_{th}}{2}\right]^{2/3}\rho_0\left(\frac{I_j}{\rho_0}\right)^{2/3},$$

where I_j is ion beam intensity, C_V - specific heat, and a_{th} - fraction of thermal component of energy.

If matter's expansion stage begins after ions range begins to depend on temperature of substance being heated, then there are following stages:

1) heating of immobile layer of a given mass (the step continues until the thermal threshold energy reaches value of ion initial energy);

2) heating of immobile layer, mass of which increases with time in accordance with increases of ion mass range(the step continues until the expansion of the layer begins);

3) heating of the layer with increasing mass during the its isothermal expansion.

The solution in this case is:

$$T = T_{h*} \begin{cases} \left(\dfrac{t_{h*}}{t_{s*}}\right)^{3/5} \left(\dfrac{t}{t_{h*}}\right), 0 \le t \le t_{s*} \\ \left(\dfrac{t}{t_{h*}}\right)^{2/5}, t \ge t_{s*} \end{cases}, P = P_{h*} \begin{cases} \left(\dfrac{t_{h*}}{t_{s*}}\right)^{3/5} \dfrac{t}{t_{h*}}, 0 \le t \le t_{s*} \\ \left(\dfrac{t}{t_{h*}}\right)^{2/5}, t_{s*} \le t \le t_{h*} \\ \left(\dfrac{t_{h*}}{t}\right)^{1/5}, t \ge t_{h*} \end{cases}, \rho = \rho_0 \begin{cases} 1, 0 \le t \le t_{h*} \\ \left(\dfrac{t_{h*}}{t}\right)^{3/5}, t \ge t_{h*} \end{cases},$$

t_{h*}, t_{s*} - are the time moments corresponding to endings of the first and second stages; T_{h*}, P_{h*} are defined by formulas:

$$T_{h*} = T_h\left(\frac{2}{5}\right)^{2/3}\left\{1+\left[1+\frac{5}{4}\left(\frac{E_j}{E_{sh}}\right)^3\right]^{1/2}\right\}^{2/3}, P_{h*} = P_h\left(\frac{2}{5}\right)^{2/3}\left\{1+\left[1+\frac{5}{4}\left(\frac{E_j}{E_{sh}}\right)^3\right]^{1/2}\right\}^{2/3},$$

where E_{sh} is the thermal threshold energy that corresponds to the temperature T_h.

The solution for the efficiency of energy transfer - the fraction of energy transferred to shock wave from heating ion flow - for the two cases are defined by formulas:

IOP Conf. Series: Journal of Physics: Conf. Series **1238** (2019) 012059 doi:10.1088/1742-6596/1238/1/012059

$$\eta = \eta_h \begin{cases} \left(\dfrac{t}{t_h}\right)^{3/2}, t \le t_h \\ 10\left(\dfrac{t_h}{t}\right)^{3/4} - 9\left(\dfrac{t_h}{t}\right), t_h \le t \le t_s \\ \left[\dfrac{90}{14}\left(\dfrac{t_s}{t_h}\right)^{1/4} - 9\right]\left(\dfrac{t_h}{t}\right) + \dfrac{25}{7}\left(\dfrac{t_h}{t_s}\right)^{9/20}\left(\dfrac{t_h}{t}\right)^{3/10}, t_s \le t \end{cases}, \quad \eta_h = \frac{3a_{th}(\gamma-1)}{5}\left(\frac{2}{\gamma+1}\right)^{1/2};$$

$$\eta = \eta_{h^*} \begin{cases} \left(\dfrac{t_{h^*}}{t_{s^*}}\right)^{9/10}\left(\dfrac{t}{t_{h^*}}\right)^{3/2}, 0 \le t \le t_{s^*} \\ \dfrac{25}{16}\left(\dfrac{t}{t_{h^*}}\right)^{3/5} - \dfrac{9}{16}\left(\dfrac{t_{s^*}}{t_{h^*}}\right)^{8/5}\left(\dfrac{t_{h^*}}{t}\right), t_{s^*} \le t \le t_{h^*} \\ \dfrac{25}{7}\left(\dfrac{t_{h^*}}{t}\right)^{3/10} - \dfrac{5}{2}\left[\dfrac{9}{40}\left(\dfrac{t_{s^*}}{t_{h^*}}\right)^{8/5} + \dfrac{45}{56}\right]\dfrac{t_{h^*}}{t}, t \ge t_{h^*} \end{cases}, \quad \eta_{h^*} = \left(\frac{P_{h^*}}{P_h}\right)^{3/2}\eta_h$$

3. Comparative analysis of the thermodynamic state of the plasma when it heated by fluxes of fast ion and electron

We introduce a time scale t_{sh} independent of the ion energy (the loading time t_h for an ion with energy equal to the energy thermal threshold at the temperature of the heated region T_h):

$$t_{sh} = \left[\frac{9}{4(\gamma-1)a_{th}}\right]^{1/3} \frac{0.002\, m_j^2 T_h^2}{Z_j^2 A_j \rho_0^{2/3} I_j^{1/3} m_e^2},$$

where m_j - is the mass of ion, m_e - is the mass of electron.

Figures 1-4 show graphs dependences of the temperature, density, and pressure normalized to T_h, P_h и ρ_h and the energy transfer efficiency on the dimensionless time t / t_{sh}.

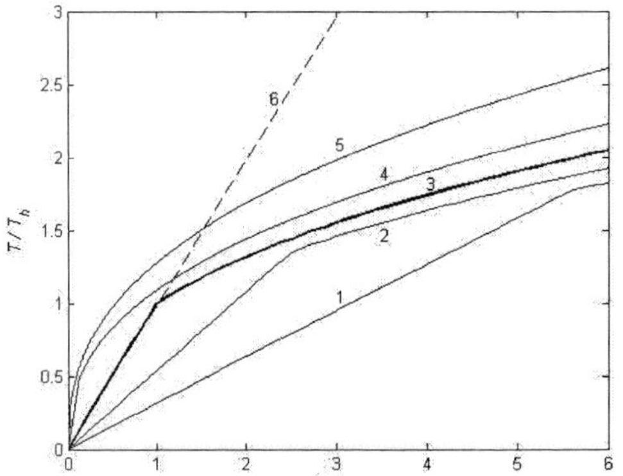

Fig. 1

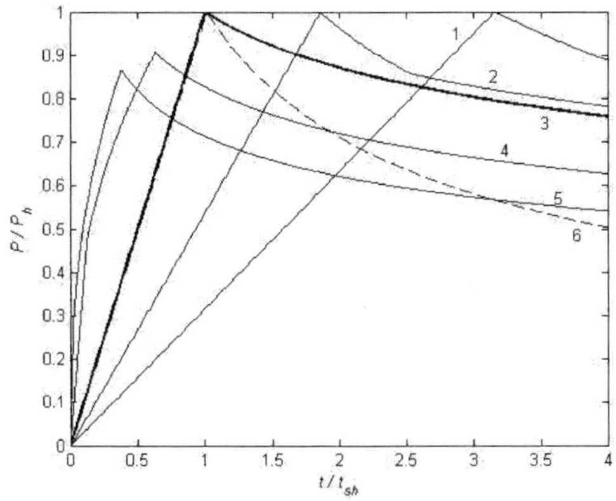

Fig. 2

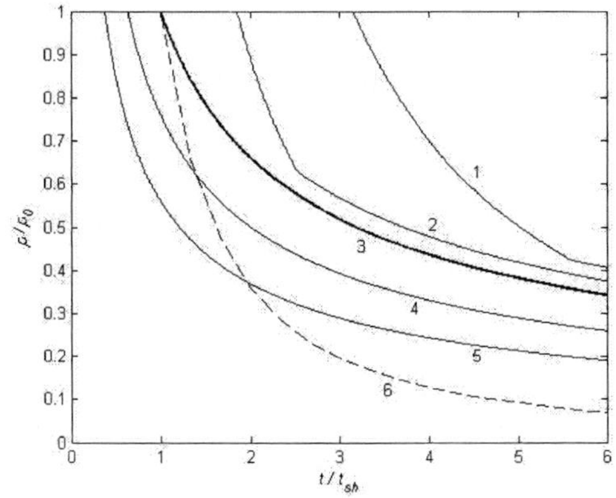

Fig. 3

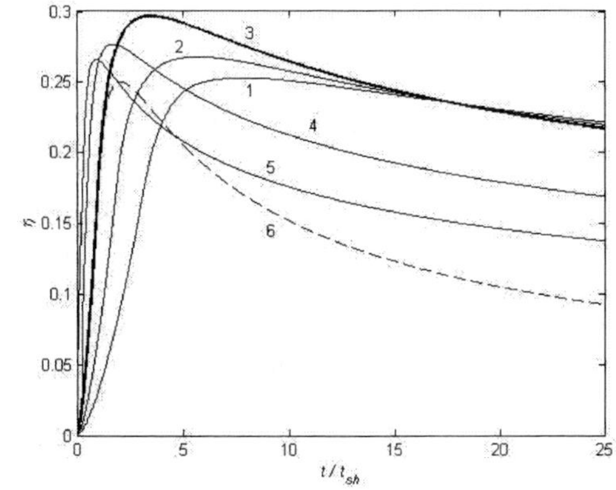

Fig. 4

Fig. 1-4. Dependence of the temperature (Fig. 1), pressure (Fig.2), density (Fig. 3) normalized to T and energy transfer efficiency (Fig. 4) on the dimensionless time. Line 1 correspond to the ion energy $E_j = 3MeV$, line 2 – $E_j = 2.3MeV$, line 3 – energy $E_j = 1.69MeV$ (energy thermal threshold), line 4 - $E_j = 1.1MeV$, line 5 - $E_j = 0.5MeV$, line 6 - the beam of fast electrons with mass range is equal to mass range of fast ion with energy equal to energy of thermal threshold.

The graphs show the features of the found solutions, connected to the variation of the dependence of the mass range of ion on the temperature of the heated region. The loading time at which the maximum pressure $P = P_h$ is reached and the expansion of the substance begins to increase with increasing of ion energy, and this growth occurs faster for ions with energy $E_j > E_{sh}$. For ions with energy $E_j \geq E_{sh}$ and fast electrons, the maximum achievable pressure is the same. For ions with energy $E_j < E_{sh}$ the maximum achievable pressure slightly decreases with decreasing ion energy.

Change to slow decrease of pressure with time ($P \propto t^{-1/5}$) for ions with energy $E_j \leq E_{sh}$ begins immediately after reaching the maximum pressure $P = P_h$. The duration this change for ions with energy $E_j > E_{sh}$ than longer that greater the ratio E_j / E_{sh}.

When heating of matter by an equivalent beam of fast electrons, there is no stage of a slow decreases pressure - after pressure reaches the maximum, it decrease like $P \propto t^{-1/2}$. The time to reach the maximum value of the energy transfer efficiency increases with increase of ions beam energy. The maximum value of energy transfer efficiency for fast electrons (25%) is 1.2 times smaller than the maximum value of the maximum for ions with an energy equal to the energy of the thermal threshold (30%).

4. Conclusion

The found solution describes the thermodynamic state of a plasma created when a flat target is heated by a stream of high-energy ions, when the free path length of ions does not depend on temperature of the heated region at the beginning of heating and and increases with the temperature as $T^{3/2}$ when the material reaches the temperature of the thermal threshold of the ion. The solution shows that the heating of substance by a stream of ions with an energy equal to the energy thermal threshold is the most advantageous regime for the formation of a quasi-stationary shock wave. Other things being equal, this regime corresponds to attainment of the limiting pressure when the substance is heated by the flow of charged particles, and the slowest subsequent pressure drop with time. Pressure of gigabar level is achieved by heating ions a flat target of aluminum by a proton beam with a particle energy about 4 MeV with a beam intensity of about $5 \cdot 10^{15}$ W/cm^2.

The paper is supported by RFBR grant # 17-02-00059_a.

References

[1] V. A. Shcherbakov, Sov. J. Plasma Phys. 9, 240 (1983).
[2] R. Betti, C. D. Zhou, K. S. Anderson, L. J. Perkins, W. Theobald, and A. A. Solodov, Phys. Rev.Lett. 98, 155001 (2007).
[3] S.Yu. Gus'kov, JEPT Letters 100, 79 (2014)
[4] S.Yu.Gus'kov, X. Ribeyre, M. Touati, J.L.Feugeas, Ph.NicolaiandV.Tikhonchuk Phys. Rev Lett. 109, 2550004 (2014).
[5] V.S. Imshenik, Sov. Phys. Dokl. 5, 253 (1960)
[6] B.A. Trubnikov, in Reviews of Plasma Physics, Ed. By M. A. Leontovich (Gosatomizdat, Moscow, 1963; Consultants Bureau, New York, 1965), Vol. 1.
[7] V.Yu. Bychenkov, W. Rozmus, A. Maksimchuk, D. Umstadter, and C. E. Capjak, Plasma Phys. Rep. 27, 1017 (2001).
[8] J.J. Honrubia, J. C. Fernandez, M. Temporal, B. M. Hegelich, and J. Meyer_ter_Vehn, Phys. Plasmas 16, 102701 (2009).
[9] S.Yu. Gus'kov, N. V. Zmitrenko, D. V. Il'in, A. A. Levkovskii, V. B. Rozanov, and V. E. Sherman, Plasma Physics Reports 35, 709 (2009).

The Development of Accelerating Structure E-type for 1 MeV Proton LINAC

N V Avreline

TRIUMF Canada's National Laboratory for Particle and Nuclear Physics, Vancouver, Canada

navreline@triumf.ca

Abstract. Experimental results and computer simulations of electrodynamic and thermodynamic characteristics are presented for an accelerating structure that is excited in the TM_{010}^{0} mode and that has the accelerating channel of URAN-1M located in the diametric plane. The idea of using this structure in the particle accelerator URAN-1M, located at the Baikov Institute of Metallurgy and Materials Science, with the goal of increasing the average beam current is explored. The experimental study and the computer simulation of accelerating structure excited in E010 mode that has accelerating channel of URAN-1M placed in the diametric plane of structure resulted in $R_{sh} = 165\ M\Omega/m$ and $Q_0 = 9736$. To form an accelerating field identical to that of URAN-1M, the structure requires less than 15 kW RF power. Thermal analysis showed that if the main parts of the structure are water-cooled, the structure could operate in CW regime.

1. Introduction

1 MeV Proton LINAC URAN-1M was developed and been in successful operation for many years at the A.A.Baikov Institute of Metallurgy and Materials Science (IMET), Russian Academy of Sciences (RAS) [1, 2]. Currently, IMET is planning to upgrade this LINAC to achieve higher average beam current. As the pulse beam current is close to the Coulomb limit, only an increase in the duty factor would allow to increase the average beam current. The high duty factor operation runs into heatsink problems. The development of a new more efficient accelerating structure would reduce the level of RF power and cool down the structure.

2. Investigation of mock-up E-type Accelerating Structure

The accelerating structure [3], excited in TM_{010}^{0} mode, with accelerating channel in the diametric plane, was selected as prototype for the new accelerating structure. Figure 1 shows the design of this structure. The two longitudinal bars, that hold the accelerating channel with drift tubes are placed on two cylindrical supports inside of the cylindrical cavity and are placed at the cavity's central axis. The diameter of the cavity is D=420 mm, the height of cavity is H=263 mm. The accelerating channel consists of 12 drift tubes, the tubes' length are 20 mm and gaps between them are 12 mm. The channel was designed for relative speed of protons β=0.022 ($\beta=v/c$, where v – speed of protons and c – speed of light in free space).

IOP Conf. Series: Journal of Physics: Conf. Series **1238** (2019) 012060 doi:10.1088/1742-6596/1238/1/012060

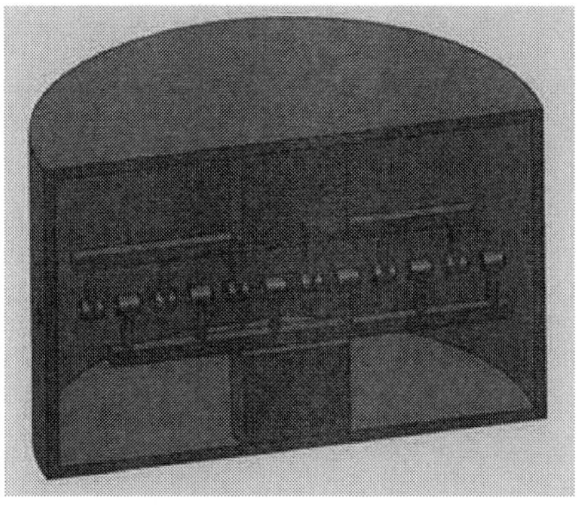

Figure 1. The model of the accelerating structure operating in TM_{010}^{0} mode.

As the result of simulating the accelerating structure in CST Microwave Studio, it was computed that $f_0 = 174.5\ MHz, Q_0 = 9457, R_{sh} = 260\ M\Omega/m$. Figure 2 presents the calculated field distribution.

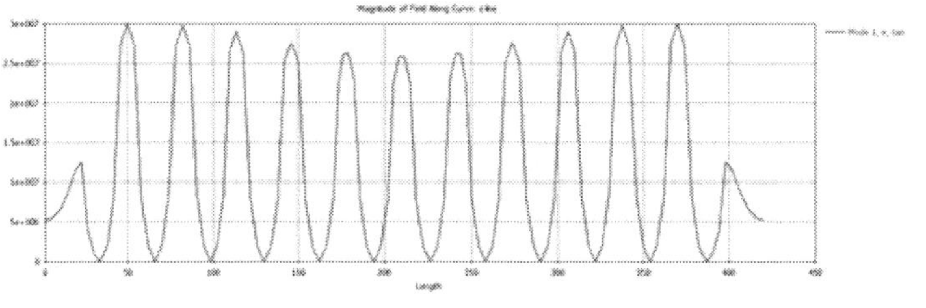

Figure 2. E_z field distribution calculated using CST Microwave Studio.

The maximum E_z field in gaps depends on the distance between these gaps to the middle of the accelerating channel. For longer distances, the field in the gaps is higher, for shorter distances, the field in the gaps is lower. Growing electric potential along cylindrical supports and further along the longitudinal bars results in a growing field distribution.

Wrapping of copper foil around the longitudinal bars and its insertion under their cylindrical supports has redistributed the electrical potential and made accelerating field uniform along the accelerating channel. Figure 3 shows the wrapped parts of the accelerating channel.

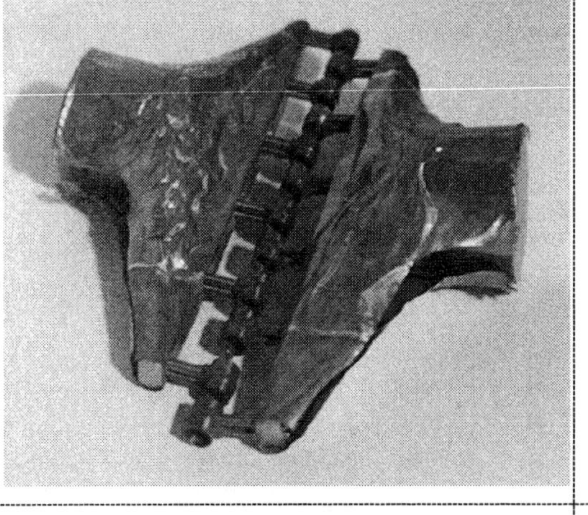

Figure 3. Acceleration channel with wrapped parts.

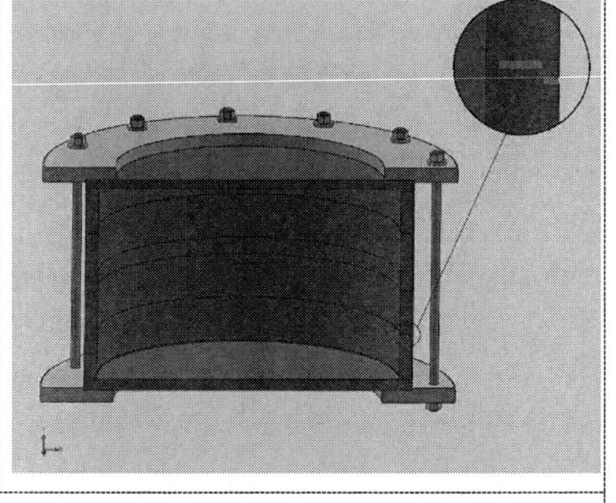

Figure 4. The design of the cavity.

IOP Conf. Series: Journal of Physics: Conf. Series **1238** (2019) 012060 doi:10.1088/1742-6596/1238/1/012060

The cavity of this structure is composed of five machined copper cylinders with 420 mm inner diameter. To provide good electrical contacts, two flanges squeeze the cavity using twelve threaded rods of 12 mm in diameter. The 2 mm wide cylindrical contact tooth in the top of the cylinder's surface also helps to provide good electrical contacts between cylinders and lids (figure 4).

The results of measurements of the mock-up structure are presented in table 1.

Table 1. Results of measurements.

f_0, MHz	$R_{sh}, M\Omega/m$	Q_0
170.613	287	9171

Figure 5 shows the measured distribution of the accelerating field.

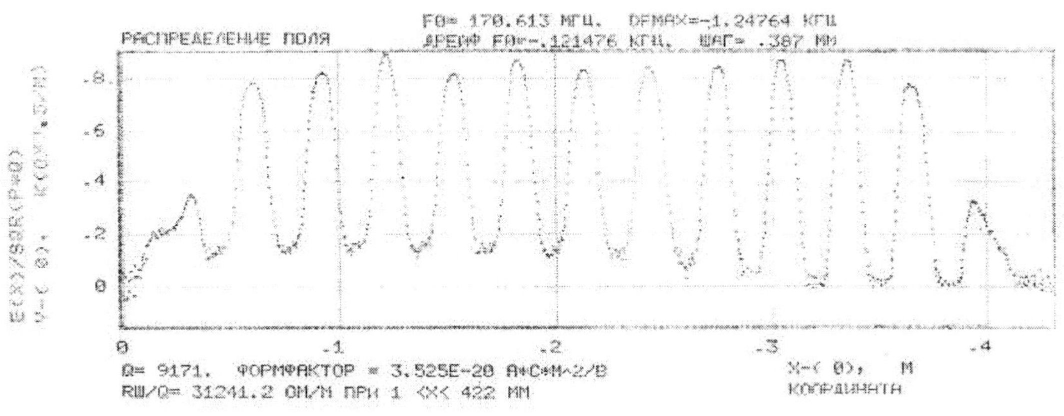

Figure 5. The tuned distribution of accelerating field measured using instrument PIRC-1.

3. Accelerating Structure of 1 MeV Proton LINAC

The next goal was to develop the new accelerating structure based on the above investigated structure and the accelerating field distribution of URAN-1M structure [4]. Figure 6 presents the field distribution of URAN-1M structure.

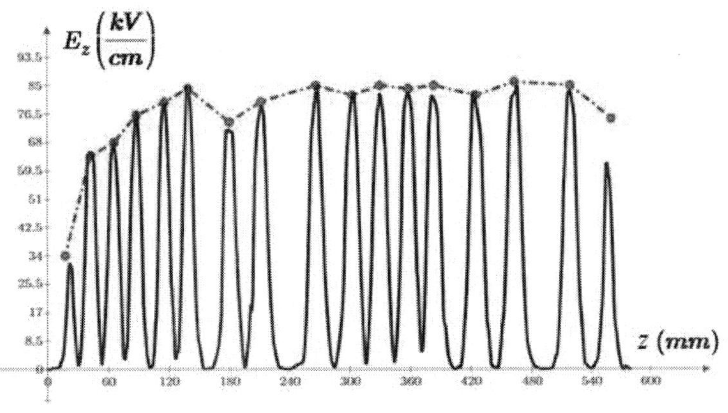

Figure 6. The distribution of the accelerating field in the accelerating structure of the accelerator URAN-1M.

IOP Conf. Series: Journal of Physics: Conf. Series **1238** (2019) 012060 doi:10.1088/1742-6596/1238/1/012060

This field distribution is unique in that the accelerating field in gaps is increasing from the first to the sixth accelerating gap. The new design uses the change of potential along the radius of the cylindrical cavity of the resonator excited in TM_{010}^{0} mode. The first six drift tubes (figure 7, pos. 1, 2) are connected to the bottom and top of the cavity at different radii to achieve a growing field. The rest of drift tubes are attached to the longitudinal bars that are sitting on the big cylindrical supports in the center of resonance cavity (figure 7, pos.3, 4) and on the shorting conductors (figure 7, pos. 5, 6). Those are used for adjustments of the field distribution.

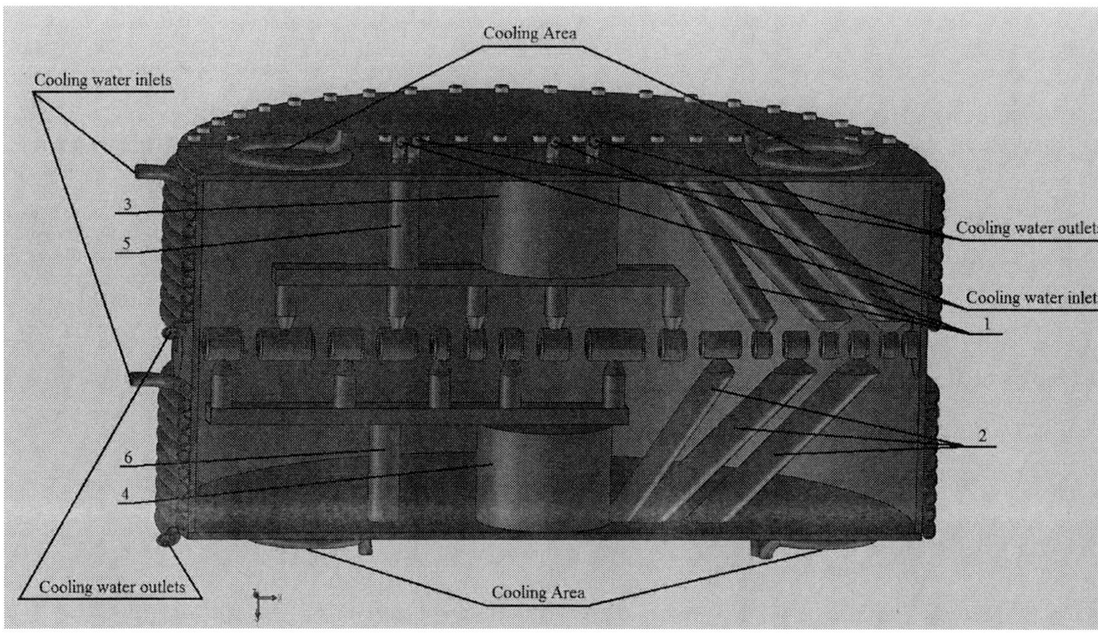

Figure 7. The new accelerating structure of URAN-1M excited in TM_{010}^{0} mode.

The simulation of electromagnetic field of the proposed accelerating structure in HFSS ANSYS Electronic Desktop showed that it is possible to tune the field distribution rather close to what is required (figure 6).

Adjustments in positions of tubes' supports (figure 6, pos. 1 and pos. 2) that attach to the lids of cavity, tune the fields in the first six accelerating gaps. Adjustments of the main cylindrical supports' position (figure 7, pos.3, 4) and the positions of the shorting conductors (figure 7, pos.5, 6) tune the field distribution along the rest of the accelerating channel. The installation of different shims around the main cylindrical support allows for frequency tuning.

4. The RF Amplifier

The high shunt impedance allows to use lower level of RF power to achieve the required field distribution. The calculation showed that 15 kW of power would be enough to reproduce the field distribution of the URAN-1M structure (figure 6). The reduction of RF power creates the condition for higher duty cycle and even for operating LINAC in CW regime. CW regime allows increasing the average beam current up to 4.0 mA.

The other benefit from using lower level RF power is that a cheaper RF amplifier could be used that has smaller dimensions. Specifically, one cabinet measuring 482 mm x 450 mm x 800 mm could contain such 15 kW solid-state amplifier.

5. Cooling of the Accelerating Structure

The simulation of thermal regime in ANSYS Work-bench using steady-state thermal solver showed (figure 8) that the supports of the drift tubes, the longitudinal bars, the supports of the accelerating channel, shorting conductors and cylindrical surface of the main chamber all require water-cooling. Additional two spots in each lids of the cylindrical cavity should be water cooled down. The design of water cooling system is presented in figure 7.

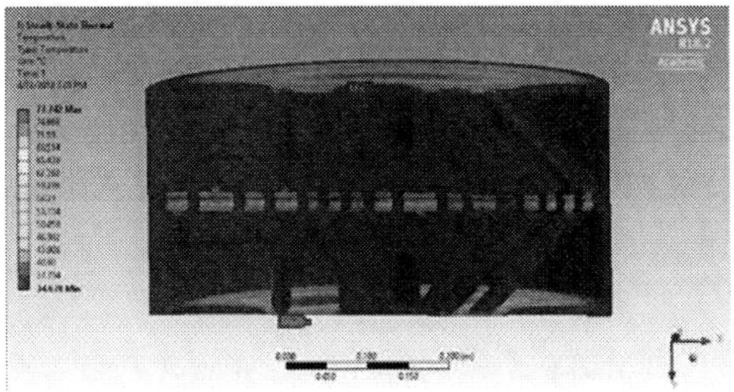

Figure 8. The steady-state thermal simulation of the accelerating structure running 15 kW CW of RF power.

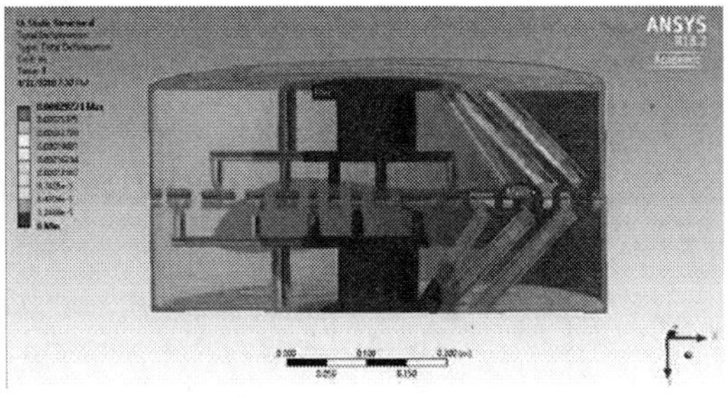

Figure 9. Static structural simulation in ANSYS Workbench of the structure.

The steady-state thermal simulation in ANSYS Work-bench showed that temperature of any part belonging to the accelerating structure will be below 78°C, if the temperature of cooling water is sustained below 35°C.

Water cooling cylindrical parts are arranged as coaxial pipe. The inlets are connected to the inner pipes, outlets are connected to the space between outer and inner pipes. 10 mm in diameter copper pipes are soldered to the outer cylindrical surface (figure 7). Longitudinal bars are water cooled. There are also four spots, optimally soldered on the top and bottom of the cylindrical cavity that are also water cooled. The cooling system requires 23 L/min of water, if the water temperature is to be supplied at 22°C.

To reduce mechanical deformation due to heat generated from the RF power, the centers of cavity lids have stainless steel fixtures to prevent deformation in the central part of the cavity. The static structural simulation in ANSYS Workbench showed that maximal deformation causes 292 μm deflection in the last drift tube position (figure 9). The deviation of the frequency due to total deformation will be below 50 kHz, the maximum deformation takes places in the position of the first and last drift tubes.

6. Conclusion

The present design shows possibility to build 1 MeV LINAC for acceleration of light ions, operating in CW mode. The average beam current could reach at least 4 mA. This LINAC could be used for imitation of influence of transmute hydrogen. It also be used to study mechanical properties of austenitic steels, to study degradation of their mechanical properties, and to study embrittlement of vanadium alloys under exposure to a proton beam.

7. References

[1] Komissarov A P, Bondarenko G G, Kostin K A 1996 Journal of nuclear materials 233-237 pp 1154-1159
[2] Bondarenko G G, Komissarov A P, Kostin K A 1995 1st International Symposium on Beam Technologies p 15
[3] Avreline N V, Gass V F, Gordeev D U, Leonov V N 1991 12th All Union Seminar of LINACs
[4] Avreline N V and etc. 1992 Tuning and testing of the resonance structure of the proton LINAC URAN-1M Preprint MEPHI 019-92

Study of collective effects in the FCC-ee collider

M Zobov[1], **E Belli**[2,3], **G Castorina**[3], **M Migliorati**[3], **S Persichelli**[4], **G Rumolo**[2] **and B Spataro**[1]

[1]LNF-INFN, Via Enrico Fermi 40, 00044 Frascati (Rome), Italy
[2]CERN, Geneva, Switzerland
[3]University of Rome "La Sapienza" and INFN Sez. Roma1, Rome, Italy
[4]LBNL, Berkeley, California, USA

E-mail: mikhail.zobov@lnf.infn.it

Abstract. The Future Circular Collider (FCC) study aims at designing different options of a post-LHC collider. The high luminosity electron-positron collider FCC-ee based on the crab waist concept is considered as an intermediate step on the way towards FCC-hh, a 100 TeV hadron collider using the same tunnel of about 100 km. Due to a high intensity of circulating beams the impact of collective effects on FCC-ee performance has to be carefully analyzed. In this paper we evaluate beam coupling impedance of the FCC-ee vacuum chamber, estimate thresholds and rise times of eventual single- and multibunch beam instabilities and discuss possible measures to mitigate them.

1. Introduction

The first electron-positron Higgs factory was proposed already several months before the official announcement of the Higgs boson discovery [1]. It was a high luminosity collider in the LHC tunnel, called "LEP3", inspired by a low emittance B-factory design [2] and lepton-proton collider LHeC design studies [3]. The proposal was aimed at reaching the luminosity of the order of 10^{34} $cm^{-2}s^{-1}$ at the energy of 120 GeV per each beam for precise measurements of the Higgs boson mass, cross-section and its decay modes. The proposal has evolved, first passing to DLEP, the e+e- collider having the double LHC circumference, and then to TLEP ("Triple LEP") hosted in a 80 to 100 km tunnel [4]. The use of such a large tunnel has opened the possibility to accommodate also a hadron machine with an unprecedented energy of 100 TeV in the center-of-mass. The overall project is being implemented at CERN under the "Future Circular Collider" (FCC) design study [5] with FCC-ee being the electron-positron collider (former TLEP) considered as a potential first step towards the 100 TeV hadron machine called FCC-hh. The electron-proton collider option, FCC-eh, is also considered in the framework of these studies. FCC-ee is foreseen to operate at different center-of-mass energies ranging from 91 GeV to 365 GeV to study the properties of the Z resonance, the W and top pair thresholds and the Higgs boson with very high precision (see Table 1). For example, the design luminosity of $2.3x10^{36}$ $cm^{-2}s^{-1}$ at Z resonance is almost 5 orders of magnitude higher than the maximum luminosity ever achieved at LEP at the same energy (see Table 2 in [6]). There are several ingredients that contribute in reaching such high luminosities in FCC-ee. These are some of them:

- Two separate rings allow to collide many bunches without their parasitic interactions.
- The longer perimeter allows to store higher beam intensities with the same synchrotron radiation (SR) loss.

- The crab waist collision scheme, proposed [7,8] and successfully tested at Frascati [9], allows to reduce the betatron functions at the interaction point, to collide beams with much lower emittances and to alleviate the nonlinear beam-beam interaction by suppressing beam-beam resonances [10].

Table 1. Relevant FCC-ee baseline parameters

Parameter	**Z**	**WW**	**H(ZH)**	**ttbar**	
Circumference C [km]	97.75	97.75	97.75	97.75	97.75
Energy E [GeV]	45.6	80	120	175	182.5
Number of bunches per beam	16640	2000	328	59	48
Bunch population N_p [10^{11}]	1.7	1.5	1.8	2.2	2.3
Beam current I [mA]	1390	147	29	6.4	5.4
SR energy loss per turn [GeV]	0.036	0.34	1.72	7.8	9.21
Bunch length with SR/BS, σ_z [mm]	3.5/12.1	3.0/6.0	3.15/5.3	2.75/3.82	1.97/2.54
Bunch energy spread, SR/BS [%]	0.038/0.132	0.066/0.131	0.099/0.165	0.144/0.196	0.150/0.192
Longitudinal damping time [turns]	1281	235	70	23.1	20
Horizontal emittance ε_x [nm]	0.27	0.84	0.63	1.34	1.46
Vertical emittance ε_y [pm]	1.0	1.7	1.3	2.7	2.9
Luminosity per IP [10^{34} cm^{-2}s^{-1}]	230	28	8.5	1.8	1.55

Analyzing the parameters shown in Table 1 we can see that the beam emittances of FCC-ee are very small, comparable to those of the modern synchrotron light sources, while the beam stored currents are close to the best current values achieved in the last generation particle factories (see Table 2 in [11] for comparison). Therefore, a careful study of collective effects is required in order to preserve the quality of the intense beams, to suppress eventual beam instabilities and to avoid excessive RF power losses leading to a damage of vacuum chamber components and accelerator hardware.
In this paper we present a preliminary study of the collective effects in FCC-ee due to parasitic electromagnetic interaction of the circulating beams with the surrounding vacuum chamber expressed in terms of wake fields and beam coupling impedances [12]. We will consider only the Z resonance option since it is more vulnerable to collective effects and instabilities because of the lower beam energy, longer damping times, higher beam intensities and highest number of bunches.

2. Beam coupling impedance

As it has been shown in [13], for the 100 km long collider, the finite conductivity of the beam vacuum chamber represents a major source of the beam coupling impedance, and wake fields strongly affect the beam dynamics and, respectively, collider design solutions and parameters choice. This fact has been taking into account while choosing the vacuum chamber shape and its dimensions. In particular, it has been proposed to use the vacuum chamber with a round cross-section in order to avoid the betatron tune shift with the beam current in multibunch operations due to quadrupolar resistive wall (RW) wake fields [14, 15]. The beam pipe radius of 35 mm has been chosen as a compromise between the beam impedance and the power required for magnet power supplies. Actually the shape of the pipe is not totally round but additional antechambers (winglets) are foreseen for pumping purposes and installation of SR absorbers, similarly to SuperKEKB design [16]. The twin dipole and quadrupole magnets of FCC-ee are being designed incorporating this vacuum chamber cross-section [17]. The constant vacuum chamber cross-section helps to avoid the use of multiple tapered transitions between different vacuum chamber sections, thus further improving the machine impedance.

The FCC-ee vacuum chambers have to be coated in order to mitigate the electron cloud effects (beam induced multipacting) in the positron ring and/or to improve the vacuum pumping in both rings. Thin layers of NEG, TiN and AC have been considered for these purposes [18]. It has been demonstrated [13] that under certain assumptions, that are valid for the FCC-ee parameters, the longitudinal and transverse impedances of a two-layer beam pipe are given by the sum of two terms, the first term representing the well-known impedance of a single layer beam pipe and the second one describing an

IOP Conf. Series: Journal of Physics: Conf. Series **1238** (2019) 012061 doi:10.1088/1742-6596/1238/1/012061

inductive perturbation proportional to the thickness of the coating:

$$\frac{Z_L(\omega)}{C} \approx \frac{Z_0\omega}{4\pi cb}\left\{[sgn(\omega) - i]\delta_2 - 2i\Delta\left(1 - \frac{\sigma_1}{\sigma_2}\right)\right\} \quad 1$$

$$\frac{Z_T(\omega)}{C} \approx \frac{Z_0}{2\pi b^3}\left\{[1 - i\,\mathrm{sgn}(\omega)]\delta_2 - 2i\Delta\,\mathrm{sgn}(\omega)\left(1 - \frac{\sigma_1}{\sigma_2}\right)\right\} \quad 2$$

Where Z_0 is the vacuum impedance, c the speed of light, b the pipe radius, δ_1, σ_1 and δ_2, σ_2 the skin depths and conductivities of the coating and the beam pipe (substrate), respectively.
As it can be seen, for the above assumptions, the real part of the impedance does not depend on the coating thickness and conductivity. The performed numerical studies [13] have confirmed that the resulting RF power losses due to the RW impedance remain almost unchanged for the coating conductivity and thickness varying in a very wide range. In turn, since the perturbation of the imaginary part due to the coating is proportional to its thickness and to the term $\left(1 - \frac{\sigma_1}{\sigma_2}\right)$, if the coating conductivity is much smaller than the beam pipe material conductivity the impedance depends only on the thickness of the coating layer. On the other hand, if the two conductivities are comparable, this term reduces the coating perturbation. It has been shown that the coating thickness plays a crucial role affecting both single and multibunch beam dynamics [13]. In particular, in order to keep longitudinal microwave instability under control the coating thickness should be as small as 50-100 nm [19]. For this reason, a campaign of dedicated measurements has been launched to study the properties of NEG thin films with thicknesses below 250 nm [19, 20], such as secondary emission yield and activation performance.

Table 2. RF power losses due to different vacuum chamber components

Component	Number	k_{loss} [V/pC]	P_{loss} [MW]
Resistive wall	97.75 km	210	7.95
Collimators	20	18.69	0.7
RF cavities	52	17.14	0.65
RF double tapers	13	24.71	0.93
Beam position monitors	4000	40.11	1.50
Bellows	8000	49.01	1.85
Total		**359.6**	**13.6**

In addition to the RW impedance there are many other impedance sources in the machine. The design of the vacuum chamber components such as RF cavities, kickers, beam position monitors (BPMs), bellows, flanges etc. has not been completed yet. In order to evaluate their possible impedance contribution, the present strategy consists in adopting the best design solutions of the accelerator components used in the modern synchrotron light sources and the particle factories. For the present impedance model we consider that at Z running the RF system consists of 52 single cell cavities operating at 400 MHz [21] and arranged in groups of four cavities connected to the beam pipe by 26 0.5 m long tapers. In order to eliminate the beam halo and to suppress the background, collimators based on PEP-II and SuperKEKB design [22, 23] are planned to be installed in the machine, for a total number of 20 (10 for each plane). The impedance contribution of the 10000 absorbers to cope with the SR has been minimized by placing them inside the two rectangular antechambers on both sides of the beam pipe. This model also includes 4000 BPMs [24] and 8000 comb-type bellows with RF shielding [25] to be allocated before and after each BPM.

The coupling impedances and wake fields for these vacuum chamber elements have been evaluated numerically. Figure 1 shows the longitudinal wake potentials of each component for the nominal bunch length of 3.5 mm. Table 2 summarizes the corresponding loss factors. As it can be seen, the resistive walls with 100 nm coating provide the dominating contribution in both the total wake potential and respective power losses that are not negligible compared with the 50 MW power lost due to SR.

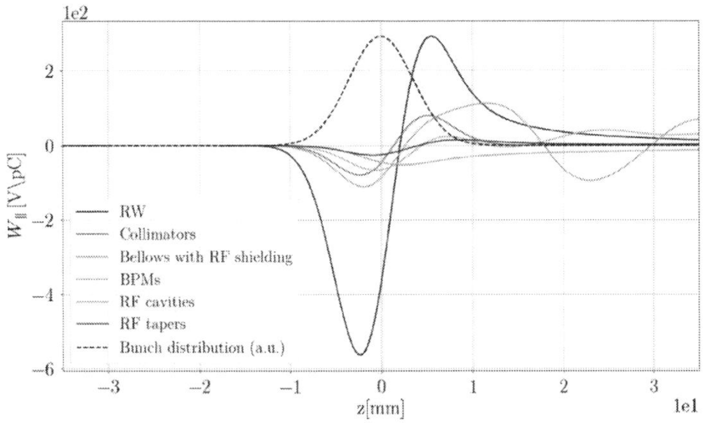

Figure 1 Longitudinal wake potential for different vacuum chamber components calculated for 3.5 mm Gaussian bunch

3. Collective effects issues

The electromagnetic beam interaction with a surrounding vacuum chamber, described in terms of wake fields and impedances, affects longitudinal and transverse beam dynamics. It can result in both single and multi-bunch instabilities and overheating of vacuum chamber components. The longitudinal wake potential (see Fig.1) modifies the RF voltage leading to a bunch length increase and a bunch shape distortion. In collisions with a large Piwinski angle, as is the case of FCC-ee, the collider geometric luminosity and beam-beam interaction depend on the bunch length, thus influencing the overall collider performance. For longer bunches the beam lifetime increases and the RF power losses are reduced. However, at a certain bunch intensity, microwave instability can take place. Typically the microwave instability does not produce a bunch loss, but the consequent energy spread growth and possible bunch internal oscillations above the instability threshold cannot be counteracted by a feedback system. In addition, the longitudinal wake fields result in the synchrotron tune reduction and a large incoherent synchrotron tune spread. Both these effects influence beam-beam performance shifting the collider working point and affecting the coherent and incoherent beam-beam resonances. The left plots in Fig.2 show the bunch length in FCC-ee as a function of the bunch population, while the right plots indicate the respective energy spread. The blue curves correspond to the bunch length and energy spread variations for non-colliding bunches. At the nominal intensity the bunch lengthens till about 7 mm, while the microwave instability threshold is about a factor of 1.5 higher than the nominal bunch population. So there is only a small margin left for eventual impedance increase. In collision the "beamstrahlung" effect [26] results in an additional energy spread increase leading to the strong bunch elongation and the microwave instability threshold increase beyond the considered bunch intensities (brown curves).

Differently from the longitudinal microwave instability, the transverse mode coupling instability (TMCI) is destructive for intense bunches. It takes place when coherent frequencies of different modes of transverse internal bunch oscillations merge. The TMCI threshold has been evaluated with the analytical Vlasov solver DELPHI [27] by considering the dominating RW impedance and by taking into account the bunch lengthening due to the longitudinal wake fields shown in Fig.2. It has been found that in the transverse case the TMCI instability threshold is affected to a lesser extent by the coating thickness due to the bunch lengthening effect. For comparison, Fig. 3 shows the real part of the frequency shift of the first coherent oscillation modes as a function of the bunch population for 50 nm (left) and 1 μm coatings (right), respectively. The dashed line represents the nominal bunch intensity. The TMCI threshold (merging lines) is about a factor 2.5 higher than the nominal intensity.

Analyzing the multi-bunch beam dynamics it has been found that the coupled bunch instability due to the transverse RW long range wake fields is another critical issue for the collider [13]. The growth rate of the fastest coupled bunch mode is estimated to be 435 s^{-1}, corresponding to about 7 revolution turns. It is worth noting here that the transverse radiation damping time is 2550 turns, i.e. it is much longer than required for the instability suppression. So a robust feedback system is necessary to mitigate the fast instability. A dedicated study is underway to develop such a challenging feedback system [28].

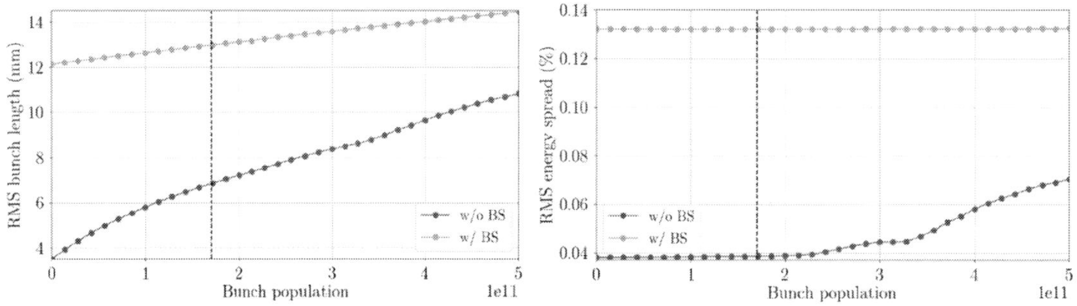

Figure 2 Bunch lengthening (on the left) and energy spread (on the right) in FCC-ee.

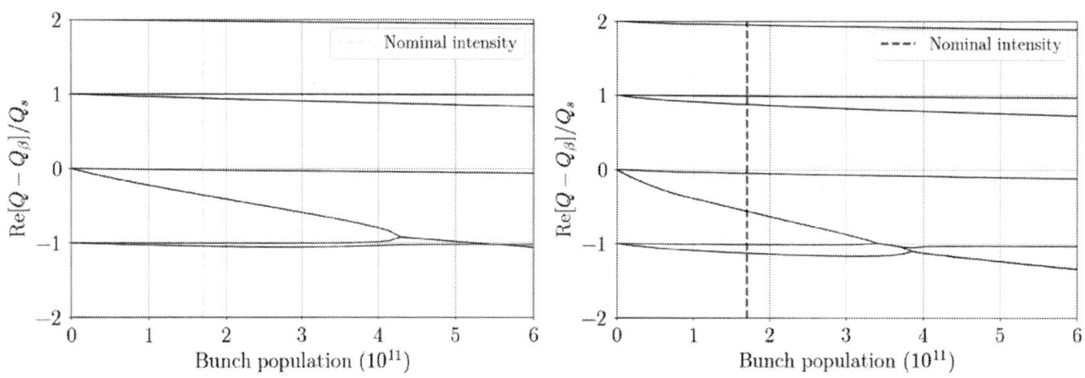

Figure 3 Real part of the frequency shift of the first transverse coherent oscillation modes for 50 nm (left) and 1 μm (right) coating thickness

The longitudinal radiation damping alone also cannot suppress the longitudinal coupled bunch instabilities due to the beam interaction with parasitic higher order modes (HOM) trapped in the vacuum chamber components. In order to cope with the instabilities special HOM damping techniques are to be applied to reduce the shunt impedances of the HOM to a harmless level, as discussed in [28]. In addition, also in this case a longitudinal feedback system has to be developed as a further safety knob.

4. Conclusions

FCC-ee beam impedance plays an important role for both the parameters choice and design solutions for the 100 km collider. In particular it has been shown that the impedance has an impact on the choice of the vacuum chamber shape, properties of the beam pipe coating, requirements of the feedback systems, and defines the microwave and TMCI instability thresholds, bunch lengthening, energy spread and RF energy loss. Further work is in progress to design low impedance vacuum chamber components and to investigate other collective effects such as the beam induced electron cloud, ion instabilities etc.

Acknowledgements

This work was in part supported by the European Commission under the HORIZON2020 Integrating Activity project ARIES, grant agreement 730871, and in part by the MICA project of INFN.

References

[1] Blondel A and Zimmermann F 2011 A high luminosity e+e- collider in the LHC tunnel to study the Higgs Boson *Preprint* arXiv: 1112.2518.
[2] Bona M *et al* 2007 SuperB: a high-luminosity asymmetric e+e- super flavor factory: conceptual design report *Preprint* arXiv: 0709.0451.
[3] Adolphsen C (LHeC Study Group)2011 A large hadron electron collider at CERN, Physics, Machine, Detector *Preprint* CERN-LHeC-Note-2011-003 GEN.
[4] Bicer M *et al* 2014 First look at the physics case of TLEP *J. High Energy Phys.* JHEP01(2014)164 (*Preprint* arXiv: 1308.6176).
[5] https://fcc.web.cern.ch
[6] Assmann R, Lamont M and Myers S 2002 A brief history of the LEP collider *Nucl. Phys. Proc. Suppl.* **109B** 17-31.
[7] Raimondi P 2006 Status on SuperB effort *in 2nd SuperB Workshop*, Frascati, Italy.
[8] Raimondi P, Shatilov D and Zobov M 2007 Beam-beam issues for colliding schemes with large Piwinski angle and crabbed waist *Preprint* physics/0702033.
[9] Zobov M *et al* 2010 Test of crab-wast collisions at DAΦNE Φ-factory *Phys. Rev. Lett.* **104** 174801.
[10] Shatilov D, Levichev E, Simonov E and Zobov M 2011 Application of frequency map analysis to beam-beam effects study in crab waist collision scheme *Phys. Rev. ST Accel. Beams* **14** 014001
[11] Zobov M 2011 Next generation of electron-positron factories *Phys.Part.Nucl.* **42** p.782.
[12] Palumbo L, Vaccaro V and Zobov M 1995 Wake fields and impedance *Preprint* physics/0309023.
[13] Migliorati M, Belli E and Zobov M 2018 Impact of the resistive wall impedance on beam dynamics in the Future Circular e+e- Collider *Phys. Rev. Accel. Beams* **21** 041001.
[14] Chao A, Heifets S and Zotter B 2002 *Phys. Rev. ST Accel. Beams* **5** 111001.
[15] Blednykh A et al.2016 Low-frequency quadrupole impedance in undulators and wigglers *Phys. Rev. ST Accel. Beams* **19** 104401.
[16] Suetsugu Y et al.2012 Design and construction of the SuperKEKB vacuum system *J. Vac. Sci. Tech.* A**30** 031602.
[17] Milanese A 2017 Progress with the design of the main FCC-ee warm magnets *in FCC Week* 2017, Berlin, Germany.
[18] Belli E et al.2017 Coupling impedances and collective effects for FCC-ee *Proc. 8th Int. Particle Accelerator Conference* (IPAC2017), p. 3734, IPAC-2017-THPAB020.
[19] Belli E et al. 2018 Single bunch instabilities in FCC-ee *Proc. 9th Int. Particle Accelerator Conference* (IPAC2018), IPAC-2018-THPAK052.
[20] Belli E et al. 2018 Electron cloud studies in FCC-ee *Proc. 9th Int. Particle Accelerator Conference* (IPAC2018), IPAC-2018-MOPMK012.
[21] Butterworth A 2017 Cavity design and beam-cavity interaction*in FCC Week* 2017, Berlin, Germany.
[22] DeBarger T et al. 2001 The PEP-II movable collimators *Preprint* SLAC-PUB-11752.
[23] Isibashi T, Terui S and Suetsugu S 2017 Low impedance movable collimators for SuperKEKB *Proc. 8th Int. Particle Accelerator Conference* (IPAC2017), p. 2929.
[24] Duarte H O C, Sanfelici L and Marques S R 2013 Design and impedance optimization of the SIRIUS BPM *Proc. of IBIC2013*, p.365.
[25] Suetsugu Y, Shirai M and Shibata K 2003 Possibility of comb-type rf shield structure for high-current accelerators *Phys. Rev. ST Accel. Beams* **6** 103201.
[26] Shatilov D 2018 IP beam parameter optimization for FCC-ee *FCC Week* 2018, Amsterdam, Netherlands.
[27] https://twiki.cern.ch/twiki/bin/view/ABPComputing/DELPHI
[28] Migliorati M et al. 2016 Collective effects issues for FCC-ee *Proc. of eeFACT2016*, p.100.

Development of the debuncher for the injector part of the accelerator complex NICA

A A Zavadtsev[2], N I Brusova[1], A V Bulanov[2,4], A V Butenko[3], M A Gusarova[2,4], D A Zavadtsev[1], J Z Kalinin[1], L V Kravchuk[1], V V Paramonov[1], I V Rybakov[1], E A Savin[2,4], A V Smirnov[3], N P Sobenin[2,4], E M Syresin[3] and D V Churanov[1]

[1]Institute for Nuclear Research of RAS, Prosp. 60 Oktyabrya 7a, Moscow 117312 Russia
[2]Nano Invest, LLC, 9-017 Acad. Chelomeya Str., Reutov, 143969 Russia
[3]Joint Institute for Nuclear Research, 6 Joliot-Curie, Dubna, 141980, Russia
[4]National Research Nuclear University MEPhI, 31 Kashirskoe Sh., Moscow 115409 Russia

azavadtsev@yandex.ru

Abstract. The developed and currently manufactured debuncher for the injector part of the accelerator facility NICA is designed to reduce the energy spread by a factor of up to ten in the ion bunches with Z/A=(0.33-1) at the output of the LU-20 linac. The debuncher includes a Split-Ring cavity, a vacuum system, a solid-state RF amplifier and an RF controller. The main design parameters are described.

1. Introduction

The Nuclotron based Ion Collider fAcility and Multi Purpose Detector (NICA/MPD) [1] aim to create an accelerator facility for research in the field of particle physics in the previously inaccessible area of parameters and experimental conditions. Among other things the injector of light ions in the NICA facility includes an accelerator LU-20 and a debuncher.

In the frame of NICA project, the light ions injector equipment is assumed to be replaced by the new one in the future. The debuncher described in the present paper will be a prototype for the new one. The debuncher is intended to reduce by a factor of up to ten the width of the energy spectrum of the accelerated ion beam after the linear accelerator LU-20 and 6 m drift space.

The ion energy at the LU-20 output is 5 MeV/nucleon. The debuncher is to be optimized for Z/A=0.5 and operate in the range Z/A =0.3-1.0 at Z=1 and the relative ion speed β=0,1028261.

The relative energy spread of the ion bunch is supposed to be $dE/E=\pm 2.5\cdot 10^{-3}$, it should not be larger than $dE/E=\pm 2.5\cdot 10^{-4}$ at the debuncher output. The total phase length of the ion bunch is of 30-35^0 at the debuncher input.

The main technical parameters of the debuncher

The debuncher consists of a cavity and vacuum, alignment and RF systems.

All the debuncher equipment will be localized close to the debuncher cavity. Location conditions of the electronic equipment provide the radiation dose rate as small as ≤2 μSv/h.

The RF master generator signal will be sent from the control system of the LU-20 accelerator.

The main parameters of the debuncher are presented in Table 1.

Table 1. The main parameters of the debuncher.

	Designation	Unit	Value
Adjustment range of the cavity accelerating voltage	UT	kV	0-200
Operating frequency	f	MHz	145.25
Tuning range of the cavity	Δf	kHz	±250
Coupling coefficient of the feeding line and the cavity	β		1.0
Instability of the input power of the cavity	δP	%	≤3
Instability of the accelerating voltage flat-top	δU	%	≤2
Operating pressure	p	mbar	$\leq 3\cdot 10^{-8}$
Adjustment range of the accelerating voltage phase	$\Delta\varphi$	degree	360
Power of the RF master generator signal	P_m	dBm	20-30
Rise/fall length of the pulse	τ_R	µsec	200
Flat-top length of the pulse	τ	µsec	100-300
Pulse repetition rate	F	Hz	≤0.5

2. Cavity types for the debuncher

Several different types of the cavity, which can be used in the debuncher potentially, have been considered and simulated. They are shown in Figure 1.

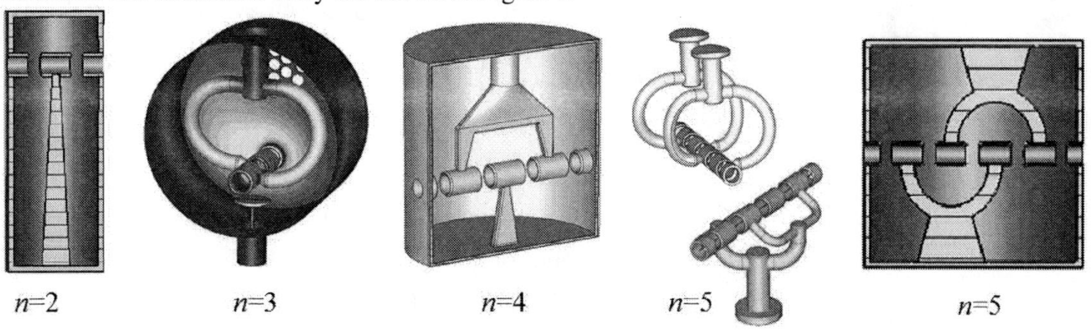

Figure 1. The cavity types considered for the debuncher (n is the number of accelerating gaps).

Electrodynamic parameters, multipacting and manufacturing technology have been investigated. The results are summarized in Table 2.

The cavity with 2 gaps is the simplest variant, but there are multipacting conditions in this cavity.

The cavities with 4 and 5 gaps are unreasonably difficult for manufacturing.

As a result, the cavity with 3 gaps (the so called Split-Ring cavity [2]) has been chosen. Its main advantage is the absence of the multipacting condition. It has rather acceptable electrodynamic parameters such as shunt impedance $R_{sh.eff}$, Q-factor Q_0, power loss P, max electric field E_{max} comparing with the Kilpatrick criterion E_k.

3. Split-Ring cavity

3.1. Field distribution

Calculated distribution of electric and magnetic fields on the cavity surface as well as electric field in the cavity volume and on its axis are shown in Figure 2.

Table 2. Calculated parameters of the cavities.

Number of gaps	n		2	3	4	5	5
Diameter×length	$D\times L$	mm	250×520	412×288	414×415	404×494	530×492
Effective voltage	UT	kV	200	200	200	200	200
Shunt impedance	$R_{sh.eff}=(UT)^2/P$	MOhm	8.42	11.68	25.5	25.5	41.6
Q-factor	Q_0		10000	9430	11700		15046
Power loss	P	kW	4.7	3.4	1.6	1.6	1.0
Max electric field	E_{max}	MV/m	10.3	8.0	5.3		
Relative max electric field	E_{max}/E_k		0.80	0.62	0.41		
Multipacting presence			yes	no	yes	no	yes
Frequency of the nearest non-operating mode	f_m	MHz	420.8	121.1	240.0	116.5	

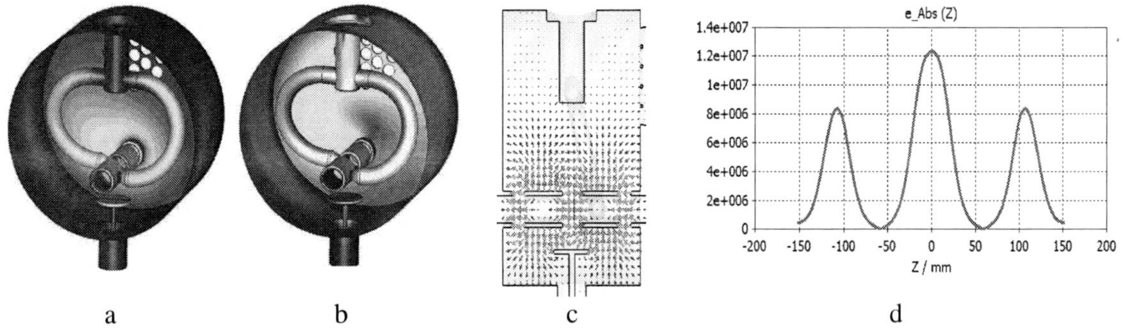

a b c d

Figure 2. Distribution of electric (a) and magnetic (b) fields on the cavity surface, electric field in the cavity volume (c) and on the cavity axis (d).

3.2. Thermal effects in the cavity
For operations with UT=200 kV, τ=400 μsec, F=1 Hz the estimated average power loss P<50 W/m^2 is rather low, resulting in a temperature rise of 0.5 C at the drift tubes with respect to the cavity walls.

3.3. Multipacting effects in the cavity
First of all, test calculations were performed for the GSI four-gap debuncher at 108 MHz [3, 4]. The results showed that the GSI debuncher is not free of multipacting. The combination of its frequency, voltage and dimensions creates conditions for the development of the medium intensity multipacting. The experimental studies of this debuncher showed that prolonged conditioning (at least a week) allowed one to get working conditions and the debuncher is currently operating successfully. The results of the performed calculations are confirmed by experimental results.

The two-gap 145 MHz cavity is not free of multipacting too, and its intensity is similar to that of the GSI debuncher.

The three-gap 145 MHz Split-Ring cavity is practically free of multipacting. The results of multipacting investigation in the cavity are shown in Figure 3: electron trajectories, increase in the number of particles as a function of the voltage UT and the number of electrons in time at different voltages UT.

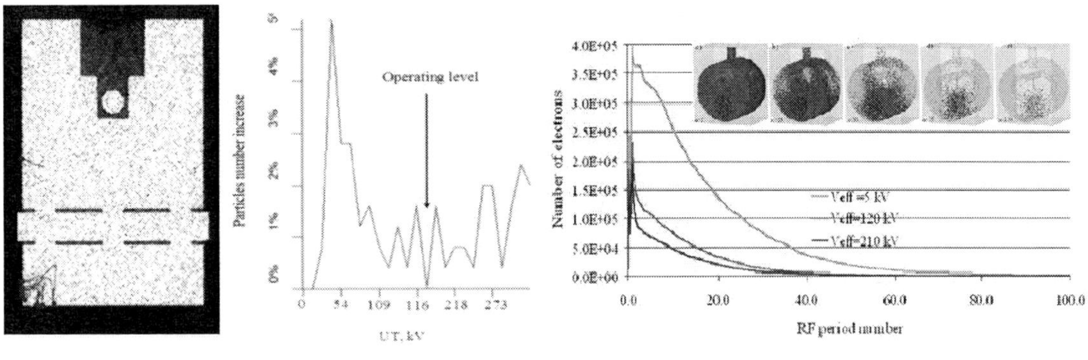

Figure 3. The results of the multipacting investigation in the cavity.

The four- and five-gap 145 MHz cavities of similar type have the multipacting conditions similar to those of the GSI debuncher.

The five-gap 145 MHz Split-Ring cavity is practically free of the multipacting development.

4. Beam dynamics

The debuncher is intended to reduce the energy spread of the accelerated ion beam after the linear accelerator LU-20 and 6 m drift space. The ion energy spectrum at the LU-20 output depends on the operation regime. The beam dynamics simulations were performed for rectangular phase space with a specified small phase length and real energy spread at the LU-20 output. Thus, any real phase space at the LU-20 output can be modelled as a sum of several rectangular phase spaces.

Figure 4 shows simulated distributions in the beam-line LU-20 – drift space – debuncher at Z/A=0.5.

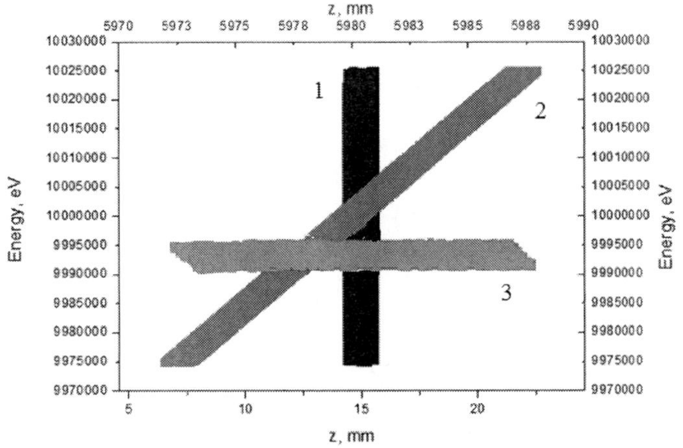

Figure 4. Beam dynamics simulation results: input distribution after LU-20 (1), distribution after 6 m drift space (2) and final distribution after debuncher (3).

To decrease the energy spread to dE/E=$\pm 2.5\cdot 10^{-4}$, the optimal voltage UT in the cavity should be UT =58 kV; 121 kV and 190 kV for Z/A=1,0; 0.5 and 0.3 respectively.

5. Engineering design

The design of the debuncher is shown in Figure 5. It includes a copper plated stainless steel cavity body, a copper vibrator with drift tubes, an input coupler with N-type connector, a pin-antenna with N-type connector, a frequency tuner with stepper-motor, input and output aperture connection pipes, two connection pipes for vacuum sensors IKR-060 and TPR-018, a connection pipe for the turbo pump HiPace 700, alignment and RF systems. All the vacuum connectors are of the ConFlat type.

Figure 5. Engineering design of the debuncher cavity, including vibrator (1) with drift tubes (2) and support (3), body (4), coupler (5), tuner (6), RF antenna (7), input aperture pipe (8), output aperture pipe (9), turbo pump (10), two flanges (11) for vacuum sensors and alignment system (12).

6. RF system

The RF system is intended to provide RF feeding of the debuncher with a peak power up to 7 kW. It includes an RF amplifier, and RF controller with a control system and a stepper-motor controller.

The master RF signal goes from the LU-20 control system, which carries out its phase control, to RF controller input.

The RF controller provides monitoring of both the RF signals from the cavity and forward and the reflected RF signals from the directional coupler located at the RF amplifier output.

Adjustment of the RF master signal power can be made from the control terminal of the LU-20 control system and ETHERNET via controlled attenuator. The accelerating voltage stabilization (i.e. RF power stabilization of the RF master signal) is realized using feed-back circuit and RF signal from the cavity.

The RF controller allows us to adjust the resonant frequency of the cavity in a such way as to make it equal to the operation frequency. This adjustment is realized via the following circuit: control terminal of LU-20 control system – ETHERNET – RF controller with the stepper-motor controller – tuner of the cavity.

The RF amplifier is made on the base of four MRFX1K80H LDMOS transistors with a max output power of 1800 W each. Input RF power comes from the RF controller. Amplified power of four transistors is summarized in the RF combiner and then passed to the cavity via coaxial directional coupler to monitor forward and reflected RF power.

The control system provides a possibility of connection to the LU-20 control system via ETHERNET and coaxial cables to monitor the signals as well as to adjust the accelerating voltage.

References

[1] Technical Project of the Accelerating Facility NICA. Under general supervision by I.N.Meshkov and G.V.Trubnikov, JINR, Dubna 2015.

[2] A. Vasyuchenko et., al., Development, Fabrication and Test of Triple Gap Bunchers for the ISAC Facility. Proc. of the 2001 PAC, p. 978, 2001.

[3] Gusarova et al., Multipacting simulation in accelerating RF structures, Nuclear Instruments and Methods in Physics Research, Section A, 599(1), p. 100-105.

[4] M. Heilmann et al., Further Upgrade Measures at New GSI CW-Linac Demonstrator Setup, Proceedings of IPAC2016, Busan, Korea, p. 892-895.

IOP Conf. Series: Journal of Physics: Conf. Series **1238** (2019) 012063 doi:10.1088/1742-6596/1238/1/012063

The Use of Interplanetary Medium as Fuel for Plasma Thrusters

A R Karimov[1,2], **O V Yakovlev**[1], **P A Murad**[3]

[1]Department of Electrophysical Facilities, National Research Nuclear University MEPhI, Kashirskoye shosse 31, Moscow, 115409, Russia
[2]Institute for High Temperatures, Russian Academy of Sciences, Izhorskaya 13/19, Moscow, 127412, Russia
[3]Morningstar Applied Physics, LLC, Vienna, USA

Sev2148@mail.ru

Abstract. Consuming long-range interplanetary medium as a fuel for driving plasma thrust is crucial. The methods of capturing the space environment depending on its density and the ship velocity are critical issues. The possible ways of creating ionization for neutral fluxes as a function of the solar radiation flux are also analyzed as a potential propellant. These processes define the permissible type of the plasma source. These options look extremely promising where a low thrust device continuously operating, could possibly deliver a celestial target close to the speed of light. Under such conditions at a considerable distance, the density of such material diminishes and it may be impossible to maintain a continuous operation mode of the plasma thruster but instead operate as a pulsed engine which might potentially be workable.

1. Introduction

The exploration of the Solar system requires a creation of a space engine with a large specific impulse. Unfortunately, such characteristics can't be provided by traditional chemical reactive engines. This follows directly from the Tsiolkovski's formula (see, for example, [1]):

$$v_f = v_{ex}\, ln\left(\frac{M_S+M_P}{M_S}\right), \qquad (1)$$

here v_{ex} is the effective exhaust velocity, M_s is the rocket mass, M_p is the propellant mass, v_f is the final velocity of the rocket. As an alternative to chemical engines, one can use plasma-driven engines. For example, the end-Hall thrusters are capable to create the exhaust velocity of the order $5 \cdot 10^5$ m/s while the maximum velocity of chemical engines is only 10^3 [1-3].

Nowadays there are several schemes of plasma thrusters: coaxial plasma thruster, stationary plasma thruster, Hall plasma thruster, plasma thruster energy/momentum exchange in crossed magnetic fields [2-9]. These devices differ from each other in the way of creating acceleration to provide thrust and performance. Nevertheless, there is a generic feature for all devices: in order, they have worked, one should obtain plasma from neutral atoms treated as propellants, and then one should accelerate this plasma flow to generate thrust. It is clear that this takes some energy and material which is continuously expended. In comparison with chemical thrusters, which can use only the reserve propellant, in plasma thrusters, one can use solar energy and interplanetary medium as a fuel. In the present piece, these issues will be addressed.

2. Interplanetary medium characteristics

As the interplanetary medium is a composition of substance and fields which fill space inside the Solar System, it appears justified to study its components. Ignoring the zero-point field and matter, the main components of the interplanetary system are the Solar wind, the high energy charged particles, the interplanetary dust, and neutral gas [10-12]. The Solar wind includes the ionized flux of the hydrogen ions of density $n_H = 5 - 6$ cm^{-3} with velocity 250—750 m/s moving from the Solar Corona [11]. Evidently, there are particles of the flow, where it is impossible to register. Taking into consideration such unknown part of the flow, in the present estimations, we will slightly increase the density of the Solar wind n_w by setting it equal to 10 particles per cm^{-3}.

A source of small particles forming a dust component is the collapsing nucleus of comets and the collisions of different bodies in the asteroid belt [11,12]. It was found that the dust concentration decreases in ecliptic as a function of $R^{-1/3}$, and at the distance $R_* \geq 3$ Astronomical Units (AU), dust is almost absent [10]. The total mass of dust in the Solar system estimates as $M_d = 10^{19} - 10^{20}$ g, and the major part of the mass (about 2/3 of the total mass), is concentrated in the dust particles of mass m_d in the mass range $10^{-3} < m_d < 10^{-5}$ g [10-12]. Using these data, we obtain an estimate of the average density for the dust component in terms of the hydrogen atoms in a sphere of a radius R_*: $n_d = 3M_d/(4\pi R_*^3 \mu_H = 10^3$ cm^{-3}, here $\mu_H = 1.66\ 10^{-24}$ g is the mass of the hydrogen atom.

Neutral gas in the Solar system was discovered by observing resonantly scattered solar radiation. At distances of $r = 5$ AU from the Sun, the neutral gas is distributed practically uniformly or homogeneous, the concentration of hydrogen atoms on the average being 0.06 cm^{-3}, helium 0.012 cm^{-3} [11,12]. As is seen, the density of this component of the interplanetary medium is much smaller than the contribution of the solar wind and dust and it can be neglected.

Proceeding from these data, we come to the conclusion that only the matter of solar wind and interplanetary dust are really suitable as a fuel for plasma thrusters. For further estimates, we shall set that the density of interplanetary medium varies in the range $n_W \leq n \leq n_d$, i.e. from 10 to 10^3 cm^{-3}.

In order to understand the possibility of using solar radiation for ionizing neutral atoms and their subsequent acceleration, we estimate the radiation power distribution as a function of the distance from the Sun. (It should be noted that here we limited our consideration to the distances on the boundary of the asteroid belt.). Taking into account that the surface temperature of the Sun at a distance of $R_S = 10^8$ m, is T_S=6000 K [11,12], from the Stephan-Boltzmann's law. We get

$$W_S = \frac{\sigma T_S^4}{(r/R_s)^2}, \qquad (2)$$

where σ is the Stefan-Boltzmann constant. This dependence is sketched in Fig. 1, where for the convenience of perception, the distances from the Sun to the planets falling into a sphere of radius R_* are labeled.

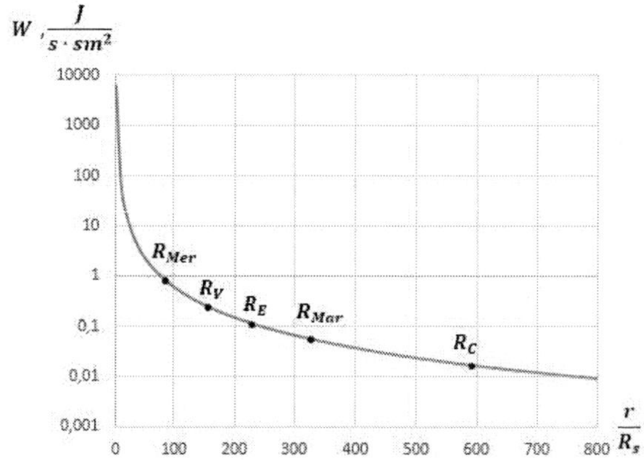

Figure 1. The radiation power with respect to the distance from the Sun. Here R_{Mer} corresponds to the Mercury orbit, R_V - the Venus orbit, R_E - the Earth orbit, R_{Mar} - the Mars orbit and R_C - the Ceres orbit.

3. The plasma production from interplanetary medium

Proceeding from information about the distribution of interplanetary medium and the Solar radiation, now we can estimate the parameters of plasma where we can produce as an interplanetary medium to use it in a plasma thruster as a fuel. Also, we have discussed the possible ways to increase the plasma density up to generate thrust at minimally required values.

As a plasma source, we shall consider the cylindrical charged chamber of volume $V = 10^3$ cm^3 where a low-pressure discharge is induced. As a rule, such types of discharges can be performed with the help of an HF microwave field at a pressure of $p{\sim}10^{-3}$ Torr that corresponds to the particle density $n_0{\sim}10^{10} - 10^{11}$ cm^{-3} under normal conditions (see, for example, [2, 13]). This indicates that it is impossible directly to use the interplanetary medium to organize a low-pressure discharge. We need to apply an additional technical parameter to obtain the required number of particles $N = n_0 \cdot V = 10^{14} - 10^{15}$ into the chamber under consideration.

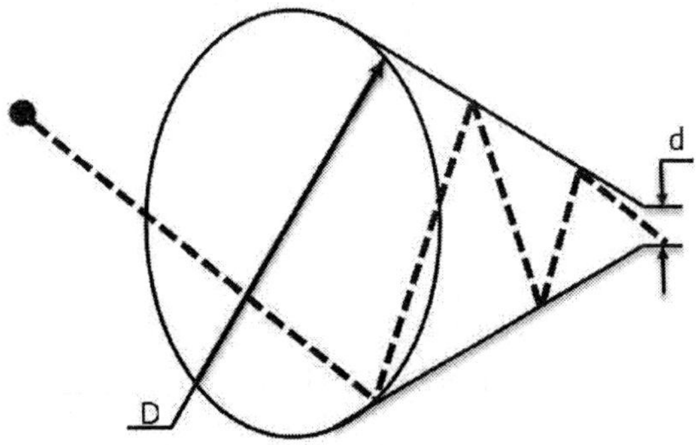

Figure 2. The scheme of trap for interplanetary substance (here, we set D=100 cm and d=10 cm).

We presume to do this with the help to trap interplanetary substances by using a special geometry (see Fig. 2) and the accumulation of particles with increasing the spacecraft's flow velocity with respect toward incoming particles. Such a conical structure should be placed on a moving spacecraft in the direction of its travel. Furthermore, the surface of such conical construction is suitable for the solar panel location.

Let us explain the simple mechanism of particle capture by assuming that the particle medium is stationary. In this case, the flux of particles that reach the surface, due to the relative velocity of the particles relative to the ship, is determined only by the ship velocity. The falling particles, repeatedly are reflected from the conical surface and fall into the region of the discharge chamber. Proceeding from the continuity of the particle flux: $nD^2 = n_c d^2$, where D is the diameter of inlet, d is the input diameter of the discharge chamber, we see that the density of particles in the narrow part of the conical funnel n_c can be increased with respect to the input density n in k $=D^2/d^2$ times. For example, for the sizes D and d of the trap shown in Fig. 2, we obtain k $= 10^4$. It should be borne in mind that this variant of density enhancement is workable as long as the medium remains collisionless, i.e. pressure effect does not work. In our case, the collisionless regime may be violated in the narrowest part of the trap. So in estimations of the mean free path λ, we put the density in the discharge chamber being equal to the desired value $n_0 = 10^{11}$ cm^{-3}. Then for the characteristic gas-kinetic cross-section $\sigma_g = 10^{-16}$ cm^2, we get $\lambda = 1/(\sigma_g n_0) = 10^5$cm, that is much larger than the characteristic size of the discharge chamber, so there is no need to take into account the initial pressure of the medium.

Now we shall pass over to estimate the accumulation of particles due to a change in the relative velocity of the interplanetary medium. It is easy to see that the flux of the interplanetary particle incident on the trap has filled the selected volume to a given number of particles N for the time τ_n:

$$\tau_n = \frac{N}{\pi R^2 v n_c}, \quad (3)$$

where v is the relative velocity of incoming flow, $n_c = k \cdot n$ is incoming particles after passing the trap with k $= 10^4$ the input radius of the discharge chamber. Here we restrict our consideration by two characteristic cases: the flux is created by the solar wind with density $n = 10$ cm^{-3} and the flux consists from dust particles with a density of $n = 10^3$ cm^{-3}. In Fig. 3 the characteristic times of particles accumulation for these both cases are plotted as a function of the relative velocity, for which we take the ship velocity.

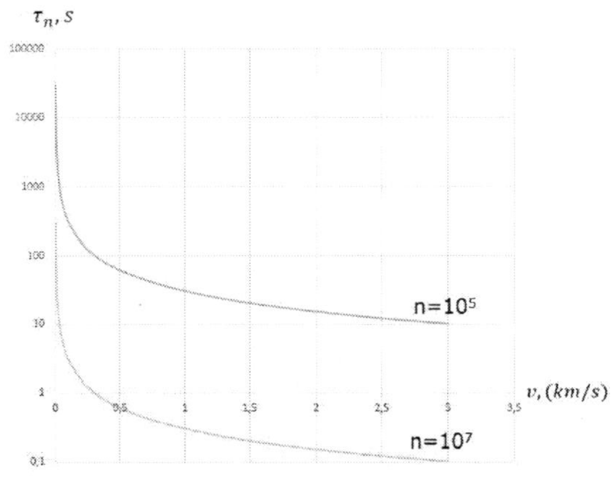

Figure 3. Accumulation time as a function of the relative velocity for the incoming flow of the solar wind particles ($n_c = 10^5$ cm^{-3}) and the dust particles ($n_c = 10^7$) cm^{-3}.

To assess the necessary energy expended on the ionization of neutral atoms and their acceleration, we shall use the data on the radiation power for the Sun presented in Fig. 1. In these estimates, we assume that the ionized medium consists only of hydrogen atoms having a minimum ionization potential of $\varepsilon_i = 13.6$ eV. For simplicity, we shall neglect to brake the ship by the oncoming flux of interplanetary medium and restrict our consideration by only elastic head-on collisions of the oncoming particles with the spacecraft surface. Despite the obvious physical fallacy of such an assumption, the result implies this is the simplest possible way to account for the contribution of particle deceleration in the energy balance:

$$WS_*\tau_\varepsilon = \varepsilon_i N + \mu_H N\left(\frac{v^2}{2} + \frac{v_{ex}^2}{2}\right), \tag{4}$$

where τ_ε is the accumulating time for ionization and accelerating particles by the sun radiation, S_* is the square of a solar panel. As previously mentioned, the elements for a solar battery may be located on the surface of the trap so in the present estimates one can use $S_* = \pi \cdot D^2/4$. The value of velocity $v_{ex} \geq 0$ is determined by the acceleration method and an additional discussion of this moment is required, but in order to obtain a minimum energy estimation in order of magnitude one can assume $v_{ex} = 0$. Then from (4), we get

$$\tau_\varepsilon = \frac{\varepsilon_i N + \mu_H N v^2/2}{W(r)\pi R^2}. \tag{5}$$

Depending on the velocity of the device, the power of solar radiation and the geometric sizes of a solar panel, this relation determines the characteristic time for the required energy accumulation. However, it should be borne in mind that these estimates of the times τ_ε should be considered as upper-bound marginal valuations.

Fig. 4 presents the dependence τ_ε for the space region between Venus and the asteroid belt between Mars and Jupiter. In these estimations, we take $S_* = 80$ m^2. As is shown from this curve, the value τ_ε changes from 0.1 s at a distance ~1 AU to several seconds for distances $r \geq 6$ AU. Under such conditions, it is impossible to maintain a continuous operation mode of the plasma thruster but a pulsed engine might be workable. Comparison of the dependencies presented in Figs. 3 and 4 show that the magnitude of pulse primarily depends on the capturing interplanetary matter that is brought about by the ship's velocity with respect to the interplanetary substance. One can try to reduce the pulse time by increasing the area of the trap, however, this will affect the dimensions of the spacecraft.

Increasing the craft's velocity makes it possible to decrease the accumulation time. In this case, a continuous mode of operation is not ruled out for high velocities.

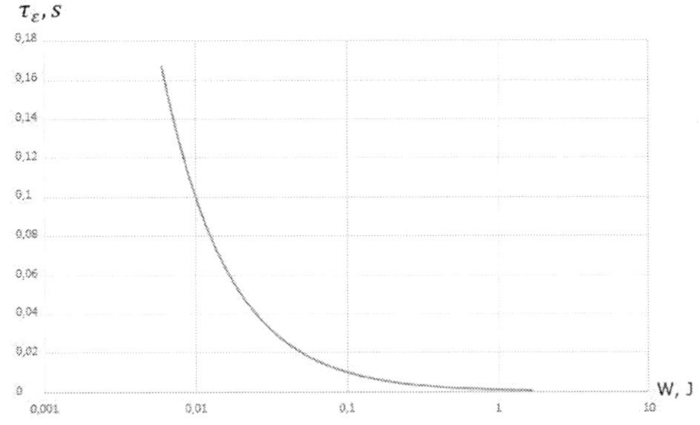

Figure 4. The value of the energy accumulation time as a function of W(r/R$_S$) under $S_* = 80$ m^2.

Finally, we may speculate about the plasma source which may be used provide the pulsed mode of a plasma source. It is clear that at low pressures, it is difficult to use a high-frequency discharge since it is caused by the particle collisions which occur rarely. In this case, there appears a problem how one should handle a required pulsed system of the discharge system. To enhance performance, one may use microwave discharges in a waveguide for the plasma production or the initiation of a discharge in a low-pressure gas. The gas may be propelled by microwave radiation with a stochastically jumping phase (see, for example, [13-15]). Although the first option is more technically simple as a breakdown power minimum value for microwaves with a stochastically jumping phase depends weakly on a working gas pressure that is caused by the anomalous nature of collisionless electron heating [14]. Such a feature allows extending the discharge existence region in a direction of overpressures. Therefore, it would be useful to combine both methods in examining this effect on the source.

4. Conclusion

In the present model, we have studied the use of the interplanetary substance as a fuel for plasma engines. For this, we used astrophysical data on the density of the cosmic medium, which we recounted relative to hydrogen atoms. It appears that the minimum and maximum density in terms of hydrogen atoms varies in the range from $n_{min} = 10$ to $n_{max} = 10^3$ cm^{-3}. The minimum estimation corresponds to taking into account only the solar wind and the maximum value that is obtained from interplanetary dust concentrated within a sphere of radius 3 AU from the Sun. Moreover, a low-pressure plasma source requires a density of the order $n_c \geq 10^7$ cm^{-3}. This means that for the low velocities of spacecraft, it is impossible to use the interplanetary medium as a fuel for plasma thrusters. However, with increasing rocket's velocity, as expected, the use of additional capturing space particles makes it possible to increase the density of the discharge volume. So in the example considered, the trap increases the particle density in k = 10^4 times irrespective of the rocket velocity only due to the ratio of geometric sizes. The dependencies presented in Fig. 3 indicate that the accumulation time of particles for the accepted value n_c depends on the device velocity. It should be noted that for simplicity, in the present estimations we have neglected by the relative velocity of particles with respect to the rocket. In the general case, this is not justified. As seen from the above dependencies, the incoming particle flux can only provide a pulsed mode of operation for the plasma source. In this case, the pulse frequency of such a device must vary with the rocket velocity and it depends on the density of the external medium.

Interestingly, the solar radiation is considered as an energy source for the ionization of accumulated particles and their further acceleration. Using the power distribution of the solar radiation, we have estimated the time needed to accumulate energy for ionization and acceleration to the required rocket velocity (see Fig. 4). As seen from the comparison of these dependencies with the accumulation time of particles, the main time will be spent on the particles accumulation regardless of the rocket velocity. In this case, the values for the accumulation time of particles and energy show that the space medium can be used as a fuel for a plasma source operating in the pulsed mode for flights in a radius of 3 AU from the Sun, even with existing technical capabilities. This may also occur for low particle values at the Ceres, where the time particles are hit for collection, the system may automatically operate like a pulsed system. The implication that a pulsed system with this requirement may be more profitable than a continuously operating system which would end up with propellant starvation.

References

[1] Sutton G P and Biblarz O 2010 *Rocket Propulsion Elements* (New York: Wiley)
[2] Goebel D M and Katz I 2008 *Fundamentals of Electric Propulsion: Ion and Hall Thrusters* (New Jersey: John Wiley & Sons)
[3] Grishin S D, Leskov L V and Kozlov H P 1975 *Electric Rocket Engines* (Moskow Russia: Mechanical Engineering)
[4] Jahn R G and Choueiri E Y 2002 Electric Propulsion *Encyclopedia of Physical Science and Technology Third Edition* vol 5 (New York: Academic Press) pp 125-141
[5] Kim V P 2015 *Tech. Phys.* **60** 362
[6] Kozlov A N 2008 *J. Plasma Phys.* **74** 261
[7] Raitses Y, Merino E and Fisch N J 2010 *J. Appl. Phys.* **108** 093307
[8] Karimov A R and Murad P A 2017 *IEEE Trans. Plasma Sci.* **45** 1710
[9] Karimov A R and Murad P A 2018 *IEEE Trans. Plasma Sci.* **46** 882
[10] Leinert C, Richter E, Pitz B and Planck B 1981 *Astron. Astrophys.* **103** 177
[11] Podgorniy I M and Sagdeev R Z 1969 *Adv. Phys. Sci.* **98** 410
[12] Beirman L 1966 *Adv. Phys. Sci.* **90**, 163
[13] Raizer Yu P 1991 *Gas Discharge Physics* (New York: Springer)
[14] Chen F F 1991 *Plasma Phys. Cont. Fusion* **33** 339
[15] Lieberman M and Lichtenberg A 2005 *Principles of Plasma Discharges and Materials Processing* (New Jersey: Wiley-Blackwell)

IOP Conf. Series: Journal of Physics: Conf. Series **1238** (2019) 012064 doi:10.1088/1742-6596/1238/1/012064

Studies of space charge dominated electron photoemission at PITZ

Mikhail Krasilnikov[1], Ye Chen and Frank Stephan

Deutsches Elektronen Synchrotron – DESY, Zeuthen, Germany

[1]mikhail.krasilnikov@desy.de

Abstract. Photoemission in modern high brightness electron sources is under studies at the PITZ photo injector. Space charge dominated photoemission in the presence of high RF field at the semiconductor photocathode is studied. By utilizing core and halo particle distributions based on measured radial laser profiles, simulations reproduce the behaviour of the measured emission curves for a wide range of RF gun parameters within the measurement uncertainties for Gaussian laser pulses. But applying this model to the case of long flattop photocathode laser pulses revealed discrepancies between experimental data and simulation results. Corresponding emittance simulations have been compared to measurements for both temporal profiles of the photocathode laser.

1. Introduction

High brightness electron sources are key components for the successful operation of modern Free Electron Lasers (FELs) like the European XFEL and FLASH. A high brightness electron source for such modern applications requires rather high bunch charge and very small transverse normalized emittance. For the European XFEL photo injector 1 nC electron bunches with ~20 ps length and normalized emittance of <1 mm mrad should be generated from the RF gun operated at high field (~60 MV/m) at the Cs_2Te photocathode. Modern RF-guns are mature technology demonstrating such a high performance. Since more than 15 years the Photo Injector Test facility at DESY in Zeuthen (PITZ) develops and optimizes high brightness photo injectors [1].

The key component of the PITZ accelerator is an L-band 1.6-cell RF-gun, as illustrated in Fig. 1. It consists of a 1.3 GHz copper cavity operated in π-mode with coaxial RF power coupler and supplied with a pair of focusing solenoids. A molybdenum cathode plug of ∅16 mm with Cs_2Te film of ∅5 mm is inserted in the cavity back wall.

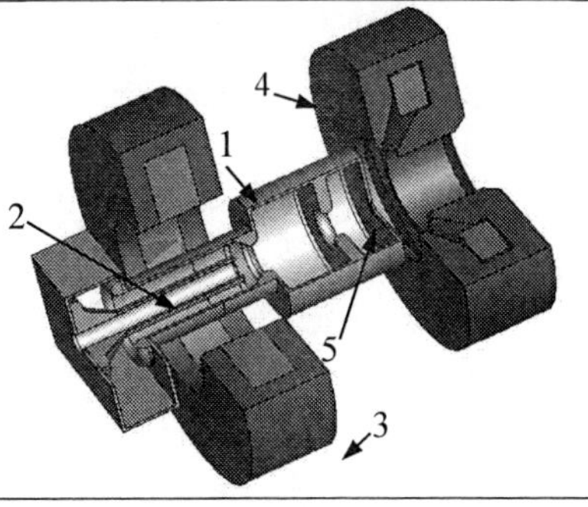

Figure 1. Sketch of the PITZ RF-gun:
1– gun cavity,
2– coaxial power coupler,
3– main solenoid,
4– bucking solenoid,
5– photocathode (Cs_2Te).

IOP Conf. Series: Journal of Physics: Conf. Series **1238** (2019) 012064 doi:10.1088/1742-6596/1238/1/012064

Experimental minimization of the transverse emittance for a wide range of electron bunch charges resulted in optimum photo injector parameters which correspond to the space charge dominated regime of photoemission from the Cs_2Te cathode. At the same time beam dynamics optimization using photo injector parameters corresponding to the experimentally tuned ones results in a more relaxed space charge effect during electron bunch photoemission. Moreover direct plugging of experimental parameters into beam dynamics simulations (ASTRA, [2]) results in strong space charge effect at the cathode and therefore to the significant dilution of the simulated beam emittance.

For more precise description of the space charge dominated dynamics in the RF-photoguns other beam measurements have to be simulated in a consistent way. One of such measurements is the so-called emission curve – dependence of the accelerated bunch charge on the photocathode laser pulse energy for fixed other parameters of the RF-gun (gradient, launch phase, laser pulse length, etc.) Discrepancies between experimental data and simulation results have been observed already for these measurements. In order to bring beam dynamics simulations closer to experimental results several improvements in the emission modelling have been implemented. A core and halo model for transverse photocathode laser distribution helped to reduce the discrepancy between measured and simulated emission curves [3]. Nevertheless, there are still remaining discrepancies, especially for the flattop temporal profile of the photocathode laser pulse. These discrepancies can be dedicated to the space charge dominated photoemission processes.

2. PITZ measurements with flattop photocathode laser pulses

The main goal of the PITZ measurements campaign of 2011 was minimization of the transverse projected normalized emittance for the flattop temporal profile of the photocathode laser pulse [1]. This optimization has been performed for a wide range of bunch charge (20 pC-2 nC) by tuning of key photo injector parameters – main solenoid current/peak field, RF-gun launch phase, laser pulse transverse spot at the photocathode. The laser pulse temporal profile of ~21 ps FWHM and the gun peak field at the photocathode of ~60.5 MV/m were fixed during these studies. A single slit scan technique has been used to measure the transverse phase space at the position of the 1st emittance measurement station (z=5.74 m) [1]. The results of emittance optimization for 1 nC bunch charge are shown in Fig. 2, where projected normalized emittance is shown as a function of the rms spot size of the photocathode laser pulse. For each laser spot size the main solenoid current and RF gun launch phase have been optimized. This yielded the optimum phase to be +6° w.r.t. the gun phase of the Maximum Mean Momentum Gain (MMMG). In contrary, the optimized simulated setup corresponds to the MMMG phase. Corresponding temporal profile and transverse distribution of the photocathode laser pulse for the optimum setup are shown in Fig. 3.

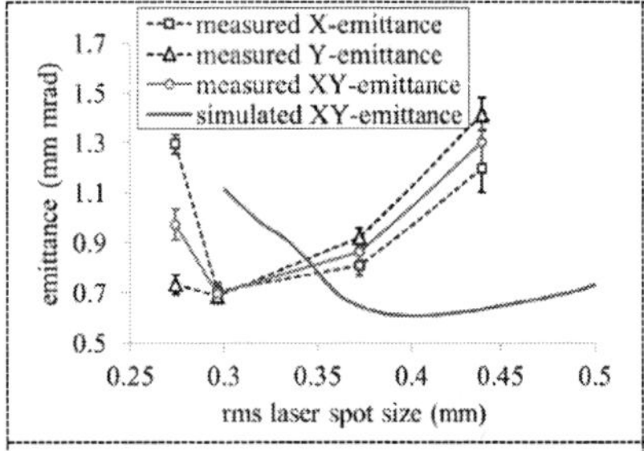

Figure 2. Transverse normalized projected emittance optimized for various rms spot sizes of the photocathode laser and for 1 nC bunch charge [1].

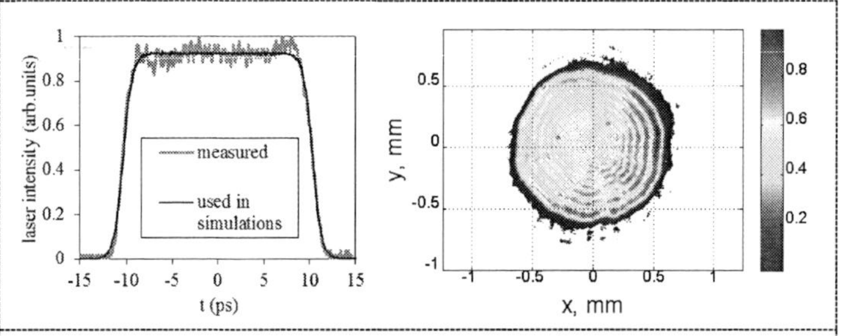

Figure 3. Left: photocathode laser temporal profile measured with an optical sampling system. Right: photocathode laser transverse distribution obtained from the image taken with CCD camera at the location optically equivalent to the photocathode position.

Beam dynamics optimization have been performed for the experimental setup assuming smooth flattop temporal profile with corresponding FWHM and rise/fall times and a radially homogeneous transverse distribution of the photocathode laser pulse. The corresponding curve is shown in Fig. 2 as well. Already from this comparison it appears a significant discrepancy between experimentally optimized rms laser spot size (~0.3 mm) and the results of ASTRA [2] beam dynamics simulations (~0.4 mm). Moreover, direct plugging of the best experimental laser rms spot size into the simulations with radial homogeneous distribution results in a very strong space charge effect already at the photocathode. Thus even the measured bunch charge cannot be reproduced and the simulated transverse phase space of the electron beam is strongly deteriorated. In contrary, the measured emittance is minimized for these settings. The measured beam transverse distribution and phase spaces [1] are shown in Fig. 4. Beam x-y asymmetry in transverse distribution as well as between horizontal and vertical phase spaces have been recently studied and the reason was found to be related to RF power coupler kick and main solenoid imperfections [4].

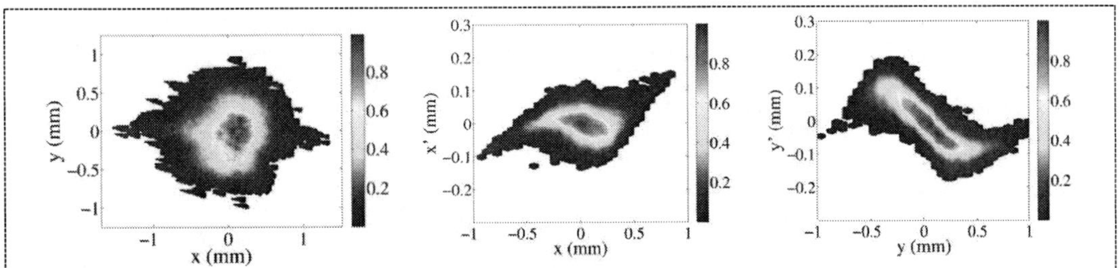

Figure 4. Left plot: 1 nC electron beam transverse distribution measured at the location of the emittance measurements (z=5.74 m from the photocathode). Measured horizontal (middle plot) and vertical (right plot) phase spaces correspond to the normalized rms emittance values: $\varepsilon_{x,n} = (0.72 \pm 0.06)$ mm mrad , $\varepsilon_{y,n} = (0.60 \pm 0.04)$ mm mrad [1].

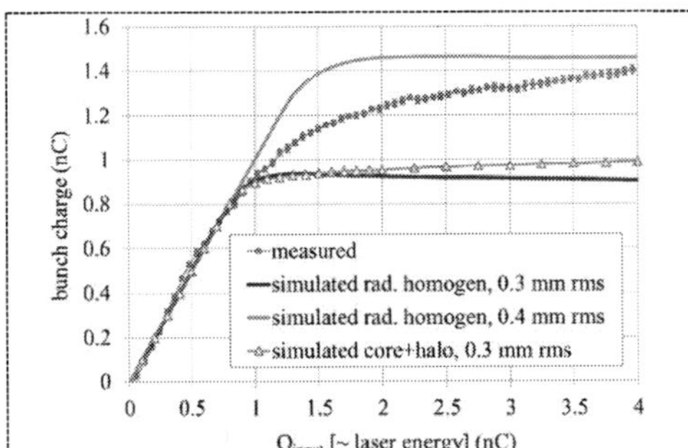

Figure 5. Dependence of the accelerated bunch charge on the photocathode laser pulse energy: measured and simulated for various transverse laser distributions.

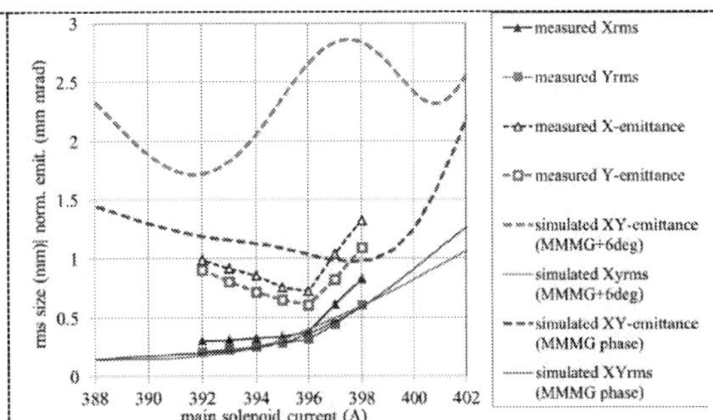

Figure 6. Measured and simulated rms transverse size and normalized transverse emittance of the electron beam versus current of the main solenoid for 1 nC bunch charge.

Besides emittance measurements several complimentary experimental curves have been taken. The emission curve for the best emittance photo injector conditions is shown in Fig. 5. The horizontal axis in this plot has been normalized w.r.t. the linear part of the emission curve (so-called QE-regime): $Q_{input} = QE \cdot E_{laser}$. Replacing the laser transverse distribution (Fig. 3, right) by radial homogeneous profile with the same rms size of 0.3 mm and applying it to ASTRA simulations yielded earlier saturation of the emission curve (blue curve). Corresponding emission curve for the laser rms size of 0.4 mm (the best simulated setup – the minimum of the green curve in Fig. 2) shows rather late saturation and working point (1 nC) would lie in the linear emission regime. The presence of a transverse halo in the photocathode laser (Fig. 3, right) could be responsible for exceeded charge measured in the saturation region. The transverse core and halo photocathode laser model [3] has been suggested in order to explain the experimental data. Unfortunately this resulted only in a small increase of the charge in the beam halo obtained from these simulations (pink curve in Fig. 5). Corresponding results of the simulated rms beam size and emittance are shown in Fig. 6 as a function of the main solenoid current compared to measurements [1]. In order to illustrate an impact of the RF gun launch phase also results of simulations for the MMMG gun phase are shown in Fig. 6. The main solenoid calibration (peak field vs. current) has been modified by ~0.5% in order to provide a better agreement between simulated and measured curves of the rms electron beam size. Transverse distribution of the photocathode laser (core and halo model) used in ASTRA simulations and results of these simulations for the optimum setup (with gun phase of MMMG+6°) are shown in Figs. 7 and 8. For results of simulations with radial homogeneous laser profile and gun MMMG phase see Fig. 8 in [1]. Double-fan shape of the simulated phase space is correlated to the two-maxima in the beam current profile. These results are distinct signatures of the strong space charge effect. Discrepancies between simulations and measurements (Fig. 4) are clearly seen and still not yet fully understood.

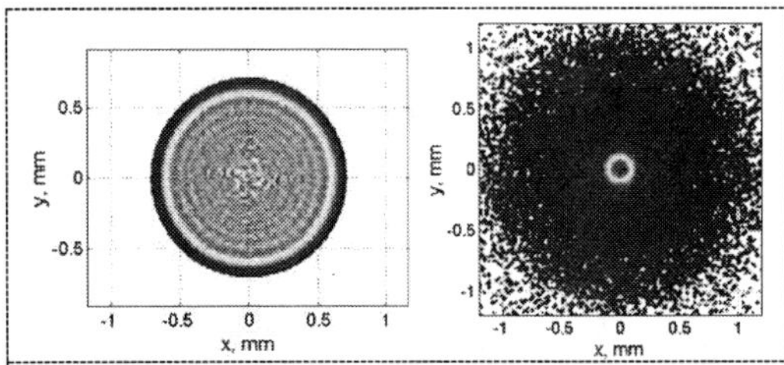

Figure 7. Left: Photocathode laser transverse distribution with applied core and halo model. Right: simulated electron beam transverse distribution at z=5.74 m.

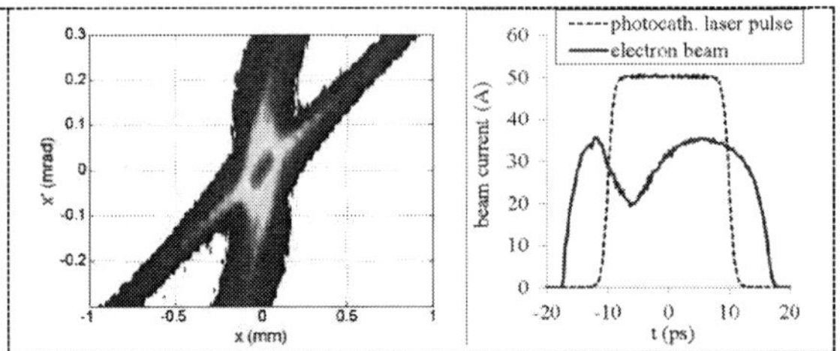

Figure 8. Left: simulated transverse phase space, $\varepsilon_{x,n} = 2.6$ mm mrad. Right: simulated beam current profile at z=5.74 m compared to the temporal profile of the photocathode laser pulse.

3. PITZ measurements with Gaussian photocathode laser pulses

Similar measurements have been performed at PITZ in 2015 but the emittance was optimized for different photo injector parameters, namely for 0.5 nC bunch charge (1 nC in 2011) and for photocathode laser pulses with Gaussian temporal profile of ~11.5 ps FWHM (in contrast to the flattop with 21 ps FWHM used in 2011). The position of the 1st emittance measurement station was shifted upstream to z=5.277 m from the photocathode. Another difference to the 2011 results is the experimental optimum gun phase was found to be close to the MMMG phase what well agrees with simulations. The measured laser transverse distribution is shown in Fig. 9 (left) together with core and halo modeled input distribution for ASTRA simulations (right plot in Fig. 9). Results of emission simulations are shown in Fig. 10. Here, the improvements from the core and halo model w.r.t. the radial homogeneous distribution are clearly observed.

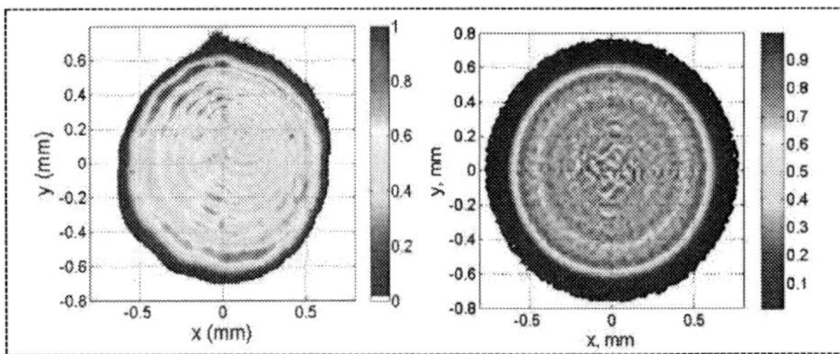

Figure 9. Photocathode laser transverse distributions: measured (left) and used in beam dynamics simulations with applied core and halo model (right).

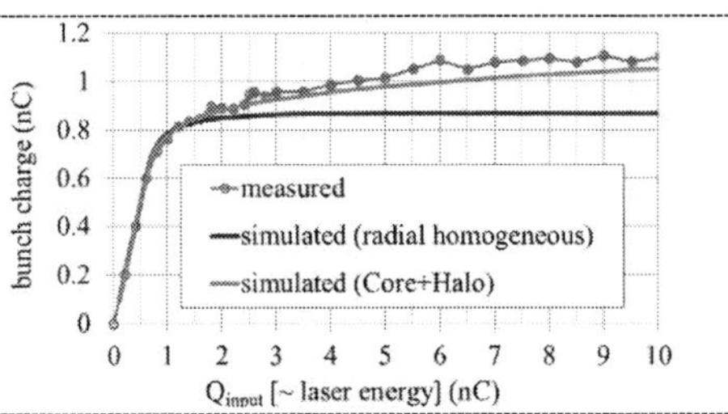

Figure 10. Emission curves: measured and simulated for various transverse laser distributions.

Simulated dependencies of the electron beam transverse rms size and emittance on the main solenoid are shown in Fig. 11 together with corresponding measurements [5]. A pair of specially designed gun quadrupoles has been used during this measurement campaign. This resulted in a more round beam (Fig. 12) and more equal horizontal and vertical phase spaces (Fig. 13). Corresponding simulated electron beam distribution and phase space are shown in Fig. 14. A better agreement is achieved for the electron beam transverse distribution whereas the simulated transverse phase space is still deviating from the measured one when the largest part of this discrepancy results from the low intensity tails of the phase space distribution. Overall, this setup corresponds to a good agreement on the transverse rms beam size in Fig. 11. The horizontal displacement between measured and simulated emittance curves still remains, but the vertical offset became smaller. The better agreement for this emittance measurement setup could be also partially related to the fact that the working point (bunch charge of 0.5 nC) lies well within the linear regime of the emission curve (Fig. 10).

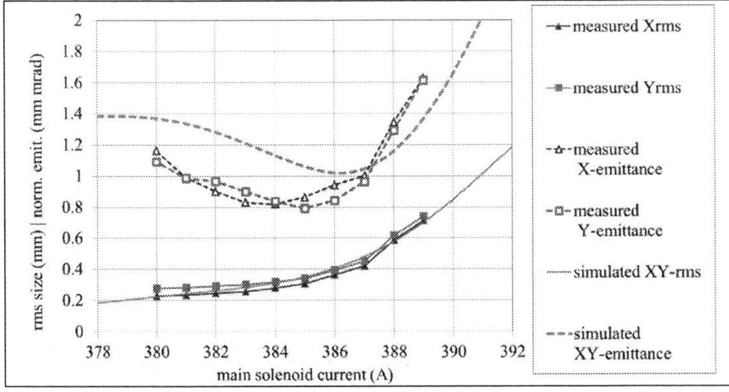

Figure 11. Measured and simulated transverse rms size and normalized emittance of the electron beam versus current of the main solenoid for 0.5 nC bunch charge and photocathode laser pulses with Gaussian temporal profile of 11.5 ps FWHM and 0.3 mm rms spot size.

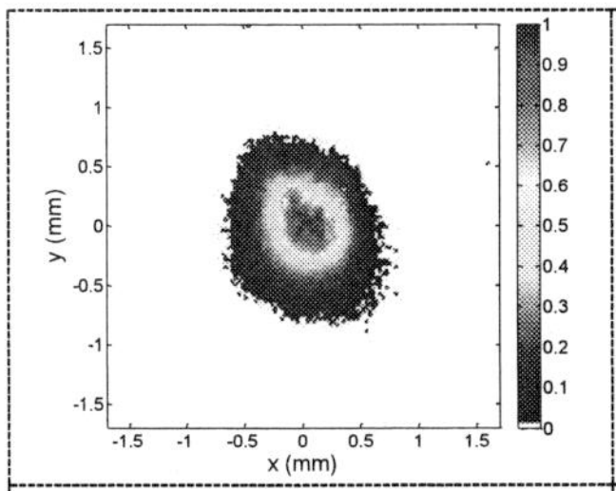

Figure 12. Measured transverse distribution of electron beam at 5.277 m from the photocathode.

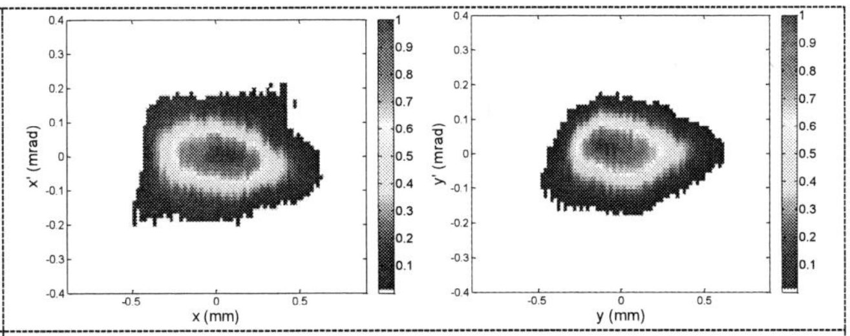

Figure 13. Measured horizontal (left) and vertical (right) phase space of the electron beam. Normalized rms emittance values: $\varepsilon_{x,n} = (0.82 \pm 0.04)$ mm mrad, $\varepsilon_{y,n} = (0.84 \pm 0.04)$ mm mrad. [5].

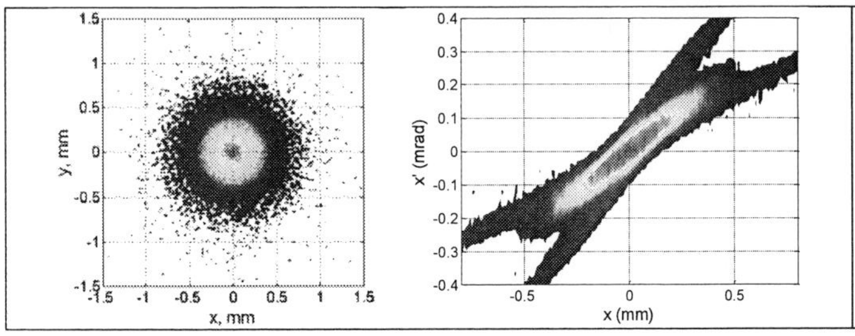

Figure 14. Transverse electron beam distribution (left) and transverse (horizontal) phase space (right) simulated for the optimum experimental conditions. Normalized rms emittance 1.1 mm mrad.

4. Conclusions

Space charge dominated photoemission is under studies at the PITZ photo injector. Discrepancies between experimental data and beam dynamics simulations have already been observed for the generated bunch charge for the optimum photo injector parameters. This also leads to strong disagreement between simulated transverse phase spaces of the electron beam and corresponding experimental data. Applying the model of the photocathode laser pulse transverse core and halo strongly improves this discrepancy for laser pulses with Gaussian temporal profiles, but for the case of flattop laser pulses it still remains significant. Further investigations are ongoing towards better emission model for the beam dynamics simulations as well as by preparing more detailed photoemission measurements.

5. References

[1] M. Krasilnikov, et al., PRST AB 15, 100701 (2012).
[2] K. Floettmann, ASTRA particle tracking code, http://www.desy.de/~mpyflo/.
[3] C. Hernandez-Garcia et al., NIM A 871 p. 97 (2017).
[4] M. Krasilnikov et al., "Investigations on electron beam imperfections at PITZ", LINAC'16, East Lansing, USA, MOPLR013, p. 165 (2016).
[5] M. Krasilnikov et al., „Electron beam asymmetry compensation with gun quadrupoles at PITZ", FEL'17, Santa Fe, USA, WEP007, p. 421 (2017).

Development of the equipment of beam channels for applied and biomedical research at the NICA complex

A A Baldin[1], A V Vishnevskiy[1], K K Laktionov[2], V A Panacik[1], S I Tyutyunnikov[1], D I Yudin[2], I P Yudin[1]

[1]Joint Institute for Nuclear Research, *Dubna, Moscow region, Russia*
[2]N.N.Blokhin National Medical Research Center of Oncology. Moscow, Russia

E-mail: yudin@jinr.ru

Abstract. The report deals with the construction of the ion beam transport line for applied and biomedical research at the NICA accelerator complex, JINR, Dubna. Calculations of the beam output channels from the Nuclotron and Booster are carried out. The used electronics is described. We also discuss the compilation and realization of the plan of treating a tumor located at a depth up to 30 cm.

1. Problem statement

The types of accelerated ions and the energies of beams at the Nuclotron and the future NICA accelerator complex are suited for applied research and certain practical applications [1] in the following fields: 1) nuclear power engineering; 2) studies of the radiation resistance of materials; 3) radiation biology and medicine.

The system for monitoring the extracted beams described in the present paper was developed within the framework of the project "Energy+Transmutation" [2].

This system should be a part of a more general set of detectors and systems that allow one to solve a number of applied problems from the list above with the use of purely electronic methods. The requirements set by radiation therapy were used as the basis for the development of the beam monitoring system. The flux of ions ranging from deuterons to carbon with an intensity ranging from 10^5 to 10^{10} s^{-1}, a beam diameter of 3–4 mm, and an energy of up to hundreds of mega-electronvolts per nucleon should irradiate an area of about 200 × 200 mm^2 with the help of scanning magnets. The system should be able to perform the following: 1) measure the position of a deflected beam at least once per millisecond in order to monitor the scanning magnet system; 2) measure the profile of a deflected beam in each position in order to monitor the quality of the beam itself; 3) measure the distribution of the target irradiation dose after each beam dump. In response to these tasks, test samples of detectors and electronics for them were constructed at department no. 5 of the Laboratory of High-Energy Physics (LHEP) JINR. Wire chambers were used in order to fulfill the first two requirements. The third requirement was satisfied with the use of a pixel chamber.

It is pertinent to note that the averaged ionization currents, which are produced by the beam particles in the detectors, and complete ionization charges are detected instead of the individual beam particles.

The constructed system that operates in real time finds practical use in nuclear engineering experiments conducted with the Nuclotron beams. It complemented the beam diagnostics system that was used until very recently and was based on the activation technique (the measurement of the integral beam dump over hours-long irradiation) and the photoplate exposure (the measurement of the beam profile in the target region and the beam position).

2. Monitoring chamber system

Functional test samples of detectors and electronics for them designed for monitoring the Nuclotron ion beam were constructed at LHEP department no. 5: 1) a wire chamber with front-end electronics and detection electronics installed in a data acquisition crate located far from the beam channel and 2) a pixel ionization chamber built with the use of electronics designed at JINR.

This system is designed to perform the following functions: 1) conduct multiple measurements of the extracted beam profile and the beam position in the *XY* plane during the beam dump in a wide range of intensities from 10^5 to 10^{10} s^{-1}; 2) determine the overall beam particle flux in the beam dump; 3) measure the distribution of the irradiation dose at the irradiated object in the *XY* plane.

2.1. Two-Coordinate Wire Beam Chambers

The chamber incorporates two signal planes for recording the projection of the beam intensity distribution onto the *X* and *Y* axes. The signal anode planes are placed between three high-voltage cathode planes. Each signal plane is basically an array of wires with a diameter of 25 μm made from gold-plated tungsten that are arranged with an interval of 2 mm. Each signal plane incorporates 96 wires. The corresponding active region size is 192 × 192 mm^2. Each wire may be connected to the detection electronics. High-voltage planes are made from stainless grid. The wire diameter is 50 μm. The grid cell size is 0.5×0.5 mm^2, and the interelectrode gap is 6.5 mm. The chambers are constantly purged with a gas mixture based on argon and carbon dioxide with the addition of propanol vapor. When the beam intensity is high, the chambers operate in the regime of collection of the ionization charges produced in the interelectrode gap. When the beam intensity is low, the chambers may operate in the gas multiplication regime at up to 10^4 s^{-1}.

Wire Chamber Electronics. The wire chamber electronics incorporates the frontend electronics installed at a distance of up to 1.5 m from the beam, the ADC unit mounted in the remote data acquisition crate at a distance of no less than 30 m, and the so-called sequencer generating a train of control pulses for synchronizing the system operation. The sequencer is also located in the remote crate. A single **the frontend electronics board** incorporates 32 channels for recording the current values from the chamber wires at specific time points. The board is basically a multichannel analog memory element and an analog multiplexer for connecting the memory channels sequentially, one after another, to the remote ADC mounted in the data acquisition crate. Each board has a single analog output that is activated if the channel being triggered is located at this board. If this is not the case, the analog output is disabled. The outputs of all boards may be tied together for readout with a single ADC channel. Each board is addressed with the use of switches located on it. Thus, all the channels tied together for readout with a single ADC channel are lined up in a single queue with consecutive numbering. The boards are mounted in twos in NIM bins (i.e., a single bin incorporates 64 channels). In response to a signal from the sequencer, the instantaneous value of the input current is saved simultaneously in all channels of the system incorporating many boards. The channels from the general queue are then sequentially connected to the ADC with the help of a control pulse train.

In accordance with the technical requirements, the decay-time constant of the signal formed in the first stage was set to be equal to 330 μs. This means that we may perform data sampling with a period down to 1 ms in order to obtain correct data on the dynamics of changes that the beam is undergoing. **The sequencer** mounted in the data acquisition crate serves to generate (on command from the crate controller or upon the arrival of an external trigger signal) a pulse train that is needed in order to control the synchronous operation of the frontend electronics and the ADC. The output signals are displayed on the front panel. The sequencer was constructed using 2K × 8 memory. Prior to the start of operation in the acceleration run, the sequencer requires loading 2048 8-bit words, which set the needed configuration of the control pulse train at the output terminals, into the memory through the crate bus. When it is triggered by an external signal or on command through the crate bus, the internal generator produces 2048 pulses and stops. The pulse counter generates the address for the sequential readout of 8-bit words. Every bit of each successive word affects the generation of a signal through one of the eight output terminals at the front panel. **Calibration dependences** for several detection channels incorporating the frontend electronics and the ADC in the data acquisition were obtained with the use of a test current supply. **The voltage dependences of the chamber gas amplification** for two gas mixtures based on argon and carbon dioxide with the addition of propanol vapor are used. The electronics for pixel chambers (to be discussed below) was used to measure these dependences. The charge passing between the chamber electrodes in unit time upon irradiation with a Sr-90 radioactive source was recorded.

The combination of the gas amplification capacity and the dynamic range of the detection electronics yields a dynamic range of the system of no less than five orders of magnitude.

Figure 1 shows examples of visualized data from the wire chambers displayed after each beam dump in the process of data acquisition:

1) beam profiles projected on the *X* and *Y* axes after the beam dump;

2) distributions of half-widths at half-maximum of the profiles for the *X* and *Y* projections in the process of data acquisition;

3) distributions of positions of the beam center in the wire chamber reference frame in the process of data acquisition.

2.2. Pixel Ionization Chamber

The basic design of this chamber is similar to the one of the wire chamber. The cathode is made from a grid of stainless wires with a diameter of 50 μm. The cell size is 0.5×0.5 mm^2. The anode plane is a removable element of the chamber. A plane made from glass-fiber laminate with a thickness of 1 mm was used in the considered case. The active region with an area of 160×160 mm^2 is divided into 256 separate sensitive pads (pixels). Each of them is connected to a terminal at the periphery of the printed board. The pixel size is 10×10 mm^2. The interelectrode gap may be varied in a wide range starting from 6.5 mm with the use of inserts.

The pixel chamber was purged in beam runs with the same gas mixture (based on argon and carbon dioxide with the addition of propanol vapor) that was used in the case of the wire chamber.

The pixel chamber electronics incorporates two unit types: the frontend electronics near the chamber and the data acquisition unit mounted in the data acquisition crate located far from the beam channel.

The frontend electronics of the pixel chamber is located up to 1.5 m from it and is based on the TERA06 microcircuit chip produced in Italy [3]. The main features of TERA06 are as follows. Each of the 64 channels converts the input current into pulse rate with continuous counting of the number of pulses by a 16-bit counter. The counter readings may be shifted into the 64-word output register at any specific time on command from the data acquisition unit. The register contents are read out word after word into the data acquisition unit mounted in the crate located far from the beam channel. The counter does not stop at this time. Thus, the system operates without dead time. The quantum of conversion of the input current into frequency is varied with the use of external potentials within the range from 100 to 800 fC. The maximum conversion frequency is 5 mHz. The average noise per channel is on the order of 1 1/s. The frontend unit and the data acquisition unit designed at JINR are used in the described system. With the chosen settings, the quantum of conversion of the input current into frequency is roughly equal to 600 fC. A total of 64 detection channels were used. This corresponded to an active area size of 80×80 mm^2 in the pixel chamber.

Figure 1 shows examples of visualized data on the two-dimensional distribution of the cumulative dose obtained in real time after each dump of the beam onto the target.

3. **The channel for biomedical research.**

One direction in the development of the JINR accelerator complex NICA is the design of a test bench for biomedical research based on the JINR Nuclotron. While designing the test bench, the general technique for manufacturing the hadronic therapy complex is tested. We present a calculation results for the optics of the charge particle transport channel of the hadronic therapy complex.

This channel is intended for the transportation of the $^{12}C^{+6}$ carbon ions with an intensity of ~2 × 10^9 and an energy of 100–550 MeV/nucleon. The channel (see fig. 2) starts near the F3 focus of the primary channel.

3.1. Setup Restrictions.

Additional restrictions of the transport problem arise in the use of specific optical elements and their mounting in the experimental hall. The following optical elements are used in the channel: dipole and deflecting magnets (SP-94) and a magnetic lens (ML-17).

The length of the transport channel is 12 m. The designed channel starts from a dipole magnet mounted in the primary transport channel at a distance of 5.25 m in front of the F3 focus of the primary channel. Optical elements (quadrupole lenses) are located over the channel. There is a beam trap at the end of the channel, the test bench in F_k (focus C, see fig. 2) is right before the trap, and the beam scanning system is before the test bench. The scanning system consists of two similar deflecting magnets rotating at an angle of 90° around the axis relative each other.

One natural restriction of beam transportation is that the diameter of the vacuum pipeline is equal to 0.25 m, which passes through all the quadupole elements of the system. A magnet aperture of 0.3 × 0.13 m is a restriction at the final stage of the scanning system. The mutual position of the scan region and the system of scanning magnets is fixed.

The maximum current is limited for each quadrupole lens; this results in a limitation of the coefficient $K < 1.5\ m^{-2}$ used for calculating the phase incursion on the lens at a preset energy range of 100–550 MeV/nucleon.

The beam cross section in the F3 focus of the primary transport channel is a circle 0.04 m in diameter [4-5], which makes it possible to define the initial conditions of beam propagation.

3.2. Results Obtained.

For a preset aperture and the magnet arrangement, we have [6]: 1) a minimum focus diameter of 2.8 mm for an emittance of 25π mm mrad; 2) a minimum focus diameter of 5.6 mm for an emittance of 50π mm mrad. The region of target scanning is restricted by an aperture of two successive magnets in a plane normal to the beam. The magnet aperture is 0.3 × 0.13 m for a specific channel; this agrees with the planned scan region of 0.1 × 0.1 m. Solutions have been found for emittances of 25π mm mrad and 50π mm mrad based on the above specified restrictions. The beam cross section in the focus C (fig.2) is close to the minimum for the given scanning system geometry: 1) 3.0 × 3.0 mm for an emittance of 25π mm mrad; 2) 5.6 × 6.0 mm for an emittance of 50π mm mrad.

4. Initial Conditions.

To check the existence of a solution in other operating modes, other initial conditions of beam propagation in an additional channel were considered. The beam cross section in the F3 focus acts as a parameter determining these initial conditions.

In addition to a solution with a focus on the primary channel target of 0.04 × 0.04 m, we considered the solution for a focus of 0.1 × 0.12 m [6]. The beam cross section in the focus on a target for configurations with initial conditions corresponding to different beam sizes in the F3 focus of the primary channel (the emittance is 50π mm mrad) is equal to 1) 5.6 × 6.3 mm for the focus 0.1 × 0.12 m and 2) 5.6 × 6.0 mm for the focus 0.04 × 0.04 m.

We suggest the following equipment for the additional channel: 1) magnetic lenses ML-17; 2) scanning and dipole magnets SP-94; 3) the pixel chamber and the wire chamber and readout electronics.

Conclusions

Test samples of the following systems were constructed: 1) two coordinate wire chambers with an active area size of 192 × 192 mm^2 and the electronics for them and 2) the pixel ionization chamber with an active area size of 160 × 160 mm^2 and 256 pads with a size of 10 × 10 mm^2 that is operated with the detection electronics based on the TERA06 microcircuit chip.

The systems ensure the monitoring of the extracted relativistic Nuclotron ion beams in real time and allow one to perform 1) multiple measurements of the beam profile, its center position, and its intensity during the beam dump; 2) measurements of the distribution of the cumulative target irradiation dose after each dump. The system was tested in a series of Nuclotron acceleration runs and performed efficiently at deuteron beam intensities of up to 10^{10} 1/s.

The construction of the carbon beam transport line for applied and biomedical research at the Nuclotron accelerator complex, JINR, Dubna is shown. We have studied the scheme and modes of magneto-optical elements of the channel. Used electronics described. We are discussed the compilation and realization of the plan of treating a tumor located at a depth up to 30 cm. Choice of beam scanning schemes and their optimization are shown.

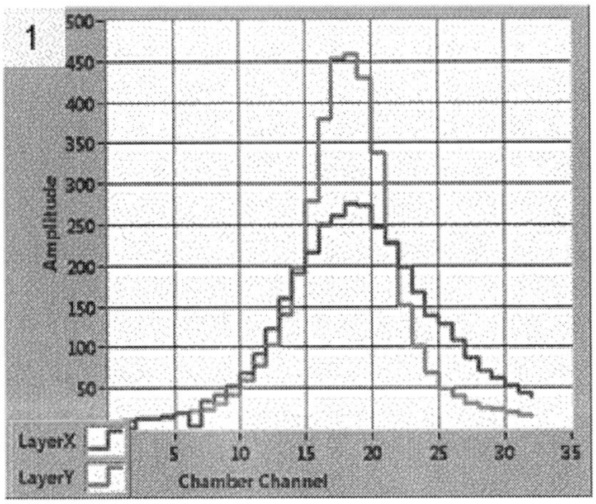

Figure 1. The view of distribution of beam is for on-line monitoring of the carbon beam.

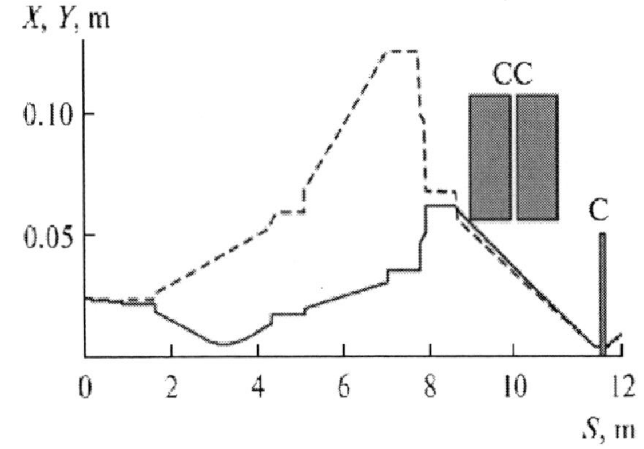

Figure 2. Position of the scanning system of magnets on a
beam (CC). The scanning region in the beam focus (C).

References

[1] N. Agapov et al., "NICA project at JINR," in Proceedings of the XXIII Russian Particle Accelerator Conference, St. Petersburg, Russia, 2012 (JACoW, 2012).FLASH project: http://flash.desy.de/.

[2] A.I. Berlev et al., The Measurement of the Neutron Flux Using Diamond Detector at the "Quinta" Setup // Physics of Particles and Nuclei Letters. 2016. Vol.13. № 3. pp.352-357.

[3] G. Mazza et al., "A 64-channel wide dynamic range charge measurement ASIC for strip and pixel ionization detectors," IEEE Trans. Nucl. Sci. 52, 847–853 (2005).

[4] A. Safronova et al., "Investigation of 4 GeV Deutron Beam Parameters at the Nuclotron," in Proceedings of the 3rd International Conference on Current Problems in Nuclear Physics and Atomic Energy, NPAE-Kyiv 2010, June 7–12, 2010 (Kiev, 2010), p. 113.

[5] V.I. Volkov, private commun.

[6] I. P. Yudin, V. A. Panasik, and S. I. Tyutyunnikov An Additional Carbon Ion Transport Channel for Biological Research at the Nuclotron of the Joint Institute for Nuclear Research // Physics of Particles and Nuclei Letters, 2012, Vol. 9, No. 3, pp. 283–287. © Pleiades Publishing, Ltd., 2012. ISSN 1547-4771.

IOP Conf. Series: Journal of Physics: Conf. Series **1238** (2019) 012066 doi:10.1088/1742-6596/1238/1/012066

Development of Proton Irradiation Facility at the INR Linac

S Bragin[1], A Feschenko[1], S Gavrilov[1], O Grekhov[1], Y Kalinin[1], Y Kiselev[1], L Latysheva[1], S G Lebedev[1], A Melnikov[1], V Serov[1], N Sobolevskiy[1], A Titov[1], O Volodkevich[1], D Arbuznikov[2], O Podgornaya[2], E Prokhorov[2], S Razinkov[2], P Tsedrik[2] and S Tsibryaev[2]

[1] Institute for Nuclear Research, Russian Academy of Sciences, Moscow, Russia
[2] Russian Federal Nuclear Center, All-Russian Research Institute of Experimental Physics, Sarov, Russia

s.gavrilov@gmail.com, lebedev@inr.ru

Abstract. The results of development, installation and tests of an experimental facility for irradiation of different products and materials at the INR RAS linear accelerator are presented. The beam energy range is 20÷210 MeV. The energy is adjusted by switching on/off the fields in accelerating cavities and with energy degraders. The intensity range is from 10^7 protons per separate pulses up to 1 μA of average beam current. Also the activation of the beam dump and the test unit as well as the radiation conditions in the vicinity of the facility are estimated.

1. Introduction

The accelerating complex based on the linear proton accelerator of Institute for Nuclear Research RAS [1] is a multipurpose facility with a wide range of objectives in neutron research [2-3], radioactive isotope production [4] and medical physics [5]. This paper describes the development and construction of the Proton Irradiation Facility (PIF) for studies of irradiation induced effects in electronics and other materials. PIF essentially expands the range of tasks performed at the INR linac. The structure of the linac is shown in Figure 1. After the analysis of possible locations of PIF the decision was made to locate it downstream of the existing 7.5° bending magnet in the 600 MeV areas at the exit of the accelerator.

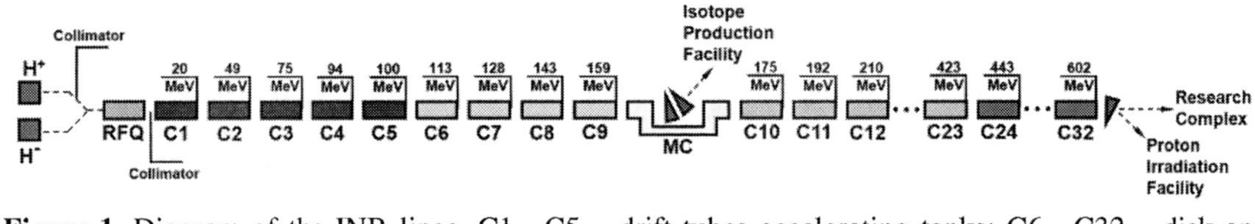

Figure 1. Diagram of the INR linac. C1...C5 – drift tubes accelerating tanks; C6...C32 – disk and washer accelerating cavities; MC – matching cavity.

As the energy range for PIF is 20÷210 MeV beam dynamics simulations with the aim to verify lower energy beam transportation to the exit of the accelerator as well as to find the focusing elements settings have been done. The simulations have been carried out with TRACE-3D code [6] and have shown that in the whole specified energy range the beam can be transported to the exit of the accelerator with minimum beam losses, no focusing elements are needed downstream of the bending

IOP Conf. Series: Journal of Physics: Conf. Series **1238** (2019) 012066 doi:10.1088/1742-6596/1238/1/012066

magnet and the size of the beam at the PIF target can be controlled with the last three accelerator quadrupole doublets.

2. PIF configuration and main parameters

The general PIF configuration is shown in Figure 2. The main elements are: bending magnet with vacuum chamber, beam pipe, beam dump, target positioning system, energy degrader and beam diagnostics.

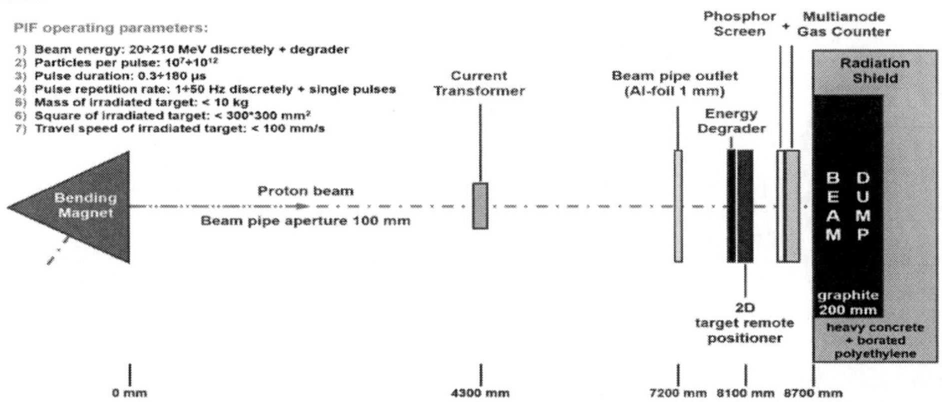

Figure 2.PIF configuration.

Beam energy at the PIF is adjusted mainly by a switching on/off the accelerating cavities, which provide discrete energy values with a full energy spread < 1%. For intermediate beam energy values the energy degrader is used. The presence of the degrader along with the outlet of 1 mm aluminium foil window and 900 mm air gap results in a significant increase of the energy spread, especially for lower beam energies. As an example figure 3 demonstrated the energy spectrum calculated with TRIM [7] for 49 MeV beam at the exit of the 4.6 mm aluminium degrader.

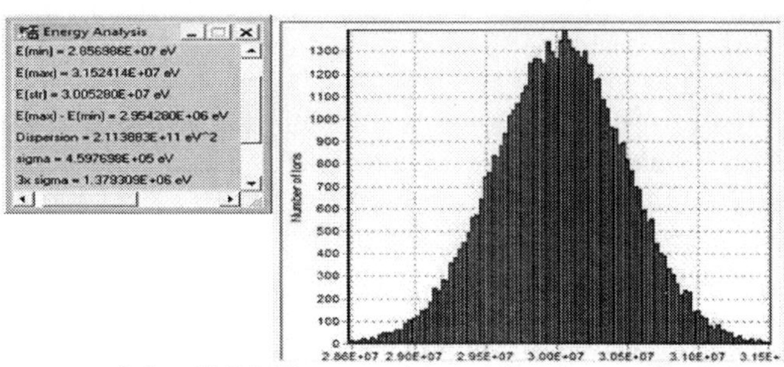

Figure 3. Energy spectrum of the 49 MeV proton beam after 1 mm aluminium window, 900 mm air gap and 4.6 mm aluminium energy degrader.

Beam intensity is defined by a combination of three parameters: pulse current, which can be adjusted in the range of $10^7 \div 10^{12}$ protons per pulse with two collimators in the linac injection line, pulse duration, variable in the range of $0.3 \div 180$ µs, and pulse repetition rate. The available repetition rates are 1, 5, 10, 20, 25, 40 and 50 Hz. The preset number of beam pulses can be produced at any of the above repetition rates.

The two-dimensional target positioner 3 (Figure 4) is intended for the irradiated target positioning in the prescribed transverse point and for moving it with the speed up to 100 mm/s in the range of ±150 mm with respect to the beam axis. The mass of the irradiated element is limited to 10 kg.

IOP Conf. Series: Journal of Physics: Conf. Series **1238** (2019) 012066 doi:10.1088/1742-6596/1238/1/012066

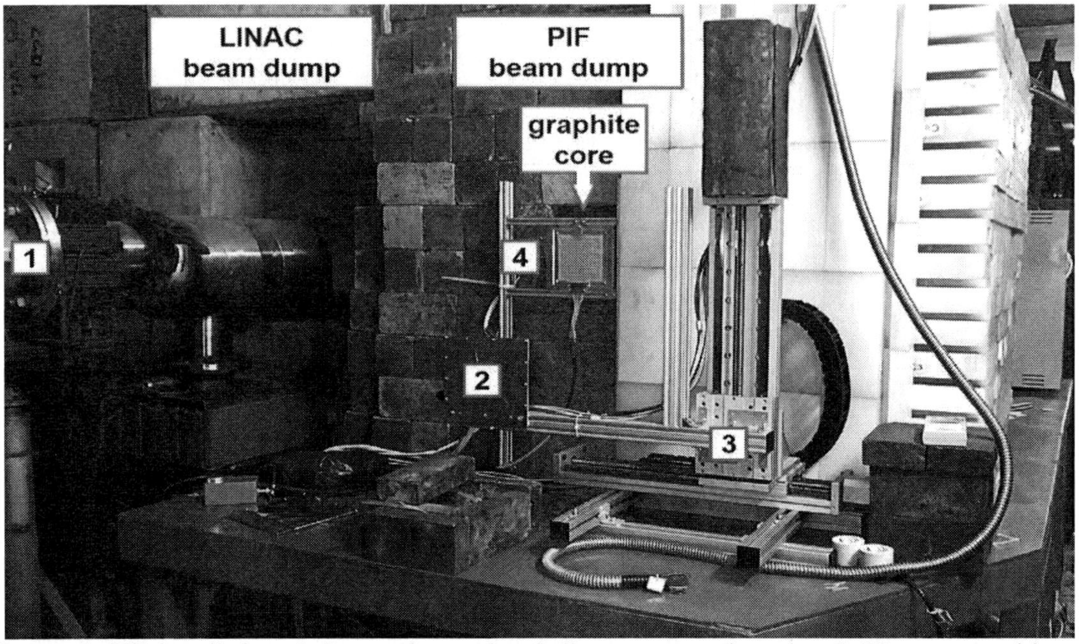

Figure 4. PIF installation. 1. Beam pipe outlet. 2. Beam energy degrader. 3. 2D target positioner. 4. Multianode gas counter with phosphor screen.

2.1. Beam dump

The beam dump consists of the graphite core surrounded by the shield of heavy concrete and borated polyethylene blocks. The calculations of shielding were carried out using the transport code SHIELD [8]. For the maximum specified beam current of 1 μA with the energy of 210 MeV the most intensive neutron fluxes are generated upwards ($2{\cdot}10^7$ n/cm^2/sec) and sideways ($5{\cdot}10^5$ n/cm^2 /sec). These fluxes are lowered to the acceptable levels by the existing 1.7 m concrete shield and 7 m soil layer. The calculations of the backward neutron flux from the dump to the irradiated unit show, that the integral flux does not exceed 10% of the neutron flux generated inside the irradiated unit and the energy of these neutrons is below 1 MeV. Hence the influence of the backward neutrons on the results of irradiation tests is negligible.

Calculations of the target activation were done with the code DCHAIN-SP [9]. The composition of the target was taken to be the following: silicon 90%, epoxy resin 5%, copper 5%. The total activity of all the radionuclides produced in the target after 8 hour irradiation with the 209 MeV, 1 μA beam is 10^9 Bq. It decreases by two orders of magnitude within one day (Fig. 5). The dose rate on the surface of the irradiated target in a day is 27.5 μSv/h

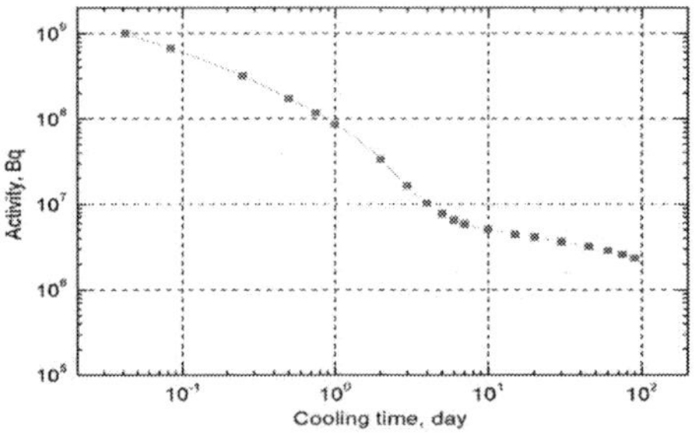

Figure 5. Integral activity of all radionuclides in the PIF target after 8 h irradiation by 210 MeV protons with 1 μA average current.

2.2. Beam diagnostics

IOP Conf. Series: Journal of Physics: Conf. Series **1238** (2019) 012066 doi:10.1088/1742-6596/1238/1/012066

PIF beam diagnostics includes three devices: beam current transformer (BCT), multianode gas counter (MGC) and phosphor screen (PS). The BCT is installed in the beam pipe about 3 m upstream of the outlet window and provides absolute nondestructive measurements of beam pulse current with the amplitude > 50 μA, which corresponds to about $>10^{10}$ protons per pulse with typical pulse durations. For smaller beam intensities a special gas counter has been developed and fabricated (Figure 6a). The counter includes several parts: single-anode ionization chamber for beam current measurements in the range of $10^{7}\div10^{11}$ protons per pulse and two proportional chambers, each containing 25 anode 100 μm gold-plated strips with 4 mm spacing (Figure 6b) for beam transverse profiles and position measurements in 100×100 mm² area. In the existing design an ambient air filling is used, so MGC must be calibrated regularly with BCT for absolute beam current measurements.

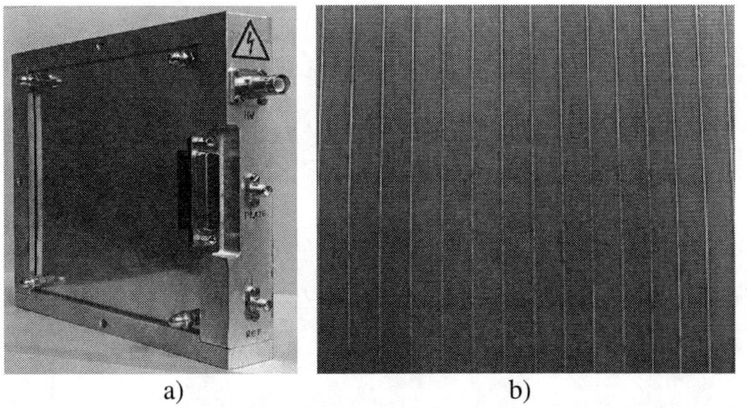

a) b)

Figure 6. a) Multianode gas counter. b) Strips for profile measurements in proportional chambers.

The phosphor screen is covered with P43 phosphor (Figure 7a). The image on the screen is visualized with CCD-camera. PS provides beam profile and beam position measurements (Figure 7b) with the resolution of about 1 mm in the 100×100 mm² area. Also it was experimentally tested, that PS can be used to measure number of particles per beam pulse independently on its duration. The threshold sensitivity was found to be near 5×10^{7} proton/cm². The linearity of the measurement was tested within the range of $10^{8}\div10^{10}$ proton/cm².

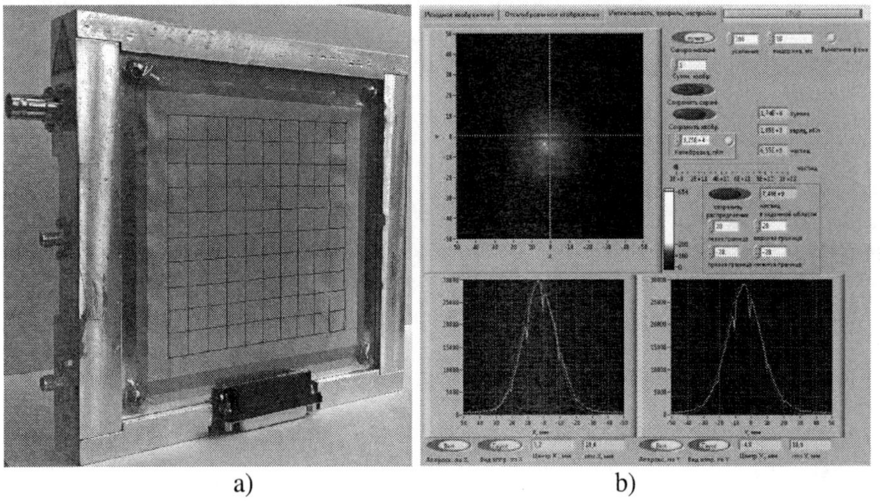

a) b)

Figure 7. a) Phosphor screen fixed at the surface of the gas counter. b) PS diagnostics software.

3. PIF beam tests

Beam tests were done within a full range of the PIF operating parameters to check all the systems. Figure 8 shows some results of irradiation with initial beam energies about tens of MeV, when protons are strongly scattered or completely stopped inside the specimen, when a "shadow" image is formed at PS.

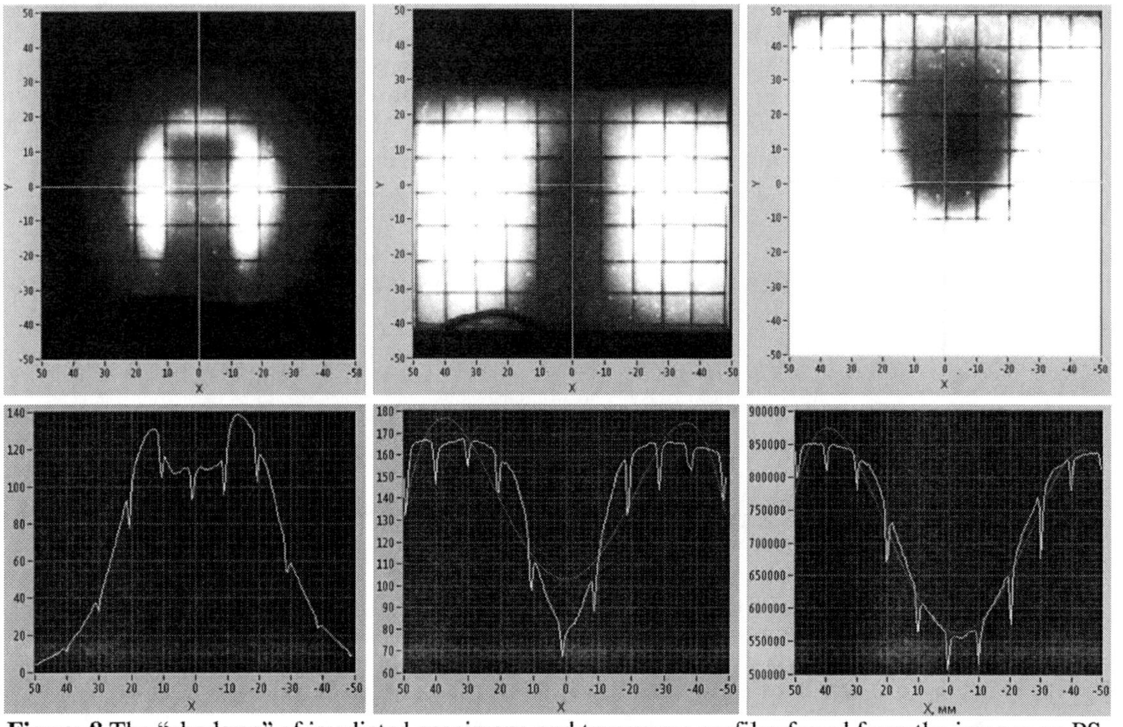

Figure 8. The "shadows" of irradiated specimens and transverse profiles found from the images on PS.

The results of the tests confirmed the correctness of the decisions taken at the stage of development of the irradiation facility and the feasibility of all the declared parameters. The experience gained in the course of the tests will subsequently speed up and optimize the irradiation procedure.

This work was done with the use of the equipment of the Shared Research Facility "Accelerator Center for Neutron Studies and Nuclear Medicine of INR RAS" of the of the Institute for Nuclear Research of the Russian Academy of Sciences.

References

[1] Kravchuk L V 2006 *Nucl. Instr. and Meth.* A **562** 932
[2] Koptelov E A, Lebedev S G, Smirnova O N, Sobolevsky N M, Stavissky Y Y, Al'tovsky I V, Orlov V V, Votinov S N 1996 *Journal of Nuclear Materials* **233** 1552
[3] Koptelov E A, Lebedev S G, Matveev V A, Sobolevsky N M, Strebkov Y S, Subbotin A V, 2002 *Nucl. Instr. and Meth.* A **480** 137
[4] Zhuikov B L 2014 *Applied Radiation and Isotopes* **84** 48
[5] Kravchuk L V 2010 *Physics-Uspekhi* **53** 635
[6] Rusthoi D, Lysenko W and Crandall K 1997 Further Improvements in TRACE 3-D *Proceedings of PAC97* pp 2574-2576
[7] www.srim.org
[8] http://www.inr.ru/shield
[9] Takada H, Kosako K 1999 Development of the DCHAIN-SP for Analysis Decay and Build-up Characteristics of Spallation Products. *JAERI-Data* / Code 99-008.

IOP Conf. Series: Journal of Physics: Conf. Series **1238** (2019) 012067 doi:10.1088/1742-6596/1238/1/012067

Beam asymmetry compensation at DESY photoelectron guns

I Isaev, M Krasilnikov, Q Zhao

Deutsches Elektronen-Synchrotron DESY, 15738 Zeuthen, Germany

E-Mail: igor.isaev@desy.de

Abstract. Quadrupole fields are considered as an option for beam asymmetry compensation for L-band photoelectron guns used at DESY. The quadrupoles dedicated for the installation at a gun coupler region have three designs. The first two designs were installed and tested with photoelectron beam. The designs of the quadrupoles and experiments results are presented.

1. Introduction

One of the most suitable electron sources for modern Free-electron Lasers (FELs) is a photoelectron gun. A photoelectron gun combines electron generation utilizing photoemission and direct acceleration in the RF cavity. Photoemission allows the generation of higher current densities than in the case of thermionic cathodes and also provides better control over the transverse and longitudinal electron pulse distributions.

The Photo-Injector Test Facility at DESY in Zeuthen (PITZ) is built to develop, operate and characterize electron sources meeting the requirements for FELs operated in the SASE regime as Free-electron LASer in Hamburg (FLASH) [1] and the European X-ray Free Electron Laser (European XFEL) [2]. The requirements include a projected transverse normalized RMS emittance below 1 mm mrad, an RMS bunch length below a few mm and a nominal bunch charge of 1 nC [3].

The tests of the photoelectron gun at PITZ showed that the gun is able to provide required emittance, but at the same time, the optimal parameters (charge, transverse laser spot size, solenoid current) for the smallest emittance were different compared to simulations.

Additionally, during emittance measurements at PITZ an electron beam asymmetry was observed. The asymmetry was found for the beam transverse profile as well as for phase space pictures. The electron beam emittance shows additional undesirable tails that disclose the beam asymmetry (see figure 1).

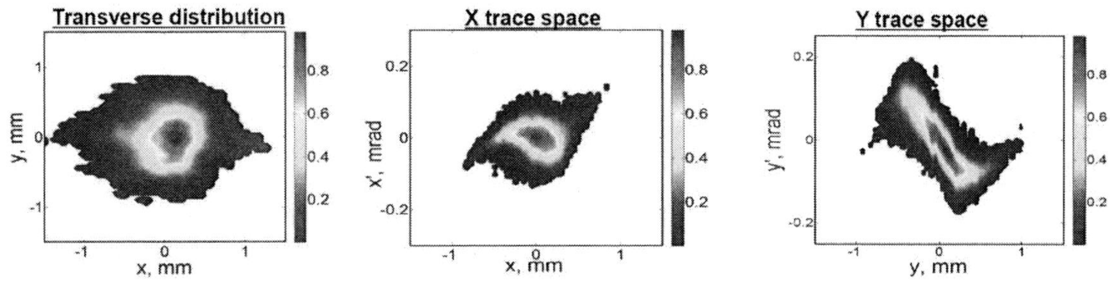

Figure 1. Asymmetric transverse electron beam shape (left), and asymmetric X and Y phase spaces (middle and right) observed during emittance measurements at PITZ.

IOP Conf. Series: Journal of Physics: Conf. Series **1238** (2019) 012067 doi:10.1088/1742-6596/1238/1/012067

2. Beam asymmetry investigations

During electron beam asymmetry investigations a number of experiments for identification of the asymmetry source were performed [4].

The experiment on Larmor angle showed that the beam asymmetry source should be located around the region of coaxial coupler transition to the full cell of the gun. In the same region, the main solenoid is installed. Additionally to this experiment, simulations on the 'back-tracking' particle were performed [5]. Similar to the Larmor experiment, the particle tracking was carried out for two polarities of the main solenoid current (positive and negative). The simulations showed that the distributions of the mentioned particles coincide at the place. That was revealed during the Larmor experiment. The orientations of the particles distributions at this place are 45 deg. This fact is a hint that the beam asymmetry source feature a quadrupole structure.

Based on the insight and experience from other laboratories [6], there were performed thorough simulations on the beam asymmetry modeling [7]. In the simulations, the beam asymmetry source was modeled by a pair of quadrupoles. Position, strength and rotation angle of the quadrupoles were optimization parameters for fitting simulated beam shapes to the beam transverse profile images obtained from the experiment.

The simulations showed that the beam asymmetry can be modeled by the pair of quadrupoles. In this case, the first quadrupole must be located at position 0.18 m (the region of a coaxial coupler to the full cell) downstream the cathode, with skew orientation (45 deg for negative current polarity or 135 deg for positive current polarity). The polarity of the first quadrupole should not vary with changing the main solenoid polarity. The second quadrupole must be located at the position of 0.36 m (around the center of the main solenoid, but slightly shifted towards the downstream end). The quadrupole must have normal orientation, and the polarity should change together with the change of the main solenoid polarity.

The simulations results lead to the idea of building a quadrupole able to compensate the asymmetrical field that produces the beam asymmetry, and might be mounted onto the gun, which was already installed in the tunnel.

3. The first design of the gun quadrupoles

As the most suitable option (among single-wire and PCB quadrupole designs), air-coil based gun quadrupole design was selected (depicted in figure 2). As the only appropriate location for the quadrupole installation was selected the coaxial coupler near the gun-coupler flange.

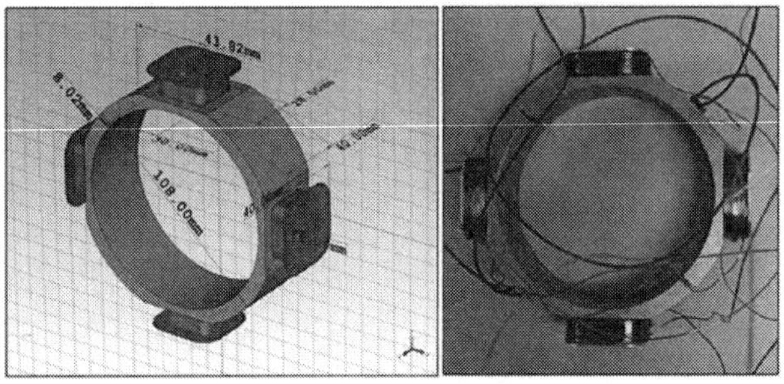

Figure 2. The view of the first design of the gun quadrupole: CAD model (left) and the corresponding produced quadrupole (right).

The coil parameters were chosen according to the magnetic field simulations performed in CST Studio [8]. Using the simulation results, there was produced a quadrupole with the following parameters: aluminium frame, four air coils made of 0.56 mm thick copper cable with 180 windings, two thermal switches with non-magnetic screws and threshold at 80 deg C to prevent overheating. The quadrupole was fixed at the coaxial coupler by a radiation hard cable tie. The quadrupole was connected to the power supply (with current stabilization) with a current limit of 3 A. The field measurements in a laboratory showed good consensus of the simulations and measurement results.

Since the quadrupole frame of the first design has only four coils, there were performed two series of the measurements, particularly with normal and skew quadrupole installations. The orientation change of the quadrupole was done by hands with the operation interruptions and tunnel access. The measurements were performed for 5 MW peak power in the gun, maximum mean momentum gain RF phase, 500 pC bunch charge, 1.2 mm laser transverse spot size, main solenoid current of 335 A (the solenoid current at which beam asymmetry mostly pronounced) at normal and opposite polarities, the bucking solenoid and the booster were switched off. The results showed that the normal-oriented quadrupole can make beam shape more round, but only for normal polarity of the main solenoid current, while the beam at opposite solenoid polarity becomes less round. The same situation was observed with the skew-oriented quadrupole.

4. The second design of the gun quadrupoles

The promising results of the first design of the gun quadrupoles tests and additional time slot for new tests allowed to create a second design of the gun quadrupoles, to produce and to test it.

The second design extends the previous design by additional air coils on an aluminium frame. The further differences are: two quadrupoles are placed on a single frame (each of the quadrupoles supplied separately), the number of copper cable (0.56 mm) windings was reduced to 140, because the geometry of the coils was changed to have as strong as possible quadrupole gradient (which is 0.0117 T/m at 1 A). In analogy to the first design, the quadrupole consists of two thermal switches with a threshold of 80 deg C. Initially, the quadrupole was fixed at the gun coaxial coupler by a radiation-hard cable tie, but later the fixation was exchanged to an aluminium clamp (because later the quadrupole was kept for standard operation). The CAD model and images of the second design of the quadrupole are presented in figure 3.

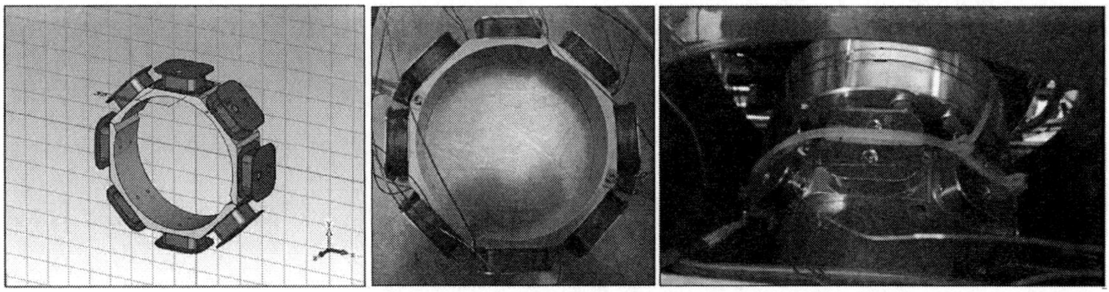

Figure 3. The view of the second design of the gun quadrupole: CAD model (left), the corresponding produced quadrupole (middle), and installed quadrupole at the gun coaxial coupler (right).

4.1. The transverse beam profile measurements

The transverse beam profile measurements were performed with the same settings as for the first quadrupole design (5 MW in the gun and switched off booster). The beam measurements with quadrupole showed the possibility to obtain symmetric beam profiles for both polarities of the main solenoid. For the case of the most round beam, the polarity of the normal-oriented quadrupole (Gun Q1) changes its polarity with changing polarity of the main solenoid, while the polarity of the skew-oriented quadrupole (Gun Q2) did not show this behavior. This result agrees with beam asymmetry modeling by two quadrupoles [7].

4.2. The emittance measurements

In order to correctly set quadrupole currents for the emittance measurements, the transverse beam shape scan adjustments were performed. During the adjustments, beam transverse profile pictures were taken for different quadrupoles settings at two screens: screen station used for slit masks, and the screen used for beamlets collection. Afterward, the quadrupoles currents were selected to have the most round beam on both screens at the same time. The resulted currents are Gun Q1 = -0.6 A and Gun Q2 = -0.5 A.

For the emittance measurements adjustments the so-called solenoid scan was performed, meaning that the emittance and beam sizes dependencies are measured vs the main solenoid current. The emittance measurements were performed at following PITZ parameters: laser transverse spot size = 1.2 mm, charge = 500 pC, gun power = 6.5 MW, booster power = 3 MW, Gaussian laser temporal profile = 11.5 ps, bunch length = 15.8 ps. Emittance measurements vs solenoid current for the case of 0 currents of the gun quadrupoles and the optimized beam shape at two screens are presented in figure 4 (left bottom plots).

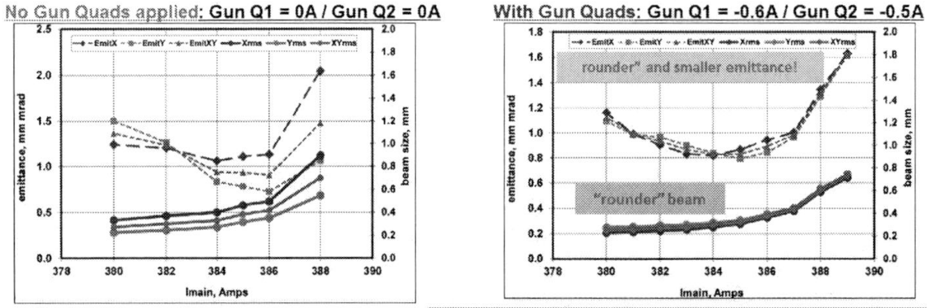

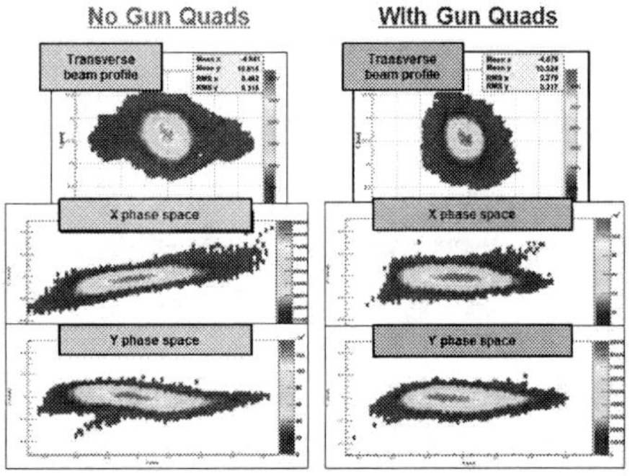

	No Gun Quads			With Gun Quads		
	x	y	xy	x	y	xy
Emittance, mm mrad	1.11	0.78	0.93	0.82	0.84	**0.83**
Beam RMS size, mm	0.46	0.32		0.28	0.32	
α	-0.99	0.39		0.06	-0.01	
β	4.85	4.37		3.18	3.24	
γ	0.41	0.26		0.32	0.31	

Figure 4. Emittance measurements results with the second gun quadrupole design: emittance values and beam sizes vs main solenoid currents (top pictures), beam profiles and phase spaces (left bottom pictures), beam parameters (table).

It is evident that in the case where the quadrupoles are applied, the values of the *X* and *Y* components of the emittance, and the beam sizes, are closer to each other, than in the case without any quadrupoles. It is also visible at the beam projection and *X* and *Y* phase spaces presented in figure 5 (left bottom). The transverse profile of the beam became symmetric as well as *X* and *Y* phase spaces approach closer to each other. Additionally to the emittance reduction by 10 %, all beam parameters became more symmetrical.

4.3. Copies of the second quadrupole design for XFEL and FLASH

The two copies of the gun quadrupoles were produced and installed at European XFEL in August 2017 and at FLASH at the end of 2017.

The emittance measurements with the gun quadrupoles at European XFEL injector were performed in October 2017. For the measurements the quadrupole scan technique was used. Due to the quick process of the emittance measurements, it was possible to perform the quadrupole scan for different combinations of the gun quadrupoles currents and produce an emittance map. The measurements results are presented in figure 5.

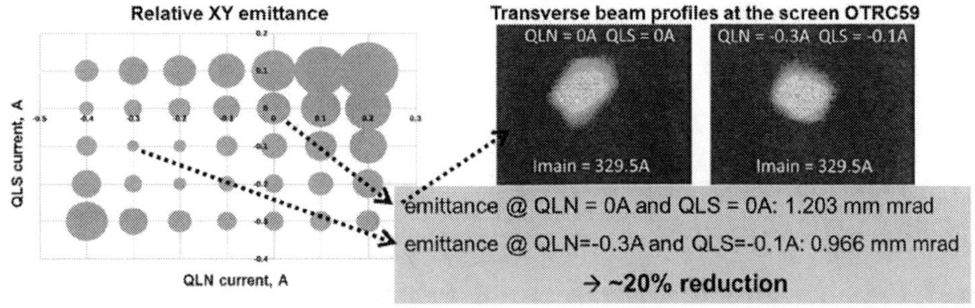

Figure 5. Results of emittance measurements with the second gun quadrupole design at European XFEL. QLN and QLS are quadrupoles names at XFEL, which correspond to Gun Q1 and Gun Q2 names at PITZ.

The smallest value of emittance was obtained at gun quadrupoles currents of -0.3 A and -0.1 A for normal and skew quadrupoles, correspondingly. The emittance value was reduced by 20 %. Also, the beam profile became more round.

5. Outlook and the third design of the gun quadrupoles

The next step of the gun quadrupoles development is the third design, which is dedicated for the installation at a location that is closer to the gun cavity (between coaxial coupler flange and the gun cavity). The design was optimized to fit the quadrupole to the more tight space in this region (see figure 6). The third design gun quadrupole is already installed at current gun cavity at PITZ.

The combination of the second and the third designs of the gun quadrupoles supposed to decrease the beam emittance even more than it was already obtained with only the second design quadrupole.

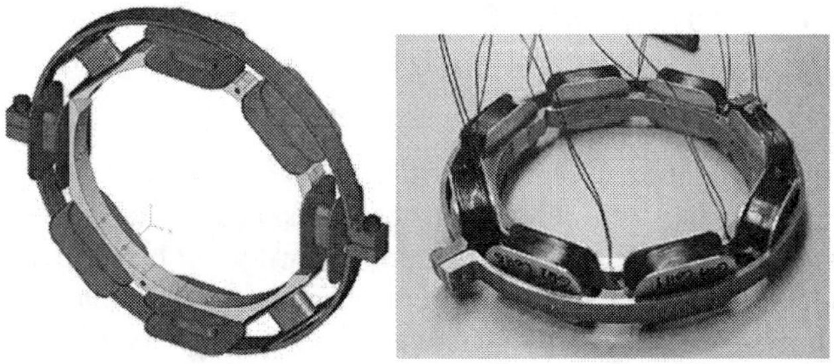

Figure 6. The view of the third design of the gun quadrupole: CAD model (left) and the corresponding produced quadrupole (right).

References

[1] FLASH project: http://flash.desy.de/

[2] XFEL project: http://xfel.desy.de/

[3] Altarelli M *et al* 2006 XFEL: The European X-Ray Free-Electron Laser. *Technical design report* DESY-06-097

[4] Isaev I 2014 Stability and performance studies of the PITZ photoelectron gun. *PhD thesis*, Universität Hamburg

[5] Krasilnikov M 2016 Update on Beam Imperfections Studies at PITZ *PITZ Collaboration Meeting*

[6] Schmerge J 2010 LCLS Gun Solenoid Design Considerations *SLAC Technical Note* SLAC-TN-10-084

[7] Quantang Z 2016 Simulations with rotation quads model for beam asymmetry studies (updated) *PITZ physics Seminar*

[8] CST software: https://www.cst.com

IOP Conf. Series: Journal of Physics: Conf. Series **1238** (2019) 012068 doi:10.1088/1742-6596/1238/1/012068

Improving the efficiency of suppression of electron conductivity in the vacuum accelerator neutron tubes with coaxial diode system

E D Vovchenko, K I Kozlowskii, M I Lisovskii, V I Rashchikov, A E Shikanov and V L Shatokhin

National Research Nuclear University (MEPhI), Moscow, Russia

E-mail: virashchikov@mephi.ru

Abstract. Discussed two options for increasing the magnetic isolation efficiency of the electron component in a vacuum neutron tube with coaxial geometry for the acceleration of deuterons. Computer analysis of the previously developed diode systems with magnetic insulation has shown that the suppression ratio does not exceed 50%. The proposed new options of suppression systems used the current ring and diaphragm, located near the edge of the cylindrical anode. Computer analysis showed that for these acceleration geometries the suppression factor could exceed 97%.

1. Introduction

Compact pulsed neutron generators based on sealed diode accelerating tubes are widely used for neutron logging of oil-gas and ore holes, express elemental analysis of compounds, neutron tomography, and detection and identification of hidden hazardous substances and objects [1].

Various methods for suppression of electron emission from the cathode of a diode accelerating system are employed to increase the energy efficiency of neutron generation in accelerating tubes [2]. Electrostatic systems cease to be efficient for the suppression of electron conduction at relatively high accelerating voltages and currents. In such systems, magnetic field can be used for isolation of the diode gap [3]. Several variants of such an approach have been studied in [4–6]. New methods for suppression electron conductivities proposed. Relativistic particle in cell (PIC) code SUMA was used for computer simulation [7].

2. Simulation results

The systems based on a permanent magnet with longitudinal magnetization that represents a hollow cylinder made of NdFeB compound and placed inside the accelerating tube working vacuum volume is rather effective for electron conductivities suppression, but their efficiency not exceed 50%. Figure 1 shows field and particle distribution in the accelerating tube. Diode anode (green area at the upper panel) is under the voltage of 300 kV and the cathode (target) is grounded (dark blue on the figure).

Black dots on the lower panel represent the deuterons emitted from plasma region near the anode. The cathode electrode is made of a magnetic material based on the NdFeB compound and represents a hollow cylinder.

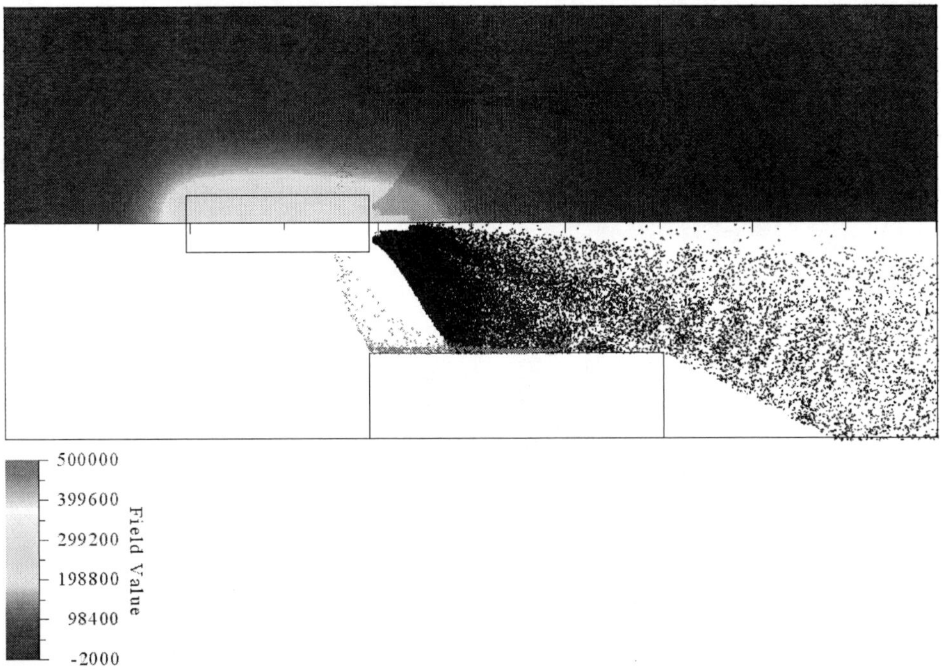

Figure 1. Distributions of (upper panel) potentials and (lower panel) charges at accelerating tube.

During acceleration by the field between anode and cathode deuterons reach the neutron-generating target (thin metal plate coated with a titanium film saturated with deuterium) and produce the neutrons. Electrons (red dots on the target and in the region between the left-hand edge of the target and anode) emerge in the diode due to the secondary emission. Most of this electrons return to the cathode because of strong longitudinal magnetic field of permanent magnet. Nevertheless significant part of them moves to the anode and decreasing neutron generation efficiency.

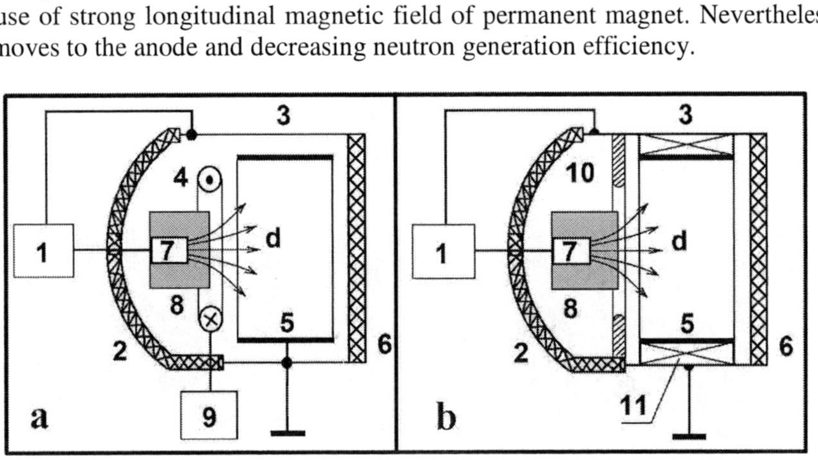

Figure 2. Two possible scheme to reduce electron conductivities: a – with current ring, b – with permanent magnet and diaphragm. 1 – high voltage power source, 2,6 – insulators, 3 – metal case, 4 - current ring, 5 - neutron-generating target, 7 – deuteron source, 8 – anode, 9 – pulse current source, 10 – diaphragm, 11 - permanent magnet.

Two new possible scheme for electron conductivity reduction for this types of neutron generators schematically shown on figure 2. At figure 2a magnetic field from thin ring with pulse current uses for

IOP Conf. Series: Journal of Physics: Conf. Series **1238** (2019) 012068 doi:10.1088/1742-6596/1238/1/012068

diminish secondary electron current from the target. At figure 2b permanent magnet and diaphragm uses for this purposes. Figure 3 and figure 4 shows the magnetic field distribution for permanent magnet and thin ring with current accordingly.

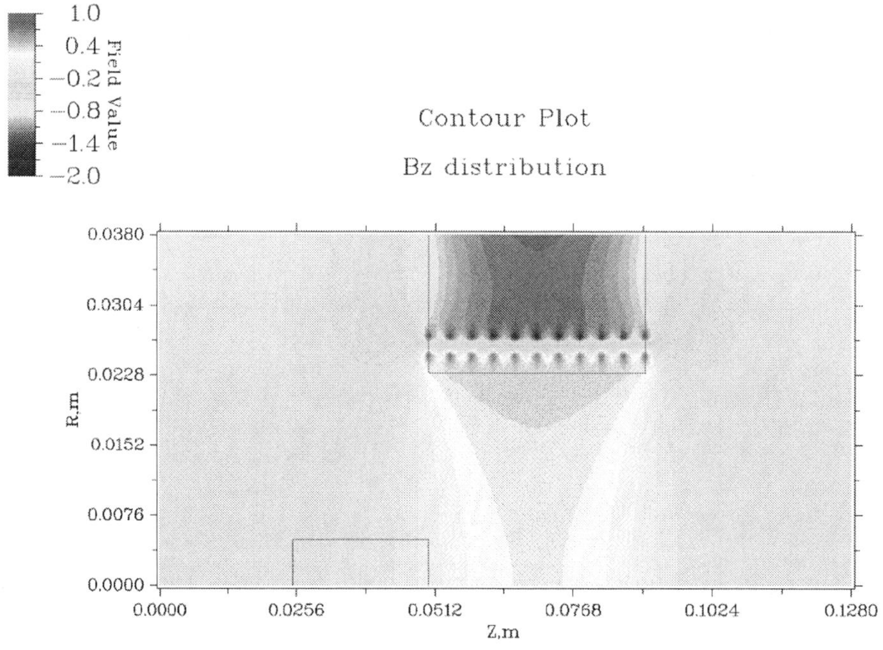

Figure 3. Magnetic field distribution of NdFeB permanent magnet.

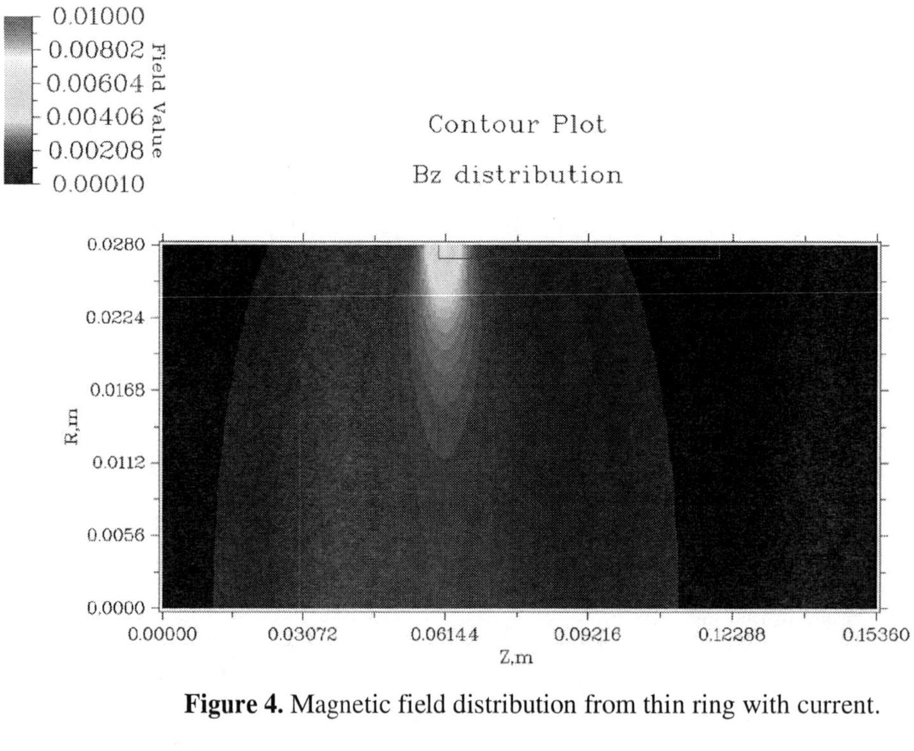

Figure 4. Magnetic field distribution from thin ring with current.

Magnetic field from thin ring with current essentially change the behavior of secondary electrons in diode (figure 5). Even with the ring current 100 A there are no more secondary electrons on the anode (left figure).

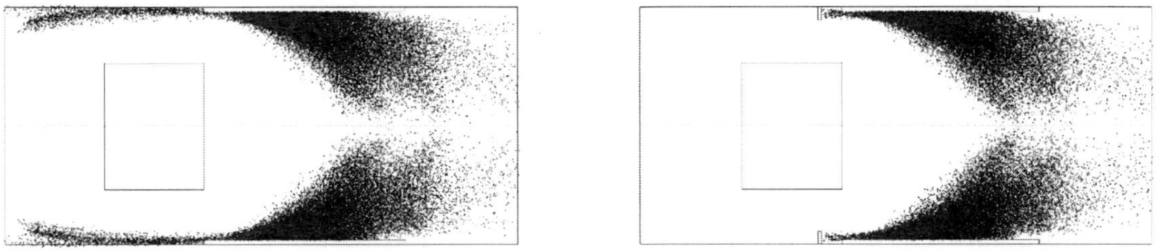

Figure 5. Secondary electrons (black dots) distribution without (left) and with (right figure) diaphragm.

Nevertheless we obtained the flow of electrons on the insulator that may leads to its destruction. This problem can be solved by inserting diaphragm in the region near the target (right figure). Secondary electron flow in the case of permanent magnet may be essentially diminished with the help of diaphragm too. Simulation shows that proper diaphragm installation reduce electron flow by the order of magnitudes.

Another way to reduce secondary electrons flow to anode without diaphragm installation is the increasing the value of thin ring current and magnetic field consequently. Figure 6 shows charge particle and their self-field distribution in accelerating tube for current value 10 kA.

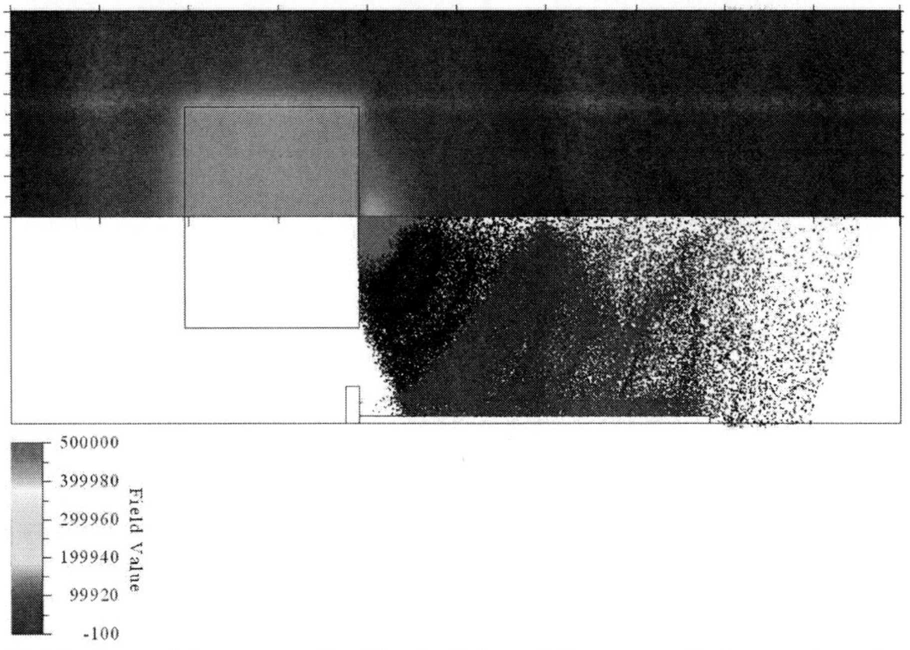

Figure 6. Distributions of (upper panel) self-potentials and (lower panel) charges at accelerating tube. Red dots – ion source plasma electrons, black dots – deuterons, blue dots – secondary electrons.

Simulation shows than less than 1% of secondary electrons can get to the diaphragm, therefore it's practically no needs for its installation.

3. Conclusion.

IOP Conf. Series: Journal of Physics: Conf. Series **1238** (2019) 012068 doi:10.1088/1742-6596/1238/1/012068

Computer simulation shows the following ways to suppress electron conductivities in vacuum neutron tubes with coaxial geometry:

- Thin current ring with diaphragm installation (low current value)
- Thin current ring without diaphragm installation (high current value)
- Long solenoid or permanent magnet with diaphragm.

References

[1] *A Proc. Int. Sci.-Tech. Conf. "Portable Neutron Generatorsand Their Application" 2012* Moscow, Russia, Ed. By Yu N Barmakov (Vseross. Nauchno-Issled. Inst. Avtom., Moscow, 2013).

[2] Bogdanovich B Yu, Nesterovich A V, Shikanov A E, Vorogushin M F and Svistunov Yu A 2009 *Remote Radiation Monitoring with Linear Charged-Particle Accelerators*, Vol. 1: *Linear Accelerators for the Generation of Bremsstrahlung and Neutrons* Energoatomizdat, Moscow

[3] Bespalov D F, Kozlovskii K I, Tsybin A S and Shikanov A E 1979 Author's Certificate SU 766048

[4] Didenko A N, Shikanov A E, Kozlovskii K I, Shatokhin V L and Ponomarev D D 2014 *Plasma Phys.Rep.* **40** 910

[5] Shikanov A E, Vovchenko E D, Kozlovskii K I and Shatokhin V L 2015 *Tech. Phys. Lett.* **41** 511

[6] Shikanov A E, Vovchenko E D and Kozlovskii K I 2015 *At. Energy* **119** 258

[7] Rashchikov V I 1990 *Vopr. At. Nauki Tekh., Ser.: Yad.-Fiz. Issled.* **10** 50

IOP Conf. Series: Journal of Physics: Conf. Series **1238** (2019) 012069 doi:10.1088/1742-6596/1238/1/012069

Design, construction and tuning of an RF deflecting cavity for the REGAE facility

V Paramonov[1], V Danielyan[2], K Floettmann[3], A Simonyan[2], V Tsakanov[2].

[1]- INR RAS, Moscow , Russia,
[2] – CANDLE SRI, Erevan, Armenia,
[3] – DESY, Hamburg, Germany.

E-Mail: paramono@inr.ru

Abstract. Extraordinary emittance requirements in the nm range (normalized) and pulse lengths down to a level of ~10 fs for REGAE bunches demand both operation at low bunch charges on the sub-pC scale and a very careful beam handling. The S-band RF deflecting cavity is intended for diagnostics of the longitudinal bunch parameters. For the first time a deflecting structure, specially developed and optimized for bunch rotation has been realized for the REGAE RF deflector. The developed cavity provides a minimized level of aberrations in the distribution of the deflecting field combined with an improved RF efficiency. The main steps in the cavity design, construction and tuning are described.

1. Introduction

The scientific program of the Relativistic Electron Gun for Atomic Exploration (REGAE) facility [1] focuses on investigations of femtosecond electron diffraction with respect to the most venerable concepts in chemistry and biology. Extraordinary emittance requirements in the nm range (normalized) and pulse lengths down to a level of ~10 fs demand both, operation at low bunch charges on the sub-pC scale and a very careful beam handling. The S-band RF deflecting cavity is designed for diagnostics of the longitudinal bunch parameters. The ultimate requirements of the bunch quality provide the special requirements to the deflecting cavity, such as a sufficient deflecting voltage at the limited length together with minimized introduced distortions. For the first time a deflecting structure, specially developed and optimized for bunch rotation has been realized for the REGAE RF deflector. Following the concept of separated control of the RF efficiency and the field quality, the developed cavity provides a minimized level of aberrations in the distribution of the deflecting field combined with an improved RF efficiency. To realize the expected cavity performances, an appropriate level of technology in the cavity construction should be supported.

2. Optimized structures for bunch rotation

Initially introduced for deflection of charged particles, in modern applications the deflecting structures operate for rotations of bunches of charged particles. They are used for luminosity enhancement in circular colliders, beam lines for emittance exchange and longitudinal bunch diagnostics. In this case the deflecting structure is applied for transformation of particles distribution in a 6D phase space and

IOP Conf. Series: Journal of Physics: Conf. Series **1238** (2019) 012069 doi:10.1088/1742-6596/1238/1/012069

should provide as minimal own distortions, as possible. The concept of special deflecting structures with a minimized level of aberrations, preferable for distribution transformation, was presented in [2].

Analysis of electron bunch dynamics and distribution of a deflecting field was presented in [3,4], indicating non linear additions mostly caused by higher spatial harmonics in the periodical deflecting field as the main source of emittance deterioration. For structures optimization a parameter $\psi_{\max}$ was proposed, [2], in the physical sense the maximal deviation of the phase of a deflecting field $E_d(r,\varphi,z)$ from the phase of a synchronous spatial harmonic

$$E_d(r,\varphi,z) = \tilde{E}_d(r,\varphi,z)e^{i\psi_d(r,\varphi,z)}, \psi_{\max} = \max(|\psi_d(0,0,z) - \frac{\theta_0 z}{d}|), \frac{-d}{2} \le z \le \frac{d}{2}, \quad (1)$$

where $\tilde{E}_d(r,\varphi,z), \psi_d(r,\varphi,z)$ are the amplitude and the phase of deflecting field, correspondingly, d is the structure period and θ_0 is the operating phase advance, see [4] for details. For the end cells design the similar procedure with a zero rise in transverse momentum during bunch rotation was adopted, [5]. The leading idea in the developed procedures is to keep the central particle in the bunch near the structure axis and simultaneously decrease the level of higher spatial harmonics in the $E_d(r,\varphi,z)$ distribution. Guided by these ideas, deflecting structures with a minimized level of aberrations have been developed, [6]. For conventional deflecting structures the conditions of equation (1) are additional requirements and result in deterioration of other parameters, in particular RF efficiency. To maintain a high RF efficiency and combine it with a minimal level of aberrations a deflecting structure with separated control of parameters was proposed, [5].

3. Cavity design parameters

The structure with a separated control of the RF efficiency and the level of aberrations was investigated and optimized more thoroughly, [7], and one option was selected for realization. The cells of the accepted structure and a childlike face - distribution of the magnetic field between irises are shown in Figure1a,b, respectively. The short cavity has three regular cells and two identical end cells, Figure 1c. In Figure 1d are shown distributions of a deflecting field along the structure axis for bunch deflection, $\phi = 0$, and bunch rotation, $\phi = 90$, see [4] for details. The main design parameters of the cavity are listed in Table 1.

Table 1. The main design parameters of the cavity.

Parameter	Definition	Unit	Value
Operating frequency	f	MHz	2997.925
Operating phase advance	θ_0	radian	π
Quality factor, calculated	Q		12550
Maximal phase deviation	$\psi_{\max}$	grad	1.8
Effective shunt impedance, structure	Z_e	MOm/m	43.2
Effective shunt impedance, total cavity	Z_{et}	MOm	7.58
Maximal input RF power		kW	5
Maximal deflecting voltage	V_d	kV	190
Particles energy		MeV	5

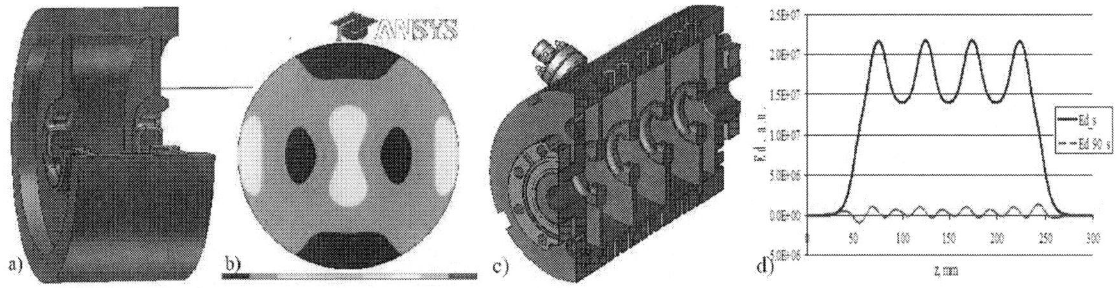

Figure 1. (a) The cells of the deflecting structure,(b) magnetic field distributions between irises, (c) the model of the deflecting cavity, (d) deflecting field E_d distribution along the axis for bunch deflection (blue curve) and bunch rotation (red curve).

4. Cavity technical design and manufacturing

As compared to a widely used deflecting structure, based on a disk loaded waveguide, the applied structure has more complicated, hence more difficult for processing, shape of cells. It is the price for better physical performances. The coupling coefficient $k_c = \frac{f_\pi^2 - f_0^2}{f_\pi^2 + f_0^2} \times 100\% = -5.04\%$ is not high, leading to a narrow passband with negative dispersion. But the cavity is rather short – the length from flange to flange is 270 mm - and has only three regular cells and two end cells. On the operating dispersion curve there are only five modes and the separation in frequency with the nearest mode is 13.5 MHz. To simplify cells processing, we can specify reasonable tolerances for cells treatment. The procedure of tolerances estimations is described in [7]:

$$\sigma_E = 2\sigma_f \left(\frac{1}{N}\sum_{m=0}^{N-1}\left(\frac{f_m^2\delta_m^2}{f_m^2 - f_\pi^2}\right)\right)^{1/2}, \sigma_f = \frac{1}{f_\pi}\left(\sum_{nd=1}^{N_d}\left(\frac{\partial f_\pi}{\partial x_{nd}}\right)^2 \sigma_{nd}^2\right)^{1/2}, \quad (2)$$

Taking into account a more complicated shape of the cell, for this short cavity the tolerances for internal cell dimensions were specified as $\pm 20\mu m$, resulting in the expected standard deviations of the cells frequency $\sigma_f \approx 700kHz$ and the related standard deviation of the deflecting field distribution $\sigma_E \approx 1.8\times 10^{-2}$.

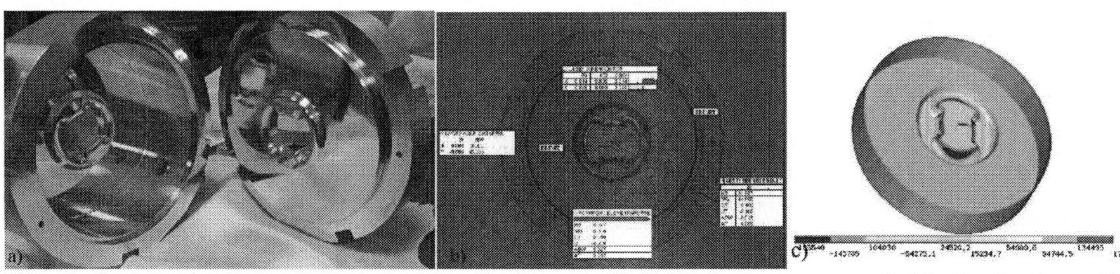

Figure 2. **(a)** Aluminum test cells, (b) diagram of CMM measurements and (c) distribution of frequency sensitivity □ $(\mu_0 H^2 - \varepsilon_0 E^2)$.

To develop and test procedures of cells processing with Numerically Controlled (NC) hardware, the test cells, shown in Figure 2a, were produced and their dimensions were measured, Figure 2b. The measured dimensions of test cells appeared to lie within tolerances. The measured frequency of operating mode, normalized for air evacuated conditions, was 3001.9 MHz. To reduce extra positive reserve in frequency at □ $2MHz$, the cells diameter was increased up to 0.1 $\hbar m$. For frequency tuning after cells brazing blind holes with central pins were introduced in the cavity design. The thread allows pins to be pushed and pulled, thus providing deformation of the cell surface and a related shift of the operating frequency in both directions.

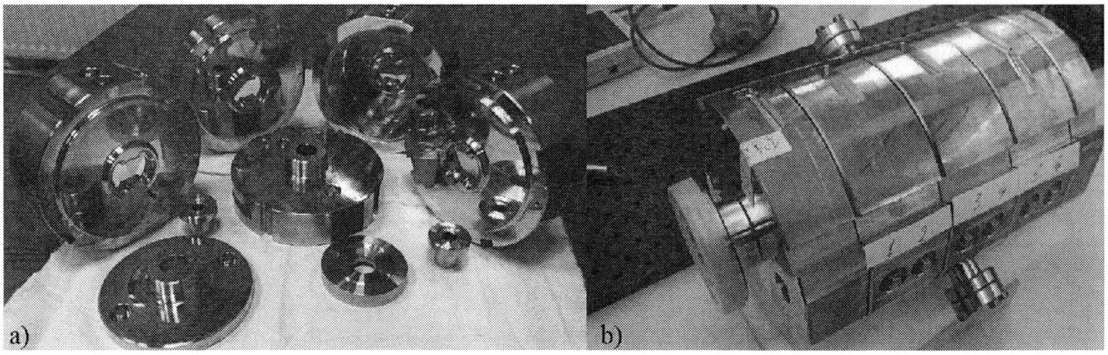

Figure 3. (a) Cavity elements after manufacturing and (b) brazed cavity before RF tuning.

Regular cells of the cavity are produced of OFHC copper by using hard tools and NC hardware, showing a high quality of surface treatment, Figure 3a. The high temperature cavity brazing was done in a vacuum oven with silver alloys. The brazed cavity before RF tuning is shown in Figure 3b.

5. RF measurements and tuning

After cavity brazing the measured operating frequency, recalculated for vacuum conditions and temperature $T = 20C$, was estimated to be $3001.5MHz$, higher than expected. Dimensions correction made after test cells measurements give the calculated value of cell frequency as $2995.86MHz$ The reason for this big deviation is not completely clear yet. The most probable reason is the shape deviation due to residual stress relaxation during high temperature brazing, which is simultaneous annealing of forged copper.

Multiple repeated bead pull measurements of field distribution, after averaging the results, indicate a very similar distribution for E_r values in the middle of four irises 100:100.08:100.04:99.86 – with a relative deviation from the average value $\approx 0.145\%$. Field values in the middle of irises are practically identical. It means very narrow spread in the own frequencies of the cells.

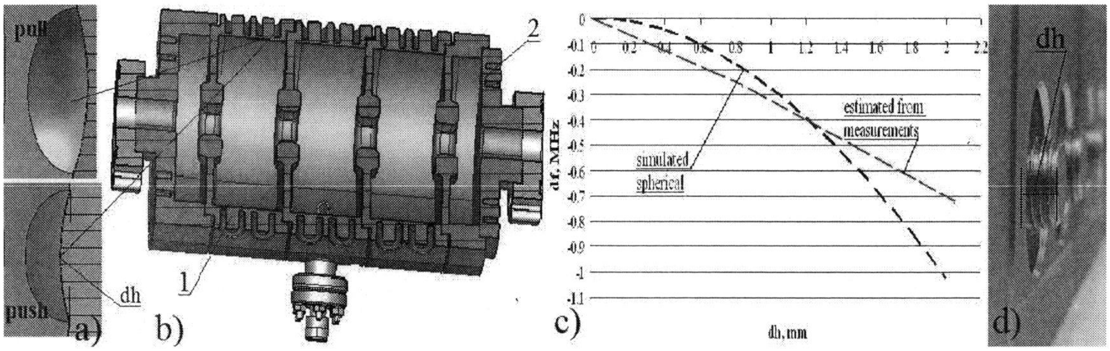

Figure 4. (a) assumed spherical surface deformations (b) with blind holes 1,2 on the cavity surface, (c) calculated for spherical deformation (blue curve) and estimated from measurements (red line) frequency shift and (d) – final pins positions after frequency tuning.

Each cell is equipped with four blind holes for frequency tuning, Figure 4b. Assuming a spherical shape for surface deformation, Figure 4a, the plot of the expected frequency shift is shown with blue curve in Figure 4c. The shape deformation was performed in two steps. The depth of pins extraction dh was defined by the angle of external nuts, because the thread M8 has a pitch of $1.25mm$. All pins were extracted at the same depth. The treatment of the results showed more linear dependence of the frequency shift on the pin extraction, red curve in Figure 4c. Finally, with a sufficiently the large depth of pins extraction $dh = 2.1mm$, see Figure 4d, the frequency value of $2997.91MHz$ was obtained assuming vacuum conditions and operating temperature $T = 35C$. In the case of this surface deformation the vacuum tightness of the cavity is not violated.

The cavity will be powered from a solid state amplifier with a rather moderate RF power value of $5kW$. Coaxial cable as a transmission line is used with turning vacuum tight feed through. The cavity has RF driving and RF signal loops. To minimize self inductance, the RF driving loop has small dimensions and is produced with 3D printing technology. Preliminary measurements with driving RF loops showed a possibility of matching $S_{11} \le -30dB$. The value of the own quality factor is estimated to be $\approx 80\%$ from value calculated in 3D approximation. The final measurements will be performed after plating the loop with a thin layer of gold and mounting the cavity at the beam line.

6. Summary

For the first time was realised the deflecting cavity specially optimized for minimal level of aberrations in the distributions of deflecting field. As compared to conventional deflecting structures, the cavity provides physical benefits in applications for transformation of particle distributions in the bunch. These cavity performances are requested in the facilities with unique bunch parameters. As the price for better physical performances this structure requires more attention during construction and RF tuning.

The experience acquired during construction and tuning of the first cavity is sure to help us replicating this deflecting structure with improved parameters.

References

[1] Manz S. et. al., Mapping atomic motion with ultrabright electrons: towards fundamental limit in space time resolution. *Faraday Discussion,* 177, p. 467-491, 2015, http://dx/doi/org/10.1039/C4FD00204K
[2] Paramonov V. Deflecting structures with minimized level of aberrations. *Proc. Linac 2012 Conference,* p. 445, 2012
[3] Floettmann K, Paramonov V, Beam dynamics in transverse deflecting rf structures. *Physical Review ST Accelerators and Beams,* v. 17, 024001, 2014
[4] Paramonov V. Field distribution analysis in deflecting structures, *DESY 18-13, arXiv:1302.5306v1, 2013*
[5] Paramonov V, Kravchuk L, Floettmann K. Standing wave RF deflectors with reduced aberrations, *Proc. RuPAC 2012 Conference,* p. 590, 2012
[6] Paramonov V, Kravchuk L, Floettmann K. Deflecting RF structures with reduced level of aberrations for transformations of particles distribution in the bunch, *Physics of Particles and Nuclei Letters,v. 11, n. 5, p. 636,* 2014
[7] Paramonov V. Effective RF deflecting structures for bunch rotation and deflection. *Proc. RuPAC 2016 Conference,* p. 201, 2015
[8] Rybakov I, Paramonov V, Skasyrskaya A. Manufacturing tolerances estimation for linac cavities, *Proc. RuPAC 2016 Conference,* p.291, 2017

IOP Conf. Series: Journal of Physics: Conf. Series **1238** (2019) 012070 doi:10.1088/1742-6596/1238/1/012070

Intense quasiperiodic beam dynamics investigation in accelerating system

D A Ovsyannikov[1], I D Rubtsova[1] and N V Lomonosova[1]

[1]St. Petersburg State University, 7/9 Universitetskaya nab., St. Petersburg, 199034, Russia

E-mail: d.a.ovsyannikov@spbu.ru, i.ribtsova@spbu.ru, st032429@student.spbu.ru

Abstract. The paper is devoted to intense quasiperiodic beam dynamics investigation on the basis of integro-differential mathematical model. Space charge density representation by trigonometric polynomial makes it possible to obtain Coulomb field analytical expression in the same form. Space charge field is presented by the integral over particle phase states domain. Beam quality criteria are suggested in the form of integral functionals. Beam dynamics optimization problem is formalized as trajectory ensemble control problem. Analytical expression of objective functional variation is obtained; it makes possible directed optimization methods using. Numerical simulation and optimization results are presented for linear waveguide accelerator.

1. Introduction

Main topics discussed in this paper are as follows: intense beam dynamics mathematical modeling with quasiperiodicity account; formalization and investigation of beam dynamics optimization problem; numerical results. We introduce further development of the research presented in [1-4].

Our study is connected with self-consistent particle distributions modeling [1-9]. Aiming at mathematical model formulating, we use Coulomb field analytical expression; such an approach is applied in [1-5,8,9]. Trigonometric polynomials provide the convenient instrument for particle interaction description and investigation [1-3,9]. Space charge field integral representation makes it possible to introduce integro-differential beam dynamics model [1-5, 9-11].

In view of Coulomb field account, independent variable in beam dynamics equations is the analogue of time; so beam dynamics quality criteria formalization presents the special problem [8,11].

Using the approach given in [5,10,12,13] we formulate the problem of beam dynamics optimization as trajectory ensemble control problem and obtain analytical expression for quality functional variation. It makes possible to apply directed optimization methods. Such an approach is successfully applied for solving different beam dynamics optimization problems [2,4,11,14-17] .

Numerical results are presented for travelling-wave linear accelerator.

2. Beam dynamics equations

Consider intense beam longitudinal dynamics model in some accelerating system:

$$\frac{dz_i}{d\tau} = p_i\left(1 + p_i^2\right)^{-1/2}, \qquad \frac{dp_i}{d\tau} = \zeta c^{-2} E^{(RF)}(\tau, z_i, \boldsymbol{u}) + |\zeta| c^{-2} E^{(int)}(\tau, z_i, p_i, z_c, p_c), \quad i = \overline{1, N}, \tag{1}$$

where $z_c = N^{-1}\sum_{i=1}^{N} z_i$, $p_c = N^{-1}\sum_{i=1}^{N} p_i$, with initial conditions

$$z_i(0) = z_{0i},\ p_i(0) = p_{0i},\ i = \overline{1,N}\ . \tag{2}$$

Here $= ct$, c is the speed of light, t is the time; z is longitudinal coordinate; p is reduced impulse; (z_i, p_i) is i-th particle phase vector; (z_c, p_c) presents bunch center phase state; ζ is particle specific charge; $E^{(RF)}$, $E^{(int)}$ characterize respectively the effect of RF field and space charge field on model particle; $\boldsymbol{u}(\tau)$ is program control.

3. Integro-differential beam dynamics model

Let us generalize beam dynamics model (1)-(2). For Coulomb field modeling we approximate particle density by trigonometric polynomial and derive the intensity $E^{(int)}(z_i, z_c, p_c)$ expression in the same form [1-3,9]. After the transformation it is represented in the form: $E^{(int)}(z_i, z_c, p_c) = N^{-1}\sum_{n=1}^{N} V(z_i, z_c, p_c, z_n)$, where the function $V(z_i, z_c, p_c, z_n)$ is determined by particle interaction account method. Applying the law of large numbers we introduce mathematical model of Coulomb field intensity:

$$\tilde{E}^{(int)}(z_i, z_c, p_c) = \int_{M_{\tau,u}} V(z_i, z_c, p_c, z_\tau)\rho(\tau, z_\tau, p_\tau)\, \mathrm{d}z_\tau\, \mathrm{d}p_\tau, \tag{3}$$

where $M_{\tau,\boldsymbol{u}}$ is particle phase states domain, $\rho(\tau, z, p)$ is phase density. The analogic integral formulae is used for average phase coordinates. Beam dynamics model is generalized with due account of (3).

Consider dynamic controlled process described by the equations [1,3]:

$$\frac{\mathrm{d}\boldsymbol{x}}{\mathrm{d}\tau} = \boldsymbol{f}(\tau, \boldsymbol{x}, \boldsymbol{x}_c, \boldsymbol{u}) = \boldsymbol{f}_1(\tau, \boldsymbol{x}, \boldsymbol{u}) + \int_{M_{\tau,u}} \boldsymbol{f}_2(\tau, \boldsymbol{x}, \boldsymbol{x}_c, \boldsymbol{y}_\tau)\rho(\tau, \boldsymbol{y}_\tau)\mathrm{d}\boldsymbol{y}_\tau, \tag{4}$$

$$\frac{\partial\rho(\tau,x)}{\partial\tau} + \frac{\partial\rho(\tau,x)}{\partial x}\boldsymbol{f}(\tau, \boldsymbol{x}, \boldsymbol{x}_c, \boldsymbol{u}) + \rho(\tau, \boldsymbol{x})div_{\boldsymbol{x}}\boldsymbol{f}(\tau, \boldsymbol{x}, \boldsymbol{x}_c, \boldsymbol{u}) = 0, \tag{5}$$

where $\boldsymbol{x}_c(\tau) = \int_{M_{\tau,u}} \boldsymbol{y}_\tau\, \rho(\tau, \boldsymbol{y}_\tau)\mathrm{d}\boldsymbol{y}_\tau$, with initial conditions

$$\boldsymbol{x}(0) = \boldsymbol{x}_0 \in M_0,\ \rho(0, \boldsymbol{x}) = \rho_0(\boldsymbol{x}). \tag{6}$$

Here $\tau \in [0, T]$ is independent variable; T is fixed; $\boldsymbol{x}$ is phase vector; $\boldsymbol{u}(\tau)$ is a control; vector function $\boldsymbol{f}_1$ is determined by external field representation; vector function $\boldsymbol{f}_2$ is determined by the method of particle interaction account; $\rho(\tau, \boldsymbol{x})$ is phase density defined on system (4) trajectories; M_0 is initial phase domain; $\rho_0(\boldsymbol{x})$ is initial phase density; $M_{\tau,\boldsymbol{u}} = \{\boldsymbol{x}_\tau = \boldsymbol{x}(\tau, \boldsymbol{x}_0, \boldsymbol{u}): \boldsymbol{x}_0 \in M_0\}$. All the functions in model (4)-(6) are supposed to be rather smooth to use mathematical optimization methods [5]. Control function $\boldsymbol{u}(\tau)$ is supposed to be piecewise-continuous and taking values in a compact.

4. Trajectory ensemble control problem

Let us give the examples of beam dynamics quality criteria and suggest control problem formulation on the basis of their generalization.

The criterion characterizing beam phase spread at device cross-section $z = \xi$ is as follows:

$$K_1(\xi, \boldsymbol{u}) = \widehat{N}^{-1}\sum_{i=1}^{\widehat{N}}\left(\varphi(\tau_i(\xi), \xi, \boldsymbol{u}) - \hat{\varphi}(\xi, \boldsymbol{u})\right)^2. \tag{7}$$

Here $\widehat{N}$ is the number of particles crossing the section $z = \xi$; $\varphi(\tau, z, \boldsymbol{u})$ is particle phase; $z_i(\tau_i(\xi)) = \xi$; $\hat{\varphi}(\xi, \boldsymbol{u}) = \widehat{N}^{-1}\sum_{i=1}^{\widehat{N}} \varphi(\tau_i(\xi), \xi, \boldsymbol{u})$ is averaged phase at cross-section $z = \xi$.

Let us introduce integral functional corresponding the criterion (7):

$$I_1(\xi, \boldsymbol{u}) = \int_0^T \int_{M_{\tau,u}} \left(\varphi(\tau, z_\tau, \boldsymbol{u}) - \bar{\varphi}(\xi, \boldsymbol{u})\right)^2 \delta_\varepsilon(z_\tau - \xi)\rho(\tau, z_\tau, p_\tau)\mathrm{d}z_\tau \mathrm{d}p_\tau \mathrm{d}\tau\,,$$

$$\bar{\varphi}(\xi, \boldsymbol{u}) = \int_0^T \int_{M_{\tau,u}} \varphi(\tau, z_\tau, \boldsymbol{u})\, \delta_\varepsilon(z_\tau - \xi)\rho(\tau, z_\tau, p_\tau)\mathrm{d}z_\tau \mathrm{d}p_\tau \mathrm{d}\tau.$$

The value of functional $\bar{\varphi}(\xi, \boldsymbol{u})$ presents average phase at $z = \xi$ cross-section; $\delta_\varepsilon(z)$ is smooth approximation of Dirac delta function $\delta(z)$.

To estimate phase scatter on the segment $[a,b]$ the following functional may be used:

$$I_2(\boldsymbol{u},a,b)=\int_0^T\int_{M_{\tau,u}}\left(\varphi(\tau,z_\tau,\boldsymbol{u})-\bar{\varphi}(z_\tau,\boldsymbol{u})\right)^2 \Pi_\varepsilon(z_\tau,a,b)\rho(\tau,z_\tau,p_\tau)\mathrm{d}z_\tau \mathrm{d}p_\tau \mathrm{d}\tau$$

Here $\Pi_\varepsilon(z,a,b)$ is smooth approximation of the function $U(z-a)\,U(b-z)$, where $U(z)$ is Heaviside function. The similar expressions may be introduced to estimate beam energy spread.

Generalizing the examples given we introduce quality criterion of dynamic control process (4)-(6):

$$I(\boldsymbol{u})=\int_0^T\int_{M_{\tau,u}}\Phi\left(\tau,\boldsymbol{x}_\tau,\boldsymbol{u},\Lambda(\boldsymbol{x}_\tau,\boldsymbol{u})\right)\rho(\tau,\boldsymbol{x}_\tau)\mathrm{d}\boldsymbol{x}_\tau \mathrm{d}\tau, \qquad (8)$$

$$\Lambda(\boldsymbol{x},\boldsymbol{u})=\int_0^T\int_{M_{t,u}}\lambda\left(t,\boldsymbol{v}_t,\boldsymbol{u},\boldsymbol{x}\right)\rho(t,\boldsymbol{v}_t)\mathrm{d}\boldsymbol{v}_t \mathrm{d}t.$$

Let us consider the problem of objective functional (8) minimization with respect to control $\boldsymbol{u}$. Assume $\Phi(\tau,\boldsymbol{x},\boldsymbol{u},\Lambda)$ and $\lambda(t,\boldsymbol{v},\boldsymbol{u},\boldsymbol{x})$ to be smooth functions. It makes possible mathematical methods [5,10,12] applying and obtaining analytical expression for quality functional variation.

5. Quality functional variation

Objective functional (8) variation is derived on the basis of the approach [5]:

$$\delta I(\boldsymbol{u},\Delta\boldsymbol{u})=\int_0^T\int_{M_{\tau,u}}\Big(\Delta_u\Phi\left(\tau,\boldsymbol{x}_\tau,\boldsymbol{u},\Lambda(\boldsymbol{x}_\tau,\boldsymbol{u})\right)-\boldsymbol{\Psi}^*(\tau,\boldsymbol{x}_\tau)\Delta_u f_1(\tau,\boldsymbol{x}_\tau,\boldsymbol{u})$$

$$+\,\partial\Phi(\tau,\boldsymbol{x}_\tau,\boldsymbol{u},\Lambda)/\partial\Lambda\int_0^T\int_{M_{t,u}}\Delta_u\lambda\left(t,\boldsymbol{v}_t,\boldsymbol{u},\boldsymbol{x}_\tau\right)\rho(t,\boldsymbol{v}_t)\mathrm{d}\boldsymbol{v}_t\mathrm{d}t\Big)\rho(\tau,\boldsymbol{x}_\tau)\mathrm{d}\boldsymbol{x}_\tau\mathrm{d}\tau. \qquad (9)$$

Here $\Delta\boldsymbol{u}$ is control $\boldsymbol{u}$ variation; $\Delta_{\boldsymbol{u}}$ designates the increment of any function with respect to argument $\boldsymbol{u}$ only; vector functions $\boldsymbol{\Psi}(\tau,\boldsymbol{x})$ satisfies on the trajectories of dynamic process (4)-(6) the auxiliary system of integro-differential equations

$$\frac{\mathrm{d}\boldsymbol{\Psi}(\tau,x(\tau))}{\mathrm{d}\tau}=\left(\frac{\partial\Phi(\tau,x,u,\Lambda(x,u))}{\partial x}\right)^*+\frac{\partial\Phi(\tau,x,u,\Lambda(x,u))}{\partial\Lambda}\left(\frac{\partial\Lambda(x,u)}{\partial x}\right)^*$$

$$+\int_0^T\int_{M_{t,u}}\frac{\partial\Phi(t,y_t,u,\Lambda(y_t,u))}{\partial\Lambda}\left(\frac{\partial\lambda(\tau,x_\tau,u,y_t)}{\partial x}\right)^*\rho(t,\boldsymbol{y}_t)\mathrm{d}\boldsymbol{y}_t\mathrm{d}t-\left(\frac{\partial f(\tau,x,x_c,u)}{\partial x}\right)^*\boldsymbol{\Psi}(\tau,\boldsymbol{x}(\tau))$$

$$-\int_{M_{\tau,u}}\left(\int_{M_{\tau,u}}\left(\frac{\partial f_2(\tau,y_\tau,x_c,v_\tau)}{\partial x_c}\right)^*\rho(\tau,\boldsymbol{v}_\tau)\mathrm{d}\boldsymbol{v}_\tau+\left(\frac{\partial f_2(\tau,y_\tau,x_c,x)}{\partial x}\right)^*\right)\boldsymbol{\Psi}(\tau,\boldsymbol{y}_\tau)\rho(\tau,\boldsymbol{y}_\tau)\mathrm{d}\boldsymbol{y}_\tau$$

with the following conditions at $\tau=T$:

$$\boldsymbol{\Psi}(T,\boldsymbol{x}(T))=0.$$

Note that * denotes transposition.

The analytical representation (9) of quality criterion variation provides the possibility of directed optimization methods applying in beam dynamics optimization problems. It may be beneficial to combine gradient optimization with some Monte Carlo method of global extremum search [18-20].

6. Numerical results

The approach presented is applied for beam dynamics investigation in linear waveguide accelerator with injection energy 40 keV, accelerating wavelength 0.1 m, accelerator length 0.78 m. Numerical experiments were performed for beam current 5A, model particles number 64 [3].

Figure 1 presents the adequate approximation of bunch charge density by trigonometric polynomial. One can detect bunching process development along the structure.

At figure 2 one can see Coulomb field intensity distribution along quasiperiod for the middle and the end of the accelerating tract. The intensity reduce is caused by particle velocity growth.

The objectives of beam dynamics optimization are as follows: obtaining average reduced energy $\hat{\gamma}=10$ at accelerator exit; phase spread minimization during beam movement; capture coefficient maximization. Some beam characteristics for initial ($\boldsymbol{u}_0$) and optimized ($\hat{\boldsymbol{u}}$) controls are presented in the table 1.

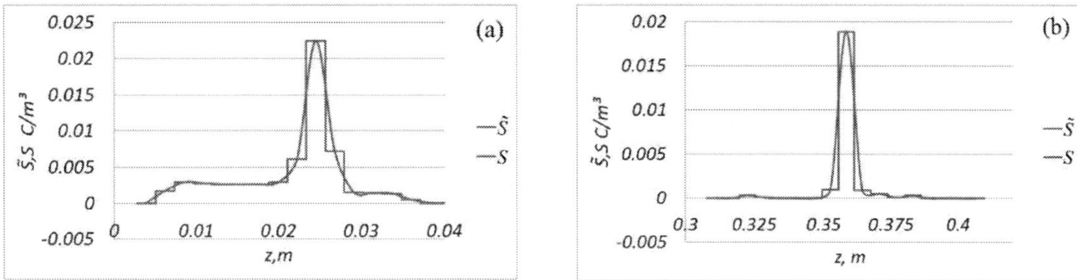

Figure 1. Piecewise-constant charge density $\tilde{S}$ and trigonometric polynomial S in bunching part (a) and accelerating part (b) of the structure

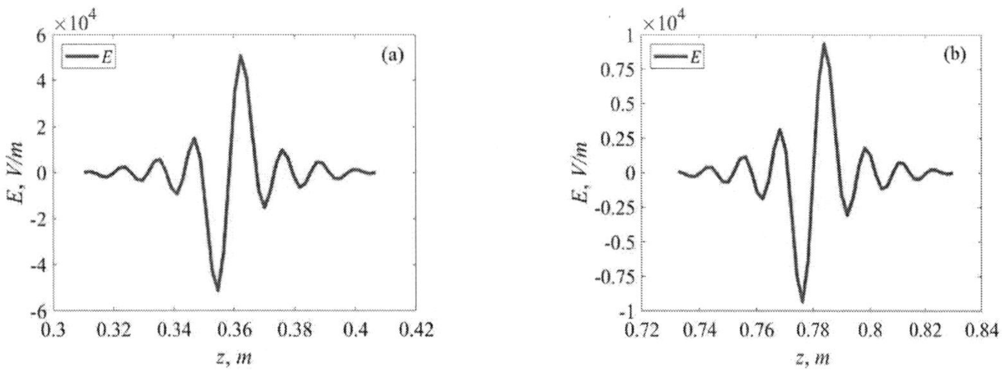

Figure 2. Space charge field intensity distribution along bunch length in the middle (a) and near the output section (b) of the accelerator

Table 1. Main beam characteristics obtained for initial ($\boldsymbol{u}_0$) and optimized ($\hat{\boldsymbol{u}}$) controls

	$\boldsymbol{u}_0$	$\hat{\boldsymbol{u}}$
Phase spread $\Delta\varphi_{exit}$ at device exit (rad)	0.400	0.389
Average reduced energy γ_{exit} at device exit	9.462	10.311
Energy spread $\Delta\gamma_{exit}$ at device exit	4.135	3.525
Capture coefficient K_c	0.922	0.938

References

[1] Rubtsova I D 2016 On modeling and optimization of intense quasiperiodic beam dynamics *Proc. XXV Russian Particle Accelerator Conf. RuPAC-2016 (21-25 November 2016 St. Petersburg)*(Geneva: JACoW http://www.JACoW.org) pp 363-66

[2] Rubtsova I D 2016 Analytical Approach to Quasiperiodic Beam Coulomb Field Modeling, *II Conference on Plasma&Laser Research and Technologies (2016), Journal of Physics: Conference Series* **747** No 1 012074 http://iopscience.iop.org/1742-6596/747/1/012074

[3] Rubtsova I D, Lomonosova N V and Chuprynina T A 2017 Investigation of quasiperiodic intense beam longitudinal dynamics in linear waveguide accelerator *Vestnik St. Petersburg State University of Technology and Design, Ser. 1: Natural and technical sciences* **3**, pp 15-23

[4] Rubtsova I D and Ovsyannikov D A 2018 Intense quasiperiodic beam dynamics in accelerating system: mathematical model and optimization method *III International Conference on Laser and Plasma Researches and Technologies (2017), Journal of Physics: Conference Series*

941 No 1 012092 http://iopscience.iop.org/article/10.1088/1742-6596/941/1/012092

[5] Ovsyannikov D A 1990 *Modeling and Optimization of Charged Particle Beam Dynamics* (Leningrad: Leningrad State University Press) p 312

[6] Drivotin O I and Ovsyannikov D A 2009 Self-consistent distributions for charged particle beam in magnetic field *Int. J. Mod. Phys.* A **24** (5) pp 816-42

[7] Bondarenko T V Polozov S M and Sumbaev A P 2016 Numerical simulation of the beam loading effect at the LUE-200 accelerator *Physics of Particles and Nuclei Letters* **13** (7), pp 919-22

[8] Rubtsova I D and Suddenko E N 2012 Investigation of program and perturbed motions of particles in linear accelerator *Proc. XXIII Russian Particle Accelerator Conf. RuPAC-2012 (24-28 September 2012 St. Petersburg)*(Geneva: JACoW http://www.JACoW.org) pp 367-69

[9] Rubtsova I D 2014 Integral-differential model of quasi-periodic beam longitudinal dynamics *Proc. 20th Int. Workshop: Beam Dynamics & Optimization (BDO) (30 June-4 July 2014 St. Petersburg)* (St. Petersburg: IEEE) p 144

[10] Ovsyannikov D A 2012 Mathematical modeling and optimization of beam dynamics in accelerators *Proc. XXIII Russian Particle Accelerator Conf. RuPAC-2012 (24-28 September 2012 St. Petersburg)*(Geneva: JACoW http://www.JACoW.org) pp 68-72

[11] Rubtsova I D 2014 Mathematical optimization model of longitudinal beam dynamics in klystron-type buncher *Proc. XXIV Russian Particle Accelerator Conf. RuPAC-2014 (6-10 October 2014 Obninsk)*(Geneva: JACoW http://www.JACoW.org) pp 66-68

[12] Ovsyannikov D A 1997 Modeling and Optimization Problems of Charged Particle Beam Dynamics *Proc. European Control Conf. ECC'97 (1-4 July 1997 Brussels Belgium)* 4 pp 1463-67

[13] Ovsyannikov D A, Ovsyannikov A D, Vorogushin M F, Svistunov Yu A and Durkin A P 2006 Beam dynamics optimization: models, methods and applications *Nuclear Instruments and Methods in Physics Research* A **558** (1) 11-19

[14] Rubtsova I D 2015 Optimization of iterative beam dynamic process *Proc. III Int. Conf. "Stability and Control Processes" in Memory of V.I.Zubov (SCP) (5-9 October 2015 St. Petersburg*) ed L A Petrosyan and A P Zhabko (St. Petersburg: IEEE) pp 198-200 DOI: 10.1109/SCP.2015.7342092

[15] Zavadsky S V, Ovsyannikov D A and Chung S-L 2009 Parametric Optimization Methods for the Tokamak Plasma Control Problem *Int. J. Mod. Phys.* A **24** No 5 pp.1040–47

[16] Ovsyannikov A D, Ovsyannikov D A, Altsybeyev V V, Durkin A P and Papkovich V G 2014 Application of Optimization Techniques for RFQ Design *Problems of Atomic Science and Technology* **91** No 3 pp 116-19

[17] Ovsyannikov D A and Altsybeyev V V 2013 Mathematical Optimization Model for Alternating-Phase Focusing (APF) Linac *Problems of Atomic Science and Technology* No 4 p 93

[18] Vladimirova L V 2014 Global Extremum Search on the Basis of Density and Its Mode Estimation *Proc. 20th Int. Workshop: Beam Dynamics & Optimization (BDO) (30 June-4 July 2014 St. Petersburg)* (St. Petersburg: IEEE) p 186

[19] Vladimirova L V 2016 Multicriterial Approach to Beam Dynamics Optimization Problem *II Conference on Plasma&Laser Research and Technologies (2016), Journal of Physics: Conference Series* **747** No 1 012070 http://iopscience.iop.org/1742-6596/747/1/012070

[20] Vladimirova L V and Fatyanova I A 2015 Construction of optimal regression experiment plan based on random search with "memory" using parallel calculations *Proc. III Int. Conf. "Stability and Control Processes" in Memory of V.I.Zubov (SCP) (5-9 October 2015 St. Petersburg*) ed L A Petrosyan and A P Zhabko (St. Petersburg: IEEE) pp 303-304.

IOP Conf. Series: Journal of Physics: Conf. Series **1238** (2019) 012071 doi:10.1088/1742-6596/1238/1/012071

Numerical simulation of longitudinal and transversal electron dynamics in classical microtron

Y A Bashmakov[1,2], V S Dyubkov[1] and Y Y Lozeev[1]

[1] National Research Nuclear University MEPhI, Kashirskoe shosse 31, Moscow 115409, Russian Federation
[2] P.N. Lebedev Physical Institute of the Russian Academy of Sciences, Leninskiy prospekt 53, Moscow 119991, Russian Federation

vsdyubkov@mephi.ru

Abstract. Numerical simulation of 3D electron dynamics in classical Lebedev institute microtron with energy 6-10 MeV is carried out. Special attention is paid to particles transversal motion investigation. A comparison of physical electric field distribution along cavity axis with that obtained by means numerical calculation is done. The obtained results are discussed.

1. Introduction

Microtron is the cyclic resonance relativistic accelerator with variable harmonic number. It is known that the mechanism of relativistic particle acceleration realized in the microtron was proposed by Veksler in 1944 [1]. Electrons dynamics in a microtron has a number of specific features that differ it from other circular accelerators and result in unique geometric and energy parameters of the accelerated beam [2]. Such electron accelerators are widely used as injectors for electron synchrotrons, as well as accelerator-drivers in experiments on generation of coherent terahertz electromagnetic radiation. In a number of experiments, a microtron is used as electron source in free-electron lasers of submillimeter band. It is important to know the parameters of the electron bunch at the microtron output to simulate the generation mode.

One of the urgent tasks of modern particle accelerators technology is a design and development of specialized synchrotron radiation sources of the fourth generation [3-11]. Such sources will become a unique tool for research tasks like ultra-small angle x-ray scattering, high resolution powder diffraction, coherent x-ray scattering, micro-macromolecular crystallography, material science, structural biology, biomedical imaging etc. Circular resonance electron accelerator, microtron, can be proposed as an injector for some of such sources.

2. Numerical simulation of electron dynamics

We investigated the three-dimensional electrons motion in 6.5 MeV microtron, that is the injector of 1.3 GeV “Pakhra” LPI synchrotron [12]. Accelerating the electrons occurs in the cylindrical cavity with operating frequency is equal to 2.856 GHz. Electron source is a thermionic cathode LaB_6 placed inside the accelerating cavity. The cavity cross section in microtron median plane is shown in figure 1. There are basic geometric dimensions of the cavity in figure: radius (R), cavity length (d), distance between center of thermal emitter and the resonator axis (r_{emit}). We consider that cavity length d was equal to 19 mm, g and R were equal to 7 mm and 40.3 mm correspondingly. The microtron is operated with the Russian first type of acceleration. Emitted electrons traverse the microtron cavity through the special windows which original dimensions ($y \times x$) are 7.3 × 25 mm (input window) and 8.2 × 14.3 (output one).

RF electromagnetic field of the resonator partially penetrates beyond the special windows and it is ending on the outer side of the cavity walls and partly on the windows surfaces. It results in field vertical component appearance. A penetration of the field beyond cavity boundaries (for $x = 0$) is shown in the figure 2.

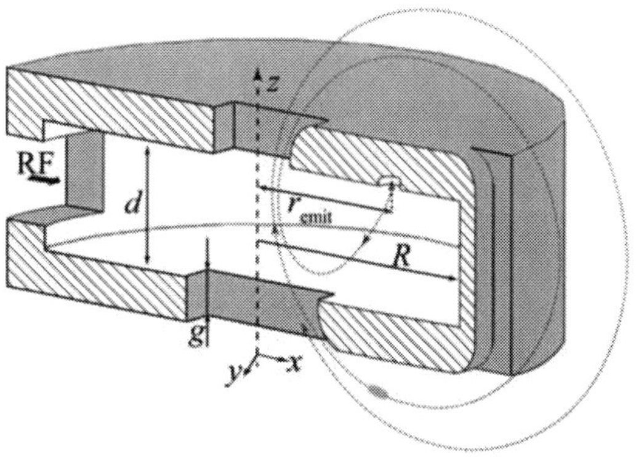

Figure 1. Cross-sectional view of RF cavity.

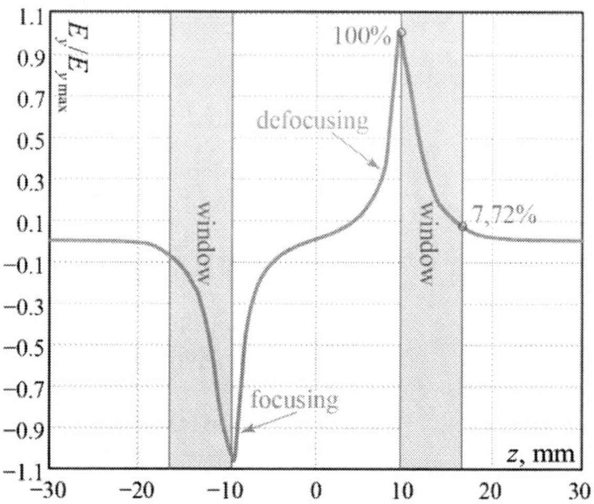

Figure 2. A penetration of the RF field beyond cavity boundaries through its windows.

One of the main requirements for electrons dynamics in a classical microtron is to provide vertical stability. Vertical stability is achieved by means of two counteracting factors: focusing and defocusing action of RF field in the area of special windows. Namely, phase shift between electric field vertical components oscillations in those areas is equal to π. It is equivalent to FODO focusing system or electrostatic focusing in cyclotron.

In order to investigate motion of electrons, and especially transverse one, numerically calculated three-dimensional distribution of the cavity RF field is used (TM_{010} mode), which coincides with the real distribution with high accuracy. The following microtron parameters are assumed: emitted electrons energy is equal to 5 eV (under Gaussian energy distribution), r_{emit} is equal to 75 cm, emitter spot diameter is equal to 4 mm, an amplitude of accelerating voltage is equal to 678 kV under equilibrium particle phase is equal to 15°, uniform magnetic field for circular motion is 0.12 T under magnet pole diameter is equal to 75 cm.

The influence of input window vertical size on particle transition was investigated as well. It was found, by means of numerous simulations, that optimal dimensions of input window are 12.5 × 25 mm. Minimum amount of particles die on the wall in that case.

Electron trajectories and electron bunch characteristics was calculated by means of numerical integration of the three-dimensional motion equation, in the approximation of non-interacting particles, as it was done in [13]. The average particle energy at the final microtron orbit (the 9th) is equal to 6.71 MeV. The calculated transverse and longitudinal dimensions of the beam, as well as the energy spread of the beam particles at the final microtron orbit are Δz = 6.9 mm, Δy = 4.2 mm, Δx = 2.3 mm and σ_W = 0.67% respectively.

Energy spectrum of electrons is shown in figure 3. One can see that it is close to the normal Gaussian distribution. A distribution of charge density in bunches on the 9th orbit in the median and vertical planes are shown in figure 4.

Vertical motion of electrons during the acceleration for a few initial orbits is shown in figure 5 where dashed rectangles indicates output cavity window. Note that the focusing and defocusing forces act alternately on the particles during each transition of electrons through the accelerating gap. The

IOP Conf. Series: Journal of Physics: Conf. Series **1238** (2019) 012071 doi:10.1088/1742-6596/1238/1/012071

first transition is specific. On the first orbit emitted electrons pass through only the one window, in which particle vertical momentum is increased significantly. It leads to significant particle losses at a few first orbits.

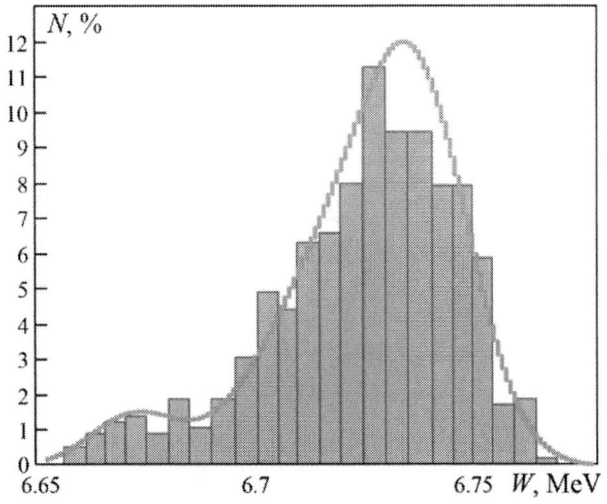

Figure 3. Particles energy spectrum.

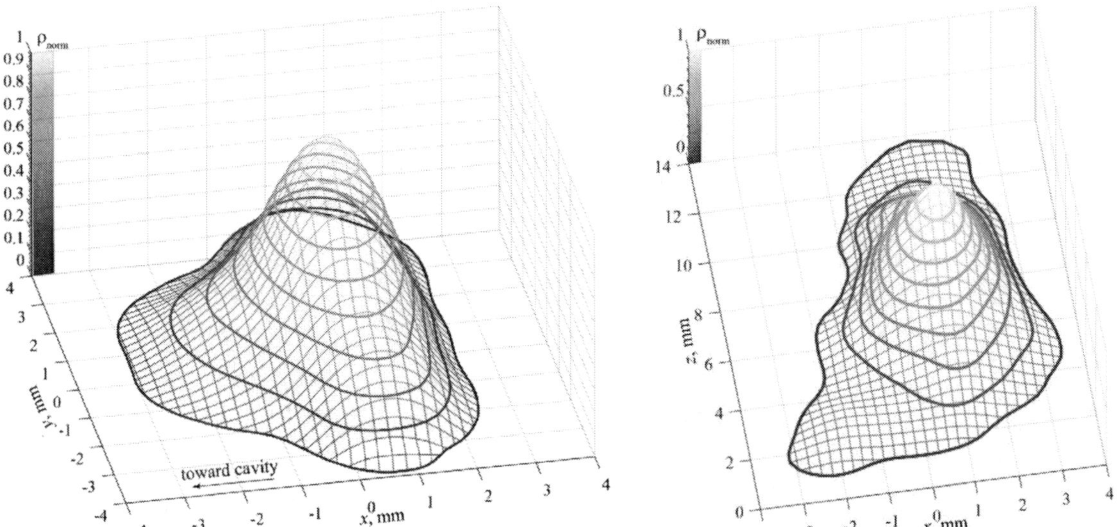

Figure 4. The distribution of charge density in bunch in vertical plane (left) and in median one (right).

Numerical simulation showed that 33% of particles were captured in stable acceleration mode, 26% of particles were lost on the inner cavity wall, 30% – on the outer cavity wall and 11% – on the other cavity parts for the RF field phase range from −90° to −50° under chosen microtron parameters. Figure 5 illustrates calculated transversal losses of electrons, which are injected in different RF field phases, on cavity walls (a – inner side of the cavity wall with output window, i.e. that one which with cathode; b – outer side of the wall with input window).

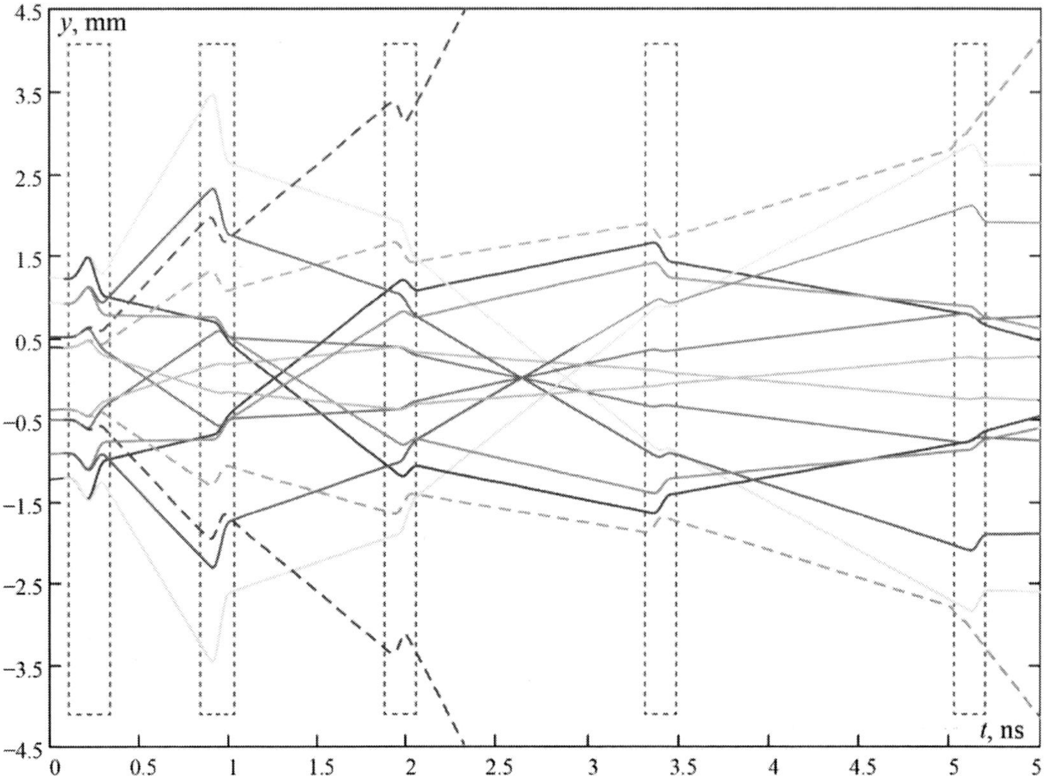

Figure 5. Vertical motion of electrons.

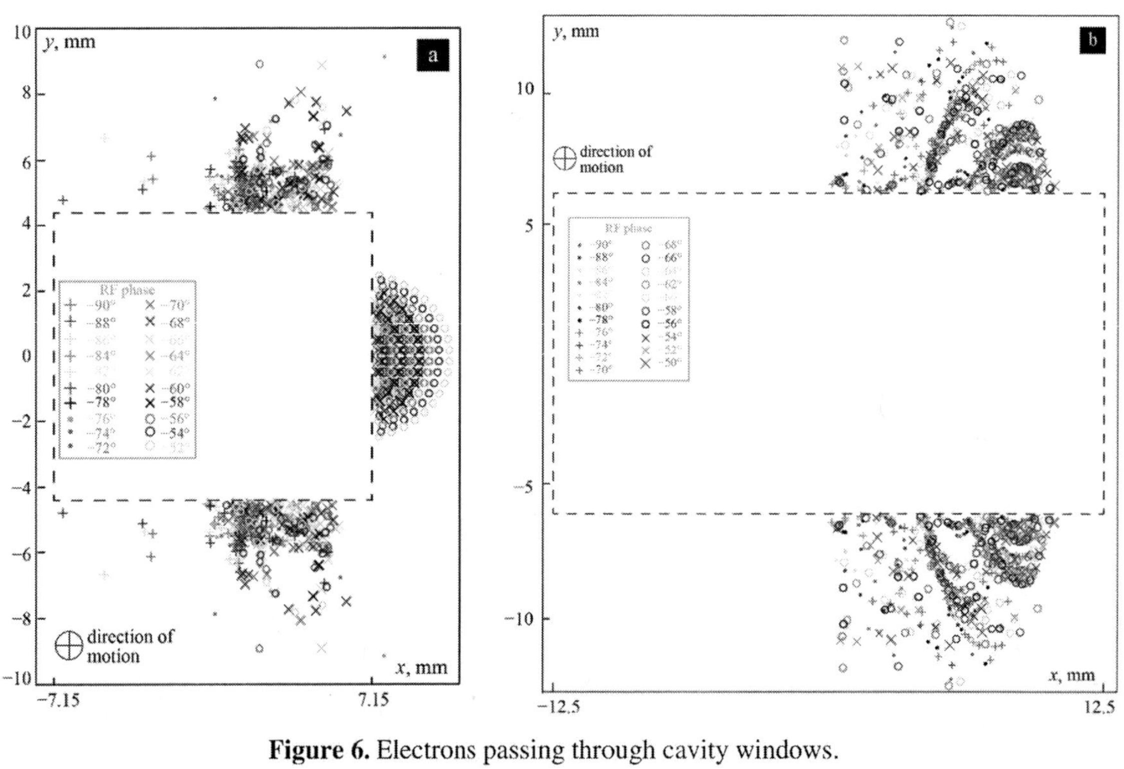

Figure 6. Electrons passing through cavity windows.

Note, that for a more efficient use of the microtron cavity it makes sense to consider using photocathode as electron source. In principle, it is possible to synchronize short laser pulses with the accelerating RF field, so that emitted particles will be accelerated effectively.

3. Conclusion

Transverse electrons dynamics in 6.5 MeV LPI microtron was investigated. It was shown that microtron cavity windows affect transverse particle motion strongly, manly on a few first orbits, leading to defocusing of particles. The optimal parameters of electron injection from the thermionic cathode were found. Good beam quality was obtained on the final orbit under used microtron parameters.

Acknowledgments

This work was supported by Competitiveness Program of National Research Nuclear University MEPhI.

References

[1] Veksler V I 1945 Journal of Physics USSR **9** 153
[2] Grinberg A P 1961 Uspekhi phyzicheskih nauk **75** 3
[3] Liu L, Resende X R and Rodrigues A R D Proc. 1st Int. Particle Accelerator Conf. (Kyoto) p 2481
[4] Williams P H et al. Proc. 2nd Int. Particle Accelerator Conf. (San Sebastian) p 3047
[5] Xiaoming J et al. 2014 Synchrotron Radiation News **27** 27
[6] In Soo Ko et al. 2017 Applied Science **7** 479
[7] Hoffstaetter G H et al. Proc. 28th Linear Accelerator Conf. (East Lansing) p 384
[8] Brinker F et al. Proc. 7th Int. Particle Accelerator Conf. (Busan) p 1044
[9] Schreiber S et al. Proc. Free Electron Laser Conf. (Shanghai) p 267
[10] Ainosuke Ando et al. Proc. 34th Int. Free-Electron Laser Conf. (Nara) 111
[11] Dowell D H et al. Proc. Free Electron Laser Conf. (Novosibirsk) p 276
[12] Bashmakov Y A, Karpov V A and Yarov A S 1984 Technical Physics **29** 539
[13] Dyubkov V S, Bashmakov Y A and Lozeev Y Y 2017 Proc. 25th Russian Particle Accelerator Conf. (St. Petersburg) p 428

IOP Conf. Series: Journal of Physics: Conf. Series **1238** (2019) 012072 doi:10.1088/1742-6596/1238/1/012072

Control and diagnostic systems for specialized synchrotron radiation source of 4th generation SSRS-4

T Kulevoy[1,2,3], D Liakin[1,2], A Tishchenko[1,3], A Blagov[1], R Senin[1], A Sitnikov[1,2], Yu Gaponov[1], A Gogin[1], A Orlov[1,2], R Kuibeda[1,2], S Barabin[1,2], M Saratovskikh[1,2], P Fedin[1,2,3], E KhabibuLlina[1,2,3], A Ziiatdinova[1,2,3], S Andrianov[1,2], A Lukashin[1,2], D Sergeeva[1,3], A Ponomarenko[1,3], L Lebedeva[1], V Stepanova[1], I Perelomova[1], N Marchenkov[1], K Funtov[1], V Vinogradov[1], E Naplekova[1], A Dmitriev[1], K-B Scheidt[4], E Plouviez[4], F Ewald[4], A Götz[4], B Roche[4], J-L Revol[4], J-M Chaize[4] and S Cardot[4]

[1]National Research Center "Kurchatov institute"
[2]Institute for Theoretical and Experimental Physics of NRC «Kurchatov Institute»
[3]National Research Nuclear University MEPhI
[4]European Synchrotron Radiation Facility,Grenoble,France
kulevoy@itep.ru

Abstract. The general layout for Specialized Synchrotron Radiation Source 4th generation SSRS 4 is under development by leading of National Research Center "Kurchatov institute". In the framework of a collaborative project with European Synchrotron Radiation Facility (ESRF) supported by the Ministry of Education and Science of the Russian Federation, the applied research on determination of the configuration and key technical parameters for the control and beam diagnostic systems for 4th generation Specialized Synchrotron Radiation Source – SSRS-4 is going on. The preliminary view of the proposed control and diagnostic system is presented and discussed. SSRS-4 inherits the experience of the scientific community including a great ESRF expertise in control system development (as founder of TANGO), robust diagnostics and long operational experience on ESRF's accelerator.

1. Introduction

Synchrotron radiation (SR) is a multipurpose science instrument, that is under permanent improvement since 1990th [1]. Now many researchers use SR at different fields of science. In the world tens of modern synchrotron complexes were built to meet researchers' needs [2]-[12]. Total number of synchrotron facilities is more than 75, where more than 20 thousand scientists are working every year. Furthermore, these numbers continue increasing each day. The synchrotron sources are used and are improved continuously at all technologically advanced countries. Moreover the international facilities with ultra-high parameters were built in international collaboration as European Synchrotron Radiation Facility in Grenoble, France, for example

Main synchrotron radiation consumers are material science and life science. These large branches of science have the greatest impact on industry based on the development of high tech devices and drugs - or "economy of knowledge". Taking into account the industrial development of various world regions and the growing role of the knowledge economy it is necessary to have the national strategies for such development, as well as the corresponding infrastructure and competence centers.

Historically sources of SR are divided into four generations.

The first generation synchrotrons were built initially for experiments with high-energy physics. That time SR was considered only as an inevitable accessory effect. The first experiments used SR started at these synchrotrons.

The second generation synchrotrons were built specially for produced SR. Only bending magnets were used for generation SR at these facilities.

The third generation synchrotrons are modern facilities operated now in more than 20 countries. Their design includes number of long (five or more meters) straight sections where wigglers or undulators are installed. Such inserted devices significantly improve the energy efficiency of the device, since most of the energy radiated by the electrons is output directly to the experimental stations, so power consumption was reduced significantly.

The fourth-generation SR sources are based originally on free electron (FEL) lasers and linear accelerators with energy recuperation. It was believed that a significant emittance reduction (less than 0.1 nm) and increase in brightness can be achieved only by switching synchrotrons to FEL. However, P. Raimondi and his collaborators showed that it is possible to create a storage ring structure where electron beam emittance can reach values much less than 100 pm [13]. Such small emittance enables X-ray sources with a characteristic transverse size of about 10 microns or less and high degree of coherence beam. For radiation with a wavelength of more than 1A, such a source can be designed even as diffraction limited and completely coherent. Now NRC "Kurchatov institute", NRNU MEPhI, Budker Institute of Nuclear Physics of SB RAS and other are developing complex of specialized synchrotron radiation source of the fourth generation - SSRS-4. It was shown that a storage ring with 6 GeV beam energy can provide the proper instruments for the key experiments. Two variants of the top-up injector in the storage ring are discussed: one use the top-up linac injection, the second one use a booster. The first approach (top-up linac) enables realization of a free-electron laser (FEL) which can work between the cycles for fulfilling the storage ring. However, it will be necessary to use two sources of the electron beam - one with a thermo cathode to fill the ring accelerator and the other one based on the photo gun to operate for the FEL. Such the accelerator complex requires the production of a reliable control system based on a powerful beam diagnostic system.

2. Beam diagnostics and control systems

The modern diagnostic and control systems have to provide the increasing quality requirements for such parameters as:

- efficiency of access to data
- velocity of experimental data processing
- uniformity in the data provided and processing quality
- fault tolerance of both hardware and software

The list of diagnostic devices for the designing complex is shown in Table 1.

It is expected that a basic configuration of the accelerator complex based on beam dynamics simulation, carried out in cooperation with ESRF experts, will be developed by the end of 2018. The simulation is under process in two directions: the development of a magnetic channel for emittance of 70-100 ppm providing the needs of all experimental works now, and for a record emittance of ~ 20 pm, in order to determine the feasibility of this solution. Speaking about beam diagnostics system, beam position sensors are under experimental development [14], [15]. Particular attention is devoted to the development of small beam emittances measurements in - SSRS-4 [16]-[21]. The work is under process in collaboration with specialists from the BNU of Kaliningrad [22], pp. **33-42**

[23]. Simultaneously the development of electronics is underway for these sensors. A scaled stand for control system synchronization system and electronic components testing is under development [24]-[32].

Table 1. List of diagnostic devices for SSRS-4.

Name	Placement location	Beam parameter
Current monitor	On all structural components of the facility	Beam charge measurement
DC current transformer	On all structural components of the facility	Beam charge measurement
screen	On injector after the buncher, The second on FEL diagnostic line	Emittance, energy, energy spread
screen	On linac, Transport line to ring Transport line to FEL	Transverse beam size (mean and end) Emittance
BPM – beam position monitor	On all structural components of the facility	Beam trajectory dispersion: horizontal and vertical betatron oscillation frequency
Faraday cup	Injector Special Faraday cup after FEL's undulator before damp	Injection phase, Microbunch charge and beam structure (approximate) after FEL's undulator (Special Faraday cup)
Streak camera	Injector Transport line to FEL Storage ring Buster ring	Bunch length
Beam loss monitor	Transport line to ring Storage ring	Beam losses
interferometer	Storage ring	Transverse beam size Emittance

The basic principles of control and diagnostics systems have been formulated and developed since the 90th of the last century. During that time the efficiency of commercial and specialized equipment has been increased continuously reducing the operation cost and simplifying communication processes between the system elements. In practice, the software and hardware components themselves are becoming increasingly complicated. Anyway the apparent simplification is reached by the development, optimization and standardization of the user interface between the application programs and the object and transport layers of the diagnostic and control systems. Modern computers allow implementing remote file systems, server name functions, and event server functions on ordinary PCs. This approach makes the system more decentralized, while reducing the load on the network segments. Effective development tools - RAD (rapid application development) compete with the tools of automatic code generation and IDL, which are add-ons over platforms of the TANGO / CORBA or EPICS classes.

The management system is based on the standard architecture of distributed systems. Figure 1 shows the main elements of the network structure of the complex. The structure shown includes workstations – 1, management terminals – 2, database server – 3, encapsulated local control systems requiring high performance – 4, modular subsystem controllers – 5, wired network routers – 6, peripheral electronics (interfaces for analogue input/output) – 7. Thus, the system is functionally a set of peripheral controllers and operating terminals connected to each other by TCP / IP and UDP network protocols. Exit to the external Internet network and use of wireless components may require additional costs to develop secure protocols with data encryption. The development of the system is due to the addition of new computers and controllers. The physical hierarchy is provided by the components of the network support: routers, network bridges, etc. Parts of the system may include isolated network sections dedicated to local communications. These sites can be protected from access from the global network of the complex. The system performance is determined by the performance of the network and its components.

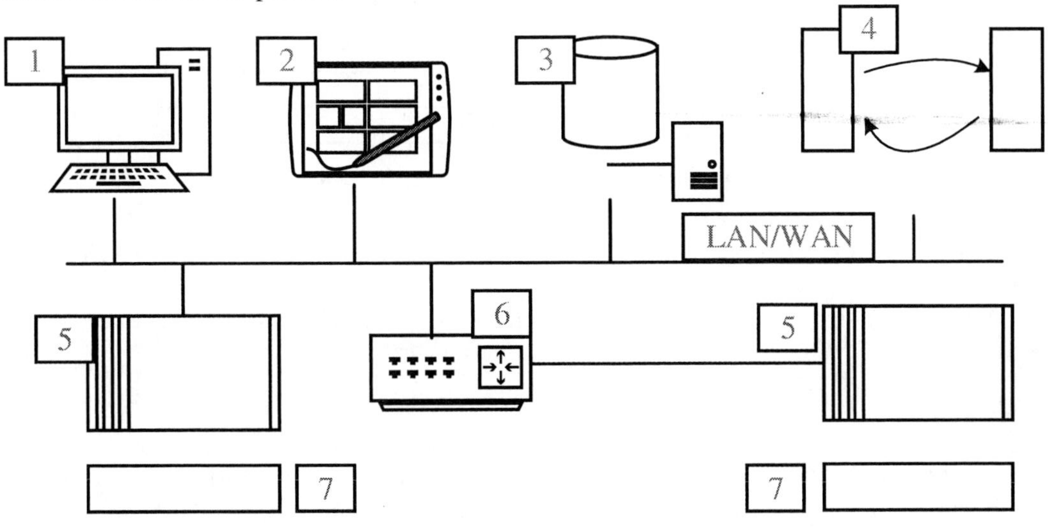

Figure 1. Distributed systems architecture of the complex.

Workstations for control system are a desktop or embedded industrial configuration PC using Windows OS (predominantly) for SCADA systems and Linux for systems using EPICS or TANGO (Windows versions also exist).

Interaction of the elements of the control system with each other is underway via a protocol called the protocol of access to the data transmission channel. As seen in Figure 1, the client and the data transfer server are provided with a network structure and a protocol procedure. As result the client can detect the servers and initiate the data transfer between themselves and the selected remote server. The server part provides access to data that describes in general a certain device – an abstract input-output controller.

Historically, the timer systems have begun with devices that distribute simple voltage pulses through the cables and each pulse corresponds to a specific event and destined for one particular device. Then, the timer events became encoding by a set of consecutive bits, which enables to transmit a set of sync pulses to different devices over a single cable. Over time, copper cables have been replaced by optical cables, which provide better stability and signal transmission for longer distances. Such synchronization systems are implemented in several laboratories that focus on the use of a relatively inexpensive, high-quality synchronization system. For the SSRS-4 the MRF standard (Micro-Research Finland) is the most practical, as the least complex providing required parameters. MRF offers a complete set of components for building a scalable synchronization system based on the encoding of a timer signal transmitted over optical cables. A typical configuration of the system used the extended MRF standard is shown on Figure 2.

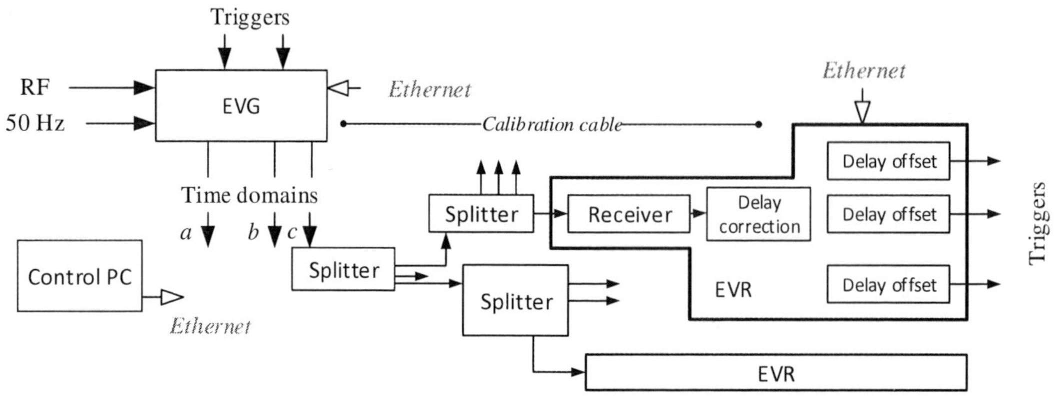

Figure 2. Layout of timer signal transmitted with MRF.

The system includes an event generator (EVG) and a set of event receivers (EVR). EVG synchronized with the phase of the high-frequency accelerating field. The signal branch can be extended by using the signal splitters. Synchronization signals can be converted directly to the form of standard TTL, LVDC, NIM, etc. pulses by the EVR at the application site.

3. Acknowledges

This work was supported by the Ministry of Education and Science of the Russian Federation. Agreement 14.616.21.0088 from 24\11\2017. ID RFMEFI61617X0088. (Federal Targeted Program "Research and development on priority directions of scientific-technological complex of Russia in 2014 -2020 year").

REFERENCE

[1] Eberhardt W 2015 Synchrotron radiation: A continuing revolution in X-ray science—Diffraction limited storage rings and beyond *Journal of Electron Spectroscopy and Related Phenomena* Volume 200 pp 31-39.

[2] Tavares P F, Leemann S C, Sjostrom M, Andersson A 2014 The MAX IV storage ring project *J. Synchrotron Rad.* 21 862–877.

[3] Rodrigues A R D et al. 2016 Sirius status report *Proceedings of IPAC2016 (Busan, Korea)* pp. 2811-2814.

[4] http://www.esrf.eu/Apache_files/Upgrade/ESRF-orange-book.pdf

[5] Henderson S 2015 Status of the APS upgrade project *Proceedings of IPAC2015 (Richmond, USA)* pp 1791-1793.

[6] Tanaka H, Ishikawa T 2016 Spring-8 upgrade project *Proceedings of IPAC2016 (Busan Korea)* pp.2867-2870.

[7] https://www.maxiv.lu.se/accelerators-beamlines/accelerators/

[8] http://lnls.cnpem.br/sirius/

[9] https://www.kek.jp/en/Facility/IMSS/PF/

[10] http://www.kek.jp/en/Facility/ACCL/LightSource/

[11] http://www-ssrl.slac.stanford.edu/content/spear3/spear3-accelerator

[12] http://www.diamond.ac.uk/Home/About.html

[13] Raimondi P, Zobov M, Shatilov D 2007 Beam-beam simulations for particle factories with crabbed waist *Proceedings of PAC07 (Albuquerque New Mexico USA)* pp 1469-1471

[14] Slater M et al 2008 Cavity BPM system tests for the ILC energy spectrometer *Nuclear instruments and Methods* V.592, Issue 3 pp.201-217

IOP Conf. Series: Journal of Physics: Conf. Series **1238** (2019) 012072 doi:10.1088/1742-6596/1238/1/012072

[15] Tobiyama M et al 2010 Development of button electrodes for SuperKEKB rings *Proceedings of BIW 2010 (Santa Fe New Mexico US)* pp.223-227.

[16] Behrens C, Gerth C, Kube G, Schmidt B, Wesch S and Yan M 2012 Electron beam profile imaging in the presence of coherent optical radiation effects *Physical Review Special Topics - Accelerators and Beams* т. 15 № 6 p 062801,

[17] Couillaud C et al. 1996 Electron Beam Transverse Emittance Measurement using Optical Transition Radiation Interferometry *Proceedings of EPAC1996 (Barselona, Spain)* pp.e-proc. 1594

[18] Ewald F et al 2013 Emittance measurement using X-ray lenses at the ESRF *Proceedings of IBIC 2013 (Oxford UK)* pp.833-836

[19] Elleaume P et al 1995 Measuring Beam Sizes and Ultra-Small Electron Emittances Using an X-ray Pinhole Camera *J. Synchrotron Rad.* 2 p.209.

[20] Ewald F et al 2011Vertical Emittance Measurement at the ESRF *Proceedings of DIPAC2011 (Hamburg Germany)* pp.188-190

[21] Ewald F 2009 Emittance Measurement Using Undulator Radiation *Proceedings of DIPAC 2009 (Basel Switzerland)* pp.182-184

[22] Snigireva I and Snigirev 2006A X-ray microanalytical techniques based on synchrotron radiation *Journal of Environmental Monitoring* т. 8 № 1 pp. 33-42

[23] Schroer C G, Kuhlmann M, Lengeler B, Gunzler T F, Kurapova O, Benner B, Rau C, Simionovici A S, Snigirev A A and Snigireva I 2002 Beryllium parabolic refractive x-ray lenses*Proceedings of SPIE - The International Society for Optical* Engineering AIP Conference Proceedings Volume 705 Issue 1 10.1063/1.1757904 .

[24] http://www.tango-controls.org/

[25] http://www.corba.org/

[26] Antonioni F et al. 2013Developments on the SCADA of CERN accelerators vacuum *Proceedings of ICALEPCS2013 (San Francisco CA USA)* pp.135-138

[27] Pieck M et al 2015LANSCE control system upgrade status and challenges *Proceedings of ICALEPCS2015 (Melbourne, Australia)* pp.464-467

[28] Stefanic R, and Geoffroy L 2007 EPICS to TANGO translator *Proceedings of ICALEPCS07 (Knoxville Tennessee USA)* pp.739-741

[29] Fleck T et al 2007 Status of the control system for HICAT at an advanced stage of commissioning: functions, restrictions and experiences *Proceedings of ICALEPCS07 (Knoxville Tennessee USA)* pp. 47-49

[30] Пономарев О П SCADA-системы, промышленные шины и интерфейсы, общие сведения о программируемых логических контроллерах и одноплатных компьютерах. Калининград 2006.

[31] Matthies S et al FESA3 integration in GSI for FAIR 2014 *Proceedings of PCaPAC2014 (Karlsruhe Germany)* pp.43-45

[32] Urakawa J and Kawamoto T 1991Improvement of TRISTAN timing system *Proceedings of PAC91* pp.1555-1557

IOP Conf. Series: Journal of Physics: Conf. Series **1238** (2019) 012073 doi:10.1088/1742-6596/1238/1/012073

Random search for global extremum of a function using Markov chains simulation

L V Vladimirova[1] and D A Ovsyannikov[1]

[1]St. Petersburg State University, 7/9 Universitetskaya nab., St. Petersburg, 199034, Russia

E-mail: l.vladimirova@spbu.ru, d.a.ovsyannikov@spbu.ru

Abstract. The problem under study is global extremum search for multiextremal function. Random search method based on Markov chains simulation is used. A comparative analysis of this method with another method – the method of random search for the global extremum with "memory", using the normal distribution – done.

1. Introduction

The problem of global extremum search is relevant for various applied problems, including optimization of physical processes [1-7].

Among the methods of search for global extremum special place is taken by the algorithms of random search. As practice shows, when solving complex extreme problems, they are almost the only way to achieve the goal, while they have the simplicity of the algorithm and can be used for large dimensions of the search area.

To increase the efficiency of random search of global extremum in this paper we use uneven distribution with density associated with the target function. This allows us to find the point of global extremum as a mode of such distribution [2].

2. Problem statement

Assume there is a nonnegative function $f(x)$, set in the region $D \subset E^n$ n-dimensional Euclidean space. The volume D is expected to be limited. The objective is to find

$$f(x^*) = \sup_D f(x). \quad (1)$$

To solve problem (1), you can enter an uneven distribution in the area D

$$p(x) = f(x)\left(\int_D f(x)dx\right)^{-1} \quad (2)$$

and find the point x^* where the mode $p(x)$ will be reached. t is clear that the realization of a random vector distributed with density (2) will appear more often near the highest value $f(x)$. Moreover, the density $p_s(x) = f^s(x)\left(\int_D f^s(x)dx\right)^{-1}$ will be when the $s \to \infty$ converged to the δ-function concentrated at the point x^* [2]. However, modeling such densities is difficult.

3. Random search for global extremum with " memory"

Let's consider a method that allows the use of uneven distribution and evaluation of the covariance matrix to model the density

$$\overline{p}(x) = Cf(x)N(x^{(0)}, B), \tag{3}$$

instead of $p_s(x)$. In (3) C is the normalization constant the density, function

$$N(x^{(0)}, B) = \frac{(2\pi)^{-n/2}}{(\det B)^{1/2}} \exp(-(x - x^{(0)})^T B^{-1}(x - x^{(0)})) \tag{4}$$

is the density of the normal distribution with parameters: vector $x^{(0)} \in D$ and covariance matrix B. The covariance matrix B to a certain extent reflects the structure of the function $f(x)$. In this sense, the described method is called the method with "memory".

It is convenient to use the generalized Neumann method for density $\overline{p}(x)$ modeling, and normally distributed random vectors with density (4) are modelled in the region D. This is a multi-iterative process. At the zero iteration, it is required to model random vectors uniformly distributed in D (the usual Neumann method for density modeling (2)). Algorithm and convergence theorem for random search of global extremum of function based on modeling of normal distribution and estimation of covariance matrix are presented in [2, 8].

4. Metropolis Method

To model density (2), you can use a rejection sampling [9]. To do this, it's needed to find a convenient simulated density $q(x)$, for which the following inequality is valid: $Cq(x) > f(x)$ for all $x \in D$, where C is a constant. However, in the algorithm of sampling variance, there is an issue with the choice the constant C (figure 1). If the constant C is too large, the algorithm will give the correct results, but the ratio of the number of test points in the strikethrough area to the total number of test points will be small, which means that the effectiveness of the method is low [9]. If an error inclines towards the lower values of the part of the area D where $f(x) > Cq(x)$ the rejection sampling cannot be applied. To resolve the issue with constant C selection, you can use the metropolis method.

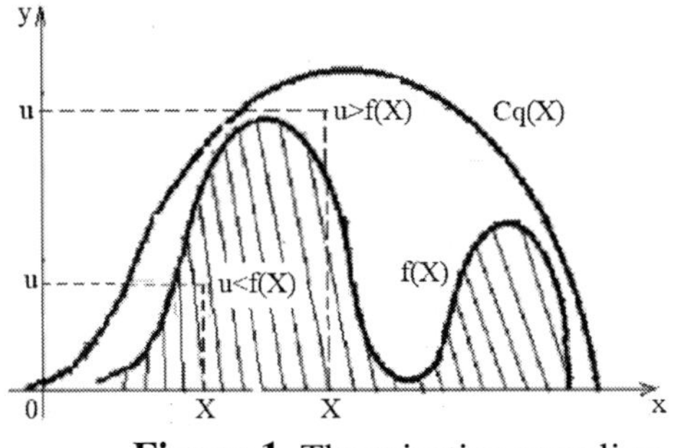

Figure 1. The rejection sampling

The algorithm of the metropolis method eliminates the problem with the choice of the constant C, using the transitive density of the Markov process as a function $q(x)$ depending on two variables $q(x, y)$. Instead of independent realizations of a random value with the density $q(x)$, the Markov chain with transition density $q(x, y)$ is modeled. Thus, the algorithm of the metropolis method serves as the basis of the Monte Carlo method on the Markov chains [10].

4.1. Metropolis method algorithm

1. Assume $k = 0$. Choose arbitrarily $X^{(0)}$ and the transition density $q(x, y)$. Modeling density $q(X^{(0)}, y)$, we obtain the realization of a random variable $Y^{(k)}$.

2. Modeling $q(Y^{(k)}, y)$, we obtain the realization of a random variable Y'.
3. The value is calculated

$$a = \left(f(Y')f^{-1}(Y^{(k)})\right)\left(q(Y^{(k)}, Y')q^{-1}(Y', Y^{(k)})\right). \quad (5)$$

4. If $a \geq 1$, we assume $Y^{(k+1)} = Y'$.
5. If $a < 1$, we assume $Y^{(k+1)} = Y'$ with probability a and $Y^{(k+1)} = Y^{(k)}$ with probability $1-a$

The random process $Y^{(k)}$ with discrete time $k = 0,1,2,3,...$ is a Markov process, and the transition density of a random process 1-5 is equal to $A(x, y) = q(x, y)\cdot \min\left(1, p(y)q(y,x)p^{-1}(x)q^{-1}(x, y)\right)$ and for all x, y the ratio $A(x, y)p(y) = A(y, x)p(x)$ is valid. Then $p(x)$ is the density of the stationary distribution for $A(x, y)$ [9]. This means that with a sufficiently large number of N the sequence

$$Y^{(1)}, Y^{(2)},..., Y^{(N)} \quad (6)$$

is the sequence of independent realizations of a random vector distributed with density $p(x)$. Thus, according to the sequence (6), the density mode can be approximately found. As the $p(x)$ is proportional to the function $f(x)$, then the maximum value of function $f(x)$ can be approximately found as well.

4.2. Special case of the metropolis method

As a transitional density, it is convenient to choose the density of the normal probabilities distribution (Gaussian)

$$q(x, y) = \frac{(\det B)^{-1/2}}{(2\pi)^{n/2}}\exp\left(-\frac{1}{2}(x-y)^T B^{-1}(x-y)\right), \quad x, y \in D \subset E^n \quad (7)$$

with covariance matrix B. Function (7) is symmetric and value search (4) is simplified: $a = f(Y')f^{-1}(Y^{(k)})$.

5. The problem of finding a locally optimal plan for the non-linear exponential regression

Let's apply the metropolis method to find the global extremum of a complex multiextremal function with 4 variables. We consider the problem of finding the locally optimal plan for the non-linear exponential regression

$$\eta(x,\theta) = \beta_1 e^{-\theta_1 x} + \beta_2 e^{-\theta_2 x}, \quad x \in [0, B] \subset R^1, \quad \beta_1, \beta_2 \neq 0, \quad \theta_2 > \theta_1 > 0, \quad (8)$$

with the vector of parameters $\theta = (\beta_1, \theta_1, \beta_2, \theta_2)$. In the ratio (8) B – is some given value.

As we can see, only the parameters θ_1, θ_2 enter nonlinearly. Functions of the type (8) form an important class of solutions of systems of differential equations and have a variety of practical applications, for example, in atomic physics.

For this particular problem, the plan x_1, x_2, x_3, x_4 consists of the minimum possible number of values $x \in [0, B]$ that is equal to the number of parameters $\beta_1, \theta_1, \beta_2, \theta_2$. To find the locally optimal plan, we consider the optimality criterion related to the information matrix $M(x,\theta)$ with fixed parameters. Such optimality criterion for regression (8) is the determinant of the following form $\det(M(x,\theta)) = \beta_1^2\beta_2^2 A(x, \theta_1, \theta_2)$ [11-13], where $x = (x_1, x_2, x_3, x_4)$ and

$$A(x,\theta_1,\theta_2)=\begin{vmatrix} e^{-\theta_1 x_1} & e^{-\theta_1 x_2} & e^{-\theta_1 x_3} & e^{-\theta_1 x_4} \\ -x_1 e^{-\theta_1 x_1} & -x_2 e^{-\theta_1 x_2} & -x_3 e^{-\theta_1 x_3} & -x_4 e^{-\theta_1 x_4} \\ e^{-\theta_2 x_1} & e^{-\theta_2 x_2} & e^{-\theta_2 x_3} & e^{-\theta_2 x_4} \\ -x_1 e^{-\theta_2 x_1} & -x_2 e^{-\theta_2 x_2} & -x_3 e^{-\theta_2 x_3} & -x_4 e^{-\theta_2 x_4} \end{vmatrix}^2 . \quad (9)$$

The function (9) depends on the points of the planning set x_1, x_2, x_3, x_4 and on the parameters θ_1, θ_2 entering nonlinearly in (8).

The task is to find the maximum of function (9) via variables x_1, x_2, x_3, x_4 from a set of scheduling $[0, B]$ with fixed values of parameters $\theta_1 = \theta_{1,u}$, $\theta_2 = \theta_{2,u}$. The plan resulting from optimization of the optimality criterion with fixed $\theta_{1,u}$, $\theta_{2,u}$ is called a locally optimal plan.

At the figure 2 locally optimal plans x_1, x_2, x_3, x_4 for 9 variants of parameters $\theta_{1,u}, \theta_{2,u}$ are presented. This result are obtained in two ways of optimization. Locally optimal plans found using the metropolis method are indicated by asterisks. Locally optimal plans found with the application of random search with "memory" are circle-marked [8,12,13].

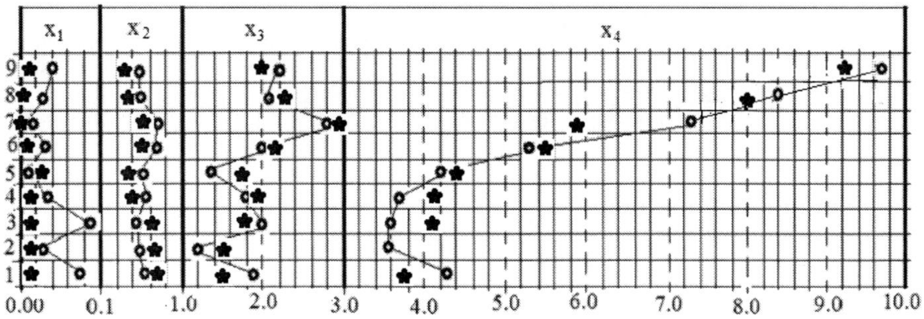

Figure 2. Four points x_1, x_2, x_3, x_4 of locally optimal plans for 9 variants of nonlinear parameters $\theta_{1,u}$, $\theta_{2,u}$ found by two optimization methods

As can be seen from figure 2 the applied simplified metropolis method gives adequate results. Due to its simplicity and convenience, it can be successfully used for solving various applied problems, including optimization problems of various accelerating structures.

6. References

[1] Ovsyannikov A D, Ovsyannikov D A, Altsybeyev V V, Durkin A P and Papkovich V G 2014 *Application of Optimization Techniques for RFQ Design* (Problems of Nuclear Science and Engineering vol 91 No 3) pp. 116-119

[2] Vladimirova L V 2014 *Global Extremum Search on the Basis of Density and Its Mode Estimation* (Proc. of 20th International Workshop: Beam Dynamics & Optimization (BDO), June-July 2014 St. Petersburg Russia IEEE) pp 186-188

[3] Vladimirova L V 2016 *Multicriterial Approach to Beam Dynamics Optimization Problem* (II Conference on Plasma&Laser Research and Technologies Journal of Physics: Conference Series vol 747 No 1) 012070 http://iopscience.iop.org/1742-6596/747/1/012070

[4] Rubtsova I D 2015 *Optimization of iterative beam dynamic process* (Proc. III Int. Conf. "Stability and Control Processes" in Memory of V.I.Zubov (SCP) (5-9 October 2015 St. Petersburg) ed L A Petrosyan and A P Zhabko (St. Petersburg: IEEE)) DOI: 10.1109/SCP.2015.7342092 pp 198-200

[5] Ovsyannikov D A and Altsybeyev V V 2013 *Optimization of APF accelerators* (Problems of Atomic Science and Technology No 88(6)) pp 119-122

[6] Zavadsky S V, Ovsyannikov D A and Chung S-L 2009 Parametric Optimization Methods for

IOP Conf. Series: Journal of Physics: Conf. Series **1238** (2019) 012073 doi:10.1088/1742-6596/1238/1/012073

the Tokamak Plasma Control Problem Int. J. Mod. Phys. A 24 No 5 pp.1040–47
[7] Kotina E D 2006 Discrete optimization problem in beam dynamics Nuclear Instruments and Methods in Physics Research, Section A: Accelerators, Spectrometers, Detectors and Associated Equipment 558(1), c. 292-294
[8] Vladimirova L V 2013 *Using random search with "memory" in the estimation of nonlinear regression parameter* (Bulletin SPbSUTD series 1 No 4) pp 30-34
[9] Ermakov S M and Sipin A S 2014 *Monte-Carlo and Parametric Rasilimali Algorithms* (SPb.: Publishing House of Saint Petersburg University) p 248
[10] Metropolis N, Rosenbluth A W, Rosenbluth M N, Teller A H and Teller E 1953 *Equations of state calculations by fast computing machines* (Journal Chemical Physics vol 21) pp 1087-1091
[11] Melas V B 1978 *Optimal designs for exponential regression* (Math. Operations forsh. Statist. Ser. Statistics. vol. 9) pp 45-59
[12] Vladimirova L V and Fatyanova I A 2015 *Construction of optimal regression experiment plan based on random search with "memory" using parallel calculations* (Sustainability and management processes: Proceedings of the III international conference (SPb, 5-9 October 2015) Ed by A P Zhabko, L A Petrosyan: Fedorova G V Publishing House) pp 303-304.
[13] Vladimirova L V and Setina I A 2017 *Locally Optimal Designs Search for Exponential Regression* (48th international scientific conference of postgraduates and students “Management Processes and sustainability" April 3-6 2017 vol 4 No 1) pp 16-21

The control of unprocessed oil stratums debit restoring by method of ultasound influence with well neutron generator

N A Berdonosova[2], B Yu Bogdanovich[1], M S Dmitriev[1], Yu F Zhuyikov[2], A V Ilyinskiy[2], A V Nesterovich[1], A E Shikanov[1] and E A Shikanov[1]

1 National Research Nuclear University MEPhI, 31 Kashirskoe shosse, Moscow, 115409, Russia
2 Institute of Geophysical and Radiation Technology of IA SHS, 4/B Atarbekova, Moscow, 107076, Russia

aeshikanov14@mail.ru

Abstract. The examples of experimental studies of oil stratum debit restoring by method of ultrasound influence to stratum are given. The method of control of such unprocessed oil stratum restoring is proposed. It uses well neutron generator with vacuum accelerating tube and neutron reagent method with pumping of neutron absorbing salt solution to stratum. The results of its successful testing are presented.

1. Catalytic action of microwave radiation

During the operation of an oil well, its production rate is reduced as a result of contamination of the fluid recovery zone. To restore it, the ultrasonic influence (UI) to stratum is applied.

Figure 1 shows a general view of the hardware complex UI in the process of operation.

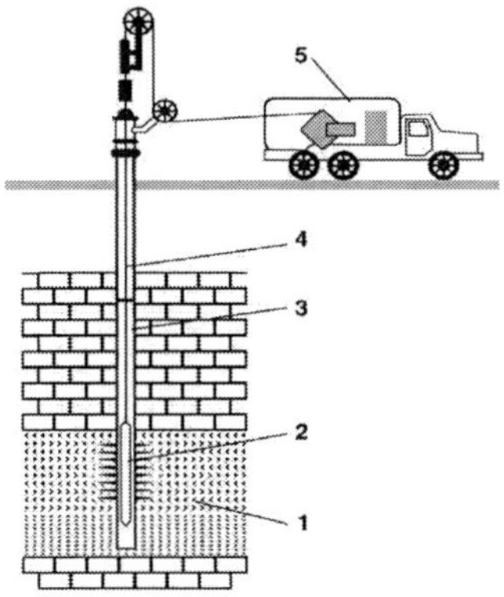

Figure 1. General view of the ultrasonic influence complex:
1 – stratum with productive fluid, 2 – well source of acoustic oscillations, 3 – cased well, 4 – geophysical load bearing cable, 5 – ground geophysical station with generator, control panel and winch.

Table 1 selectively presents the data on the increase in the production rate Δ and additionally produced oil Q for 8 wells, illustrating the efficiency of the UI in the example of the Fedorovsky oil field.

Table 1. The increase in the production rate Δ and additionally produced oil Q.

	Well No.							
	1063	1876	1886	1914	1967	2612	2634	2850
Δ, t / day	6.0	5.6	8.0	0	8.7	13.0	0	40.6
Q, t	500	846	720	0	783	720	0	14846

UI method was used to process 74 wells with a success rate of about 76%. The increase in production rate per one well was 9.9 tons per day.

As can be seen from the table, there are objects (wells No. 1914, 2634) for which the application of the UI technology does not give a positive result. In work [1] a method was proposed for identification of such wells to exclude them from the sphere of preventive actions of production rate recovery. To implement it, it is necessary to use a controlled well emitter of fast neutrons [2] generated in the pulse-periodic mode when the accelerated deuteron flux interacts with a metal target saturated with tritium, where the nuclear reaction of synthesis T (d, n)^{4}He is carried out. The target is located on the cathode electrode of a sealed vacuum accelerator tube. One of the versions of its design developed in VNIIA [3] is shown in figure 2.

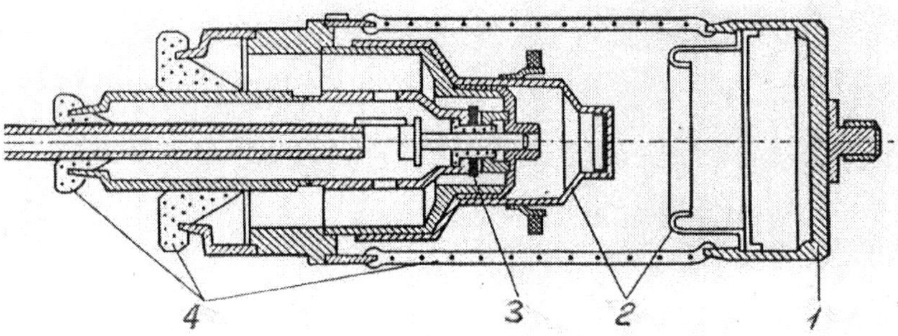

Figure 2. Schematic section of the accelerator tube:
1 target; 2-diode accelerating system; 3- source of deuterons; 4- insulators.

The change in permeability of the collector was analyzed by neutron logging methods [4]. The general scheme of the hardware complex implementing any of the method modifications is shown in figure 3.

When replacing quasi-cement formations in the pores of the oil collector, the environment under investigation begins to effectively slow down the fast neutrons generated by the emitter. Their radiative capture increases at the same time. As a result the fluxes of γ-quanta to the detector and the counting rate of their registration are increased. On its growth it is possible to judge success of UI carrying out. The dependence of the integral count of the γ-quantum detector $A(x)$ on the coordinate delayed along the wellbore in the region from the beginning of the oil stratum with a thickness of about 8 m taken before and after the realization of UI is shown in figure 4.

The lower curve corresponds to the measurement of the integral count of γ-quantum before UI, the upper curve after it. The relative error in the measurement did not exceed 15%. A comparison of the curves obtained indicates an increase in the integral count of the γ-quantum recording events during UI, and, consequently, an increase in the oil saturation, as well as the permeability of the well bottom zone.

IOP Conf. Series: Journal of Physics: Conf. Series **1238** (2019) 012074 doi:10.1088/1742-6596/1238/1/012074

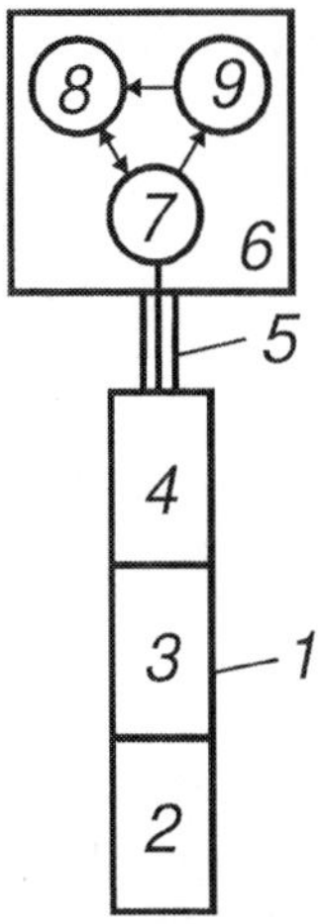

Figure 3. Schematic diagram of the hardware complex:
1 – well device with a protective bar-resistant housing, 2 – neutron source based on accelerating tube, 3 – detection system, 4 well executive and telemetry systems, 5 – load carrying cable, 6 – complex of ground equipment (logging station), 7 – ground control and telemetry systems, 8 – onboard computer, 9 - depth indicator of well device.

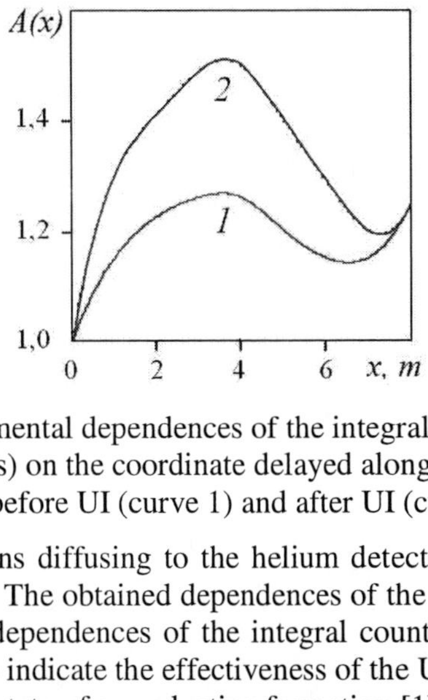

Figure 4. The typical experimental dependences of the integral count of the γ-quantum detector (relative units) on the coordinate delayed along the wellbore taken before UI (curve 1) and after UI (curve 2).

Registration of thermal neutrons diffusing to the helium detector located in the well device was carried out in another experiment. The obtained dependences of the integral count of thermal neutrons are similar to the corresponding dependences of the integral count of γ-quantum shown in figure 4. The results of this experiment also indicate the effectiveness of the UI.

A method for determining the state of a productive formation [1] by a pulsed neutron method with injecting a neutron absorbing substance into the reservoir (NaCl, $CdCl_2$, $GdCl_3$, etc.) was implemented in the third experiment. During the experiment, the time spectrum of thermal neutrons produced in the process of slowing down fast neutrons generated by the emitter in a pulsed mode was analyzed. The spectrum is characterized by the following time dependence of the thermal neutron flux detected by the detector:

$$n(t) \cong n(t_0)\exp[-\lambda(t-t_0)]$$

where t_0 is the registration start time, λ is the decrement of the neutron flux decay proportional to the macroscopic cross section of their radiative capture. 6 measurements were taken at one well. At the 1st (background) measurement, the initial decrement λ was determined. Then a NaCl solution was pumped into the formation, and the 2nd measurement was carried out. Further, the formation was exposed to ultrasonic radiation at a frequency of 20 kHz, and another 4 measurements were carried out every two hours. The results of measurements of λ_1, λ_2 for two points of the productive formation, separated by a distance 2 m along the wellbore, are presented in table 2.

Table 2. The results of measurements of λ_1, λ_2 for two points of the productive formation.

	Measurement No.					
	1	2	3	4	5	6
λ_1, ms^{-1}	4.0	5.5	5.3	5.2	4.5	4.1
λ_2, ms^{-1}	3.0	5.0	4.6	4.5	4.1	3.1

Table data indicate that after UI the decrement is restored to the initial value within 8 hours. This indicates a rapid releasing of the reagent from the formation, which means its good hydrodynamic connection with the well, which determines the effectiveness of the UI method. A solution of GdCl3 can be used to increase the contrast of measurements.

This project is supported in part by RFBR grant 15-19-00151 and by the MEPhI 5/100 Program of the Russian Academic Excellence Project.

References

[1] N. V. Berdonosova, B. Yu. Bogdanovich, S. Yu. Voronchikhin, et al., Patent No, 2517824 RF, "The method for determining the state of a productive formation by a pulsed neutron method," subm. Dec. 28, 2012, Byull. Izobret. Polezn. Modeli, No. 15 (2014)

[2] Bogolyubov E P and Ryzkov V I 2004 *Instruments and Experimental Techniques* **2** 160-163

[3] Bessarabskiy Yu G, Bogolyubov E P, Kurdyumov I G, et al. 1994 *Instruments and Experimental Techniques* **5** 206-207

[4] Meyer V A and Vaganov P A 1985 *Fundamentals of nuclear Geophysics* (Leningrad: Leningrad university)

IOP Conf. Series: Journal of Physics: Conf. Series **1238** (2019) 012075 doi:10.1088/1742-6596/1238/1/012075

Power combining scheme of solid-state RF amplifier

T V Bondarenko, S Y Ilynski, S A Polikhov, G B Sharkov

Scientific Research Institute of Technical Physics and Automation (NIITFA), Varshavskoe shosse 46, Moscow, 115230, Russia

E-mail: tvbondarenko@niitfa.ru, sapolihov@niitfa.ru, gbsharkov@niitfa.ru

Abstract. Solid state amplification technology had developed a lot in recent years. This article is focused on a power combining system of a new power amplifier stage for ISIS neutron source (Rutherford Appleton Lab., STFC, UK). The amplifier is based on a modular block-built architecture that intends using of amplifier modules and combination of their output power (of >kW levels) using high power RF combiners. The power combining system for high power RF amplifier is discussed in this article. The system scheme together with results of modeling, low power measurements of key electrodynamic characteristics and high power operation test are presented.

1. Introduction

RF high power combining system is a part of solid-state RF power amplifier [1] (fig. 1) that is being produced in NIITFA to replace the high power feeding system of ISIS neutron source in Rutherford-Appleton Laboratory (STFC, UK) [2]. Nominal output power level of the system is 250 kW (4 % duty cycle) at frequency of 202.5 MHz. Amplifier is based on a modular system that incorporates 64 amplifying modules with up to 4.3 kW output power each that are connected to the first stage of power combining (PC) system [3].

2. General system design

The high power combining system is based on a two stage architecture with four 16:1 power combiners (PC) as a first stage, and one 4:1 power combiner as a second stage.

Both types of power combiners are box shaped coaxial structures. Power combiners has built-in Teflon tuning elements that allow to change the PC operation frequency ±5 MHz and to tune it's impedance to compensate possible manufacturing intolerances. 16:1 PC (fig. 2) has RF switch on each input port that can operate as a through or short on our operating frequency independently for each port. RF switches has built-in control system that controls its state (through/short) and gives command to change state in order to turn on and off the input port. Power combiner together with RF switches can provide effective power combination of incoming power with up to 3 (of 16) simultaneously turned off channels.

The power combining system is extremely flexible: the number of input ports for both combining cascades is limited only by the case size. This feature opens the potential to build the amplifier with megawatt power range without strong design changes.

IOP Conf. Series: Journal of Physics: Conf. Series **1238** (2019) 012075 doi:10.1088/1742-6596/1238/1/012075

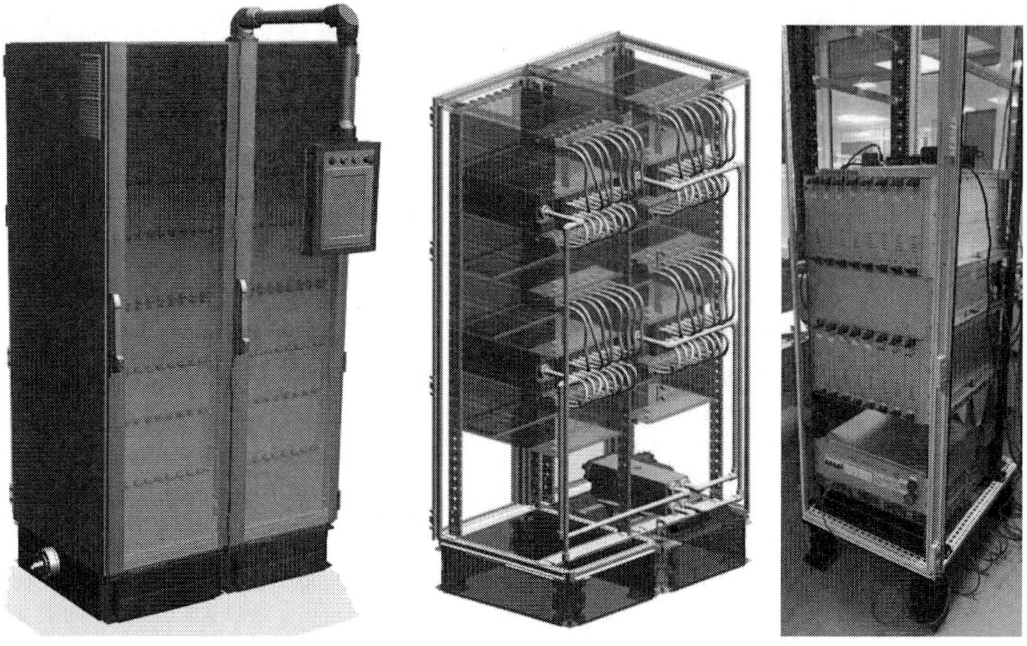

Figure 1. General view of two 19" solid state amplifier racks for ISIS neutron source (left), highlighted power combining system of the amplifier (middle) and rack with installed 16 power amplifier modules and one 16:1 power combiner during high power tests (right).

RF switches are manufactured in a housing that incorporates 4 RF switches. RF switch housing has a quick-disconnect design that together with control systems allow using of the RF amplifier modules hot-swap principle in the amplifier. The 16:1 combiner sizes are 560x447x177 mm (LxWxH) without fixation flanges.

Figure 2. 16:1 PC assembled and ready for low level tests (7/8"-N and N-SMA adapters are on output port and N-SMA adapters are on input ports).

Second stage of RF combining system is one 4:1 power combiner that has much simpler design without RF switches or other operational elements (fig. 3). Four 16:1 power combiners are connected to 4:1 power combiner using the 7/8" coaxial waveguide. Output of 4:1 power combiner is connected

IOP Conf. Series: Journal of Physics: Conf. Series **1238** (2019) 012075 doi:10.1088/1742-6596/1238/1/012075

to the next amplification cascade of RF power feeding system of ISIS facility through the gas barrier and bidirectional coupler. The overall dimensions of the 4:1 Power Combiner are 446x140x149 mm (LxWxH) without supporting frame and 446x210x153 mm with supporting frame. Bullets of the flanges are not taken into account.

Figure 3. 16:1 PC inner guide and outer view during assembling process

3. Measurement

Low level RF tests were performed using the Agilent E5072A vector network analyzer. All input ports were terminated with 50 Ohm loads except for the one under test at the moment (fig. 4-7).

All Power Combiner characteristics meet requirements except for slight mismatch of insertion loss. The insertion loss is 0.5% higher that it was predicted in high power RF combining system design. However this excess of power loss is compensated by lower power losses in 4:1 Power Combiner (0.35% instead of predicted 0.9%) so the total insertion losses of the high power RF combining system are now 3.14% that is within predicted 3.2%.

Results show that 16:1 power combiner is tuned 0.25 MHz down from operation frequency: to 202.25 MHz. This is in very precise agreement with CST calculation. Power transmission deviation for all of the ports lies in +0.34..-0.27% band from mean transmission value that is also in fine agreement with CST.

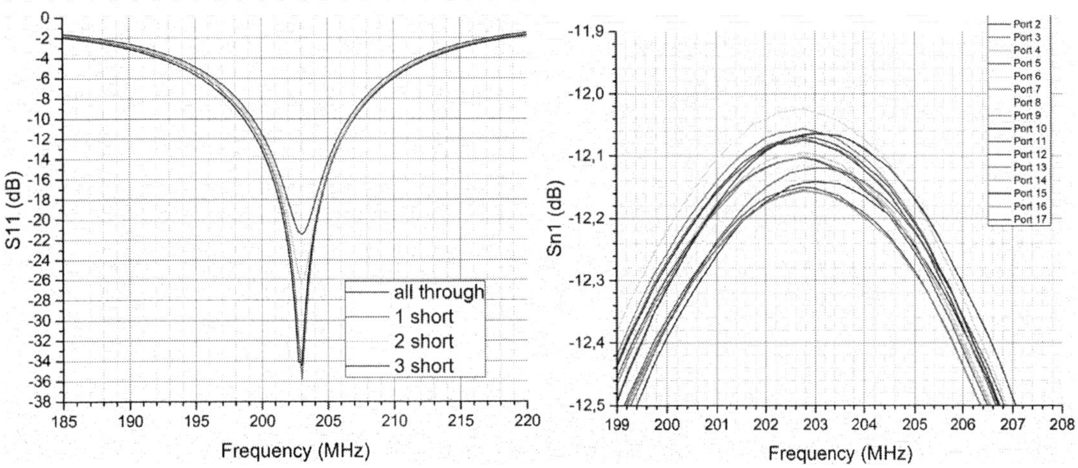

Figure 4. S11 parameter (reflection from output port) for different amount of shorted RF switches and Sn1 parameters (transmission from output to one of the input ports) of 16:1 PC.

IOP Conf. Series: Journal of Physics: Conf. Series **1238** (2019) 012075 doi:10.1088/1742-6596/1238/1/012075

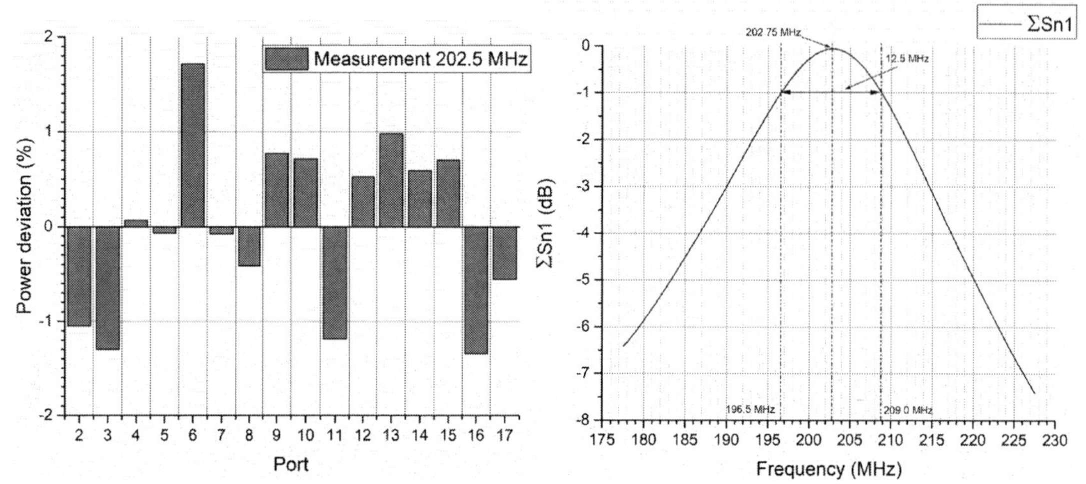

Figure 5. Deviation of power transmission coefficients (output-input transmission) of input ports and -1 dB operational frequency bandwidth of 16:1 PC.

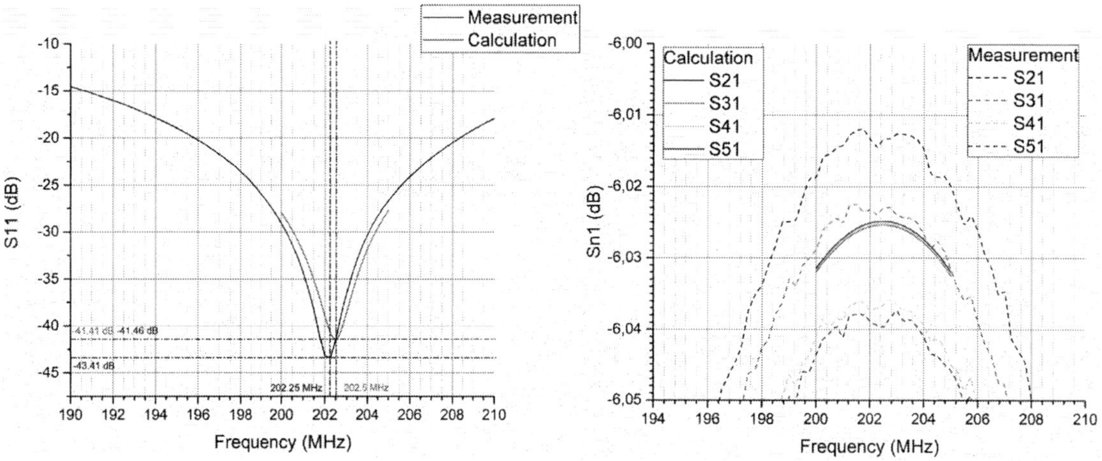

Figure 6. S11 parameter (reflection from output port) and Sn1 parameters (transmission from output to one of the input ports) of 4:1 PC.

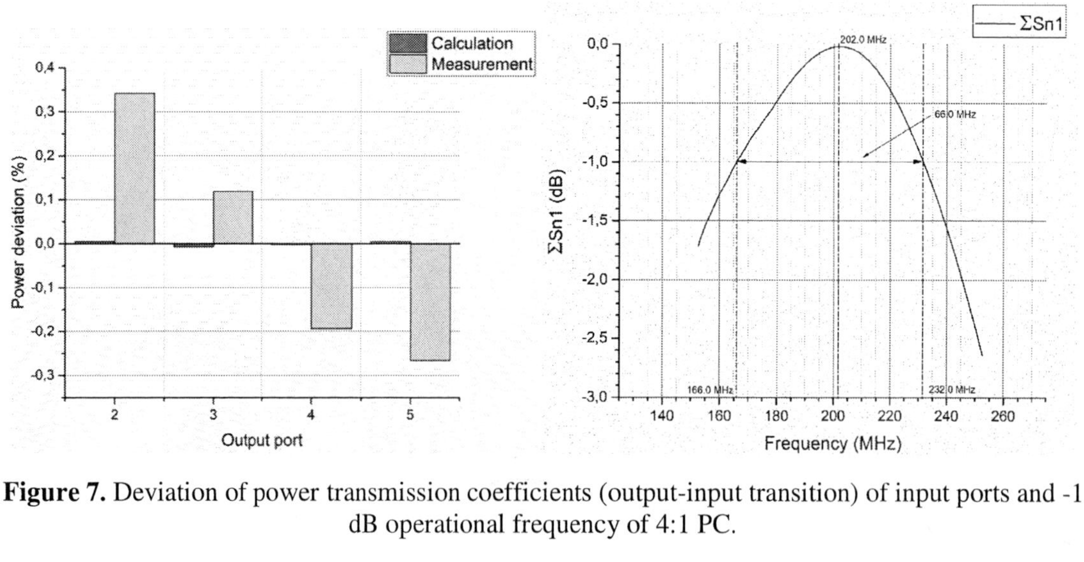

Figure 7. Deviation of power transmission coefficients (output-input transition) of input ports and -1 dB operational frequency of 4:1 PC.

Table 1. Measured parameters of 16:1 and 4:1 power combiners and initial requirements

Parameter	16:1 PC Requirements	16:1 PC Measurements	4:1 PC Requirements	4:1 PC Measurements
S11, dB	<-20 dB	-29.13	<-25.0 dB	-41.5
<Sn1>, dB	>-12.10	-12.10	>-6.04	-6.03
Power division equality, %	<5% span	+1.72..-1.35	<5% span	+0.34..-0.19
Insertion loss, %	<2.3	2.79	<0.90	0.35
1 dB bandwidth, MHz	>10.1	12.5	>10.1	66.0

4. High power level RF tests

After ensuring that both power combiners operate correctly first high power test were held. 16 power amplifier modules were connected to 16:1 power combiner's inputs, water cooled matched load was connected to combiner's output. Tests had shown stable operation with nominal output power of 64 kW with up to 2 shorted RF switches for more than one hour. System operation with more than two shorted RF switches on nominal output was not tested.

The operating mode was: 50 Hz pulse repetition rate with 2 % duty cycle. Duty cycle was set lower than nominal value due to limited power throughput of cable connection between power combiner and matched load. Nevertheless duty cycle value doesn't affect much the power combiner functionality due to exceptionally low power dissipation in power combiner. Now therefore we can conclude that the power combiner is fully functioning.

5. Conclusion

Power combining scheme for a solid-state power amplifier was described in this article. Power combining scheme has two stages with four 16:1 power combiners on the first and one 4:1 power combiner on the second stage. The first power combiner has RF switch on each input port that allow to turn on and off any power amplifying module of the amplifier, up to three of 16 RF switches can be turned off simultaneously without efficiency loss. This feature also allows using the hot-swap to change any power amplifier module on the run.

Both power combiners were manufactured and tested on low RF levels: results shows fine matching between calculations and assembled units. High power tests were held for one 16:1 power combiner. They confirmed correct operation of 16:1 power combiner on nominal output pulse power of 64 kW for more than one hour without any failures with up to two turned off RF inputs.

6. References

[1] Бондаренко Т.В., Краснов А.А., Полихов С.А., Шарков Г.Б. ВАНТ Серия: Техническая физика и автоматизация, выпуск 74, с. 3-13, 2016

[2] D J S Findlay, ISIS — PULSED NEUTRON AND MUON SOURCE, Proceedings of PAC07, p. 695-699, 2007.

[3] G.B. Sharkov, A.A. Krasnov, S.A. Polikhov, et al. Proceedings of NAPAC, p. 873-876, 2016

Influence of the Frequency Detuning to Electrodynamics Parameters of an Electron Linac

S M Polozov[1,2], **A I Pronikov**[1,2]

[1] National Research Nuclear University - Moscow Engineering Physics Institute, Moscow 115409, Russia
[2] National Research Center "Kurchatov Institute", Moscow, 123182, Russia
SMPolozov@mephi.ru

Abstract. It is very important task to define the tolerances necessary for correct operation of an electron linac. The detuning of linac main parameters as frequency, RF field amplitude and phase can sufficiently influence to the beam parameters. It is highly important for linacs with high brightness beams which are necessary for light sources and colliders. It is also sufficient for industrial linacs. A linac section consists of regular and bunching parts. The influence of one cell detuning on electrodynamics parameters of the whole section was studied. Results of such simulation will discuss in this paper.

1. Introduction

The standing wave biperiodic accelerating structures (BAS) with coupling windows are one of the popular structures for research and industrial electron linacs. Such structures are used in the new industrial 10 MeV/ 20 kW accelerators developed by the joint team of NRNU MEPhI (Moscow) and Corad Ltd. (St. Petersburg) [1-3]. Optimized omega – shape accelerating cells provide the maximum RF field on the structure axis and high shunt impedance in these linacs. New linac has high electrical efficiency, narrow beam energy spectrum within energy range of 5-10 MeV. The high coupling coefficient was obtained also to achieve the maximal efficiency of RF-pulse power usage. The gentle buncher was used to provide high capturing coefficient and narrow energy spectrum in the wide energy band.

The same structure was proposed as the first accelerating section (gentle buncher) for new electron injector of the CERN Future Circular Collider (FCC) [4, 5]. This injector will consist of two e-guns and up to 150 regular sections with output energy up to 14 GeV. One of RF-guns with photo-cathode is proposed to use for injection of short and high-brightness electron bunches in the FCC-ee main booster and the second RF-gun with the thermionic cathode is proposed to use for 6 nC bunches acceleration necessary for e- / e+ conversion.

Starting 2017 the new project of 4th generation synchrotron light source called Specialized Synchrotron Radiation Source (SSRS4) is under development in Russia. Current version of SSRS4 general layout includes 6 GeV main storage ring and top-up injection linac. Free-electron laser will also include in SSRS and top-up linac will used both for injection in the storage ring and for generation of the drive beam for FEL. An additional 1.5 GeV ring is also proposed as an option to generate photons in vacuum UV band. Such layout leads to three linac operation modes: 6 and 1.5 GeV beams for injection and 6-7 GeV high-brilliance bunches for FEL. It leads to the same for-injection scheme as it was used for SuperKEK-B and MAX-IV and is proposed for FCC-ee: two RF-

IOP Conf. Series: Journal of Physics: Conf. Series **1238** (2019) 012076 doi:10.1088/1742-6596/1238/1/012076

guns with photocathode and thermionic cathode and one main linac with large number of identical regular sections.

The high-quality manufacturing of the buncher cells and its frequency tuning is one of the main problems for BAS. Investigation of the influence of one cell frequency detuning on the whole structure electrodynamics parameters is very sufficient for e-linac R&D. Results of such effect simulations on the whole structure frequency and RF-field amplitude distribution will be presented bellow.

2. Results of the detuning simulations

BAS consisting of 28 accelerating and 27 coupling cells was used for simulations [1, 3]. This structure was early proposed for the industrial linac with energy of 10 MeV, the structure total length is 143 cm. The bunching part consists of six accelerating cells with variable phase velocity and RF-field amplitude. One of the middle cells of BAS is used as RF-power coupler. Similar structure is proposes to use in for-injector of SSRS4. The key problem of the development and testing of the linac is the accuracy of bunching cells manufacturing. Frequently some dimensions of one or more bunching cells can be out of tolerance. It can leads to the variation of main electrodynamics parameters of the whole structure as the operating resonant frequency or the phase and the amplitude RF-field. It is very interesting problem to study the influence of the one accelerating cell detuning to electrodynamics parameters of the whole structure.

BAS model was developed to study this effect. As an example, the iris diameter of the sixth bunching cell was varied in the band of $D_6=D_{opt}\pm0.1$ mm, where D_{opt} is the optimum 6th cell diameter. The distribution of RF-field amplitudes in centers of bunching and regular cells is presented in Table 1. The distribution for D_{opt} is marked by the green color. Frequencies of the whole structure for the operation mode $\mu=\pi$ ($f\pi/2$) are also presented in Table 1. It is clear that the operation frequency shift is equal only ±0.6 MHz for the whole structure. Comparatively, the variation parameter $df_{\pi/2}$ / $dD_6 \approx$ 70 MHz/mm for separate sixth accelerating cell. The RF-field amplitude distribution on the linac axis is presented in Figure 1 for three different D_6 values. It was found that detuning of one bunching cell leads to the variation of the RF-field amplitudes in other bunching and regular cells. Shown that the error less than 0,1 mm give the error in the RF-field amplitude about 3,6 %. The frequency of the whole structure shifts up and the amplitude of RF field in bunching cells decreases if the diameter of 6th cell is lowers than optimal value. In the opposite case, when $D_6>D_{opt}$ the frequency shifts down and amplitudes for bunching cells increase.

Table 1. Results of the sixth cell detuning simulation

$f_{\pi/2}$, MHz	$A_{regular}$	A_6	A_5	A_4	A_3	A_2	A_1
			$D_6=D_{opt} - 0.1$ mm				
2859.32	1	0.94	0.78	0.63	0.64	0.52	0.58
			$D_6=D_{opt} - 0.05$ mm				
2859.15	1	0.94	0.78	0.63	0.63	0.51	0.57
			$D_6=D_{opt}$				
2858.98	1	0.94	0.78	0.,63	0.63	0.50	0.56
			$D_6=D_{opt} + 0.05$ mm				
2858,82	1	0,.93	0.78	0.62	0.62	0.49	0.55
			$D_6=D_{opt} + 0.1$ mm				
2858,66	1	0.93	0.78	0.62	0.61	0.48	0.54

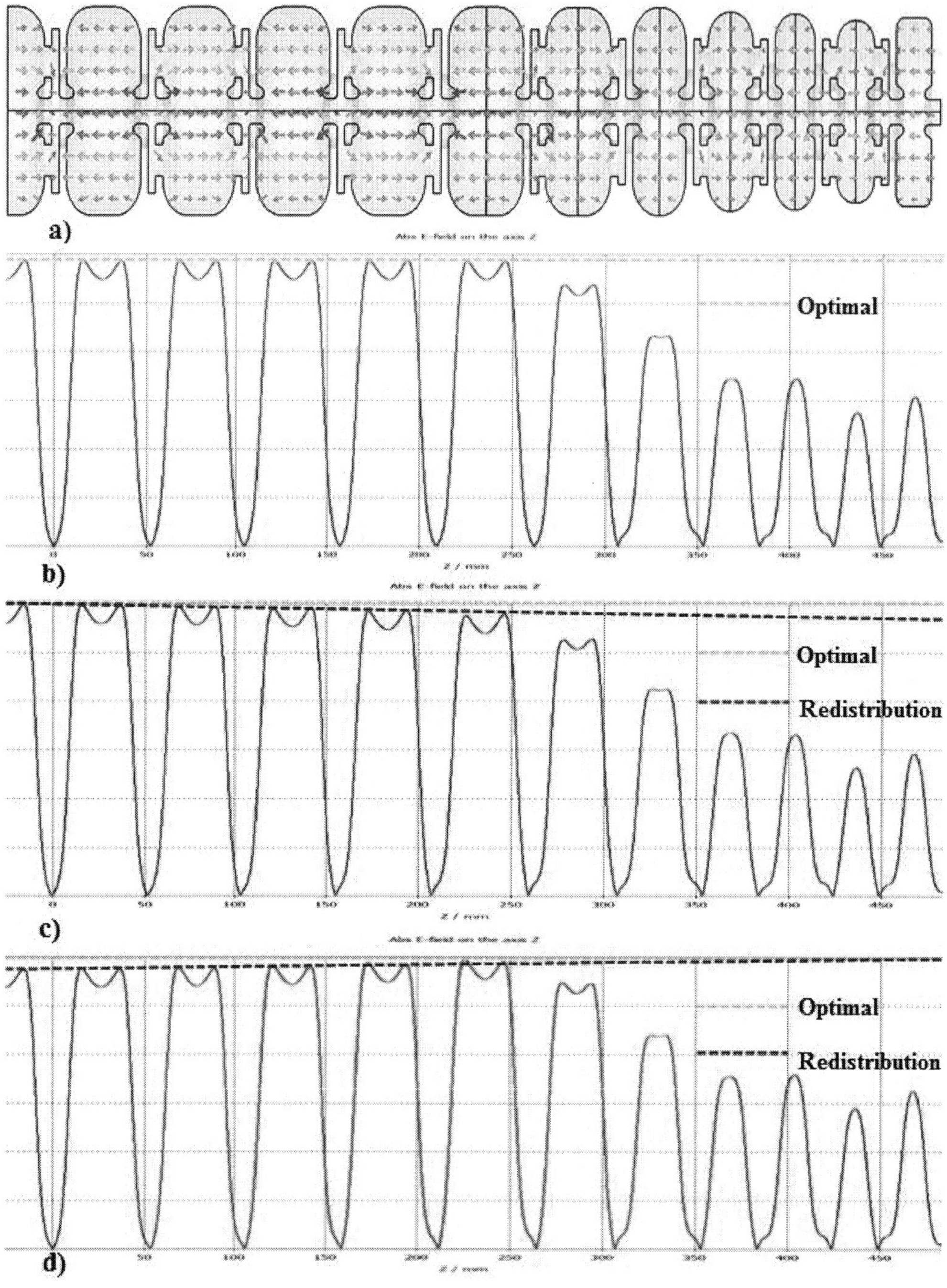

Figure 1. The RF-field amplitude distribution (a) on the linac axis for three different D6 values: $D_6=D_{opt}$ 0 mm (b), + 0.1 mm (c) and – 0.1 mm (d)

3. Conclusions

Results of the one cell detuning simulation in the accelerating structure of an electron linac were studied in the frame of its effect on electrodynamics parameters of the whole section. The operation frequency shift is equal to ±0.6 MHz for the whole structure with the one bunching cell diameter error of 0.1 mm. The detuning of one bunching cell leads to the variation of the RF-field amplitudes in other

IOP Conf. Series: Journal of Physics: Conf. Series **1238** (2019) 012076 doi:10.1088/1742-6596/1238/1/012076

bunching and regular cells but in the RF-field amplitude is about 3.6 % for the diameter error noted above. The coupling coefficient is not depends of the error.

This work was supported by the Ministry of Science and Education of Russian Federation, Agreement No 14.616.21.0086 from 24.12.2017, ID RFMEFI61617X0086 (Federal Targeted Program "Research and development on priority directions of scientific-technological complex of Russia in 2014 -2020 year").

References

[1] Demsky M.I., Lalayan M.V., Polozov S.M. et al., 2016. New 10 MeV High-power Electron Linac for Industrial Application. Proc. of IPAC'16, 1794-1796.

[2] Demsky M.I., Lalayan M.V., Polozov S.M. et al., 2016. Commissioning and First Tests of the New Standing Wave 10 Mev Electron Accelerator. Proc. of RuPAC'16, 173-175.

[3] Demsky M.I., Polozov S.M., Rashchikov V.I., 2016. Beam Dynamics in New 10 MeV High-Power Electron Linac for Industrial Application. Proc. of RuPAC'16, 493-495.

[4] Benedikt M., Zimmermann F., 2016. Status of the Future Circular Collider Study. Proc. of RuPAC'16, 34-38.

[5] Bondarenko T.V., Polozov S.M., 2017. Beam Dynamics Simulation in Two Versions of New Photogun for FCC-ee Electron Injector Linac. Proc. of IPAC'17, 1326-1329.

The light nuclei stopping in metal targets of accelerators

V I Rashchikov, A E Shikanov

National Research Nuclear University (MEPhI), Moscow, Russia

E-mail: virashchikov@mephi.ru

Abstract. The report discusses processes of light nuclei stopping in a solid-state barrier. Accounting algorithm of energy losses of light nuclei for (0÷20) MeV - range was considered. Calculated functions of the energy losses for various materials were presented.

1. Introduction

In this paper, we focus on the process of bombarding the surface of a solid-state barrier with light nuclei with energies up to 20 MeV, which are the products of fusion reactions. According to the considerations given in [1], in order to estimate the energy losses of nuclei in the surface layer of a solid in the region of kinetic energies of the nuclei satisfying condition

$$T >> 10^{-6} \frac{e^3 M}{2h^2 \varepsilon_0^2} Z_1^2 A_1 \cong 0.1 Z_1^2 A_1 \text{ [MeV]} ,$$

the formula Bethe-Bloch can be used:

$$F_+(w) = -\frac{dT}{dx} = 10^{-6} \frac{2\pi e^3 M}{m} \frac{\rho Z_1^2 Z_2}{A_2 w} \ln\left[10^6 \frac{4m}{M} \frac{w}{I(Z_2)}\right] \text{ [MeV/m]}, \quad (1)$$

where $w = \frac{T}{A_1}$ - the kinetic energy per nucleon in the stopping nucleus (reduced kinetic energy), e- the elementary charge, M- the mass of a proton, m – the mass of an electron, A_1, Z_1- the atomic masses and numbers of the stopping nuclei, A_2, Z_2, ρ, – the atomic numbers and the density of a stopping substance respectively, $I(Z) \cong 10(Z+0.5)$ -[V]- the mean ionization potential [2].

The guaranteed lower limit for the applicability of the Bethe-Bloch formula in this case is several MeV.

The purpose of this report is obtained of accounting algorithm of energy losses of light nuclei for (0÷20) MeV – range.

2. Accounting algorithm

At lower kinetic energies, one can use the empirical formula obtained on the basis of the data from [3], which gives a good approximation to the function of the energy losses in the energy range of (0 ÷ 5):

$$F(w) = 10^{-6} \alpha \frac{\rho Z_1^2 Z_2 w^{0.5}}{A_2} \frac{1+\beta w^{0.5}}{w^{1.275} + \gamma} . \quad (2)$$

The parameters in the equation (2) for some possible elements are determined in the table 1.

Table 1. Parameters in equation (2).

Element	Be	C	Al	Ti	Fe	Cu	Zr
α	50.2	62.15	31.55	28.10	32.00	26.80	17.00
β	0.006	0.085	0.104	0.110	0.020	1.124	0.063
γ	0.046	0.049	0.044	0.038	0.123	0.103	0.041

For computer modeling of the processes of the nuclei stopping in a solid, it is necessary to specify a continuous function of the energy losses in the entire energy range of $[0 \div w_{max}]$, where w_{max} is the maximum value of reduced kinetic energy in the spectrum of the stopping nuclei. This requires matching the equations (1) and (2). In the simplest case, this can be done by linear interpolation using the function

$$\tilde{F}(w) = a_0 - a_1 w \tag{3}$$

where
$$a_1 = \frac{F_-(w_c - \delta) - F_+(w_c + \delta)}{2\sigma}$$

$$a_1 = \frac{(w_c + \delta)F_-(w_c - \delta) - (w_c - \delta)F_+(w_c + \delta)}{2\sigma}.$$

The value w_c is determined by the intersection of the curves $F_-(w)$ and $F_+(w)$; δ – the arbitrary parameter selected in the range of $[0.1 \div 1]$ MeV. As a result of such constructions, a continuous, sufficiently smooth function of energy losses is obtained

$$F(w) = \begin{Bmatrix} F(w),\ 0 < w < w_c - \delta \\ \tilde{F}(w),\ w_c - \delta < w < w_c + \delta \\ F_+(w),\ w_c + \delta < w < w_{\max} \end{Bmatrix}. \tag{4}$$

The degree of its smoothness can be improved by using in the proposed matching procedure a cubic spline

$$\tilde{F}(w) = \sum_{n=0}^{3} a_n (w - \delta)^n$$

instead of a linear function (2). Vector $\mathbf{a} = \{a_0, a_1, a_2, a_3\}$ is the solution of next linear algebraic equation:

$$\mathbf{a}\begin{Vmatrix} 1 & 0 & 1 & 0 \\ w_0 - \delta & 1 & (w_0 + \delta) & 1 \\ (w_0 - \delta)^2 & 2(w_0 - \delta) & (w_0 + \delta)^2 & 2(w_0 + \delta) \\ (w_0 - \delta)^3 & 3(w_0 - \delta)^2 & (w_0 + \delta)^3 & 3(w_0 + \delta)^2 \end{Vmatrix} = \begin{Vmatrix} F_-(w_0 - \delta) & F_-^{'}(w_0 - \delta) & F_+(w_0 + \delta) & F_+(w_0 + \delta) \end{Vmatrix}.$$

3. The calculated results

Figure 1 presents the calculated functions of the energy losses obtained on the basis of the described algorithm for various materials.

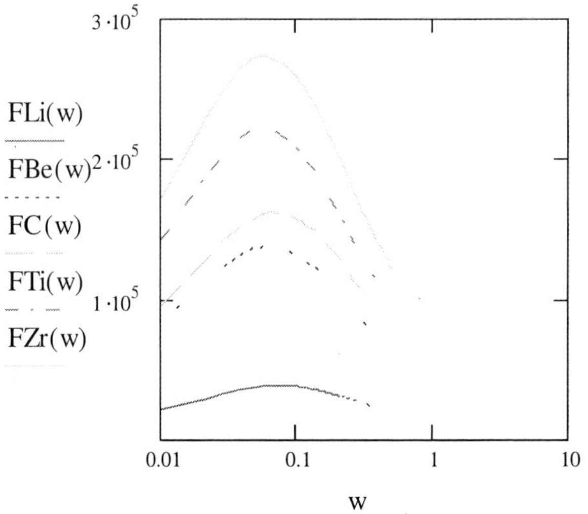

Figure1. The calculated dependences of the energy losses of a nucleus moving in zirconium (the upper curve), in titanium (the second curve from above), in carbon (the third from above), in beryllium and lithium (the lower curve)

Figure 2 shows the characteristic calculated dependences of the kinetic energy of the nucleus per nucleon from the depth of its penetration into a solid.

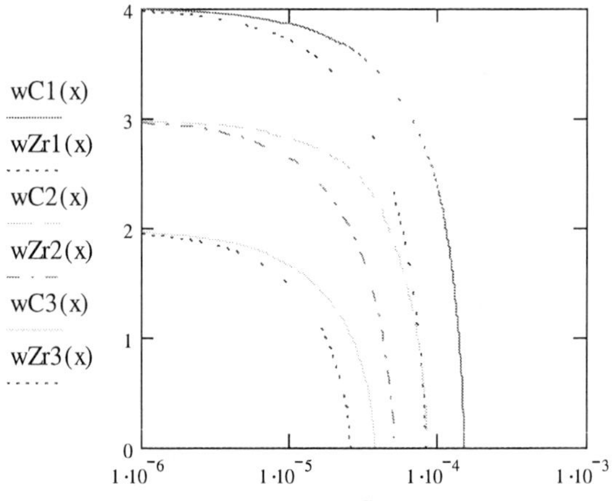

Figure2. Three pairs of calculated dependences of the kinetic energy of the nucleus per nucleon from the depth of penetration into carbon (upper curves in pairs) and zirconium. The pair of curves $wC1(x)$ and $wZr1(x)$ corresponds to the initial energy w_0=4MeV; the pair $wC2(x)$, $wZr2(x)$ to w_0=3MeV; the pair $wC3(x)$, $wZr3(x)$ to w_0=2MeV

These curves were obtained by were obtained by the numerical solution of a differential equation

$$\frac{dw}{dx} = \frac{F(w)}{A_1},$$

with initial conditions $w(0) = \{4;3;2\}$ MeV.

The path length of the nucleus will be the root of the equation $w(x) = 0$ and can be estimated for the case in question directly from figure 2.

4. Case of multiply component barrier

The elaborated algorithm can be generalized to the case when a solid under consideration represents a composition of *N* different elements with stoichiometry coefficients k_n normalized to 1 ($\sum_{n=1}^{N} k_n = 1$). In this case, according to the principle of additivity, the loss function can be represented as the sum:

$$F(w) = \sum_{n=1}^{N} k_n F_n(w) ,$$

where the partial loss functions corresponding to different stopping components are calculated by equations (1-3), in which the values of A_2, Z_2 are replaced by the corresponding values for various components. Similarly, the coefficients α, β, γ should be replaced according to their table values.

Thus, the algorithm has been developed for calculating the functions of energy losses in solids based on the interpolation matching of the empirical formula for the energy losses in the low-energy region with the analytical Bethe-Bloch formula, which accurately describe the stopping in the medium-energy region. The proposed analytical formulae make it possible to increase the accuracy and reliability of calculations of heat at stopping the nuclei in solids, and can also be used in calculating the output parameters of the projected neutron generators.

5. Conclusion

1. Accounting algorithm of energy losses of light nuclei for (0÷20) MeV - range was considered.
2. Calculated functions of the energy losses for various materials were presented.

References

[1] Lehmann C 1979 *Interaction of Radiation with Solids and Elementary Defect Production (on Russian)* (Moscow: Atomizdat) p 296.

[2] Gusev N G , Klimanov V A, Mashkovich V P, Suvorov A P 1989 *Protection from ionizing radiation (on Russian). Vol.1. Physical fundamentals of radiation protection* (Moscow: Energoatomizdat) p 514

[3] Gott YuV 1978 *Interaction of particles with substances in plasma research (on Russian)* (Moscow: Atomizdat) p 272

IOP Conf. Series: Journal of Physics: Conf. Series **1238** (2019) 012078 doi:10.1088/1742-6596/1238/1/012078

Applied calculations of accelerator characteristics for sterilization installation with local radiation shielding

N E Rozanov

Moscow Radiotechnical Institute of Russian Academy of Sciences, Moscow 117519, Russia

nrozanov@mail.ru

Abstract. The calculations of electron beam characteristics in the accelerator of sterilization installation with the local radiation shielding in relation to the use of a magnetron with a larger pulse power are made. It is shown that a simple method on the basis of "moving" of cathode-heating unit toward the anode plane of the injector may be used for increasing of current and power of accelerated beam.

1. Introduction

In relation to the use of magnetron MI-456AM with increased approximately to 3.5 MW pulse power, the possibilities of effective application the existing accelerating structure in the sterilization installation with the local radiation shielding the calculations [1] are made. The application of such self-shielded accelerators makes it possible to avoid construction of special buildings and makes less expensive the creation of radiation technology systems. This existing accelerating structure on a standing wave was developed and created using the magnetron MI-456A with a pulse power of about 2.5 MW. In this structure the electron beam is accelerated to an averaged, over the energy spectrum, energy of about 5 MeV, has a pulse current of 250 mA (up to 270 mA), i.e. the pulse power of 1.25 MW, the efficiency of the accelerating structure - 58%. The injected beam current is 1.6 amperes [2]. For effective use of microwave power from the upgraded magnetron it is necessary to increase the injection current. A simple way to achieve this purpose may be the "moving" of cathode-heating unit toward the anode plane of the beam injector. The calculations show the reality of this possibility. By means of variation of the new position of the cathode-heating unit, the optimal position of this unit is found.

2. Computer codes description

Several variants of computer codes DINA [1] were used for calculations.

DINA_RoZ is a 2.5-dimensional computer code to describe the processes of bunching and acceleration of electron beams in the accelerators in a steady-state, periodical in time, regime. It allows you to assign the initial parameters of the beam both with arbitrary distribution of different types and obtained in calculations of the electron injector with the help of specialized computer codes. 2D-distributions of radio-frequency (RF) fields in accelerating structures can also be set both as arbitrary ones and entered from calculations by specialized computer codes. The DINA_RoZ code calculates and takes into account the flow of electrons to the walls of the channel in accelerators with RF field focusing. This possibility allows explaining the power balance in linear accelerators of electrons on a standing wave. Only taking into account these power losses it is possible to significantly optimize the operation of such accelerators. In this code, to find the solutions of equations in the state with a given RF power, "pumped" into the resonator of the accelerating structure by an external RF generator, the method of "power balance" is used. It consists in calculating, first of all, the total cost of microwave power in the accelerating structure. Firstly, it is the microwave power transmitted to the accelerated beam, derived from the accelerator. Secondly, it is the power spent on the partial acceleration of the electron flow, which is not derived from the accelerator, but settles on the walls of the accelerating resonator. And, thirdly, it is

IOP Conf. Series: Journal of Physics: Conf. Series **1238** (2019) 012078 doi:10.1088/1742-6596/1238/1/012078

the power of ohmic losses of microwave energy in the walls of the resonator. Then this total amount of "spent" power is compared with the RF power "pumped" into the resonator by an external RF generator. The latter is the difference between the power of the RF generator and the loss of RF power in the microwave channel connecting the RF generator and the accelerating resonator. After a series of calculations of beam dynamics in the accelerator with different, but fixed at every calculation, values of the amplitude of the electric field generated by the microwave generator, and calculating the above amounts of "spent" microwave power, the code selects the regime in which the above "balance of power" is observed. This method determines the steady-state, periodic-in-time with the microwave field period, the state in the accelerating structure.

The computer code DINA_TIME simulates the 2.5-dimensional time evolution of electron beams in transient modes without the assumption of the periodicity of processes in time. In addition to the processes of bunching and acceleration, it describes the dynamics of reflected electrons moving in the direction to the injector. Such electrons appear in the regimes when the voltage on the injector is noticeably less than the voltages on the first accelerating gaps. In addition, it simulates the operation of the injector (in the regime with a given current) taking into account the own fields of these reflected electrons and calculates the power of the flux falling on the cathode of the injector. The characteristics of the flow of the injected beam distorted by these fields may be used in a self-consistent way to calculate the processes of bunching and acceleration. Calculation of the time interval in several times exceeding the electrons transit time of the accelerating structure leads to the establishment (if the effect of the field of the reflected electrons is small enough) of the periodic-in-time (with the frequency of the RF field) regime described by the code DINA_RoZ. The calculation time of such process is several times larger than the calculation time of the established process.

The computer code DINA_3D describes the three-dimensional dynamics of the electron beam in accelerators in periodic-in-time regime. It allows taking into account the non-alignment of the injected beam and the accelerating structure, azimuthal inhomogeneity of the beam self-fields and RF fields in the structure, and other three-dimensional effects.

The computer code DINA_RoZ_BL is created to describe the process of bunching and acceleration of electron beams with taking into account the effect of "beam loading". The general technique of describing the effect of "beam loading", i.e. the effect of excitation of electromagnetic waves in accelerating structure by electron bunches of the intense beam, has been developed by E. S. Masunov [3]. The equations describing the time evolution of the complex amplitude of the electric field radiated by a modulated electron beam in an accelerating structure with a known longitudinal field distribution of the resonator' own mode have been derived. The paper [4] describes a package of three-dimensional computer codes that use the solution of such equations to describe the beam loading effect in conjunction with the equations for the dynamics of the beam in the accelerator, taking into account the quasi-static fields of the beam' space charge and current. Taking into account the beam loading, the Masunov' equations describe the growth of the radiation field with the time calculated from the beginning of the beam current pulse. Given that this process is slow compared to the microwave field period, it is possible to find the quasi-stationary state of the accelerated beam at fixed radiation field characteristics at each such moment of time using, in our case, the DINA_RoZ code. As a result, either stationary, time-periodic solutions in the case of long beam current pulses or values at the end of short beam current pulses are determined. To find the total, taking into account the field emitted by the beam, the electromagnetic field in the accelerating resonator with a given, fixed microwave power, the equations for it are supplemented by two terms in the right side [5, 6]. First of them is a "source" that simulates the "pumping" of RF power into the resonator from an external microwave generator, a second term accounts the loss of microwave power in the resonator, taking place either because of its departure into the external waveguide, or due to ohmic losses in the resonator walls or in the absorber of microwave power. These two terms compensate each other in the absence of beam radiation, for example, at low injection current or in the absence (for any reason) of beam current modulation. The main result [7] of account for the effect of radiation of the electron beam (i.e., the effect of beam loading) is some frequency change of total microwave accelerating field (as compared with the case of absence of the beam radiation) under the condition of "pumping" in the accelerating structure of the same total microwave power. At the same time, there is no change (reduction) in the efficiency of the accelerator in this case.

3. Results of the calculations

Calculations of the beam formation processes in the injector with different positions of the cathode-heating unit relative to the anode plane have been carried out using computer code [8]. Calculations of beam dynamics in the accelerating structure using information about the beam in the injector have shown the possibility of its effective acceleration. Provided that the previous energy, averaged over the energy spectrum, of about 5 MeV is maintained, the pulse accelerated current rises to a value of about 400 mA, which gives a pulse beam power of 2 MW and the efficiency of the accelerating structure 68% (figure 1).

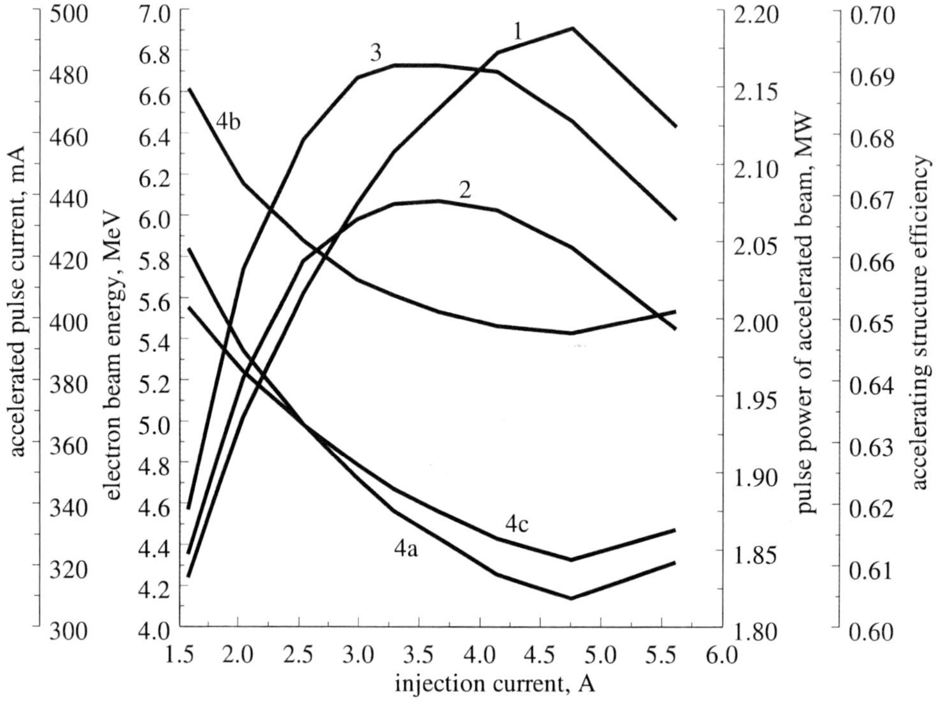

Figure 1. Dependences on the injection current: of accelerated pulse current (curve 1) and pulse beam power (2), of efficiency of accelerating structure (3), of beam energies: averaged over the energy spectrum (4a), maximal energy in the spectrum (4b) and the energy found by using the coefficient of beam current transmission through the 6-mm aluminum plate in the method using Faraday cylinder (4c).

Figure 1 shows the dependences of the main characteristics of the accelerated beam on the injection beam current. In particular, it can be seen that there are optimal values of injection currents to achieve the maximum of different beam characteristics. Also can be seen, that the use of the method with a cylinder of Faraday of finding the energy by using the coefficient of beam current transmission through the 6-mm aluminum plate, allows enough accurately determine the averaged over the spectrum beam energy. Indeed, these energies, as can be seen from a comparison of curves 4a and 4c in the figure 1, are different not strong.

It is supposed in these calculations, that the part of lost microwave power in RF channel remains the same (about 14%, as with the old magnetron MI-456A), so the pulse power about 3 MW is entered to the accelerating structure.

In the above regime with the average, over the spectrum, energy of about 5 MeV, the injected beam current is equal to 2.5 amperes as can be seen from the curve 4a in the figure 1. To achieve this beam current, the offset of cathode-heating unit toward the anode plane of the injector must be equal to 7 mm.

Further increase of beam injection current by this method, as can be seen on figure 1, still more increases accelerated beam current - up to 500 mA, but it decreases the averaged over the spectrum energy - up to 4 MeV, which may be unacceptable for the practical use.

This method of increasing the injection current, in general, is not optimal, because, in particular, the beam characteristics at the entrance of the accelerating structure may be not optimal. Therefore, the above results are not optimal, but relate, we believe, to the simple method of utilizing the existing accelerating structure, leading to increase the accelerated beam current and the structure efficiency. It is probably that by some correcting the characteristics of accelerating structure a better result can be achieved, if, in particular, to provide an improved conditions of beam entering to the accelerating structure. But the calculations with beam characteristics at the entrance from the injector to the accelerating structure, similar to the used in the calculations with applying to old magnetron MI-456A (which can be considered as optimal characteristics), show that the result is not improved practically.

Recall that in the sterilization installation with local radiation shielding, the use of a focusing magnetic field is unacceptable. Therefore, the focusing of the electron beam is realized by the radial component of the electric field of the microwave accelerating wave.

4. Conclusion

The calculations of electron beam characteristics in the accelerator of sterilization installation with the local radiation shielding in relation to the use of the upgraded magnetron MI-456AM with a larger, then in used magnetron MI-456A, pulse power are made. It is shown that the proposed simple method of increasing the injection current by means of "moving" the cathode-heating unit toward the anode plane of the injector is very promising.

References

[1] Rozanov N.E. 2002. Computer codes DINA for calculation of intense beams dynamics in linear accelerators. In: Proc. of MEPhI-2002 7, p.124.

[2] Belugin V.M., Pirozhenko V.M., Rozanov N.E. 2009. Self-shielded electron linear accelerators designed for radiation technologies. Physical Review Special Topics - Accelerators and Beams 12, issue 9 (090101), pp.1-8.

[3] Masunov E.S. 1999. Beam loading effects in charged particle accelerators. MEPhI, 112 p.

[4] Bondarenko T.V., Masunov E.S., Polozov S.M. 2013. Computer codes BEAMDULAC-BL for three-dimensional simulation of electron beams with account of self electrostatic field and beam loading. Problems of Atomic Science and Technology. 6 (18), pp.114-118.

[5] Rozanov N.E. 2014. On the question of beam loading in standing-wave electron accelerators. In: Proc. of MEPhI-2014 2, p.146.

[6] Rozanov N.E. 2015. Development of the theory of beam loading in standing-wave electron accelerators. In: Proc. of MEPhI-2015 2, p.203.

[7] Rozanov N.E. 2016. An influence of beam loading in standing-wave electron accelerators. In: Proc. of MEPhI-2016, p.180.

[8] Vasiljev A.E., Vetrov V.V., Pikunov V.M., Smirnov V.S. Account of near-cathode phenomena in calculations of electron injector for TWT. Preprint of Lomonosov Moscow state University, 2005, №11, 10 p.

IOP Conf. Series: Journal of Physics: Conf. Series **1238** (2019) 012079 doi:10.1088/1742-6596/1238/1/012079

Modeling of spin-orbital dynamics in a storage ring

Alexander Aksentev[1,2]

[1] National Research Nuclear University "MEPhI," Moscow, Russia
[2] Forschungszentrum Jülich, Jülich, Germany

a.aksentev@fz-juelich.de

Abstract. Two distinct storage ring designs have been proposed for the deuteron Electric Dipole Moment (dEDM) search experiment: a Frozen Spin, and a Quasi-Frozen Spin ring. The FS-ring allows for the most accurate measurement, but the continuous fulfilment of the FS condition is unrealizable. The QFS design relaxes the condition to only hold on average, at a slight cost of precision. [1] The goal of the present work is to investigate which design is preferable for the attainment of the target precision. This task requires one to model the spin-orbital dynamics of the beam in both rings.

1. The Frozen and Quasi-frozen Spin concepts

There exist two design approaches to the problem of measuring the deuteron Electric Dipole Moment (dEDM) inside a storage ring: the Frozen Spin (FS) lattice, and the Quasi-Frozen Spin (QFS) lattice.

In the FS ring design, the spin of the reference particle is aligned with its momentum at any point in time; this allows the maximization of the useful signal, but requires that the energy of the particle be its so-called "magic energy" – a condition that cannot be fulfilled exactly for an ensemble of particles. Also, in this design, combined E+B field elements in the arcs must be used, which further complicates the construction of such a lattice. The QFS design does not require the continuous fulfillment of the Frozen Spin condition; the cost for this is a slight (on the order of percents) degradation of the EDM signal. It can also be implemented in two versions: one in which the E- and B-field elements are separated, and another in which the E+B elements are located in the straight sections of the lattice. [1]

In order to decide which design solution is preferable for the attainment of the target accuracy (upped bound on the dEDM at 10^{-29} e·cm) one has to model the spin-orbital dynamics of the beam inside both structures. The present work deals with some problems that are inherent in this enterprise, on the example of an FS lattice design with E+B elements in the arcs.

2. Experiment systematics

In the original BNL proposal, a longitudinally polarized beam is injected in the storage ring, and the EDM is measured by the growth, over a long period, of its vertical polarization component. [2] The spin dynamics of a particle put into an electro-magnetic field are determined by the Thomas-Bargmann-Michel-Telegdi equation:

$$d\vec{S}/dt = \vec{\Omega}\times\vec{S},$$

$$\vec{\Omega} = -\frac{e}{m}\left[G\vec{B} + \left(\frac{1}{\gamma^2-1} - G\right)\vec{\beta}\times\vec{E} + \frac{\eta}{2}\left(\vec{E} + \vec{\beta}\times\vec{B}\right)\right], \quad (1)$$

where $G = (g-2)/2$ is the anomalous magnetic moment, g the gyromagnetic ratio, the first two terms are due to the Magnetic Dipole Moment (MDM), and the last two to the EDM.

As is readily observed from the above equation, the presence of a parasitic horizontal magnetic, or vertical electric, field caused by the imprecision of accelerator element positioning (specifically, tilts about the optical axis), induces a faking MDM signal in the measurement plane. To solve this problem, a variation on the CW/CCW procedure [2] was proposed in Jülich. [3] The idea behind this method is to estimate the net MDM+EDM spin precession frequency when the beam in injected into the ring counter-/clockwise, and then add the results. Since the magnetic field is reversed going from the CW to the CCW case, the MDM frequency changes sign, and hence cancels in addition. One requirement of this method, though, is that the MDM frequency must be much higher than the one caused by the EDM; the sensitivity of MDM precession to the element positioning inaccuracy has to be analyzed by simulation.

Another problem encountered in the EDM search is the finite spin coherence time (SCT). Because the beam particles are injected into the ring at different energies and positions, and hence have different orbit lengths, their spins precess at different frequencies; due to this fact the net beam polarization vector dissipates over time. By Taylor-expanding the expressions for the spin tune in the electric and magnetic fields around the value γ_0 of the reference particle, we arrive at: [3]

$$\Delta\nu_s^E = G_6 \cdot \Delta\gamma + O(\Delta\gamma),$$
$$\Delta\nu_s^B = G \cdot \Delta\gamma,$$
$$\Delta\gamma(t) = \Delta\gamma_m cos(\Omega_s t) + f(\alpha_1, \eta, \gamma)\Delta\gamma_m^2 + g\left(\eta, \beta, \frac{\Delta L}{L}\right)\gamma^2, \quad (2)$$

where G_6 is a function of G and γ_0, $\Delta\gamma_m$is the synchrotron oscillation amplitude of the particle, $\alpha_1$is the second-order momentum compaction factor, $\eta$the slip factor, $\Omega_s$the synchrotron oscillation frequency, and $\Delta L/L$is the orbit lengthening due to betatron motion. The first term in equation (2) is averaged out as a result of synchrotron oscillation; the orbit-lengthening-related decoherence effect is mitigated by the introduction of sextupole fields.

3. Tracking code

For the analysis of the lattice designs a tracking code was written in python. Tracking is performed via element by element integration of the equations of motion, utilizing a conventional integrator LSODA from Scipy's Integrate package. Major elements used in the construction of a lattice, such as the magnetic dipole, quadrupole, electric cylindrical deflector and Wien filter are implemented, with two ways to simulate element positioning error: 1) via the computation of a tilt matrix, which is applied to the computed field vector at run time (which is the more general, but computationally inefficient approach), and 2) customized tilting of E+B and quadrupole elements. The computation of the right-hand side of the system of ODEs is vectorized, due to the use of Numpy arrays.

Sensitivity and decoherence tests

In order to analyze the sensitivity of the MDM precession to positioning error, the simulation was set up as follows:

- The E+B elements were tilted about the optical axis, the tilt angles were picked from a normal distribution with zero mean and 10^{-4} rad standard deviation.
- An ensemble of 1000 initial conditions, taken from a normal distribution with $\Delta\gamma(0) \sim N(0, 10^{-4})$, was tracked through the lattice for 100 turns (for reasons of run time)
- The test statistics (spin precession frequencies in the vertical and horizontal planes) were estimated by $\Omega_x = S_y(100)/\Delta t(100); \Omega_y = S_x(100)/\Delta t(100)$.

IOP Conf. Series: Journal of Physics: Conf. Series **1238** (2019) 012079 doi:10.1088/1742-6596/1238/1/012079

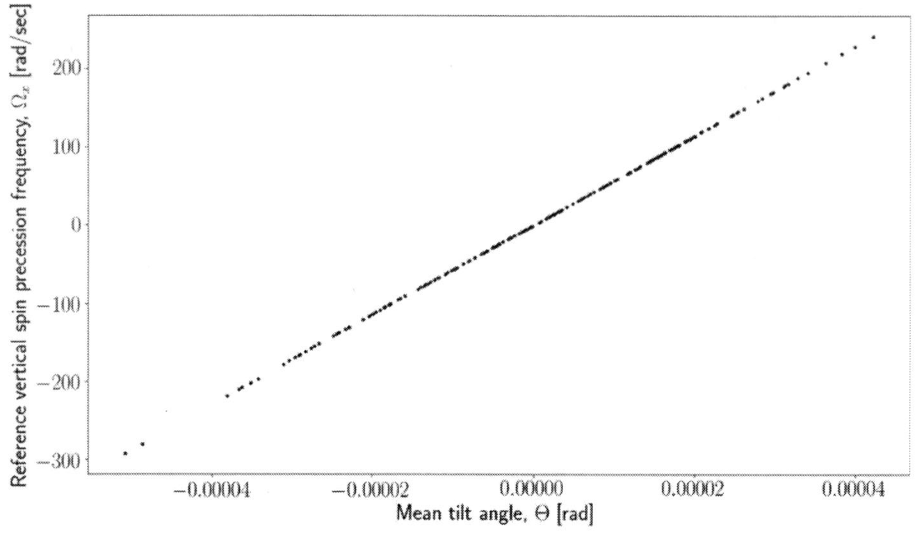

Figure 1. Reference particle spin precession frequency in the vertical plane as a function of the average E+B element tilt angle. The average tilt angle is linearly proportional to the net radial magnetic field. Linear dependence is thus in accordance with the T-BMT equation. Slope: $5.72 \cdot 10^6$ 1/sec, intercept $4.19 \cdot 10^{-4}$ rad/sec.

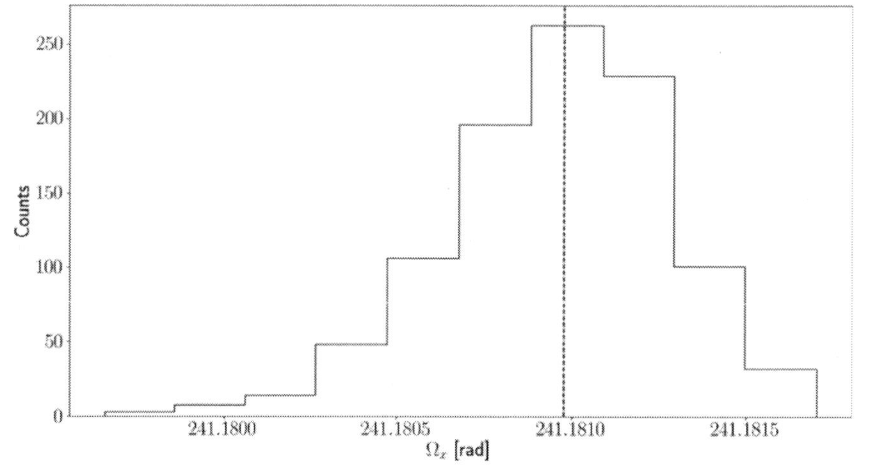

Figure 2. Decoherence in the vertical plane. Dashed line: frequency of the reference particle. At higher turn numbers this should become a chi-square distribution with the reference frequency being the lowest. This isn't observed here due to the low number of synchrotron oscillations.

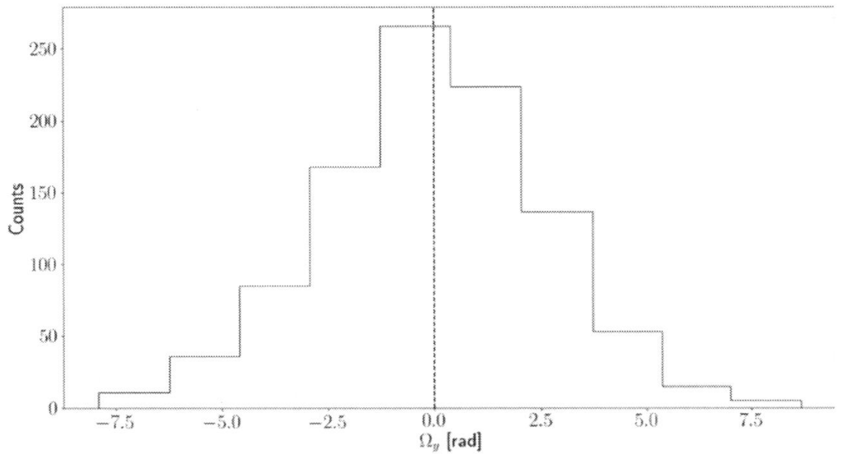

Figure 3. Decoherence in the horizontal plane. Dashed line: reference particle. It can be observed that the frozen spin condition is fulfiled (reference frequency is zero), but decoherence in this plane is four orders of manitude higher.

Major problems

As was observed at the end of section 2, the linear decoherence terms vanish as a result of synchrotron oscillations. This averaging out, however, requires a significant number of synchrotron oscillations (in the hundreds) to have occurred. As the ratio of synchrotron to revolution frequencies is significantly less than unity, any reasonable simulation must track the initial particle ensemble for tens of thousands of revolutions in the accelerator. This poses a significant problem in terms of run time, not only to any interpreted code, but also to any code using conventional integration in general.

As an example, we estimate the run time required for a single decoherence test run, if a Runge-Kutta integrator is used with performance on the level of MSURK89. [4] MSURK89 is an 8-th order, Runge-Kutta-Verner, step-size adaptive integrator, with initial step size calibrated for optimal speed-precision ratio. Use of an 8-th order stepper yields better run times due to fewer required evaluations of the right hand side. That code requires approximately $1.23 \cdot 10^{-4}$ sec per homogenous dipole field, and $6.25 \cdot 10^{-4}$ sec per spherical deflector (tracking was performed for 10 initial conditions simultaneously). An FS lattice design we used in simulations contains 20 E+B cylindrical deflectors, out of 397 elements. MSURK89 would thus take $5.88 \cdot 10^{-2}$ sec/turn to track ten particles through a single realization of the lattice, 20,000 turns would take 1176 seconds, and an ensemble of 1,000 particles would further require 20.4 hours of run time on a 16-core machine. In order to give certainty bounds on the MDM precession frequency, one has to repeat this experiment a reasonable number of times. Unless the code is massively vectorized, using technologies like OpenCL, the use of a conventional integrator for this type of analysis is not a tenable option; use of mapping methods, like COSY INFINITY [4] should be preferred.

References

[1] Senichev Y, Andrianov S, Ivanov A, Chekmenev S, Berz M, Valetov E 2015 *Investigation of lattice for deuteron EDM ring*. Proc. ICAP2015 (Shanghai, China).

[2] Anastassopoulos D et al. 2008 *AGS Proposal: Search for a permanent electric dipole moment of the deuteron nucleus at the* 10^{-29} *e · cm level.* BNL Report.

[3] Senichev Y 2016 *Search for the Charged Particle Electric Dipole Moments in Storage Rings.*

Proc. RuPAC2016 (St. Petersburg, Russia).
[4] Valetov E 2017 *Field modeling, symplectic tracking, and spin decoherence for EDM and muon G-2 lattices*. Fermilab thesis, pp 156—165.

AUTHOR INDEX

AUTHOR INDEX

AUTHOR INDEX

AUTHOR INDEX